Borehole Imaging: applications and case histories

Geological Society Special Publications

Series Editors

It is recommended that reference to all or part of this book should be made in one of the following ways:

LOVELL, M. A., WILLIAMSON, G. & HARVEY, P. K. (eds) 1999. *Borehole Imaging: applications and case histories.* Geological Society, London, Special Publications, **159**.

RIDER, M., GOODALL, T. & DODSON, T. 1999. A pre-development turbidite reservoir evaluation using FMS electrical images. *In*: LOVELL, M. A., WILLIAMSON, G. & HARVEY, P. K. (eds) *Borehole Imaging: applications and case histories.* Geological Society, London, Special Publications **159**, x–xx.

GEOLOGICAL SOCIETY SPECIAL PUBLICATION NO. 159

Borehole Imaging: applications and case histories

EDITED BY

MIKE LOVELL
University of Leicester, UK

GAIL WILLIAMSON
Gail Williamson Associates

PETER HARVEY
University of Leicester, UK

1999

Published by

The Geological Society

London

THE GEOLOGICAL SOCIETY

The Geological Society of London was founded in 1807 and is the oldest geological society in the world. It received its Royal Charter in 1825 for the purpose of 'investigating the mineral structure of the Earth' and is now Britain's national society for geology.

Both a learned society and a professional body, the Geological Society is recognized by the Department of Trade and Industry (DTI) as the chartering authority for geoscience, able to award Chartered Geologist status upon appropriately qualified Fellows. The Society has a membership of 8600, of whom about 1500 live outside the UK.

Fellowship of the Society is open to persons holding a recognized honours degree in geology or a cognate subject and who have at least two years' relevant postgraduate experience, or not less than six years' relevant experience in geology or a cognate subject. A Fellow with a minimum of five years' relevant postgraduate experience in the practice of geology may apply for chartered status. Successful applicants are entitled to use the designatory postnominal CGeol (Chartered Geologist). Fellows of the Society may use the letters FGS. Other grades of membership are available to members not yet qualifying for Fellowship.

The Society has its own Publishing House based in Bath, UK. It produces the Society's international journals, books and maps, and is the European distributor for publications of the American Association of Petroleum Geologists (AAPG), the Society for Sedimentary Geology (SEPM) and the Geological Society of America (GSA). Members of the Society can buy books at considerable discounts. The Publishing House has an online bookshop (*http://bookshop.geolsoc.org.uk*).

Further information on Society membership may be obtained from the Membership Services Manager, The Geological Society, Burlington House, Piccadilly, London W1V 0JU (Email: *enquiries@geolsoc.org.uk*; tel: +44 (0)171 434 9944).

The Society's Web Site can be found at *http://www.geolsoc.org.uk/*.The Society is a Registered Charity, number 210161.

Published by The Geological Society from:
The Geological Society Publishing House
Unit 7, Brassmill Enterprise Centre
Brassmill Lane
Bath BA1 3JN, UK

(*Orders*: Tel. +44 (0)1225 445046
Fax +44 (0)1225 442836)
Online bookshop: *http://bookshop.geolsoc.org.uk*

First published 1999

British Library Cataloguing in Publication Data
A catalogue record for this book is available from the British Library.

ISBN 1-86239-043-6

Typeset by Aarontype Ltd, Bristol, UK

Printed by Aldens

Distributors

USA
AAPG Bookstore
PO Box 979
Tulsa
OK 74101-0979
USA
Orders: Tel. +1 918 584-2555
Fax +1 918 560-2652
Email bookstore@aapg.org

Australia
Australian Mineral Foundation Bookshop
63 Conyngham Street
Glenside
South Australia 5065
Australia
Orders: Tel. +61 88 379-0444
Fax +61 88 379-4634
Email bookshop@amf.com.au

India
Affiliated East-West Press PVT Ltd
G-1/16 Ansari Road, Daryaganj,
New Delhi 110 002
India
Orders: Tel. +91 11 327-9113
Fax +91 11 326-0538

Japan
Kanda Book Trading Co.
Cityhouse Tama 204
Tsurumaki 1-3-10
Tama-shi
Tokyo 206-0034
Japan
Orders: Tel. +81 (0)423 57-7650
Fax +81 (0)423 57-7651

Contents

Preface

Borehole images now form a major part of the interpretation strategy for subsurface geological interpretation in a wide variety of scenarios. This volume attempts to document some case histories together with image acquisition and analysis techniques. It by no means covers the full breadth of the subject but is intended to provide a valuable source of information within one volume. We hope the broad spectrum of this text will encourage geologists to look outside their immediate areas of interest with the potential to learn from complementary work in other areas.

This publication grew out of an SPE (Society of Petroleum Engineers) forum held at Breckenridge where the need for documenting the use of borehole images was identified. Subsequently, a conference entitled 'Borehole Imaging: Case Histories', held at the Geological Society of London, provided a forum for presenting real examples of the application and use of borehole images. Much of the contents of this book were discussed at that meeting.

The original meeting included discussion workshops which proved exciting and stimulating but were difficult to document in a worthwhile manner. There was a technical exhibition demonstrating the latest advances in data manipulation and interpretation and other forums which also emphasised the case histories at the expense of technological developments. However, comments from colleagues identified the need to document how borehole images are obtained, how images may be misinterpreted, and the range of published material existing in this subject area. This final text has evolved through time with additional papers solicited to cover areas with previously limited documentation.

Mike Lovell, Gail Williamson & Peter Harvey

Advances in borehole imaging technology and applications

STEPHEN E. PRENSKY

US Minerals Management Service, 1201 Elmwood Park Blvd., New Orleans, LA, 70123, USA (e-mail: stephen_prensky@mms.gov)
Currently: Prenskey Consulting Services, 11800 Auth Lane, Silver Spring, MD 20902 (e-mail: steve@sprensky.com)

Abstract: Borehole imaging devices are currently available for both wireline and logging-while-drilling (LWD) environments for use in both openhole and cased wells. Most wireline, and all LWD devices provide indirect images that are derived from a high-density grid of electrical or ultrasonic acoustic measurements. Optical techniques, such as television and photography, provide direct images of the borehole. Imaging devices consist of (1) a rotating sensor (acoustic or electrical) that scans the borehole with each revolution, (2) multiple, azimuthally placed sensors, or (3) a downhole television or photographic camera. Advances in digital acquisition and processing permit real-time display and evaluation of images of the open borehole or the inside of casing. The resolution and borehole coverage of microelectrode devices has increased with newer designs that allow additional measurement sensors per pad, additional pad segments on a four-arm tool, and the introduction of six-arm imaging tools. In nonconductive borehole fluids, a four-arm microinduction device can provide crude images. Macroelectrode imaging tools (laterolog-type; wireline) provide lower-resolution electrical images from either an azimuthal arrangement of sensors around a mandrel (wireline), or from azimuthally sensitive sensors (logging-while-drilling), that scan the borehole during rotation of the bottomhole assembly. In acoustic imaging, recent devices incorporate design changes that provide improved image resolution and operate in a wider range of borehole conditions. Ultrasonic transducers now use lower transmitting frequencies and spherical focusing and these tools also offer surface-selectable acquisition parameters. Acoustic and microresistivity imaging techniques measure different physical properties and these data are complementary. One recently introduced wireline service combines microresistivity and acoustic imaging devices in a single tool for simultaneous, single-pass acquisition. Recent advances in cable and sonde design permit downhole video in a wider range of petroleum environments, in flowing wells, and in tubing. Openhole image interpretation involves qualitative (visual) identification and quantitative characterization of strike and dip on planar features, lithology, bedding, fractures, and vugs. When core is available, features identified on log-derived images can be correlated or calibrated to it, or when core is absent, images may serve as a substitute. Borehole images are useful in formation evaluation (when calibrated), sedimentology, stratigraphy, and structural analysis. Current applications include core orientation, 'reconstructing' missing core, identification and characterization of fractures and depositional features (e.g. bedforms), reservoir architecture, bedding, grain size, porosity, permeability, net sand and net pay counts, faults, folds, and fractures (hydrocarbon, water, geothermal wells, and in rock engineering), evaluation of borehole stress (borehole stability), directional placement of wellbore for optimal drainage (fractured reservoirs), evaluation of fracture stimulations, and perforation control. Cased hole applications include cement evaluation, casing inspection, detection of fluid entry, identifying production blockages, and assisting in fishing operations.

Borehole imaging developed from industry's need for borehole inspection devices, for fracture detection and for use in well remediation. Advances in technology (Table 1) represent evolutionary improvements on, or adaptations of television cameras, the acoustic televiewer, wireline microlog, dipmeters and laterologs. The first devices, introduced in the 1950s, consisted of photographic cameras rigged for borehole use. These were followed in the mid-1960s by the introduction of downhole television devices and, in the late 1960s, by the first acoustic imaging device, the borehole televiewer (BHTV). Resistivity-based borehole imaging devices were introduced in the mid-1980s and represent the evolution of high-resolution dipmeter (microelectrode) and laterolog (macroelectrode) designs.

Recent developments in borehole imaging technology have been driven, in large part, by changes in petroleum exploration strategies.

From: LOVELL, M. A., WILLIAMSON, G. & HARVEY, P. K. (eds) 1999. Borehole Imaging: applications and case histories. Geological Society, London, Special Publications, **159**, 1–43. 1-86239-043-6/99/$15.00.

Table 1. *Summary of significant advances in borehole imaging technology*

Company	Tool name	Year	Borehole coverage (%)	Logging speed (ft/hr)	Comments	Reference
Optical						
Photography						
Birdwell	–	1958	100	–	16-mm lens	Dempsey & Hickey 1958
—	–	1964	na	–	high-resolution	Kotyakhov & Serbrennikov 1964
Layne Texas	–	1965	na	–	35-mm stereo	Jensen & Ray 1965
Laval Surveys	–	1966	na	—	packer assembly	Mullins 1966
Television						
Shell		1964	100	?	black and white	Briggs 1964
SW Res./GRI		1985	100	?	color, VCR, UV light	Darilek 1985, 1986
Shimizu Const.		1987	100	>100	high-resolution colour	Iizuka *et al.* 1987
Raax	BIPS	1989	100	180–240	high-resolution colour, 3 display modes	Kamewada *et al.* 1989, 1990; Raax 1993
Westech/DHV	DHV	1992	100	600–1800	fiber-optic cable	Cobb & Schultz 1992; Cobb 1993
Halliburton	DHV	1993	100	600–1800	coiled tubing	Rademaker *et al.* 1992
Halliburton	DHV	1993	100	600–1800	flowing wells	Allen *et al.* 1993; Maddox *et al.* 1995; Maddox 1996*b*
Kyoto Univ.	BSS	1995	100	250	digital color scanner, 0.1 mm resolution, shallow wells (<650 ft)	Tanimoto *et al.* 1995

Company	Tool name	Year	Tool dia. (inches)	Operating frequency (kHz)	Rotational speed (rpm)/ samples per revolution	Borehole coverage (%)	Recommend logging speed (ft/hr)	Sampling rate vert./horiz. (inches) (8-inch hole)	Comments	Reference
Acoustic										
Borehole televiewer										
Mobil	BHTV	1968	3.375	2000	180/?	100	900	<1.0	analog, oscilloscope, Polaroid film	Zemanek *et al.* 1968
Mobil	BHTV	1971	1.75	1300	180/?	100	900	<1.0	slimhole tool, can run in 2-inch tubing	Glenn *et al.* 1971
Amoco	BHTV	1980	3.375	1300	180/485	100	300	0.33/<0.1	analog-to-raster conversion, CRT display, continuous VCR, digital reprocessing	Wiley 1980; Jaegler 1980

Amoco	BHTV	1981	3.375	1300	?/512	100	?	?	volumetric scanning, 3-D presentation	Broding 1981
Los Alamos	BHTV	1981							high-temperature (535 deg F) tool	Heard 1981; Heard & Bauman 1983
Arco	BHTV	1983	3.375	400/1300	180/256	100	300	0.33/0.1	digitization of analog recording, digital reprocessing	Pasternack & Goodwill 1983
	SABIS	1983	1.8	1400	180/256	100	300 [?]	0.33/0.1	digital tool, video display, AGC, slim hole	Hinz & Schepers 1983
Shell	BHTV	1984	3.375	1300	180/256	100	300 [?]	0.33/0.1	analogue-to-digital conversion, AGC, video display	Rambow 1984
Arco	DUST	1988	3.25	1000–2700	180/200	100	300	n/a	digital, no protective window, casing thickness	Katahara *et al.* 1988
Atlas	CBIL (Star)	1989	3.625	250	360/250	100	600	0.33/0.1	dual spherically focused transducers wbm $\leq$16 lb/gal	Faraguna *et al.* 1989; McDougall & Howard 1989; Zemanek *et al.* 1990
	DBT	1990	?	500/1500	300/512	100		≈0.3/0.1	dual transducers, openhole and cased-hole modes	Xie 1990
DMT (Germany)	FACSIMILE	1989	2.37, 3.5	?	720/512	100	600	0.2/0.25	focused transducer, high temperature and pressure	Schepers 1991
Halliburton	CAST	1990	3.625	450	?/100–500	100		/0.2	no transducer window	Seiler *et al.* 1990
Schlumberger cased hole	USI	1991	3.375	195–650	390–450	100	400–800 1600–3200	0.35/0.33 0.7/	cement evaluation, casing inspection	Catala *et al.* 1991; Hayman *et al.* 1991*b*
	UCI	1995	3.375	2000	450/180	100	425	0.2/0.15	focused transducer casing corrosion	Hayman *et al.* 1995
openhole	UBI	1993	3.375	250 500	450/180 450/180	100	800 400	0.4/0.4 0.2/0.2	WBM <16 lb/gal OBM <11.6 lb/gal	Schlumberger 1993; Hayman *et al.* 1994
BPB	SAS	1996	2.25	500/1000	480/200	100	500	0.5/0.2	slimhole tool	Elkington 1996
Tohoku Univ.	DBHTV	1995		2390					doppler shift (fracture permeability) measurement	Niitsuma & Inagaki 1995
Halliburton/ GRI		1997							images $\leq$6 inches beyond casing	Birchak *et al.* 1997

Table 1. (*continued*)

	Tool	Year	Arms (pads)	Total electrodes	Electrodes per row/rows	Borehole coverage percent	Logging speed (ft/hr)	Sampling rate vertical/ horizontal (inches)	Comments	Reference
Electrical										
Microelectrical									pad-type	
Schlumberger	FMS	1986	2	54	2/27	20	1600		2-pads imaging, 2-pads SHDT	Ekstrom *et al.* 1986; Lloyd *et al.* 1986
Schlumberger	FMS	1988	4	64	4/16	40	1600	0.1/0.2	4-pads imaging	Boyeldieu & Jeffreys 1988
BRGM (France)	ELIAS	1989	16	16 +16 guard	2/1	100	1200	0.1	2.5-inch diameter, focused electrodes	Straub *et al.* 1989
Schlumberger	FMI	1991	4+4	192	4/48	80	1800	0.1/0.1	4 pads plus 4 flaps	Safinya *et al.* 1991
Halliburton	EMI	1994	6	150	6/25	60	1800	0.1/	6-arm tool, calibrated electrode	Seiler *et al.* 1994
Western Atlas	Star	1995	6	144	6/24	60	600–2400	0.1/?	6-arm tool; powered standoff	Tetzlaff & Paauwe 1997
BRGM (France)	ELIAS	1995	16	32	2/1	100	1200	0.1	2.0-inch diameter	Straub *et al.* 1995
Macroelectrical									mandrel-type	
CNRS (France)	–	1987	–	24	24/1	100	?	3.0/4.0	focused laterologue-type	Mosnier 1982, 1987
CNRS (France)	–	1989	–	160	16/10	100	?	?	image acquisition during frac job	Mosnier & Cornet 1989
Schlumberger	ARI	1992	–	12	12/1	100	1800/3600	8.0/	focused laterologue, calibrated image; dip estimates	Davies *et al.* 1992
	RAB	1994	–	1	–	100	30	2.0/2.0	LWD device, real-time images possible	Bonner *et al.* 1994; Rosthal *et al.* 1995
	HALS	1995	–	12	12/1	100	1800/3600	8.0/	software focused laterologue	Smits *et al.* 1995

Hydrocarbon reservoirs previously considered non-conventional or difficult, such as, deep-water, low permeability, fractured, thin bed, and laminated, are now standard exploration targets. Borehole imaging now plays a central role in modern reservoir description and characterization upon which successful exploration and production strategies depend. Improvements in borehole logging, in general, and in imaging technology, in particular, were made possible by concurrent advances in data acquisition, telemetry, and data processing (see review by Prensky 1994*a*, *b*).

The purpose of this paper is to summarize both the historical and current status of borehole imaging technology and applications. Image interpretation is not covered and readers should consult the cited references for discussions on this topic. Tables summarizing tool development, comparison of imaging types, and applications are provided for easy reference. A comprehensive bibliography of published literature on borehole imaging accompanies this paper. Additional references specifically related to borehole stress analysis, can be found in Prensky (1992). For more detailed discussions of the historical development, tool operation, data processing, and image interpretation, readers are referred to papers by Paillet *et al.* (1990*a*), Serra (1989), Rider (1996), and Lofts *et al.* 1997.

General discussion

Modern borehole imaging logging tools provide high-resolution images of the borehole wall, in openhole, or of tubulars (casing and tubing) in cased wells. These devices create images either directly, through optical technologies (photographs or video) or indirectly, through a high-density of geophysical measurements (electrical resistivity or acoustic reflectivity). The extent of borehole coverage and image resolution are functions of the area measured, which is in turn, related to sensor size and number (in the case of electrical tools), or transducer size and focusing (in the case of acoustic devices), and the horizontal and vertical sampling rates (Table 1). Magnetometers, inclinometers, and accelerometers comprising part of the tool or located elsewhere in the tool string, measure tool azimuth, inclination, and speed.

Early acoustic tools and borehole television devices acquired analogue images that were stored on film or video tape and subsequently digitized for analysis, if so desired. In contrast, modern acoustic and electrical imaging tools acquire digital data directly, as raster scans or as a matrix of individual measurements. Electrical and acoustic devices record data in the form of continuous helical scans, as the tool moves upwards. Computers convert the helical scans into circumferential scans and process the matrix of measurements acquired at each depth into pixels. Standard digital processing techniques that include signal conditioning, image enhancement, and image analysis are applied to borehole images (Clavier 1994). The measurements of tool azimuth and inclination are used to provide the proper borehole orientation to the images. Errors in processing or the inappropriate use of a technique may introduce image artifacts that must be recognized prior to interpretation and feature extraction (Bourke 1989).

Signal conditioning corrects the raw sensor signals for noise and distortion introduced during acquisition (e.g. synchronization of adjacent sensors, speed corrections, geometrical corrections for tool eccentering, amplitude equalization). Image enhancement techniques (e.g. filtering, sharpening, normalization, and false coloring) improve overall image quality and can be used to emphasize specific features (Ekstrom 1984; Burns 1985; Serra 1989; Jain 1989; Goetz *et al.* 1990; Wong *et al.* 1989*a*, *b*; Pratt 1991; Walbe 1992; Rao & Sunder 1995).

Image analysis techniques are used for qualitative recognition and quantitative extraction of structural features (strike and dip), fracture features (e.g. apertures), and depositional features (e.g. vugs and bedding). The choice of techniques may vary with the intended application of the image data. The resulting images are displayed in gray or color scales to represent the range of electrical conductivity or resistivity and acoustic traveltime or amplitude.

Digital analysis of static or single-frame images derived from borehole imaging tools includes fracture and petrophysical characterization techniques originally designed for use with core and core thin sections. These techniques can provide both qualitative and quantitative information on fracture orientations (Thapa *et al.* 1997; Xu & Jacobi 1997) lithology, sand/shale ratios in thin beds, and in some cases, quantitative evaluation of a reservoir's 3-D pore system (porosity, permeability, pore structure, and distribution) (Ehrlich & Davies 1989; Phillips *et al.* 1991; Georgi *et al.* 1992; Gerard *et al.* 1992; Ruzyla 1992; Bigelow 1993; Standen *et al.* 1993; Oyno *et al.* 1996) and can be used to correlate and calibrate borehole images to core for quantitative applications (Berger & Roestenburg 1989; Harker *et al.* 1990; Jackson *et al.* 1990, Bourke 1992, 1993; Sovich & Newberry 1993; Sullivan 1994; Sullivan & Schepel 1995; Sovich *et al.* 1996). Using images alone,

Table 2. *Comparisons and guidelines, for fracture identification tools (modified from Kubik* et al. *1992)*

Technique	Advantages	Limitations/disadvantages	Guidelines
Core	direct observation; most accurate fracture ID and characterization	expensive	continuous (electronic) orientation preferred
	ID of other fracture types	possible sample loss especially in heavily fractured intervals	detailed reconstruction critical
	ID of mineralization type and extent; aperture estimates	variable reliability of orientation data	procedures to minimize coring/handling induced fractures
	not dependent upon spell, borehole coverage, tool noise, etc.		
	can be cored with air		
Microresistivity	high-resolution	requires conductive muds	use interpretation windows of 4 and 30 ft
	good strike/dip quantification of natural and induced fractures	limited borehole coverage on single pass of 4-pad tool may result in poor fracture ID	make multiple passes with a 4-pad tool, or use 6- or 8-pad devices for higher borehole coverage
	thin-bed detection	joint ID dependent largely upon degree of borehole spalling	cross-correlate with core and/or other tools
	data routinely corrected for borehole deviation and magnetic declination	large spalls difficult to identify as joints and to quantify	many studies indicate acoustic techniques image fewer (undersample) fractures compared with microresistivity techniques
	low operational sensitivity	resolution decreased by pad standoff	
		irregular logging speed (stick/pull) causes layer misalignment	

Acoustic	essentially full borehole coverage	reduced effectiveness in heavy muds; can't operate in air/gas-filled holes	take all precautions possible to minimize eccentering and tool noise
	good strike/dip quantification of natural and induced fractures	rugosity, eccentricity, and motor noise may locally limit effective borehole coverage	use directionally corrected tool package if available
	breakout detection and orientation (stress tensor)	joint identification dependent largely upon degree of borehole spalling	cross-correlate with core and/or other tools
	high-resolution borehole caliper (geometry)	large spalls difficult to identify as joints and to quantify	
		some tools are not routinely corrected for borehole deviation and magnetic declination	
Television	quick and inexpensive; can log in air-and gas-filled holes	light intensity often varies around borehole; slight obstruction by tool arms	needs gas or clear borehole fluid; pre-certify that directional gyro is functional
	high-resolution visual image	poor dip quantification	correction should be made for gyro drift
	easily reviewed and interpreted with TV and VCR	poor ID of induced fractures	multiple runs may be useful, particularly with active and/or variable shows
	full borehole coverage	joint ID dependent largely upon degree of borehole spalling	cross-correlate to core and/or other tools
	gas/fluid entry detection	limited ID of bedding	
	can properly identify and quantify large spalls as joints	active gas/oil production may limit visibility	
	good/fair strike quantification; only qualification of dips		

quantitative fracture-characterization methods have been proposed (Svor & Meehan 1991).

Due to the physics of the two-electrode arrangement of microelectrode imaging devices, these measurements represent only relative values (Lovell *et al.* 1997*a*, *b*). True quantitative interpretation of borehole images requires either absolute calibration of sensor response (Fam *et al.* 1995, 1996), calibration of sensor response to the small-scale features of interest in image interpretation, or both. Calibration of qualitative image data is accomplished through modeling: using analog or numerical techniques, or both; synthetic-rock boreholes (Williams *et al.* 1994, 1997; Lovell *et al.* 1997*a*, *b*; Yu *et al.* 1997, 1998); or a combination of core and other wireline logs (Adams *et al.* 1990; Williams & Sharma 1991; Bourke 1992; Sullivan 1994; Fam *et al.* 1995, 1996; Mathis *et al.* 1995; Sullivan & Schepel 1995). A device designed to make direct microelectrode measurements on core, at the same scale as microelectrode imaging devices, is being used to relate qualitative borehole images to quantitative petrophysical parameters (Jackson *et al.* 1990; review of Lovell *et al.* 1997*b*).

The resolution of an imaging tool, that is, the ability to recognize objects, is primarily a function of the tool's sample rate. This, in turn is dependent on sensor size (resistivity) or beam spot size (acoustic). Each measured sample is converted into a graphic pixel. However, since features the size of a single pixel cannot be resolved, the actual resolution is a multiple of the sample rate (Rider 1996). For example, a tool with horizontal and vertical sampling rates of 1-inch by 1-inch will be able to resolve features that are >0.2 in.

Borehole images are presented in a variety of 2D and 3D visualizations (e.g. Barton & Zoback 1989; Seiler 1995). The standard 2D presentation for acoustic and electrical borehole images is in the form of an unrolled cylinder (the borehole) split along magnetic north. On this 2D presentation, planar features (e.g. bed boundaries, faults, and fractures) that intersect the cylindrical borehole at an angle from the horizontal, i.e. have a dip component and appear as sinusoidal features, the amplitude of the sine curve is a function of the dip angle (the low side of the sine curve amplitude equals the apparant dip which is converted to true dip). Optical (video) images which are normally presented in circular downward-looking or side-view formats, can also be displayed in this manner. Acoustic traveltime data are also commonly displayed as polar cross sections. Three-dimensional presentations provide more accurate information regarding spatial relationships than do 2D displays, particularly in non-circular boreholes (Seiler 1995). As with conventional well logs, the interpretation of borehole images in horizontal wells requires a paradigm shift in thinking of the geologic conditions that can produce the observed response (image) (Lofts *et al.* 1997).

Modern processing and interpretation systems are commonly Unix- or PC workstation-based, are interactive, and, in addition to manual techniques, generally incorporate automatic techniques for feature-extraction (bed boundaries, dip-analysis, rock texture, lithofacies) (e.g. Barton 1988; Antoine & Delhomme 1990; Cheung & Heliot 1990; Torres *et al.* 1990; Schmitt 1991, 1993; Barton *et al.* 1992; Ma *et al.* 1993; Luthi 1994; Faivre & Catala 1995; Verhoeff & Frost 1995; Hall *et al.* 1996; Glossop *et al.* 1997; Ye *et al.* 1997, 1998).

Table 2 compares the advantages and limitations of the three types of borehole imaging technologies. For some applications, either microresistivity and acoustic devices provide acceptable images and the choice of technique depends on borehole mud type. For other applications, differences in the desired petrophysical parameters may be the determining factor. Microresistivity imaging tools are sensitive primarily to resistivity contrasts (fluid contrasts) in rocks and, due to sensor size, number, and arrangement, and they can resolve finer-scale features than acoustic devices. However, due to the sensor design, these electrical measurements are relative (uncalibrated), rather than absolute and do not represent quantitative measurements of formation resistivity. Microresistivity devices are better suited to applications related to lithology and fracture characterization (large and small, open and filled) than are acoustic methods. In contrast, acoustic amplitude is very sensitive to geometrical variations in the borehole surface, e.g. rugosity. Acoustic imaging techniques are better suited to evaluating rock properties such as porosity (e.g. vugs), borehole failure (breakouts), and large open fractures (Laubach *et al.* 1988) and acoustic traveltime provides a very high-resolution borehole caliper. Acoustic imaging is most effective where there are high contrasts in acoustic impedance, generally in hard formations (Paillet *et al.* 1990*b*). In formations where variations in lithology are accompanied by variations in hole size, acoustic images can be used for discriminating changes in lithology. In fractured formations (water-based mud), microresistivity and acoustic images respond to different fracture attributes and electrical and acoustic data are therefore complementary. Combined image interpretation and display of acoustic and electrical images allows discrimination between open and closed fractures, deep and shallow fractures, and natural and induced fractures (e.g. Dennis *et al.* 1987; Laubach *et al.* 1988; Kubik *et al.* 1992; Schlumberger 1993). However, separate logging runs with acoustic and electric tools are generally not made in the same borehole interval due to the rigtime and expense involved. To overcome this disincentive, a new commercial service provides simultaneous acquisition of both acoustic and microresistivity images. The objective is to make use of the synergy derived from combined image interpretation (acoustic images provide full borehole coverage while microresistivity images offer higher resolution and dynamic range) (Tetzlaff & Paauwe 1997; Yu *et al.* 1997, 1998).

The primary applications for borehole imaging derive from their inherent ability to provide more detail of the features present in the formations surrounding the borehole or of the casing. These applications fall under the following categories: (1) fracture identification and evaluation (both natural and induced through drilling or formation stimulation), (2) diagnosing completion and production problems, (3) analysis of sedimentology, bedforms, and depositional structures, (4) net pay counts in thinly bedded formations, and (5) identification of wellbore failures such as borehole breakouts, tensile wall cracks, and shear surfaces. Borehole breakouts, i.e. spalling of the borehole wall in a preferential direction due to regional stress patterns (Cox 1970; Babcock 1978), are readily detected by all forms of borehole imaging (see reviews by Bell 1990, 1996; Prensky 1992; Barton *et al.* 1997). Knowledge of the horizontal stress patterns is used for inferring natural fracture patterns, designing fracture stimulation programs, developing drilling programs to maximize borehole stability, and to study regional crustal stresses. Acoustic and microresistivity imaging devices can also be run in reduced data-set, 'dipmeter mode', to obtain structural and stratigraphic dips at higher logging speeds. Additional, or specialized applications of borehole images are discussed below.

Optical imaging devices

The need for visual inspection and recording of downhole conditions led to the use of photographic cameras in openhole (to provide direct images of fractures) and in cased hole (for completion evaluation, e.g. verification of water inflow, for tubular inspection, and to aid in fishing operations, The development of miniature television cameras in the 1950s and 1960s enabled downhole video, which subsequently replaced downhole photography. Although the need and potential applications were ever present, until recently, direct-imaging technology has had only limited use because of the operational requirement for a clear (transparent) borehole fluid. In most cases, either adverse borehole conditions (e.g. formation instability), cost, or the time required to displace conventional drilling mud (opaque) with a clear fluid, were prohibitive. Furthermore, the quality and resolution of coaxially transmitted downhole video is adversely affected by high pressures and temperatures. Consequently, early optical imaging techniques were used primarily in shallow (<5000 ft) air-, gas-, or water-filled boreholes.

Table 1 summarizes the development of optical imaging techniques. Photographic equipment has included both standard and wide-angle lenses, stereoscopic photo pairs, and the use of an inflatable packer to enable displacing a limited interval of borehole fluid. Both photographic- and television-camera borehole devices include a downhole light source and use either a super-wide-angle (fisheye) lens or a mirror (stationary or rotating) to provide up to 100% borehole coverage (downward-looking or side-views). Downhole television devices use miniature television cameras to acquire black-and-white or color images of the open or cased borehole. Directional orientation is provided either by a magnetic compass mounted on the mirror, where it can be included in the image, or a by gyroscope run in conjunction with the imaging tool. Borehole video images are acquired in real-time as a series of analogue static images and recorded in continuous mode using conventional VCR technology. The effect of motion can be simulated by short image refresh rates and relatively slow logging speeds. The recorded analogue images can be converted to digital images at the surface either during logging or off the video recordings, at a later time. Once in digital form, image processing techniques can be applied to image enhancement (Walbe 1992) and fracture image analysis (Xu & Jacobi 1997). Currently, borehole video images are only qualitative in nature but future development will include quantitative image analysis. The conventional downhole video image is a forward projection looking down a circular borehole. One system also records images in the standard 2D flattened (unwrapped) cylindrical projection and in a detailed image of a small portion of the borehole (Raax 1992, 1993). Current downhole video systems primarily use plain light illumination, although some systems also offer an ultraviolet (UV) light source (Darilek 1985, 1986).

Recent improvements in camera-lens technology, the introduction of surfactants to prevent oil adhesion to camera lens and lighting, and in data telemetry have resulted in high-resolution borehole video systems capable of operating at the depths (≤20 000 ft), temperatures (121°C), and pressures (≤10 000 psi) of deep petroleum environments (Cobb & Schultz 1992). Fiber-optic telemetry systems offer the greater bandwidth needed for high-resolution real-time video and improved lens protection permits operation during oil inflow (Maddox 1996*b*; Whittaker & Linville 1996). Advances in microelectronics and camera construction have resulted in smaller-diameter (1–11/16 inches) and more reliable tools that can be run on wireline or coiled tubing, in vertical or horizontal wells (Rademaker *et al.* 1992; Maddox *et al.* 1995).

Table 3 summarizes the applications of borehole video. At the present time, the primary commercial application is diagnosing well production problems such as inspection of tubulars for wear, damage, collapse, leaks, scale buildup, or corrosion; identifying fluid entry and phase; inspecting blocked or damaged perforations, gravel packs, or bridges; and in fishing operations. Downhole television has also been used in the United States for evaluation of formation lithology and fractures (natural, drilling induced, and hydraulically induced for well stimulation), particularly in water wells and in low-permeability gas reservoirs (Devonian shales and coalbed methane). Video surveys have also been used for developing post-logging production profiles (Allen *et al.* 1993), the study of openhole formation porosity (Hickey 1993; Warner 1996) and pressure compartmentalization (Bilodeau *et al.* 1997) and glacier research (Harper & Humphrey 1994). High-resolution borehole video is also widely used for geotechnical and environmental evaluations in shallow boreholes (<5000 ft). High-resolution and portable colour systems have been developed for rock engineering applications, e.g. fracture evaluation, in shallow wells (Tanimoto *et al.* 1995; Raax 1992, 1993) and also in 'non-borehole' applications such as the examination of tunnel wall condition (the surface equivalent of horizontal boreholes) (Taniguchi *et al.* 1993).

Acoustic imaging devices

Acoustic imaging tools are mandrel devices that consist of a rotating (4–12 rps) piezoelectric transducer that operates in a pulse-echo mode at an ultrasonic frequency. The transducer, which acts as both transmitter and receiver, measures the traveltime and amplitude of an acoustic pulse reflected off the borehole wall or casing as the transducer scans the borehole with each rotation. Magnetometers provide azmuthal information for each scan. Acoustic devices operate in both conductive (water-based) and non-conductive (oil-based) muds. Because acoustic energy does not travel in a gas, televiewer-type tools cannot operate in air- or gas-filled boreholes.

Acoustic traveltime (velocity) is a function of the distance to the borehole wall, or casing, and the velocity of the borehole fluid. This in turn is related to the density of the borehole fluid. Signal amplitude is a function of the acoustic impedance (density and velocity) of the borehole fluid. Loss in signal amplitude (image quality) results from conditions that either scatter, absorb, or spread the acoustic energy, such as, tool eccentering, irregularities in the borehole shape and surface, high-density and some oil-based drilling muds, contrasts in acoustic impedance between the borehole fluid and borehole wall or casing. Tool centering is necessary for obtaining optimum image quality. Algorithms have been developed to correct for geometrical problems such as eccentering and to extract high-resolution borehole caliper data (e.g., Georgi 1985; Lysne 1986; Barton 1988; Menger & Schepers 1988; Menger 1994; Priest 1997). Eccentering problems in horizontal wells due to gravity can be reduced by use of powered tool standoff or special stabilizers. Signal scattering and attenuation in heavy muds and in oil-based muds is reduced through use of mud excluders, by eliminating or changing transducer window material, use of focused transducers, and by lowering the operating frequency.

Vertical image resolution is a function of acoustic beam spot size (that is, through focusing or, where unfocused transducers are used, their size), azimuthal sampling rate and logging speed. However, since acoustic beam spot size, i.e., image resolution, is directly proportional to signal frequency, there is a necessary compromise in the operating frequency to balance the desire for high image resolution and the need to operate in high density muds. The logging speed is determined, in part, by the logging objective, high-resolution openhole applications and cased-hole applications, such as casing inspection (thickness and corrosion) or cement evaluation, require slower logging speeds. For borehole imaging in open and cased hole, a high-resolution focused transducer is used and recognition of features as small as 1 mm are possible under good operating conditions. When used for cement evaluation, an unfocused plane-faced transducer may be used (Schlumberger 1993). The designs, materials, and operating characteristics (power and frequency) of the current generation of acoustic imaging tools represent a compromise between the need for spatial resolution, mud penetration, circumferential coverage and logging speed (Paillet *et al.* 1990*a*).

The acoustic imaging tool, the borehole televiewer, developed in the 1960s (Zemanek *et al.* 1968), eliminated the need for a transparent borehole fluid, the primary factor that, until that time, had limited the usefulness of optical devices. The earliest acoustic devices were analogue systems, the reflected acoustic pulses were converted electrical signals which modulated the electron beam on an oscilloscope; a camera recorded the image off the oscilloscope. Limitations, such as poor image display and high signal attenuation (lower signal-to-noise ratio) in heavy muds, resulted in a rapid decline in its use within a few years after its introduction (Rambow 1984). However, the need for a reliable fracture evaluation tool led to continued tool development in the late 1970s and early 1980s. Modification and enhancement of the basic televiewer design which had been licensed by Mobil to and manufactured by Simplec, resulted in a number of significant advances. The introduction of (1) automatic gain control (AGC) resulted in improved image quality, (2) analog-to-digital data conversion (raster scan images) led to improved video displays, videotape recording, and enabled digital image processing, and (3) 'volumetric scanning' which added a depth element and enable imaging beyond the borehole face (see Table 1). In the late 1980s and 1990s, development of true digital acquisition and telemetry, together with improvements in transducer design (lower frequencies and focusing) that offer better resolution and higher signal-to-noise ratios, and digital image processing techniques, resulted in a broad expansion in use of the acoustic imaging service.

Compared with earlier designs, the current generation of acoustic imaging tools (see Table 1) have lower broadcast frequencies (195–650 kHz) to permit operation in heavier muds and higher rotational rates that enable increased logging speeds at previous image resolution. The use of focused transducers enable these tools to provide image resolution up to 0.2 inches at 500 kHz and 0.3 inches at 250 kHz, despite the decrease in broadcast frequencies. To optimize logging speed and image resolution in different operating conditions (e.g. different and changing borehole size and mud density), some tools offer selectable transducer size (i.e. focusing) (McDougall & Howard 1989; Seiler *et al.* 1990), rotational speed, and vertical and horizontal sampling rates (Seiler *et al.* 1990). Modern tools also measure the downhole acoustic velocity of the borehole fluid (mud) for calibrating acoustic traveltime. Future tool developments may include the use of dynamic, electronically focused transducers that can provide optimal focal distance for any borehole size and eliminate the need for multiple transducers (Maki *et al.* 1991).

Televiewer-type, pulse-echo devices are also used in cased hole for perforation control, casing inspection (wear and damage) and cement evaluation (distribution and quality). In these applications, the acoustic transducer excites a thickness-mode resonance in the casing and then measures the decay of this signal. The attenuation of the signal is related to the acoustic properties of the adjacent materials, e.g., mud, casing, and cement. Analysis of these data provides measurements of internal surface roughness, casing thickness, and cement acoustic impedance (Schlumberger 1993). Thickness of the cement sheath evaluated by current tools is approximately one inch. A new tool, currently in development, will image formation boundaries up to six inches behind casing and cement thicknesses of one inch to four inches (Birchak *et al.* 1997).

Table 3 lists common applications for acoustic borehole imaging and references to published examples. The primary applications of acoustic imaging are identification and evaluation of fractures (e.g. Dudley 1993) and breakouts, borehole geometry, and cement evaluation. High-resolution caliper measurements are obtained from traveltime measurements either by assuming a constant mud velocity (density) or by obtaining actual measurement of downhole mud properties. Recently, efforts have focused on deriving petrophysical values, e.g. porosity, permeability and formation velocity, from televiewer amplitude and traveltime data (e.g. Davidson *et al.* 1992; Bigelow 1993). In one case, acoustic amplitude images are calibrated to core porosity leading to generation of maps of amplitude-derived porosity which can define the distribution formation anisotropy (Verhoeff & Chelini 1993). In another, an experimental doppler-shift televiewer measures the traveltime of backscattered waves from particles in the borehole fluid, in addition to the borehole wall reflection signal. If the fluid is moving, a doppler shift in the backscattered waves is detected. This technique can provide an estimate of fluid velocity in fractures, i.e. fracture hydraulic permeability (Niitsuma & Inagaki 1995).

Table 3. *Applications of borehole imaging*

Application	Lithology			Category			References
	Clastics	Carbonates	Crystalline	Acoustic	Electrical	Optical	
Openhole							
lithology/lithofacies	×					×	Darilek 1985
	×			×			Rambow 1984; Rubel *et al.* 1986
		×		×			Zemanek *et al.* 1968; Clerke & Van Akkeren 1986
		×			×		Badr & Ayoub 1989; Lovell *et al.* 1997*a*
			×		×		Brewer *et al.* 1995; Lovell *et al.* 1997*a*
			×			×	Hawkins *et al.* 1989
thin beds, tubidites, and net sand	×					×	Kotyakhov & Serbrennikov 1964; Walbe 1986
	×	×		×			Zemanek *et al.* 1968; Jageler 1980; Rambow 1984; Clerke & Van Akkeren 1986; Plumb & Luthi 1986; Hackbarth & Tepper 1988; Luthi & Banavar 1988; Goetz *et al.* 1990; Seiler *et al.* 1990; Ma & Bigelow 1993; Gautama 1994; Firmansjah *et al.* 1996
	×				×		Plumb & Luthi 1986; Hackbarth & Tepper 1988; Luthi & Banavar 1988; McGann *et al.* 1988; Bourke *et al.* 1989; Serra 1989; Trouiller *et al.* 1989; Adams *et al.* 1990; Harker *et al.* 1990; McNaboe 1991; Heine 1993; Sovich & Newberry 1993; Deri & Chiarabelli 1994; Hurley 1994; Slatt *et al.* 1994; Sullivan 1994; Eubanks *et al.* 1995; Fam *et al.* 1995 1996; Sullivan & Schepel 1995; Ramamoorthy *et al.* 1995; Carr *et al.* 1996; Prilliman *et al.* 1997; Sovich *et al.* 1996; Coll *et al.* 1997; Lovell *et al.* 1997a; Pirmez *et al.* 1997; Spang *et al.* 1997; Taha 1997; Reid and Enderlin 1998
stress analysis	×			×			Hickman *et al.* 1985; Plumb & Hickman 1985; Zoback *et al.* 1985; Frikken 1996; Barton *et al.* 1997
	×				×		Pezard *et al.* 1992; Fett *et al.* 1994; Messent and Yacopetti 1997
			×	×			Paillet 1985; Stock *et al.* 1985; Barton *et al.* 1988, 1992; Barton & Zoback 1988, 1990; Paillet & Kim 1985; Burns

fracture characterization	×			×			Taylor 1983; Hickman *et al.* 1985; Paillet *et al.* 1985; Morin & Barrash 1986; Barton 1988, Laubach *et al.* 1988; Barton & Zoback 1989; Glowka *et al.* 1990; Heliot *et al.* 1990; Seiler *et al.* 1990; Paillet & Goldberg 1991; Schaar 1992; Ma & Bigelow 1993; Taylor 1991; Kubik *et al.* 1992; Dudley 1993; Johnston & Wachi 1994; Knight 1995; Frikken 1996
	×				×		Laubach *et al.* 1988; Bourke *et al.* 1989; Serra 1989; Heliot *et al.* 1990; Svor & Meehan 1991; Hornby & Luthi 1992; Kubik *et al.* 1992; Lambertini 1992; Schaar 1992; Fett *et al.* 1994; Hurley *et al.* 1994; Johnston & Wachi 1994; Sullivan 1994; Eubanks *et al.* 1995; Onions & Whitworth 1995; Sullivan & Schepel 1995; Evans *et al.* 1996; Kawamoto *et al.* 1996; Roestenburg 1996; Carmona *et al.* 1997
	×					×	Dempsey & Hickey 1958; Smith *et al.* 1982; Darilek 1985; Overbey *et al.* 1988; Palmer & Sparks 1990; Walbe & Collart 1991; Hollub 1993; Thapa 1994; Choi & Saul *et al.* 1995; Thapa *et al.* 1997
		×		×			Zemanek 1968; Wiley 1980, Rambow 1984; Paillet *et al.* 1985; Dennis *et al.* 1987; Hornby *et al.* 1990; Paillet & Goldberg 1991; Svor & Meehan 1991
		×			×		Ekstrom *et al.* 1986; Dennis *et al.* 1987; Lehne 1988; Casarta *et al.* 1989; Serra 1989; Stang 1989; Gonfalini & Anxionnaz 1990; Hornby *et al.* 1990; Svor & Meehan 1991; Hornby & Luthi 1992; Faivre 1993; Standen *et al.* 1993; Bullwinkel *et al.* 1994; Friedman & McKiernan 1994; Hurley *et al.* 1994; Roca-Ramisa *et al.* 1994; Xu & Yun 1994; Akbar *et al.* 1995; Mendoza 1996; Sanders & Burkhart 1996; Sanders & Fuchs 1996; Hammes 1997; McDonald *et al.* 1997
		×				×	Owen & Darilek 1987
			×	×			Paillet 1985; Paillet *et al.* 1985; Stock *et al.* 1985; Lau *et al.* 1987; Paillet *et al.* 1987; Hornby *et al.* 1990; Genter *et al.* 1991; Tenzer *et al.* 1992; ; Morin *et al.* 1993; Hayman *et al.* 1994; Barton *et al.* 1995; Dezayes *et al.* 1995, 1996; Genter *et al.* 1995; Tenzer 1994; Genter *et al.* 1997*a, b*
			×		×		Hornby *et al.* 1990; Luthi & Souhaite 1990; Genter *et al.* 1991; Straub *et al.* 1990; Hornby & Luthi 1992; Genter *et al.* 1995; Faivre 1993; Deri *et al.* 1994; Tenzer 1994; Dezayes *et al.* 1995, 1996; Kato *et al.* 1995; Celerier *et al.* 1996; Ogawa 1996; Walthan *et al.* 1996; Genter *et al.* 1997; Lovell *et al.* 1997*a*

Table 3. *(continued)*

Application	Lithology			Category			References
	Clastics	Carbonates	Crystalline	Acoustic	Electrical	Optical	
	×				×		Plumb & Luthi 1986; Serra 1989; Safinya *et al.* 1991; Hurley 1994; Eubanks *et al.* 1995; Haase 1996; Messent and Yacopetti 1997
	×					×	Smith *et al.* 1982; Darilek 1985; Walbe 1986; Overbey *et al.* 1988; Walbe & Collart 1991; Palmer & Sparks 1990
		×			×		Hurley 1994; Xu & Yun 1994
			×	×			Barton *et al.* 1995; Ikeda *et al.* 1995
			×			×	Hawkins *et al.* 1989
horizontal wells	×				×		Svor & Meehan 1991; Hurley *et al.* 1994; Evans *et al.* 1996; Coll *et al.* 1997; Follows 1997; Lofts *et al.* 1997
	×					×	Overbey *et al.* 1988
		×			×		Svor & Meehan 1991; Hurley *et al.* 1994; Ishak 1994; Al-Kharusi & Binbeck 1995
structure (dips)	×			×			Frikken 1996; Voight & Haberland 1996
	×				×		Pezard *et al.* 1992; Hurley 1994; Lovell *et al.* 1995; Favire & Catala 1995; Roestenburg 1996; Rider 1996; Messent and Vacopetti 1997; Prilliman *et al.* 1997; Rosthal *et al.* 1997; Taha 1997
		×			×		Al-Waheed *et al.* 1994
depositional features, fabric, image facies	×			×			Luthi & Banavar 1988; Berger & Roestenburg 1989; Seiler *et al.* 1990; Verdur *et al.* 1991; Gautama *et al.* 1994
	×				×		Grace *et al.* 1986; Harker *et al.* 1990; Luthi 1990; Lloyd *et al.* 1998; Luthi & Banavar 1988; Bourke *et al.* 1989; Serra 1989; Trouillier *et al.* 1989; Adams *et al.* 1990; McNaboe 1991; Bourke 1992; Molinie & Ogg 1992; Pezard *et al.* 1992; Salimullah & Stow 1992*a*, *b*; Englet & Sweet 1993; Bullwinkel *et al.* 1994; Connally & Wiltse 1994; Dueck & Paauwe 1994; Slatt *et al.* 1994; Eubanks *et al.* 1995; Mathis *et al.* 1995; Carr *et al.* 1996; Rider 1996; Carmona *et al.* 1997; Lovell *et al.* 1997*a*, Messent and Yacopetti 1997; Taha 1997; Anxionnaz *et al.* 1998

		×			×		Ekstrom *et al.* 1986; Badr & Ayoub 1989; Serra 1989; Gonfalini & Anxionnaz 1990; Nurmi *et al.* 1990; Roestenburg 1994; Akbar *et al.* 1995; Cooper *et al.* 1995; Mathis *et al.* 1995; Hammes 1997; Jurado-Rodriguez 1997
			×			×	Hawkins *et al.* 1989; Bennecke 1994
			×		×		Walthan *et al.* 1996
reservoir architecture	×				×		Plumb & Luthi 1986; Luthi & Banavar 1988; Carr *et al.* 1996; Ye & Kerr 1996; Carr-Crabaugh *et al.* 1996; Thompson & Snedden 1996; Mancini *et al.* 1997; Pirmez *et al.* 1997; Taha 1997
	×			×			Plumb & Luthi 1986; Luthi & Banavar 1988
		×			×		Akbar *et al.* 1995
palaeocurrent analysis	×				×		Pezard *et al.* 1992; Carr *et al.* 1996; Haase 1996; Roestenburg 1996; Mancini *et al.* 1997
sequence correlations	×				×		Hurley 1996
core orientation	×			×			Goetz *et al.* Seiler 1990; MacLeod *et al.* 1992
	×				×		MacLeod *et al.* 1992, 1994, 1995; Mathis *et al.* 1995; Lovell *et al.* 1997*a*
		×			×		Mathis *et al.* 1995
porosity evaluation	×			×			Verhoeff & Chelini 1993
		×		×			Zemanek *et al.* 1968; Clerke & Van Akkeren 1986; Bigelow 1993; Ma & Bigelow 1993
		×			×		Ekstrom *et al.* 1986; Casarta *et al.* 1989; Nurmi *et al.* 1990; Standen *et al.* 1993; Delhomme *et al.* 1996; Newberry *et al.* 1996; Sanders & Fuchs 1996
		×				×	Hickey 1993; Warner 1996
permeability and flow capacity				×			Tang & Martin 1994; Niitsuma & Inagaki 1995
	×				×		Thomas *et al.* 1996; Messent and Yacopetti 1997
		×			×		Delhomme *et al.* 1996; Fitzsimmons *et al.* 1997*a*, *b*
pressure compartments	×					×	Bileodeau *et al.* 1997
rock engineering		×	×			×	Iizuka *et al.* 1987; Kamewada *et al.* 1989, 1990; Taniguchi *et al.* 1989, 1993; Tanimoto *et al.* 1995

Table 3. *(continued)*

Application	Lithology			Category			References
	Clastics	Carbonates	Crystalline	Acoustic	Electrical	Optical	
fluid entry	×	×	×			×	Darilek 1985; Owen & Darilek 1987; Walbe & Collart 1991
Cased hole							
perforation control	–	–	–			×	Darilek 1985; Walbe 1986; Walbe & Collart 1991
	–	–	–	×			Hayman *et al.* 1994; Schlumberger 1993
well completion	–	–	–			×	Darilek 1985, Walbe 1986; Walbe & Collart 1991; Allen *et al.* 1993; Olsen *et al.* 1995; Maddox 1997
cement evaluation	–	–	–	×			Broding 1984; Broding & Buchanan 1986; Catala *et al.* 1991; Hayman *et al.* 1991*a, b*; Schlumberger 1993; Uswak & McLafferty 1993; Rouillac 1994; Butsch 1995; Birchak *et al.* 1997
tubular inspection (wear, corrosion)	–	–	–			×	Mullins 1966; Jensen & Ray 1965; Darilek 1985; Allen *et al.* 1993; Rademaker *et al.* 1992; Hollub 1993; Borchers *et al.* 1994; Maddox *et al.* 1995; Olsen *et al.* 1995
	–	–	–	×			Zemanek *et al.* 1968; Schaller *et al.* 1972; Carson & Bauman 1987; Cryer *et al.* 1987; Katahara *et al.* 1988; Graham 1989; Hayman *et al.* 1991*b*; Bettis *et al.* 1993; Schlumberger 1993; Brondel *et al.* 1994; Hayman *et al.* 1994; Hayman *et al.* 1995; Woods 1995; Ananto & Schnorr 1996
fluid entry	–	–	–			×	Jensen & Ray 1965; Darilek 1985; Walbe & Collart 1991; Allen *et al.* 1993; Starcher *et al.* 1995; Olsen *et al.* 1995; Maddox *et al.* 1995; Maddox 1996*b*; Whittaker *et al.* 1997
production profile						×	Allen *et al.* 1993; Maddox *et al.* 1995
fishing operations						×	Darilek 1985; Walbe 1986, Allen *et al.* 1993; Rademaker *et al.* 1992

Electrical imaging devices

Electrical imaging devices are electrode-type resistivity devices that measure resistivity and require a conductive borehole fluid. As with conventional resistivity logging devices, the resistivity measurement is a function of porosity, pore fluid, pore geometry, cementation, and clay content (Archie relationship) and influenced by mineralogy. Electrical imaging devices in current use, fall under three categories (Table 1), (1) multi-arm (4, 6, 16) wireline pad devices consisting of a matrix array of microelectrodes or microinduction sensors, (2) wireline mandrel tools consisting of multiple azimuthally arranged laterolog-type sensors, and (3) an azimuthally sensitive laterolog-type electrode on a rotating drill collar. In all cases, each sensor makes a resistivity measurement of the borehole wall as a function of azimuth and depth.

The resistivity measurements of early devices were uncalibrated and only qualitative in nature. However, some of the newer devices, including both macroelectrode and microelectrode tools, provide calibrated (absolute) resistivity measurements which can be used for traditional formation evaluation purposes, such as determination of fluid saturation.

Resistivity logging measurements, in general, represent a rock volume some distance into the formation, beyond the borehole wall. In contrast, an acoustic imaging measurement represents the borehole wall or casing surface only. The practical consequence of this difference is that electrical devices are capable of resolving smaller features than acoustic devices. Normal drilling conditions (overbalanced) force borehole fluid (mud) into openings in the borehole wall, especially into fractures, thereby creating a conductivity contrast with the adjacent rock formations. These contrasts are measured by electrical imaging devices which makes them excellent tools for fracture detection and characterization.

Table 3 summarizes applications of electrical borehole imaging and provides references to published examples. The two primary applications of electrical imaging, fracture characterization and reservoir characterization, derive from the high-resolution of microelectrode devices which enables qualitative identification and quantitative analysis of planar features (e.g. fractures and bedding). The larger features can also be characterized by macroelectrode devices. Fracture characterization includes estimation of fracture density, orientation (dip and strike), aperture width, and porosity. Reservoir characterization involves qualitative identification and quantitative evaluation of bed boundaries, thin-beds, lithology and net sand and net pay, litho- or image facies, sedimentology and depositional fabric, reservoir architecture and heterogeneity, and carbonate porosity and heterogeneity (anisotropy). Other applications include fault evaluation (based on analysis of associated fractures), stress analysis (breakouts, tensile wall cracks, shear failure), core orientation, 'reconstructing' missing core, and permeability estimation.

Microresistivity devices

These are pad-type resistivity devices comprised of multiple, generally independent arms (four, six, and sixteen) each of which has one or more pads and each pad contains several or an array of closely spaced microelectrodes. The pads inject current into the formation and the microelectrodes measure the current density across the pad which is kept at a constant potential relative to the return electrode. Variations in current density are due to variations in local formation resistivity. The measurement is a passively focused (in most cases) electrode resistivity, i.e. micrologue-type, measurement. The sensor depth of investigation, approximately ten inches, falls between that of a microresistivity tool (e.g. microlaterologue) and a shallow laterologue (LLS) (Roca-Ramisa *et al.* 1994; Seiler *et al.* 1994; Williams *et al.* 1994). However, the actual depth of investigation represented by the image is much less, approximately one inch. As with all pad-type logging devices, the measurements are highly sensitive to conditions that reduce reliable pad contact with the borehole wall, e.g. mudcake and rugosity, and tool eccentering.

Microelectrode imaging devices were introduced in the mid- and late 1980s (Ekstrom *et al.* 1986; Boyeldieu & Jeffreys 1988) and represent a natural evolution of the multi-electrode, high-resolution four-arm dipmeters that immediately preceded them (Chauvel *et al.* 1984). During this period the six-arm dipmeter was also introduced (Morrison & Thibodaux 1984) and one service company developed a 'resistivity map' processing, which generates low-resolution borehole images from the dipmeter resistivity measurements. In the early 1990s, responding to the needed for increased borehole coverage, the six-arm designs evolved into multisensor-array imaging devices (Seiler *et al.* 1994) and additional electrodes were added to a four-arm device through the use of pad extensions ('flaps') (Safinya *et al.* 1991). During this time interval, a sixteen-arm slimhole imaging device was also developed for fracture characterization in small-diameter geotechnical boreholes (Straub *et al.* 1989, 1995).

The small sensor size and the spacing arrangement of current commercial microelectrode imaging devices provide an image resolution as high as 0.1 inches by 0.1 inches. Borehole coverage in an 8-inch diameter hole ranges from 40 to 80%, depending on the number of arms and pads. In highly deviated and horizontal wells, wireline tools are generally conveyed on drillpipe (e.g. Hurley *et al.* 1994). In these cases, hydraulically powered arms improve pad contact.

The use of conventional microresistivity imaging devices is precluded in the nonconductive oil-based and synthetic muds now in common use. However, it is still possible to obtain electrical borehole images in these wells by use of an oil-based mud dipmeter (a conventional four-arm dipmeter design where the four microelectrodes are replaced by microinduction sensors) (Stuart-Bruges 1984; Dumont *et al.* 1987; Kleinberg *et al.* 1987). Modern processing methods can generate low-resolution images from the four azimuthal measurements and these images can be used to derive dips off bed boundaries and to obtain other stratigraphic or sedimentological information, as the resolution allows.

Macroresistivity devices

There are both wireline and LWD devices in current use. Both types are focused and calibrated electrode (laterologue-type) resistivity devices that provide 100% borehole coverage by averaging (binning) the measurements into azimuthal sectors. Although the resulting images have a lower resolution than those from microresistivity devices, if strong conductivity contrasts are present, then relatively large-scale features such as lithology, bedding, fractures (>1-inch aperture), faults, and unconformities are recognizable and calculation of structural dip is possible. In addition, the larger transmitter-to-receiver spacings of macroresistivity designs results in deeper depths of investigation. Consequently, these measurements are less sensitive to borehole rugosity, washout, and mudcake. Furthermore, electrode calibration enables these measurements to also be used in quantitative formation evaluation.

The wireline devices consist of multiple electrodes placed azimuthally around a mandrel and, depending on the design, the number of electrodes in a single ring may range from 12 to 24. The first such devices were developed in the early and mid-1980s for characterization of induced hydrofractures in geothermal wells and for regional stress analysis (Mosnier 1982, 1987; Mosnier & Cornet 1989). Vertical resolution and depth of investigation vary with tool design. A recent commercial design, with 12 electrodes, has a vertical resolution of 8 to 12 inches and a depth of investigation that may equal that of the conventional deep laterolog (LLD), approximately 80 inches (Davies *et al.* 1992; Roca-Ramisa *et al.* 1994).

The LWD device uses data from three azimuthally sensitive, 1-inch diameter electrodes that scan the borehole as the bottomhole assembly rotates during drilling. Measurements are binned into 6° azimuthal sectors (<0.5 inches, in 8.5-inch diameter borehole). The vertical image resolution is 2.0 inches, an order of magnitude less than the 0.1 to 0.2 inches of microelectrode devices. The depths of investigation of the three sensors, a function of the sensor-receiver spacings, are 1-, 3-, and 5-inches. Images are processed from downhole memory data stored in the tool while dips can be calculated from a reduced data set transmitted in real-time while drilling. Image quality is a function of bit penetration rate (<60 ft/hr) and bit rotation speeds (>30 rpm) (Bonner *et al.* 1994). Real-time LWD images can be used to assist in geosteering (Prilliman *et al.* 1997; Rosthal *et al.* 1997) and the LWD image resolution may be sufficient to replace a wireline imaging tool run (Prilliman *et al.* 1997).

Other imaging measurements

In general, curent processing techniques permit borehole images to be generated from any type of logging measurement that is obtained from either a series of azimuthally oriented sensors or from a single azimuthally oriented sensor in a rotating drillstring by means of sophisticated data processing techniques that interpolate between the individual measurements. An example of a different type of image is a bulk density image obtained by a new azimuthally sensitive LWD density device that can provide real-time images of the borehole wall (Evans *et al.* 1995; Holenka *et al.* 1995; Carpenter *et al.* 1997; Bornemann *et al.* 1998). Applications of these images are similar to the conventional resistivity and acoustic borehole images: bed boundaries, structural dip, net pay, and stratigraphic features.

Future directions

Advances in well logging technology, in general, and borehole imaging in particular, have been and will continue to be driven by the need for more detailed and precise reservoir characterization. Improvements in existing logging services, as well as the introduction of new services, have largely paralleled and been made possible by advances in solid-state electronics and data telemetry. The high quality and reliability of current borehole imaging tools enables them to operate in a wider range of borehole environments and with greater resolution than previous generation tools.

In the future, we can anticipate the development of microelectrode imaging devices that provide quantitative measurements as well as acoustic devices that use dynamically focused transducers. Over the long term, higher tool sampling rates, made possible by advances in data telemetry, such as the routine use of fiber-optic cable (wireline) and drill-pipe data transmission (LWD), will enable tool designs that offer further increases in image resolution. In the near term, the greatest advances in borehole imaging technology will likely occur in LWD as industry's use of and reliance on this method of well logging in conventional, directional, and geosteered wells continues to increase. Higher data telemetry rates may allow real-time transmission of LWD images and, as additional LWD azimuthal-type sensors are developed, a wider variety of logging measurements will become available for use in borehole images. At the present time, because of high logging costs and operational concerns, borehole image logs are not routinely included in the logging programs for most exploration and development wells. This may change with future improvements in LWD image quality and the eventual introduction of real-time LWD borehole imaging. LWD image logging may eventually replace wireline services in some logging situations. Improvements in computing hardware and interpretation software, combined with the continual downwards trend in cost (especially in personal computers), will open up the field of borehole image interpretation to a broader segment of geoscientists and engineers.

References

ADAMS, J., BOURKE, L. & BUCK, S. 1990. Integrating Formation MicroScanner images with core. *Schlumberger Oilfield Review*, **2**(1), 52–65.

AKBAR, M., PETRICOLA, M., WATFA, M. *et al.* 1995. Classic interpretation problems – evaluating carbonates. *Schlumberger Oilfield Review*, **7**(1), 4–22.

AL-KHARUSI, H. & BINBEK, N. 1995. An FMI and 3D seismic combined interpretation of the Haima (Ghudun) reservoir along a horizontal well trajectory in the south Oman Qaharir field, ADSPE-44. *In*: *6th ADNOC/SPE Abu Dhabi International Oil Conference Proceedings*. Society of Petroleum Engineers, 199–207.

AL-WAHEED, H. H., AUDAH, T. M. & CAO MINH, C. 1994. Application of the Azimuthal Resistivity Imager tool in Saudi Arabia, SPE-28439. *In*: *SPE Annual Technical Conference And Exhibition Proceedings, v. Omega, Formation Evaluation And Reservoir Geology*. Society of Petroleum Engineers, 811–818.

ALLEN, T. T., WARD, S. L., CHAVERS, R. D., ROBERTSON, T. N. & SCHULTZ, P. K. 1993. Diagnosing production problems with downhole video logging at Prudhoe Bay, SPE-26043. *In*: *Western Regional Meeting Proceedings*. Society of Petroleum Engineers, 149–158. Later published in 1994. *In*: *European Production Operations Conference And Exhibition Proceedings*. Society of Petroleum Engineers, 7–16. Later published in 1994, as, WARD, S. L., ALLEN, T. T., CHAVERS, R. D., ROBERTSON, T. N. & SCHULTZ, P. K. *Journal of Petroleum Technology*, **46**(11), 973–978.

ANANTO, D. & SCHNORR, D. R. 1996. Determining drill pipe wear inside casing using ultra sonic measurements, paper IPA-96-2.5-149. *In*: *25th Annual Convention Proceedings*, **2**. Indonesian Petroleum Association, Jakarta, 411–423.

ANTOINE, J. N. & DELHOMME, J. P. 1990. A method to derive dips from bed boundaries in borehole images, SPE-20540. *In*: *SPE Annual Technical Conference Exhibition Proceedings, v. Omega, Formation Evaluation And Reservoir Geology*. Society of Petroleum Engineers, 121–130. Later published in 1993. *SPE Formation Evaluation*, 96–102.

ANXIONNAZ, H. & DELHOMME, J. 1998. Near-wellbore 3D reconstruction of sedimentary bodies from borehole electrical images, paper N. *In*: *39th Annual Logging Symposium Transactions*. Society of Profesional Well Log Analyts, 14.

BABCOCK, E. A. 1978. Measurement of subsurface fractures from dipmeter logs. *AAPG Bulletin*, **62**(7), 1111–1126. Later reprinted in 1990. *In*: FOSTER, N. H. & BEAUMONT, E. A. (eds) *Formation Evaluation II – Log Interpretation*. AAPG Treatise of Petroleum Geology Reprint Series No. 17, 457–472.

BADR, A. R. & AYOUB, M. R. 1989. Study of a complex carbonate reservoir using the Formation MicroScanner (FMS), SPE-17977. *In*: *SPE Middle East Oil Technical Conference And Exhibition Proceedings*. Society of Petroleum Engineers, 507–516. Later reprinted in 1990. *In*: PAILLET, F. L., BARTON, C., LUTHI, S., RAMBOW, F. & ZEMANEK, J. (eds) *Borehole Imaging*. Society of Professional Well Log Analysts Reprint Volume, 345–354.

BARTON, C. A. 1988, *The development of in-situ stress measurement techniques for deep drillholes*. PhD thesis, Stanford University.

—— & MOOS, D. 1988. Analysis of macroscopic fractures in the Cajon Pass scientific drillhole; over the interval 1829–2115 meters. *Geophysical Research Letters*, **15**(9), 1013–1016. Later reprinted in 1990. *In*: PAILLET, F. L., BARTON, C., LUTHI, S., RAMBOW, F. & ZEMANEK, J. (eds) *Borehole Imaging*. Society of Professional Well Log Analysts Reprint Volume, 287–290.

—— & ZOBACK, M. D. 1989. Utilization of interactive analysis of digital borehole televiewer data for studies of macroscopic fracturing, paper GG. *In*: *3rd International Symposium On Borehole Geophysics For Minerals, Geotechnical, And Groundwater Applications Proceedings*. Society of Professional Well Log Analysts, Minerals and Geotechnical Logging Society, Chapter-at-Large, 623–653.

——, —— & BURNS, K. L. 1988. In-situ stress orientation and magnitude at the Fenton geothermal site, New Mexico, determination from wellbore breakouts. *Geophysical Research Letters*, **15**(5), 467–470. Later reprinted in 1990. *In*: PAILLET, F. L., BARTON, C., LUTHI, S., RAMBOW, F. & ZEMANEK, J. (eds) *Borehole Imaging*. Society of Professional Well Log Analysts Reprint Volume, 439–442.

—— & ——1990. Self-similar distribution of macroscopic fractures at depth in crystalline rock in the Cajon Pass scientific drillhole. *In*: BARTON, N. & STEPHANSSON, O. (eds) *Rock Joints*. A. A. Balkema, Rotterdam, 163–170. An expanded version later published in 1992. *Journal of Geophysical Research*, **97**(B4), 5181–5200.

——, TESLER, L. G. & ZOBACK, M. D. 1992. Interactive image analysis of borehole televiewer data. *In*: PALAZ, I. & SENGUPTA, S. K. (eds) *Automated Pattern Analysis In Petroleum Exploration*. Springer-Verlag, New York, 223–248.

——, ZOBACK, M. D. & MOSS, D. 1995. Fluid flow along potentially active faults in crystalline rock. *Geology*, **23**(8), 683–686.

——, MOOS, D., PESKA, P. & ZOBACK, M. D. 1997. Utilizing wellbore image data to determine the complete stress tensor–application to permeabilty anisotropy and wellbore stability. *The Log Analyst*, **38**(6), 21–33.

BELL, J. S. 1990. Investigating stress regimes in sedimentary basins using information from oil industry wireline logs and drilling records. *In*: HURST, A., LOVELL, M. A. & MORTON, A. C. (eds) *Geological Applications Of Wireline Logs*. Geological Society, London, Special Publications No. **48**, 305–325.

——1996. In situ stresses in sedimentary rocks, part 1 measurement techniques; part 2 applications of stress measurements. *Geoscience Canada*, **23**(2), 85–100, **23**(3), 135–153.

BENNECKE, W. 1994, The identification of basalt flow features from borehole television logs. *In*: LINK, P. K. (ed.) *Hydrogeology, Waste Disposal, Science And Politics* [30th symposium on engineering, geology, and geotechnical engineering proceedings]. Idaho State University, Pocatello, 371–383.

BERGER, P. & ROESTENBURG, J. 1989. High resolution in-situ geochemical and image analysis applied to a well in the South China Sea, paper N. *In*: *12th International Formation Evaluation Symposium Transactions*. Society of Professional Well Log Analysts, Paris Chapter (SAID). Also published in 1990 as paper FF. *In*: *International Well Logging Symposium Transactions*. Society of Professional Well Log Analysts, Beijing Chapter.

BETTIS, F. E., CRANE, L. R. II, SCHWANITZ, B. J. & COOK, M. R. 1993. Ultrasound logging in cased boreholes pipe wear [USI], SPE-26318. *In*: *SPE Annual Technical Meeting And Exhibition Proceedings, v. Delta, Drilling*. Society of Petroleum Engineers, 15–27.

BIGELOW, E. L. 1993. The petrophysical and geological impact of borehole images, SPE-26064. *In*: *Western Regional Meeting Proceedings*. Society of Petroleum Engineers, 345–354.

BILODEAU, B. J., SMITH, S. C. & JULANDER, D. R. 1997. Comprehensive reservoir characterization using open-hole wireline logs, core, and downhole video, Buena Vista Hills field, California [abstract]. *AAPG Bulletin*, **81**(4), 680.

BIRCHAK, B., MANDAL, B., MASINO, J., MINEAR, J. & RITTER, T. 1997. Looking behind casing – developing a cement imaging and caliper tool. *GasTips* [Gas Research Institute] **3**(2), 37–44.

BONNER, S., BAGERSH, A., CLARK, B. *et al.* 1994. A new generation of electrode resistivity measurements for formation evaluation while drilling, paper OO. *In*: *35th Annual Logging Symposium Transactions*. Society of Professional Well Log Analysts.

BORCHERS, J. W., GERBER, M., WILEY, J., SMITH, D. & MITTEN, H. T. 1994. Use of a television camera to evaluate damage to well casings caused by land subsidence in the Sacramento Valley, California. *EOS Transactions* [American Geophysical Union], **75**(44), Supplement, 234.

BORNEMANN, E., HODENFIELD, K., MAGGS, D., BOURGEOIS, T. & BRAMLETT, K. 1998. The application and accuracy of geological information from a logging-while-drilling density tool, paper L. *In*: *39th Annual Logging Symposium Transactions*. Society of Profesional Well Log Analyst.

BOURKE, L. T. 1989. Recognizing artifact images of the Formation Microscanner, paper WW. *In*: *30th Annual Logging Symposium Transactions*. Society of Professional Well Log Analysts. Later reprinted in 1990. *In*: PAILLET, F. L., BARTON, C., LUTHI, S., RAMBOW, F. & ZEMANEK, J. (eds) *Borehole Imaging*. Society of Professional Well Log Analysts Reprint Volume, 191–215.

——1992. Sedimentological borehole image analysis in clastic rocks – a systematic approach to interpretation. *In*: HURST, A., GRIFFITHS, C. M. & WORTHINGTON, P. F. (eds) *Geological Applications Of Wireline Logs II*. Geological Society, London, Special Publications No. **65**, 31–42.

——1993. Core permeability imaging – its relevance to conventional core characterization and potential application to wireline measurement. *Marine and Petroleum Geology*, **10**(4), 297–408.

BOURKE, L., DELFINER, P., TROUILLER, J-C. *et al.* 1989. Using Formation Microscanner images. *The Technical Review*, **37**(1), 16–40. Reprinted in 1990. *In*: FOSTER, N. H. & BEAUMONT, E. A.

(eds) *Formation Evaluation I – Log Evaluation*. AAPG Treatise of Petroleum Geology Reprint Series No. **16**, 610–634.

BOYELDIEU, C. & JEFFREYS, P. 1988. Formation Microscanner – new developments, paper WW. *In*: *11th European Formation Evaluation Symposium Transactions*. Society of Professional Well Log Analysts. Later reprinted in 1990. *In*: PAILLET, F. L., BARTON, C., LUTHI, S., RAMBOW, F. & ZEMANEK, J. (eds) *Borehole Imaging*. Society of Professional Well Log Analysts Reprint Volume, 175–190.

BREWER, T., LOVELL, M., HARVEY, P. & WILLIAMSON, G. 1995. Stratigraphy of the ocean crust in ODP Hole 896A from FMS images. *Scientific Drilling*, **5**(2), 87–92.

BRIGGS, R. O. 1964. Development of a downhole television camera, paper N [abstract only]. *In*: *5th Annual Logging Symposium Transactions*. Society of Professional Engineers, 1.

BRODING, R. A. 1981. Volumetric scan well logging, paper B. *In*: *22nd Annual Logging Symposium Transactions. Society Of Professional Well Log Analysts*. Later published in 1982. *The Log Analyst*, **23**(1), 14–19. Also published in 1982 as, Volumetric scanning allows 3-D viewing of the bore hole. *World Oil*, **194**(6), 190–196. Later reprinted in 1990. *In*: PAILLET, F. L., BARTON, C., LUTHI, S., RAMBOW, F. & ZEMANEK, J. (eds) *Borehole Imaging*. Society of Professional Well Log Analysts Reprint Volume, 93–97.

——1984. Application of the sonic volumetric scan log to cement evaluation, paper JJ. *In*: *25th Annual Logging Symposium Transactions*. Society of Professional Well Log Analysts.

—— & BUCHANAN, L. K. 1986. A sonic technique for cement evaluation, paper GG. *In*: *27th Annual Logging Symposium Transactions*. Society of Professional Well Log Analysts.

BRONDEL, D., EDWARDS, R., HAYMAN, A., HILL, D., MEHTA, S. & SEMERAD, T. 1994. Corrosion in the oil industry. *Schlumberger Oilfield Review*, **6**(2), 4–18.

BRUDY, M., ZOBACK, M. D., FUCHS, K., RUMMEL, F. & BAUMGARTNER, J. 1997. Estimation of the complete stress tensor to 8 km depth in the KTB scientific drill holes – implications for crustal strength. *Journal of Geophysical Research*, **102**(B8), 18, 453–18, 475.

BULLWINKEL, R. J., BERRY, S. H. & FISCHER, J. A. 1994. The Formation MicroScanner – its use and verification in the Passaic Formation. *In*: *8th National Outdoor Action Conference And Exposition Proceedings. Ground Water Management* [National Ground Water Association], **18**, 449–462.

BURNS, K. L. 1985. Information extraction from noisy televiewer logs of inclined wellbores in hard rock. *Geothermal Resources Council Transactions*, **9**, part 2, 441–445.

——1987. *Geological Structures From Televiewer Logs Of GT-2 Fenton Hill, New Mexico, Part 1– Feature Extraction; Part 2–Rectification; Part 3 – Quality Control*. Los Alamos National Laboratory Report LA-10619-HDR.

——1988. *Televiewer Measurement Of The Orientation Of In Situ Stress At The Fenton Hill Hot Dry Rock Site, New Mexico*. Los Alamos National Laboratory Report LA-UR-88-1775, 6. Also published in 1988. *Geothermal Resources Council Transactions*, **12**, 229–235.

——1991. *Orientation Of Minimum Principal Stress In The Hot Dry Rock Geothermal Reservoir At Fenton Hill, New Mexico*. Los Alamos National Laboratory Report No. LA-UR-91-1990.

BUTSCH, R. J. 1995. Overcoming interpretation problems of gas-contaminated cement using ultrasonic cement logs, SPE-30509. *In*: *SPE Annual Technical Conference And Exhibition Proceedings, v. Delta, Drilling*. Society of Petroleum Engineers, 607–622.

CARMONA, R., GRATEROL, J., PINA, N. *et al.* 1997. Revitalization of Mara field, SPE-38665. *In*: *Annual Technical Conference And Exhibition Proceedings*, omega, *Formation Evaluation And Reservoir Geology*, part 1. Society of Petroleum Engineers, 241–256.

CARPENTER, W., BEST, D. & EVANS, M. 1997. Applications and interpretation of azimuthally sensitive density measurements acquired while drilling, paper EE. *In*: *38th Annual Logging Symposium Transactions*. Society of Professional Well Log Analysts.

CARR, D., JOHNS, R., ELPHICK, R. & FOULK, L. 1996. High-resolution reservoir characterization of Mid-continent sandstones using wireline resistivity imaging, Boonsville (Bend Conglomerate) gas field, Fort Worth basin, Texas, paper YY. *In*: *37th Annual Logging Symposium Transactions*. Society of Professional Well Log Analysts. Later published in 1997. *The Log Analyst*, **38**(6), 54–70.

CARR-CRABAUGH, M., HURLEY, N. F. & CARLSON, J. 1996. Interpreting eolian reservoir architecture using borehole images. *In*: PACHT, J. A., SHERIFF, R. E. & PERKINS, B. F. (eds) *Stratigraphic Analysis Utilizing Advanced Geophysical, Wireline, And Borehole Technology For Petroleum Exploration And Production* [17th annual research conference proceedings]. SEPM [Society of Sedimentary Geology] Foundation, Gulf Coast Section, 39–50.

CARSON, C. C. & BAUMAN, T. 1987. Use of an acoustic borehole televiewer to investigate casing corrosion in geothermal wells. *Materials Performance*, **26**(6), 53–59.

CASARTA, L. J., MCNAUGHTON, D. A., BORNEMANN, E. & BETTIS, F. E. 1989. Fracture identification and matrix characterization using a new borehole imaging device in the Lisburne carbonate, Prudhoe Bay, Alaska, paper XX. *In*: *30th Annual Logging Symposium Transactions*. Society of Professional Well Log Analysts.

CATALA, G., DE MONTMOLLIN, V., HAYMAN, A. *et al.* 1991. Modernizing well cementing design and evaluation [USI]. *Schlumberger Oilfield Review*, **3**(2), 55–71.

CELERIER, B., MACLEOD, C. J. & HARVEY, P. K. 1996. Constraints on the geometry and fracturing of Hole 894G, Hess Deep, from Formation MicroScanner logging data. *In*: MEVEL, C. G.,

ALLAN, J. F. & MEYER, P. S. (eds) *Proceedings Of The Ocean Drilling Program, Scientific Results*, **147**. Ocean Drilling Program, Texas A&M University, College Station, TEXAS, 329–345.

CHAUVEL, Y., SEEBURGER, D. A. & ORJUELA, A. C. 1984. Applications of the SHDT Stratigraphic High Resolution Dipmeter to the study of depositional environments, paper G. *In*: *25th Annual Logging Symposium Transactions*: Society of Professional Well Log Analysts.

CHEUNG, P. S. & HELIOT, D. 1990. Workstation based fracture evaluation using borehole images and wireline logs, SPE-20573. *In*: *Annual Technical Conference And Exhibition Proceedings, v. Omega, Formation Evaluation And Reservoir Geology*. Society of Petroleum Engineers, 465–470.

CHOI, D. R. & SAUL, G. R. S. 1995. A comparison between the borehole image processing system [BIP] and logging of oriented drill-core in determining fracture characteristics within coal seams. *In*: *Bowen Basin Symposium Proceedings*.

CLAVIER, C. 1994. Image logging – a new dimension in interpretation. *In*: *2nd International Symposium On Well Logging Transactions*. Society of Professional Well Log Analysts, Beijing Chapter, 204–218.

CLERKE, E. A. & VAN AKKEREN, T. J. 1986. Borehole televiewer improves completion results in a Permian basin San Andres reservoir, SPE-15033. *In*: *1986 SPE Permian Basin Oil And Gas Recovery Conference Proceedings*. Society of Petroleum Engineers. Later published in 1988. SPE Production Engineering, **3**(1), 89–95.

COBB, C. C. & SCHULTZ, P. K. 1992. A real-time fiber optic downhole video system, OTC-7046. *In*: *24th Annual Offshore Technology Conference Proceedings*, **4**. Society of Petroleum Engineers, 575–582. Later published in 1992 as, SCHULTZ, P. K. & COBB, C., Fiber optics improve downhole video. *Oil and Gas Journal*, **90**(19), May 11, 46–50.

——1993, Downhole videos extend their reach. *American Oil and Gas Reporter*, **36**(9), 27–33.

COLL, C., DE VILLARROEL, H. G., LOZADA, T., CHACARTEGUI, F., SUAREZ, O. & RONDON, L. 1997. Horizontal well images help describe thinly bedded reservoirs. *The Search for Oil and Gas in Latin America and the Carribean* [Schlumberger Surenco], no. **5**, 37–51.

CONNALLY, T. C. & WILTSE, E. W. 1994. Correlation of Ordovician sandstones in central Saudi Arabia. *In*: AL-HUSSEINI, M. I. (ed.) *Geo'94, The Middle East Petroleum Geosciences*, **1**. Gulf PetroLink, Manama, Bahrain, 321–333. Later published in 1996 as, Saudi sandstone correlation. *Schlumberger Middle East Well Evaluation Review*, no. **16**, 27–41.

COOPER, P., ARNAUD, H. M. & FLOOD, P. G. 1995. Formation Microscanner logging responses to lithology in guyot carbonate platforms and their implications – sites 865 and 866. *In*: WINTERER, E. L., SAGER, W. W., FIRTH, J. V. & SINTON, J. M. (eds) *Proceedings Of The Ocean Drilling Program, Scientific Results*, **143**. Texas A&M University, Ocean Drilling Program, College Station, Texas, 329–372.

COX, J. W. 1970. The high resolution dipmeter reveals dip-related borehole and formation characteristics paper D. *In*: *11th Annual Logging Symposium Transactions*. Society of Professional Well Log Analysts.

CRYER, J., DENNIS, B., LEWIS, R., PALMER, K. & WATFA, M. 1987. Logging techniques for casing corrosion. *The Technical Review*, **35**(4), 32–39.

DARILEK, G. T. 1985. A borehole television system for developing enhanced gas production from tight formations. *In*: *International Gas Research Conference Proceedings*. Government Institutes Inc., Rockville, Maryland, 122–127. Later published in 1986 as A color borehole television system for developing gas production from Devonian shales, SPE-15219. *In*: *Unconventional Gas Technology Symposium Proceedings*. Society of Petroleum Engineers, 145–152.

——1986. *Borehole Television, final report*. Gas Research Institute Report GRI-86/0128.

DAVIDSON, J. A., MORRISS, S. L. & PODIO, A. L. 1992. Estimates of formation sound speed from ultrasonic reflections, SPE-24688. *In*: *SPE Annual Technical Conference And Exhibition Proceedings, v. Omega, Formation Evaluation And Reservoir Geology*. Society of Petroleum Engineers, 287–298.

DAVIES, D. H., FAIVRE, O., GOUNOT, M-T. *et al.* 1992. Azimuthal resistivity imaging – a new generation laterolog, SPE-24676. *In*: *SPE Annual Technical Conference And Exhibition Proceedings, v. Omega, Formation Evaluation And Reservoir Geology*. Society of Petroleum Engineers, 143–153. Later published in 1994. *SPE Formation Evaluation*, **9**(3), 165–174.

DELHOMME, J. P., BEDFORD, J., COLLEY, N. M. & KENNEDY, M. C. 1996. Permeability and porosity upscaling in the near-wellbore domain – the contribution of borehole electrical images, SPE36822. *In*: *SPE European Petroleum Conference [EUROPEC] Proceedings*, **1**. Society of Petroleum Engineers, 89–101.

DEMPSEY, J. C. & HICKEY, J. R. 1958. Use of a borehole camera for visual inspection of hydraulically-induced fractures. *Producers Monthly*, April, 18–21.

DENNIS, B. 1988. Casing corrosion evaluation using wireline techniques, paper 88-39-118. *In*: *39th Annual Technical Meeting Preprints*, **3**. Petroleum Society of CIM, Calgary.

——, STANDEN, E., GEORGI, D. T. & CALLOW, G. O. 1987. Fracture identification and productivity predictions in a carbonate reef complex, SPE-16808. *In*: *SPE Annual Technical Conference And Exhibition Proceedings, v. Omega, Formation Evaluation And Reservoir Geology*. Society of Petroleum Engineers, 579–588. Also published in 1987 as Fracture identification techniques in carbonates, paper Y. *In*: *11th Formation Evaluation Symposium Transactions*. Canadian Well Logging Society. Later reprinted in 1990. *In*: PAILLET, F. L., BARTON, C., LUTHI, S., RAMBOW, F. & ZEMANEK, J. (eds) *Borehole Imaging*. Society of Professional Well Log Analysts Reprint Volume, 277–286.

DERI, S., CHIARABELLI, A. & MOUSSA, A. 1994. Borehole imaging analysis – a technique to enhance petrophysical evaluation of siliciclastic and weathered basement reservoirs. *In*: *12th Conference Proceedings*. Egyptian General Petroleum Corporation, Cairo.

DEZAYES, C., VILLEMIN, T., GENTER, A., TRAINEAU, H. & ANGELIER, J. 1995. Analysis of fractures in boreholes of Hot Dry Rock project at Soutz-sous-Forets (Rhine graben, France). *Scientific Drilling*, **5**(1), 31–41.

——, GENTER, A. & VILLEMIN, T. 1996. Inherited and induced fractures characterized from acoustic and electric borehole images [in French]. *Bulletin Centres de Recherches et Exploration et Production Elf Aquitaine*, 197–212.

DUDLEY, J. W., II 1993. Quantitative fracture identification with the borehole televiewer, paper CC. *In*: *34th Annual Logging Symposium Transactions*. Society of Professional Well Log Analysts.

DUECK, R. N. & PAAUWE, E. F. W. 1994. The use of borehole imaging techniques in the exploration for stratigraphic traps – an example from the Middle Devonian Gilwood channels in north-central Alberta. *Bulletin of Canadian Petroleum Geology*, **42**(2), 137–154.

DUMONT, A., KUBACSI, M. & CHARDAC, J. L 1987. The Oil-Based Mud Dipmeter tool, paper LL. *In*: *28th Annual Logging Symposium Transactions*. Society of Professional Well Log Analysts.

EHRLICH, R. & DAVIES, D. K. 1989 Image analysis of pore geometry; relationship to reservoir engineering and modeling, SPE-19054. *In*: *SPE Gas Technology Symposium Proceedings*. Society of Petroleum Engineers, 15–30.

EKSTROM, M. P. (ed.) 1984. *Digital Image Processing Techniques*. Academic Press, Inc., Orlando, Florida, Computational Techniques, **2**.

——, DAHAN, C. A., CHEN, M., LLOYD, P. M. & ROSSI, D. J. 1986. Formation imaging with microelectrical scanning arrays, paper BB. *In*: *27th Annual Logging Symposium Transactions*. Society of Professional Well Log Analysts, 21. Also published in 1986 as, EKSTROM, M. P., CHEN, M.-Y., ROSSI, D. J., LOCKE, S. & ARON, J. 1986. High-resolution microelectrical borehole wall imaging, paper K. *In*: *10th European Formation Evaluation Symposium Transactions*: Society of Professional Well Logging Analysts. Later published in 1987. *The Log Analyst*, **28**(3), May–June, 294–306. Later reprinted in 1990. *In*: PAILLET, F. L., BARTON, C., LUTHI, S., RAMBOW, F. & ZEMANEK, J. (eds) *Borehole Imaging*. Society of Professional Well Log Analysts Reprint Volume, 163–174.

ELKINGTON, P. 1996. Acquisition and analysis of Slim Acoustic Scanner log data. *Australian Society of Exploration Geophysicists (ASEG) Preview Publication*, issue 65, 21–32.

ENGLET, K. T. & SWEET, M. L. 1993. Image analysis characterization of small-scale heterogeneities in fluvial reservoirs – a case study from the Gypsy sandstone of Central Oklahoma, paper EE. *In*: *34th Annual Logging Symposium Transactions*. Society of Professional Well Log Analysts.

EUBANKS, D., SEILER, D. & RUSSELL, B. 1995. Geological applications using an electrical micro imaging tool. *Oil and Gas Journal*, **93**(47), November 20, 84–89.

EVANS, L. W., THORN, D. & DUNN, T. L. 1996. Formation Microimager, Microscanner, and core characterization of natural fractures in a horizontal well in the Upper Almond bar sand, Echo Sprngs field, Wyoming. *In*: PACHT, J. A., SHERIFF, R. E. & PERKINS, B. F. (eds) *Stratigraphic Analysis Utilizing Advanced Geophysical, Wireline, And Borehole Technology For Petroleum Exploration And Production* [17th annual research conference proceedings]. SEPM [Society of Sedimentary Geology] Foundation, Gulf Coast Section, 89–95.

EVANS, M., BEST, D., HOLENKA, J., KURKOSKI, P. & SLOAN, W. 1995. Improved formation evaluation using azimuthal porosity data while drilling, SPE-30546. *In*: *Annual Technical Conference And Exhibition Proceedings, v. Omega, Formation Evaluation And Reservoir Geology*. Society of Petroleum Engineers, 137–146.

FAIVRE, O. 1993. Fracture evaluation from quantitative azimuthal resistivities, SPE-26434. *In*: *SPE Annual Technical Conference And Exhibition Proceedings, v. Omega, Formation Evaluation And Reservoir Geology*. Society of Petroleum Engineers, 179–192.

—— & CATALA, G. 1995. Dip estimation from azimuthal laterolog tools, paper CC. *In*: *36th Annual Logging Symposium Transactions*. Society of Professional Well Log Analysts.

FAM, M., CHEMALI, R., HAUGLAND, M., SEILER, D. & STEWART, W. 1996. Integrating imaging logs in formation evaluation, paper ZZ. *In*: *37th Annual Logging Symposium Transactions*. Society of Professional Well Log Analysts.

FAM, M. Y., CHEMALI, R., SEILER, D., HAUGLAND, M. & STEWART, W. F. 1995. Applying electrical micro-imaging logs to reservoir characterization, SPE-30608. *In*: *SPE Annual Technical Conference And Exhibition Proceedings, v. Omega, Formation Evaluation And Reservoir Geology*. Society of Petroleum Engineers, 843–854.

FARAGUNA, J., CHACE, D. M. & SCHMIDT, M. G. 1989. An improved borehole televiewer system – image acquisition, analysis, and integration, paper UU. *In*: *30th Annual Logging Symposium Transactions*. Society of Professional Well Log Analysts. Also published in 1989 as paper Q. *In*: *12th International Formation Evaluation Symposium Transactions*. Society of Professional Well Log Analysts, Paris Chapter (SAID). Later reprinted in 1990. *In*: PAILLET, F. L., BARTON, C., LUTHI, S., RAMBOW, F. & ZEMANEK, J. (eds) *Borehole Imaging*. Society of Professional Well Log Analysts Reprint Volume, 149–159.

FETT, T., BRYANT, I. & GREENSTREET, C. 1994. Integration of state-of-the-art wireline reservoir delineation devices with core data from an offset well to optimize reservoir management – Maui field, New Zealand. *In*: *New Zealand Petroleum Conference Proceedings*. Petroleum exploration association of New Zealand, 231–239.

FIRMANSJAH, K., LOMBARD, C., MERCIER, F., CHAPPE, J. L. & MONTAGGIONI, P. 1996, New sand count using core calibrated UBI (Ultra-Sonic Borehole Imager) data, paper IPA96-1-1-087. *In*: *25th Annual Convention Proceedings*, **1**. Indonesian Petroleum Association, Jakarta, 191–213.

FITZSIMMONS, T. E., KACIR, T. P. & WADLEIGH, E. E. 1997*a*. Cautious utilization of Formation Micro-Imaging logs, SPE-37480. *In*: *Production Operations Symposium Proceedings*. Society of Petroleum Engineers, 721–728.

——, WADLEIGH, E. E., CURRAN, B. C. & KACIR, T. P. 1997*b*. Flowing features identification for massive carbonates – a real opportunity, SPE-37741. *In*: *10th SPE Middle East Oil Show And Conference Proceedings*, **1**. Society of Petroleum Engineers, 533–540.

FRIEDMAN, M. & MCKIERNAN, D. E. 1994. Extrapolation of fracture data from outcrops of the Austin Chalk in Texas to corresponding petroleum reservoirs at depth, paper HWC-94–36. *In*: *SPE/CIM/CANMET International Conference On Recent Advances In Horizontal Well Applications, Preprints*. CIM Petroleum Society, Calgary, Alberta.

FRIKKEN, H. W. 1996. CBIL logs – vital for evaluating disappointing well and reservoir performance, K15-FG field, central offshore Netherlands. *In*: *Geology Of Gas And Oil Under The Netherlands*. Kluwer Academic Publishers, 103–114.

GAUTAMA, A. B., FIRMANSJAH [sic], LEANEY, W. S. & HANSEN, S. M. 1994. Ultrasonic images – new applications – examples from Kalimantan, paper IPA94-2.2-082. *In*: *23th Annual Convention Proceedings*, **2**. Indonesian Petroleum Association, Jakarta, 65–82.

GENTER, A., MARTIN, P. & MONTAGGIONI, P. 1991. Application of FMS and BHTV tools for evaluation of natural fractures in the Soultz geothermal borehole GPK1. *Geothermal Science and Technology*, **3**(1–4), 1–29.

——, TRAINEAU, H., DEZAYES, C., ELSASS, P., LEDESERT, B., MEUNIER, A. & VILLEMIN, T. 1995. Fracture analysis and reservoir characterization of the granitic basement in HDR Soultz project (France): *Geothermal Science and Technology*, **4**(3), 189–214.

——, CASTAING, C., DEZAYES, C., TENZER, H., TRAINEAU, H. & VILLEMIN, T. 1997*a*. Comparative analysis of direct (core) and indirect (borehole imaging tools) collection of fracture data in the host dry rock Soultz reservoir (France). Journal of Geophysical Research, **102**(B7), 15 419–15 431.

——, FIRMANSJAH [sic] & MARTIN, P. 1997*b*. Assessment of reservoir fracturing from boreholes–comparison between core and wall-image data [in French]. *Revue de l'Institut Francais du Petrole*, **52**(1), 45–60.

GEORGI, D. T. 1985. Geometrical aspects of borehole televiewer images, paper C. *In*: *27th Annual Logging Symposium Transactions*. Society of Professional Well Log Analysts. Later reprinted in 1990. *In*: PAILLET, F. L., BARTON, C., LUTHI, S., RAMBOW, F. & ZEMANEK, J. (eds) *Borehole Imaging*. Society of Professional Well Log Analysts Reprint Volume, 129–147.

——, PHILLIPS, C. & HARDMAN, R. 1992. Applications of digital core image analysis to thin-bed evaluation, paper SCA-9206. *In*: *Society Of Core Analysts Preprints [33rd SPWLA Annual Logging Symposium Transactions*, **3**]. Society of Professional Well Log Analysts, Society of Core Analysts.

GERARD, R. E., PHILIPSON, C. A., MANNI, F. M. & MARSCHALL, D. M. 1992. Petrographic image analysis – an alternative method for determining petrophysical properties. *In*: PALZ, I. & SENGUPTA, S. K. (eds) *Automated Pattern Analysis In Petroleum Exploration*. Springer-Verlag, New York, 249–263.

GLENN, E. E., BALDWIN, W. F., SLOVER, V. R., STRUBHAR, M. K. & ZEMANEK, J. 1971. New borehole televiewer can be run through tubing. *World Oil*, **172**(1), 63–65.

GLOSSOP, K. J., SIDDANS, A. W. B., LISBOA, P. J. G. & RUSSELL, P. C. 1997. Robust and efficient PC-based software for the semi-automatic interpretation of borehole televiewer data, WS 2.1. *In*: *1997 Technical Program Expanded Abstracts With Authors' Biographies*, **2**. Society of Exploration Geophysicists, Tulsa, 2046–2049.

GLOWKA, D. A., LOEPPKE, G. E., LYSNE, P. C. & WRIGHT, E. K. 1990. Evaluation of a potential borehole televiewer technique for characterizing lost circulation zones. *Geothermal Resources Council Transactions*, **14**, part 1, 395–402.

GOETZ, J. F., SEILER, D. D. & EDMISTON, C. S. 1990. Geological and borehole features described by the Circumferential Acoustic Scanning Tool, paper Z. *In*: *13th European Formation Evaluation Symposium Transactions*. Society of Professional Well Log Analysts, Budapest Chapter. Also published in 1990, as paper OSEA-90113. *In*: *8th Offshore South East Asia Conference Preprints* [Singapore]. Society of Petroleum Engineers, 1–18.

GONFALINI, M. & ANXIONNAZ, H. 1990. A complete use of structural information from borehole imaging techniques [FMS]; a case history for a deep carbonate reservoir, paper J. *In*: *31st Annual Logging Symposium Transactions*. Society of Professional Well Log Analysts.

GRACE, L. M., LUTHI, S. M. & PIRIE, R. G. 1986. Stratigraphic interpretation using formation imaging and dipmeter analyses, SPE-15611. Society of Petroleum Engineers, 61st Annual Meeting [New Orleans] Preprint.

GRAHAM, P. D. 1989. Advances in downhole corrosion evaluation. *In*: *36th Annual Short Course Proceedings*. Southwestern Short Course Association, Lubbock, Texas, 217–240.

HAASE, G. M. 1996. Formation MicroScanner interpretation and caliper measurements in Nubian Sandstones, southeast Sirt Basin. *In*: SALEM, M. J., EL-HAWAT, A. S. & SBETA, A. M. (eds) *The Geology Of The Sirt Basin*, **2**. Elsevier Scientific Publishers, Amsterdam, 275–285.

HACKBARTH, C. J. & TEPPER, B. J. 1988. Examination of BHTV, FMS, and SHDT images in very thinly bedded sands and shales, SPE-18118. *In*: *SPE Annual Technical Conference And Exhibition Proceedings, v. Omega, Formation Evaluation And Reservoir Geology*. Society of Petroleum Engineers, 119–127. Later reprinted in 1990. *In*: PAILLET, F. L., BARTON, C., LUTHI, S., RAMBOW, F. & ZEMANEK, J. (eds) *Borehole Imaging*. Society of Professional Well Log Analysts Reprint Volume, 307–315.

HALL, J., PONZI, M., GONFALINI, M. & MALETTI, G. 1996. Automatic extraction and characterization of geological features and textures from borehole images and core photographs, paper CCC. *In*: *37th Annual Logging Symposium Transactions*. Society of Professional Well Log Analysts.

HAMMES, U. 1997. *Electrical Imaging Catalog – Microresistivity Images And Core Photos From Fractured, Karsted, And Brecciated Carbonate Rocks*. Texas Bureau of Economic Geology, Geological Circular 97-2.

HARKER, S. D., MCGANN, G. J., BOURKE, L. T. & ADAMS, J. T. 1990. Methodology of Formation MicroScanner image interpretation in Claymore an Scapa fields (North Sea). *In*: HURST, A., LOVELL, M. A. & MORTON, A. C. (eds) *Geological Applications Of Wireline Logs*. Geological Society, London, Special Publications No. **48**, 11–25.

HARPER, J. T. & HUMPHREY, N. F. 1994. Borehole video observations within a temperate valley glacier–implications for englacial and subglacial processes [abstract]. EOS Transactions [American Geophysical Union], **74**(44), supplement, 224.

HAWKINS, W. L., OLIVER, R. D. & LAVELLE, M. J. 1989. Borehole inspection system for large diameter holes. *The Log Analyst*, **30**(1), 26–30. Later reprinted in 1990. *In*: PAILLET, F. L., BARTON, C., LUTHI, S., RAMBOW, F. & ZEMANEK, J. (eds) *Borehole Imaging*. Society of Professional Well Log Analysts Reprint Volume, 47–51.

HAYMAN, A. J., GAI, H. & TOMA, I. 1991*a*. A comparison of cementation logging tools in a full-scale simulator, SPE-22779. *In*: *Annual Technical Conference And Exhibition Proceedings, v. Pi, Production*. Society of Petroleum Engineers, 197–207.

——, HUTIN, R. & WRIGHT, P. V. 1991*b*. High-resolution cementation and corrosion imaging by ultrasound, paper KK. *In*: *32nd Annual Logging Symposium Transactions*. Society of Professional Well Log Analysts.

——, CHEUNG, P. & BERGES, P. 1994. Improved borehole imaging by ultrasonics, SPE-28440. *In*: *SPE Annual Technical Conference And Exhibition Proceedings*, omega, *Formation Evaluation And Reservoir Geology*. Society of Petroleum Engineers, 977–992. Later published in 1998. *SPE Production Facilities*, **13**(1), 5–13.

——, PARENT, P., ROUAULT, G. *et al.* 1995. Developments in corrosion logging using ultrasonic imaging, paper W. *In*: *36th Annual Logging Symposium Transactions*. Society of Professional Well Log Analysts.

HEARD, F. E. 1981. A high temperature acoustic borehole televiewer. *In*: *High Temperature Electronics And Instrumentation Conference[Houston] Proceedings. IEEE.*

—— & BAUMAN, T. J. 1983. *Development Of A Geothermal Acoustic Borehole Televiewer*. Sandia National Laboratory Report, SAND83-0681.

HEINE, C. J. 1993. Reservoir characterization integrating borehole imagery and conventional core data, Unayzah Formation, Hawtah field, central Saudi Arabia, SPE-25638. *In*: *8th Middle East Oil Show And Conference Proceedings*, **2**. Society of Petroleum Engineers, 353–361.

HELIOT, D., ETCHECOPAR, A. & CHEUNG, PH. 1990. New developments in fracture characterization from logs. *In*: MAURY, V. & FOURMAINTRAUX, D. (eds) *Rock At Great Depth*. A. A. Balkema, Rotterdam, The Netherlands, **3**, 1471–1478.

HICKEY, J. J. 1993. Characterizing secondary porosity of carbonate rocks using borehole video data [abstract]. Geological Society of America, Abstracts with Programs, **25**(4), Southeastern Section.

HICKMAN, S. H., HEALY, J. H. & ZOBACK, M. D. 1985. In situ stress, natural fracture distribution, and borehole elongation in the Auburn geothermal well, Auburn, New York. *Journal of Geophysical Research*, **90**(B7), 5497–5512. Later reprinted in 1990. *In*: PAILLET, F. L., BARTON, C., LUTHI, S., RAMBOW, F. & ZEMANEK, J. (eds) *Borehole Imaging*. Society of Professional Well Log Analysts Reprint Volume, 371–386.

HINZ, K. & SCHEPERS, R. 1983. SABIS – The digital version of the borehole televiewer, paper E. *In*: *8th European Formation Evaluation Symposium Transactions*. Society of Professional Well Log Analysts, London Chapter.

HOLENKA, J., BEST, D., EVANS, M., KURKOSKI, P. & SLOAN, W. 1995. Azimuthal porosity while drilling, paper BB. *In*: *36th Annual Logging Symposium Transactions*: Society of Professional Well Log Analysts.

HOLLUB, V. A. 1993. Results of camera surveys at Rock Creek, paper 9329. *In*: *1993 International Coalbed Methane Symposium Proceedings*, **2**. University of Alabama, Birmingham, 659–665.

HORNBY, B. E. & LUTHI, S. M. 1992. An integrated interpretation of fracture apertures computed from electrical borehole scans and reflected Stoneley waves. *In*: HURST, A., GRIFFITHS, C. M. & WORTHINGTON, P. F. (eds) *Geological Applications Of Wireline Logs II*. Geological Society, London, Special Publications No. **65**, 179–184.

——, LUTHI, S. M. & PLUMB, R. A. 1990. Comparison of fracture apertures computed from electrical borehole scans and reflected Stoneley waves–an automated interpretation, paper L. *In*: *31st Annual Logging Symposium Transactions*. Society of Professional Well Log Analysts, 25. Later published in 1992.*The Log Analyst*, **33**(1), 50–66.

HUBER, K., FUCHS, K., PALMER, J. *et al.* 1997. Analysis of borehole televiewer measurements in the Vorotilov drillhole, Russia – first results. *Tectonophysics*, **275**, 261–272

HURLEY, N. F. 1994. Recognition of faults, unconformities & sequence boundaries using cumulative dip plots. *AAPG Bulletin*, **78**(8), 1173–1185.

——1996. Parasequence-scale stratigraphic correlations in deep-marine sediments using borehole images. *In*: PACHT, J. A., SHERIFF, R. E. & PERKINS, B. F. (eds) *Stratigraphic Analysis Utilizing Advanced Geophysical, Wireline & Borehole Technology For Petroleum Exploration And Production* [17th annual research conference proceedings]. SEPM [Society of Sedimentary Geology] Foundation, Gulf Coast Section, 147–152.

——, THORN, D. R., CARLSON, J. I. & EICHELBERGER, S. L. W. 1994. Using borehole images for target-zone evaluation in horizontal wells. *AAPG Bulletin*, **78**(2), 238–246.

IIZUKA, Y., ISHII, T., NAGATA, K., MATSUMOTO, Y., MURAKAMI, O. & NOGUCHI, K. 1987. New system for borehole TV logging and evaluation of weathered granites according to their colour indices. *In*: HERGET, G. & VONGPAISAL, S. (eds) *6th International Congress On Rock Mechanics Proceedings*, **1**. International Society for Rock Mhanics, 393–396.

IKEDA, R., YAMAMOTO, T. & OMURA, K. 1995. Fault fracture zone evaluation by well logging, paper J. *In*: *1st Annual Well Logging Symposium Of Japan Proceedings*. Society of Professional Well Log Analysts, Japan Chapter.

ISHAK, I. B. 1994. FMS applications on reservoir modelling and horizontal drilling in Marmul area, south Oman, ADSPE-25. *In*: *6th ADNOC/SPE Abu Dhabi International Oil Conference Proceedings*. Society of Petroleum Engineers, 37–47.

JACKSON, P. D., LOVELL, M. A., PITCHER, C., GREEN, C. A., EVANS, C. J., FLINT, R. & FORSTER, A. 1990. Electrical resistivity imaging of core samples. *In*: WORTHINGTON, P. F. (ed.) *Advances In Core Evaluation, Accuracy, And Precision In Reserves Estimation* [Reviewed proceedings of the first Society of Core Analysts European core analysis symposium, London, UK]. Gordon and Breach Science Publishers, New York, 365–378. Also published in 1990 as, High-resolution electrical resistivity imaging of core samples, paper G. *In*: *13th European Formation Evaluation Symposium Transactions*. Society of Professional Well Log Analysts, Budapest Chapter.

JAGELER, A. H. 1980. New well logging tools improve formation evaluation. *World Oil*, **190**(4), 89–103.

JAIN, A. K. 1989. *Fundamentals Of Digital Image Processing*. Prentice Hall, Englewood Cliffs, New Jersey.

JENSEN, O. F., JR. & RAY, W. 1965. Photographic evaluation of water wells. *The Log Analyst*, v, **5**(4), 15–26.

JOHNSTON, P. & WACHI, N. 1994. Estimation of natural fracture orientation using borehole imaging logs and vertical seismic profiles at Orcutt oil field, California, USA. *In*: *14th World Petroleum Congress Proceedings*, **2**. John Wiley and Sons, New York, 147–148.

JURADO-RODRIGUEZ, M.-J. 1997. Characterization of carbonate facies in eastern Mediterranean – results from an integrated interpretation of Formation Microscanner data and conventional logs (ODP Leg 160 data). *In*: PAWLOWSKY-GLAHN, V. (ed.) *3rd Annual Conference Of The International Association For Mathematical Geology [IAMG '97] Proceedings*. International Center for Numerical Methods in Engineering, Barcelona, Spain, 208–211.

KAMEWADA, S., ENDO, T., KOKUBU, H. & NISHIGAKI, Y. 1989. The device and features of BIP system [in Japanese]. *In*: *21st Symposium On Rock Mechanics Proceedings*. *Committee On Rock Mechanics*. Japanese Society of Civil Engineers, 196–200.

——, GI, H. S., TANIGUCHI, S. & YONEDA, H. 1990. Application of borehole image processing system to survey for tunnels. *In*: BARTON, N. & STEPHANSSON, O. (eds) *Rock Joints* [international symposium on rock joints, June 4–6, 1990, Loen Norway, proceedings]. A. A. Balkema, Rotterdam, The Netherlands, 51–58.

KATAHARA, K. W., KYLE, D. G., SIEGFRIED, R. W., GARD, M. F., GOODWILL, W. P., SCHASTEEN, T. & PETERMANN, S. G. 1988. Detection of external pipe defects with a modified borehole televiewer, paper UU. *In*: *29th Annual Logging Symposium Transactions*. Society of Professional Well Log Analysts. Later reprinted in 1990. *In*: PAILLET, F. L., BARTON, C., LUTHI, S., RAMBOW, F. & ZEMANEK, J. (eds) *Borehole Imaging*. Society of Professional Well Log Analysts Reprint Volume, 251–263.

KATO, O., DOI, N., AKAZAWA, T., SAKAGAWA, Y., YAGI, M. & MURAOKA, H. 1995. Characteristics of fractures based on FMI logs and cores in well WD-1 in the Kakkonda geothermal field, Japan. *Geothermal Resources Council Transactions*, **19**, 317–322.

KAWAMOTO, T., ARATO, H. & ANNAKA, H. 1996. Geological interpretation of the extension fractures in core and borehole image, Niigata Tertiary basin, northeastern Japan, paper I. *In*: *2nd Annual Well Logging Symposium Of Japan Proceedings*. Society of Professional Well Log Analysts, Japan Chapter.

KLEINBERG, R. L., CHEW, W. C., CHOW, E. Y., CLARK, B. & GRIFFIN, D. D. 1987. Micro-induction sensor for the oil-based mud dipmeter, SPE-16761. *In*: *Annual Technical Meeting And Exhibition Proceedings*, omega, *Formation Evaluation And Reservoir Geology*. Society of Petroleum Engineers. Later published in 1988. *SPE Formation Evaluation*, **3**(6), 733–742.

KNIGHT, J. L. 1995. In-situ fracture state from core, torctec, and a televiewer – three very different results but which if any are correct? *Scientific Drilling*, **5**(2), 61–68.

KOEPSELL, R. J., JENSEN, F. E. & LANGLEY, R. L. 1989. Gulf Coast fault orientation determined by formation imaging techniques, paper VV. *In*: *30th Annual Logging Symposium Transactions*. Society of Professional Well Log Analysts. Also published in 1989 in condensed form as Formation imaging, part 1 Formation imaging yields precise fault orientation, minimizes dry offsets; part 2 Imaging aids visualization of faults. *Oil and Gas Journal*, **87**(49), 55–58; **87**(50), 85–86.

KOTYAKHOV, F. I. & SEREBRENNIKOV, S. A. 1964. Evaluation of the distribution of fractures in oil and gas reservoirs by subsurface photography. *Geologia Nefti I Gaza*, 26–30. Later published in English translation in 1969. *Petroleum Geology*, **8**, 626–630.

KRAMMER, A., MENGER, S., CHABERNAUD, T. & FUCHS, K. 1994. Borehole televiewer data analysis from the New Hebrides island arc–the state of stress at hole 829A and 831B. *In*: GREENE, H. G., COLLOT, J-Y., TOKKING, L. B. *et al.* (eds) *Proceedings Of The Ocean Drilling Program, Scientific Results*, **134**. Ocean Drilling Program, Texas A&M University, College Station, 565–576.

KRAMMER, A., PEZARD, P., HARVEY, P. K. H. & FUCHS, K. 1995. Borehole televiewer data analysis of hole 504B from legs 137 and 140. *In*: ERZINGER, J., BECKER, K., DICK, H. J. B. & STOKKING, L. B. (eds) *Proceedings Of The Ocean Drilling Program, Scientific Results*. Texas A&M University, Ocean Drilling Program, College Station, Texas, **137/140**. 293–304.

KUBIK, W., LORENZEN, J. & WALBE, K. 1992. Fracture identification and characterization using cores, FMS, CAST, and borehole camera – Devonian shale, Pike County, Kentucky. *Gas Shales Technology Review*, **8**(1), December, 20–31. Later published in 1993 as KUBIK, W. & LOWRY, P., SPE-25897. *In*: *SPE Rocky Mountain/Low Permeability Reservoirs Symposium Proceedings*. Society of Petroleum Engineers, 543–554. Later reprinted in 1996. *In*: *Production From Fractured Shales*. Society of Petroleum Engineers Reprint Series No. **45**, 121–132.

LAMBERTINI, R. 1992. Fracture identification and quantification using borehole images – Maracaibo Basin, Venezuela, SPE-23637. *In*: *2nd Latin American Petroleum Engineering Conference Proceedings*. Society of Petroleum Engineers, 89–95.

LAU, J. S. O., AUGER, L. F. & BISSON, J. G. 1987. Subsurface fracture surveys using a borehole television camera and acoustic televiewer. *Canadian Geotechnical Journal*, **24**(4), 499–508. Later reprinted in 1990. *In*: PAILLET, F. L., BARTON, C., LUTHI, S., RAMBOW, F. & ZEMANEK, J. (eds) *Borehole Imaging*. Society of Professional Well Log Analysts Reprint Volume, 37–46.

LAUBACH, S. E., BAUMGARDNER, R. W., JR., MONSON, E. R., HUNT, E. & MEADOR, K. J. 1988. Fracture detection in low-permeability reservoir sandstone – a comparison of BHTV and FMS logs to core, SPE-18119. *In*: *SPE Annual Technical Conference And Exhibition Proceedings, v. Omega, Formation Evaluation And Reservoir Geology*. Society of Petroleum Engineers, 129–139. Later reprinted in 1990. *In*: PAILLET, F. L., BARTON, C., LUTHI, S., RAMBOW, F. & ZEMANEK, J. (eds) *Borehole Imaging*. Society of Professional Well Log Analysts Reprint Volume, 265–275.

LEHNE, K. A. 1988. Fracture detection from logs of North Sea chalk, paper E. *In*: *11th European Formation Evaluation Symposium Transactions*. Society of Professional Well Log Analysts, Norwegian Chapter. Later published in 1990. *In*: HURST, A., LOVELL, M. A. & MORTON, A. C. (eds) *Geological Applications Of Wireline Logs*. Geological Society, London, Special Publications No. **48**, 263–271.

LLOYD, P. M., DAHAN, C. & HUTIN, R. 1986. Formation imaging with micro-electrical scanning arrays – a new generation of stratigraphic high resolution dipmeter tool. *In*: *10th European Formation Evaluation Symposium Transactions*. Society of Professional Well Log Analysts, Aberdeen Chapter.

——, KOCH, R., DES AUTELS, D., ZAIN, A. M. & DAVIS, R. 1998. Chasing channel sands in South East Asia, paper I. *In*: *4th Well Logging Symposium of Japan Proceedings*. Sociey of Professional Well Log Analysts, Japan Chapter.

LOFTS, J. C., BEDFORD, J., BOULTON, H., VAN DOORN, J. A. & JEFFREYS, P. 1997. Feature recognition and the interpretation of images acquired from horizontal wellbores. *In*: LOVELL, M. A. & HARVEY, P. K. (eds) *Developments In Petrophysics*. Geological Society, London, Special Publications No. **133**, 345–365.

LOVELL, J. R., ROSTHAL, R. A., ARCENEAUX, C. L., JR., YOUNG, R. A. & BUFFINGTON, L. 1995. Structural interpretation of resistivity-at-the-bit images, paper TT. *In*: *36th Annual Logging Symposium Transactions*. Society of Professional Well Log Analysts. Later published in 1996 as YOUNG, R. A., LOVELL, J. R., ROSTHAL, R. A., BUFFINGTON, L. & ARCENEAUX, C., JR., LWD borehole images/dips aid offshore California evaluation. *World Oil*, **217**(4).

LOVELL, M. A., HARVEY, P. K., BREWER, T. S., WILLIAMS, C., JACKSON, P. D. & WILLIAMSON, G. 1997*a*. Application of FMS images in the Ocean Drilling Program – an overview. *In*: CRAMP, A., MCLEOD, C. J., LEE, S. V., & JONES, E. J. W. (eds) *Geological Evolution Of Ocean Basins – Results From The Ocean Drilling Program*. Geological Society, London, Special Publications No. **131**, 287–303.

——, ——, WILLIAMS, C. G., JACKSON, P. D., FLINT, R. C. & GUNN, D. A. 1997*b*. Electrical resistivity core imaging – a petrophysical link to borehole images. *The Log Analyst*, **38**(6), 45–53.

LUTHI, S. M. 1990. Sedimentary structures of clastic rocks identified from electrical borehole images. *In*: Hurst, A., LOVELL, M. A. & MORTON, A. C. (eds) *Geological Applications Of Wireline Logs*. Geological Society, London, Special Publications No. **48**, 3–10.

——1994. Textural segmentation of digital rock images into bedding units using texture energy and cluster labels. *Mathematical Geology*, **26**(2), 181–196.

—— & BANAVAR, J. R. 1988. Application of borehole images to three-dimensional geometric modeling of eolian sandstone reservoirs, Permian Rotliegende, North Sea. *AAPG Bulletin*, **72**(9), 1074–1089. Later reprinted in 1990. *In*: PAILLET,

F. L., Barton, C., Luthi, S., Rambow, F. & Zemanek, J. (eds) *Borehole Imaging*. Society of Professional Well Log Analysts Reprint Volume, 317–332.

—— & Souhaite, P. 1990. Fracture apertures from electrical borehole scans. *Geophysics*, **55**(7), 821–833.

Lysne, P. 1986. Determination of borehole shape by inversion of televiewer data. *The Log Analyst*, **27**(3), 64–71. Later reprinted in 1990. *In*: Paillet, F. L., Barton, C., Luthi, S., Rambow, F. & Zemanek, J. (eds) *Borehole Imaging*. Society of Professional Well Log Analysts Reprint Volume, 419–426.

Ma, T. A. & Bigelow, E. L. 1993. Borehole imaging tool detects well bore fractures. *Oil and Gas Journal*, **91**(2), January 11, 33–36.

——, Lincecum, V., Reinmiller, R. & Mattner, J. 1993. Natural and induced fracture classification using image analysis, paper J. *In*: *34th Annual Logging Symposium Transactions*. Society of Professional Well Log Analysts.

MacLeod, C. J., Parson, L. M., Sager, W. W. & Scientific Shipboard Party OPDP Leg 135. 1992. Identification of tectonic rotations in boreholes by the integration of core information with Formation MicroScanner and borehole televiewer images. *In*: Hurst, A., Griffiths, C. M. & Worthington, P. F. (eds) *Geological Applications Of Wireline Logs II*. The Geological Society, London, Special Publications No. **65**, 235–246.

——, Parson, L. M. & Sager, W. W. 1994. Reorientation of cores using the Formation Microscanner and borehole televiewer – application to structural and paleomagnetic studies with the Ocean Drilling Program. *In*: Hawkins, J. W., Parson, L. M., Allan, J. F. *et al.* (eds) *Proceedings Of The Ocean Drilling Program, Scientific Results*, **135**. Ocean Drilling Program, Texas A&M University, College Station, Texas, 301–311.

——, Celerier, B. & Harvey, P. K. 1995. Further techniques for core reorientation by core-log integration – application to structural studies of lower oceanic crust in Hess Deep, Eastern Pacific. *Scientific Drilling*, **5**(2), 77–86.

Maddox, S. 1996*b*. Visualizing production in flowing oil wells, paper EEE. *In*: *37th Annual Logging Symposium Transactions*. Society of Professional Well Log Analysts, 6. Also published in 1996, Systems visualize flow in real time. *American Oil and Gas Reporter*, **39**(11), **74**, 76–77. Later published in 1997 as paper A. *In*: *3rd Well Logging Symposium Of Japan Proceedings*: Society of Professional Well Log Analysts, Japan Chapter.

Maddox, S. D. 1997. Application of downhole video technology to multilateral well completions, SPE-38546. *In*: *Annual Technical Conference and Exhibition Proceedings*, delta, *Drilling And Completion*. Society of Petroleum Engineers, 29–34. Also published in 1997. *In*: *SPE Offshore Europe Conference Proceedings*. Society of Petroleum Engineers, 517–522. Later published in 1998, as paper OTC-8801. *In*: *Offshore Technology Conference Proceedings*, **3**, *Construction and Installation/Field Drilling and Development Systems*. Society of Petroleum Engineers, 521–527. Also published in 1998 in synopsis form. *Journal of Petroleum Technology*, **50**(6), 34, 36.

——, Gibling, G. R. & Dahl, J. 1995. Downhole video services enhance conformance technology, SPE-30134. *In*: *European Formation Damage Conference Proceedings*. Society of Petroleum Engineers, 535–549. Similar papers also published in 1995 as Maddox, S. & Dahl, J., Conformance technology uses for downhole video cameras. *In*: *1st Annual International Conference On Reservoir Conformance Technology Proceedings*. 6. And as, Maddox, S. 1995. Applications of high-resolution, real-time video. *In*: *7th Annual International Conference On Horizontal Well Technologies And Applications Proceedings*. Later published in 1996. *In*: *Permian Basin Oil And Gas Recovery Conference Proceedings*. Society of Petroleum Engineers, 49–64.

Maki, V., Gianzero, S., Strickland, R., Kepple, H. N. & Gianzero, M. V. 1991. Dynamically focused transducer applied to the CAST imaging tool, paper HH. *In*: *32nd Annual Logging Symposium Transactions*. Society of Professional Well Log Analysts.

Mancini, E. A., Epsman, M. L. & Stief, D. D. 1997. Characterization and evaluation of the Upper Jurassic Frisco City Sandstone reservoir in southwestern Alabama utilizing fullbore Formation MicroImager technology. *Gulf Coast Association of Geological Societies Transactions*, **47**, 329–335.

Mathis, B., Haller, D., Ganem, H. & Standon, E. 1995. Orientation and calibration of core and borehole image data, paper JJJ. *In*: *36th Annual Logging Symposium Transactions*. Society of Professional Well Log Analysts.

McDonald, P., Lorenz, J. C., Sizemore, C., Schechter, D. S. & Sheffield, T. 1997. Fracture characterization based on oriented horizontal core from the Spraberry trend reservoir – a case study, SPE-38664. *In*: *Annual Technical Conference And Exhibition Proceedings*, omega, *Formation Evaluation And Reservoir Geology, Part 1*. Society of Petroleum Engineers, 231–239.

McDougall, J. G. & Howard, M. G. 1989. Advances in borehole imaging with second-generation CBIL [circumferential borehole imaging log] borehole televiewer instrumentation, paper 12. *In*: *28th Annual Conference Proceedings*. Ontario Petroleum Institute Inc., Lambeth, Ontario, Canada.

McGann, G. J., Riches, H. A. & Renoult, D. C. 1988. Formation evaluation in a thinly bedded reservoir; a case history, Scapa field, North Sea, paper V. *In*: *29th Annual Symposium Transactions*. Society of Professional Well Log Analysts.

McNaboe, G. J. 1991. Comparison of Formation MicroScanner images to cores in sandstone reservoirs, Saudi Arabia, SPE-21436. *In*: *SPE 7th Middle East Oil Show Conference Proceedings*. Society of Petroleum Engineers, 831–839.

MENDOZA, J. R. 1996. The contribution of wellbore imaging to interval selection in naturally fractured reservoirs, SPE-35292. *In*: *International Petroleum Conference And Exhibition Of Mexico Proceedings*. Society of Petroleum Engineers, 389–399.

MENGER, S. 1994. New aspects of the borehole televiewer decentralization correction. *The Log Analyst*, **35**(4), 14–20.

—— & SCHEPERS, R. 1988. Method to derive high-resolution caliper logs from borehole televiewer traveltime data. *In*: *1988 Annual Meeting Expanded Abstracts With Biographies*, Society of Exploration Geophysicists, **1**, 554–556.

MESSENT, B. E. J. & YACOPETTI, C. M. 1997. The value of integrating borehole resistivity images with geological data – relevance to hydrocarbon exploration. *Australian Petroleum Production and Exploration Association Journal*, 301–314.

MOLINIE, A. J. & OGG, J. G. 1992. Formation Microscanner imagery of Lower Cretaceous and Jurassic sediments from the western Pacific (site 801). *In*: LARSON, R. L., LANCELOT, Y. *et al.* (eds) *Proceedings Of The Ocean Drilling Program, Scientific Results*, Texas A&M University, Ocean Drilling Program, College Station, Texas, **129**, 671–691.

MOOS, D. & ZOBACK, M. D. 1990. Utilization of observations of well bore failure to constrain the orientation and magnitude of crustal stresses – Application to continental, Deep Sea Drilling Project, and Ocean Drilling Program boreholes. *Journal of Geophysical Research*, **95**(B6), 9305–9325.

MORIN, R. H. 1990. Information on stress conditions in the oceanic crust from oval fractures in a deep borehole. *Geophysical Research Letters*, **17**(9), 1311–1314.

—— & BARRASH, W. 1986. Defining patterns of ground water and heat flow in fractured Brule Formation, western Nebraska, using borehole geophysical methods. *In*: *Surface And Borehole Geophysical Methods And Ground Water Instrumentation Conference And Exposition Proceedings*. National Water Well Association, Dublin, Ohio, 545–569.

——, ——, PAILLET, F. L. & TAYLOR, T. A. 1993. *Geophysical Logging Studies In The Snake River Plain Aquifer At The Idaho National Engineering Laboratory – Wells 44, 45, And 46*. US Geological Survey Water Resources Investigations Report 92-4184.

MORRISON, R. & THIBODAUX, J. 1984. The six-arm dipmeter, a new concept by Geosource, paper MMM. *In*: *25th Annual Logging Symposium Transactions*. Society of Professional Well Log Analysts, 22.

MOSNIER, J. 1982. Electrical detection of natural and artificial fractures in a borehole [in French]. *Annales Geophysique*, **38**, 537–540.

——1987. Electrical method for detecting fractures in borehole walls [in French]. *Bulletin de la Societe Geologique de France*, series 8 **3**(6), 1049–1054.

—— & CORNET, F. 1989. Apparatus to provide an image of the wall of a borehole during a hydraulic fracturing experiment. *In*: K. LOUWRIER, K., STAROSTE, E., GARNISH, J. D. & KARKOULIAS, V. (eds) *4th International Seminar On The Results Of EC Geothermal Energy Research And Demonstration Proceedings*, Commission of the European Communities, Kluwer Academic Publishers, Dordrecht, **4F**, 205–212.

MULLINS, J. E. 1966. Stereoscopic deep well photography in opaque fluids, paper NN. *In*: *7th Annual Logging Symposium Transactions*. Society of Professional Well Log Analysts. Also published in 1966 in condensed form as, New tool takes photos in oil and mud-filled wells. *World Oil*, **164**(6), 91–94. Later reprinted in 1990. *In*: PAILLET, F. L., BARTON, C., LUTHI, S., RAMBOW, F. & ZEMANEK, J. (eds) *Borehole Imaging*. Society of Professional Well Log Analysts Reprint Volume, 27–35.

NEWBERRY, B. M., GRACE, L. M. & STIEF, D. D. 1996. Analysis of carbonate dual porosity systems from borehole electrical images, SPE-35158. *In*: *Permian Basin Oil And Gas Recovery Conference Proceedings*. Society of Petroleum Engineers, 123–129.

NIITSUMA, H. & INAGAKI, Y. 1995. Pulsed doppler borehole televiewer to estimate permeability of subsurface fractures, paper K. *In*: *1st Annual Well Logging Symposium Of Japan, Proceedings*. Society of Professional Well Log Analysts, Japan Chapter.

NURMI, R., CHARARA, M., WATERHOUSE, M. & PARK, R. 1990. Heterogeneities in carbonate reservoirs; detection and analysis using borehole electrical imagery. *In*: HURST, A., LOVELL, M. A. & MORTON, A. C. (eds) *Geological Applications Of Wireline Logs*. Geological Society, London, Special Publications No. **48**, 95–111.

OGAWA, T. 1996. FMI imaging in geothermal field, paper J. *In*: *2nd Annual Well Logging Symposium Of Japan Proceedings*. Society of Professional Well Log Analysts, Japan Chapter.

OKABE, T., SHINOHARA, N., TAKASUGI, S. & HAYASHI, K. 1996. Estimation of a stress field using drilling-induced tensile fractures initiated on a borehole wall, paper K. *In*: *2nd Annual Well Logging Symposium of Japan Proceedings*. Society of professional Well Log Analysts, Japan Chapter.

OLSEN, J. E., KRISTENSEN, R. & TAYLOR, R. W. 1995. Case histories in the Europe/Africa area demonstrate improved capabilities of fiber-optic video camera technology, SPE-29300. *In*: *SPE Asia Pacific Oil And Gas Conference Proceedings*. Society of Petroleum Engineers, 421–432.

ONIONS, K. R. & WHITWORTH, K. R. 1995. Applications of electrical borehole imaging to mining design. *Scientific Drilling*, **5**(2), 69–76.

OVERBEY, W. K., JR., YOST, L. E. & YOST, A. B., II 1988. Analysis of natural fractures observed by borehole video camera in a horizontal well, SPE-17760. *In*: *SPE Gas Technology Symposium Proceedings*. Society of Petroleum Engineers, 9–16.

OWEN, T. E. & DARILEK, G. T. 1987. Borehole color television imagery, paper C. *In*: *2nd International*

Symposium on Borehole Geophysics for Minerals, Geotechnical, and Groundwater Applications Proceedings. Society of Professional Well Log Analysts, Minerals and Geotechnical Logging Society Chapter-at-Large, 25–33.

OYNO, L., TJETLAND, B. G., ESBENSEN, K. H., SOLBERG, SCHEIE, A. & LARSEN, T. 1996. Prediction of petrophysical parameters based on digital core images, SPE-36853. *In*: *SPE European Petroleum Conference [EUROPEC] Proceedings*, Society of Petroleum Engineers, **1**, 341–347.

PAILLET, F. L. 1985. *Acoustic-Televiewer And Acoustic-Waveform Logs Used To Characterize Deeply Buried Basalt Flows, Hanford Site, Benton County, Washington*. US Geological Survey Open-File Report, 85–419.

—— & KIM, K. 1985. *The Character And Distribution Of Borehole Breakouts And Their Relationship To In Situ Stresses In Deep Columbia River Basalts. Rockwell Hanford, Richland, Washington*, Operations Report RHO-BW-CR-155. Later published in 1987. *Journal of Geophysical Research*, **92**(B7), 6223–6234. Later reprinted in 1990. *In*: PAILLET, F. L., BARTON, C., LUTHI, S., RAMBOW, F. & ZEMANEK, J. (eds) *Borehole Imaging*. Society of Professional Well Log Analysts Reprint Volume, 387–398.

—— & GOLDBERG, D. 1991. Acoustic televiewer log images of natural fractures and bedding planes in the Toa Baja borehole, Puerto Rico. *Geophysical Research Letters*, **18**(3), 501–504.

——, KEYS, W. S. & HESS, A. E. 1985. Effects of lithology on televiewer-log quality and fracture interpretation, paper JJJ. *In*: *27th Annual Logging Symposium Transactions*. Society of Professional Well Log Analysts. Later reprinted in 1990. *In*: PAILLET, F. L., BARTON, C., LUTHI, S., RAMBOW, F. & ZEMANEK, J. (eds) *Borehole Imaging*. Society of Professional Well Log Analysts Reprint Volume, 99–128.

——, HESS, A. E., CHENG, C. H. & HARDIN, E. A. 1987. Characterization of fracture permeability with high-resolution vertical flow measurements during borehole pumping. *Ground Water*, **25**(1), 28–40. Later reprinted in 1990. *In*: PAILLET, F. L., BARTON, C., LUTHI, S., RAMBOW, F. & ZEMANEK, J. (eds) *Borehole Imaging*. Society of Professional Well Log Analysts Reprint Volume, 251–263.

——, BARTON, C., LUTHI, S., RAMBOW, F. & ZEMANEK, J. (eds) 1990*a*. *Borehole Imaging*. Society of Professional Well Log Analysts Reprint Volume.

——, ——, ——, —— & ——1990*b*. Borehole imaging and its application in well logging – an overview. *In*: *Borehole Imaging*. Society of Professional Well Log Analysts Reprint Volume, 1–23.

PALMER, I. D. & SPARKS, D. P. 1990. Measurement of induced fractures by downhole television camera in coalbeds of the Black Warrior basin, SPE-20660. *In*: *SPE Annual Technical Conference And Exhibition Proceedings, v. Pi, Production Operations And Engineering*. Society of Petroleum Engineers, 459–472. Later published in 1991. *Journal of Petroleum Technology*, **43**(3), 270–275, 326–328. Later reprinted in 1992. *In*: *Coalbed Methane*. Society of Petroleum Engineers, Reprint Series No. 35, 167–175.

PASTERNACK, E. S. & GOODWILL, W. P. 1983. Applications of digital borehole televiewer logging, paper X. *In*: *24th Annual Logging Symposium Transactions*. Society of Professional Well Log Analysts, 12. Later reprinted in 1990. *In*: PAILLET, F. L., BARTON, C., LUTHI, S., RAMBOW, F. & ZEMANEK, J. (eds) *Borehole Imaging*. Society of Professional Well Log Analysts Reprint Volume, 427–438.

PEZARD, P. A., LOVELL, M. A. & HISCOTT, R. N. 1992. Downhole electrical images in volcaniclastic sequences of the Izu-Bonin forearc basin, western Pacific. *In*: TAYLOR, B., FUJIOKA, K. *et al.* (eds) *Proceedings Of The Ocean Drilling Program, Scientific Results*, **126**. Texas A&M University, Ocean Drilling Program, College Station, Texas, 603–624. Also published as PEZARD, P. A., HISCOTT, R. N., LOVELL, M. A., COLLELA, A. & MALINVERNO, A. 1992. Evolution of the Izu-Bonin intraoceanic forearc basin, western Pacific, from cores and FMS images. *In*: HURST, A., GRIFFITHS, C. M. & WORTHINGTON, P. F. (eds) *Geological Applications Of Wireline Logs II*. The Geological Society, London, Special Publications No. **65**, 43–69.

PHILLIPS, C., DIFOGGIO, R. & BURLEIGH, 1991. Extracting information from digital images of core, SCA-9125. *In*: *5th Annual Technical Conference Proceedings*, Society of Professional Well Log Analysts, Society of Core Analysts Chapter-at-Large, **3**.

PIRMEZ, C., HISCOTT, R. N. & KRONEN, J. D., JR. 1997. Sandy turbidite succession at the base of channel-levee systems of the Amazon fan revealed by FMS logs and cores – unraveling the facies architecture of large submarine fans *In*: FLOOD, R. D., PIPER, D. J. W., KLAUS, A. & PETERSON, L. C. (eds) *Proceedings of the Ocean Drilling Program, Scientific Results*, Ocean Drilling Program, Texas A&M University, College Station, **155**, 7–33.

PLUMB, R. A. & HICKMAN, S. H. 1985. Stress-induced borehole elongation – a comparison between the four-arm dipmeter and the borehole televiewer in the Auburn geothermal well. *Journal of Geophysical Research*, **90**(B7), 5513–5521. Later reprinted in 1990. *In*: PAILLET, F. L., BARTON, C., LUTHI, S., RAMBOW, F. & ZEMANEK, J. (eds) *Borehole Imaging*. Society of Professional Well Log Analysts Reprint Volume, 407–415.

—— & LUTHI, S. M. 1986. Application of borehole images to geologic modeling of an eolian reservoir, SPE-15487. Society Of Petroleum Engineers, Annual Technical Conference And Exhibition, Preprint. Later published in 1989. *SPE Formation Evaluation*, **3**(4), 505–514. Later reprinted in 1990. *In*: PAILLET, F. L., BARTON, C., LUTHI, S., RAMBOW, F. & ZEMANEK, J. (eds) *Borehole Imaging*. Society of Professional Well Log Analysts Reprint Volume, 333–343.

PRATT, W. K. 1991. *Digital Image Processing*, 2nd ed. John Wiley and Sons, New York.

PRENSKY, S. E. 1992. Borehole breakouts and in-situ stress – a review. *The Log Analyst*, **33**(3), 304–312.

——1994*a*. A survey of recent developments and emerging technology in well logging and rock characterization. *The Log Analyst*, **35**(2), 15–45.

——1994*b*. A survey of recent developments and emerging technology in well logging and rock characterization – A supplement. *The Log Analyst*, **35**(5), 78–84.

PRIEST, J. 1997. Computing bore hole geometry and related parameters from acoustic caliper data, paper G. *In*: *38th Annual Logging Symposium Transactions*. Society of Professional Well Log Analysts.

PRILLIMAN, J., BEAN, C., HASHEM, M., BRATTON, T., FREDETTE, M. & LOVELL, J. 1997. A comparison of wireline and LWD resistivity images in the Gulf of Mexico, paper DDD. *In*: *38th Annual Logging Symposium Transactions*. Society of Professional Well Log Analysts.

RAAX AUSTRALIA PARTY LTD. 1993. New Borehole Image Processing System. *New South Wales Mining and Exploration Quarterly*, **38**, 10–13.

RAAX CO., LTD. 1992. *Borehole Image Processing System*. Sapporo, Japan.

RADEMAKER, R. A., OLSZEWSKI, K. K., GOIFFON, J. J. & MADDOX, S. D. 1992. A coiled-tubing-deployed downhole video system, SPE-24794. *In*: *Annual Technical Conference And Exhibition Proceedings, v. Pi., Production Operations And Engineering*. Society of Petroleum Engineers, 291–299. Also published in 1993. *In*: *8th Middle East Oil Show And Conference Proceedings*, Society of Petroleum Engineers, **1**, 67–75. Also published in 1993. *In*: *Offshore Europe 93 Proceedings*, Society of Petroleum Engineers, **1**, 11–20.

RAMAMOORTHY, R., FLAUM, C. & COLL, C. 1995. Geologically consistent resolution enhancement of petrophysical analysis using image log data, SPE-30607. *In*: *SPE Annual Technical Conference And Exhibition Proceedings, v. Omega, Formation Evaluation And Reservoir Geology*. Society of Petroleum Engineers, 833–841. Later published in 1997. *SPE Formation Evaluation*, **12**(2), 95–100.

RAMBOW, F. H. K. 1984. The borehole televiewer – some field examples, paper C. *In*: *25th Annual Logging Symposium Transactions*. Society of Professional Well Log Analysts. Later published in 1985 as Unique logging tool 'Sees' into wellbore. *World Oil*, **201**(1), 79–88. Later reprinted in 1990. *In*: PAILLET, F. L., BARTON, C., LUTHI, S., RAMBOW, F. & ZEMANEK, J. (eds) *Borehole Imaging*. Society of Professional Well Log Analysts Reprint Volume, 71–91.

RAO, T. V. S. & SUNDER, K. R. 1995. Improved reservoir definition in wells through borehole imaging. *Journal of the Association of Exploration Geophysics*, **16**(2), 71–83.

REID, R. & ENDERLIN, R. 1998. True pay thickness determination of laminated sand and shale sequences using borehole resistivity image logs, paper GGG. *In*: *39th Annual Logging Symposium Transactions*. Society of Profesional Well Log Analyst.

RIDER, M. 1996. Image logs, chapter 13. *In*: *The Geologic Interpretation Of Well Logs*, 2nd edn. Gulf Publishing Company, Houston 199–225.

ROCA-RAMISA, L. 1994. Carbonate characterization and classifi-cation from in-situ wellbore images, paper IPA94-1-1-115. *In*: *23rd Annual Convention Proceedings*, Indonesia Petroleum Association, Jakarta, **1**, 181–188.

——1996. The combined application and utilisation of CMR and FMI wireline technology in Wonnich-1. *Australian Petroleum Production Exploration Association Journal*, **36** (1) 202–208.

——, MENDOZA, J. & SANCHEZ-GALINDO, M. A. 1994. Evaluation of fracture distribution and continuity in carbonate reservoirs using wellbore imaging tools, SPE-28676. *In*: *SPE International Petroleum Conference And Exhibition Of Mexico Proceedings*. Society of Petroleum Engineers, 81–96.

ROSTHAL, R. A., YOUNG, R. A., LOVELL, J. R., BUFFINGTON, L. & ARCENEAUX, C. L., JR. 1995. Formation evaluation and geological interpretation from the Resistivity-at-the-Bit tool, SPE-30550. *In*: *SPE Annual Technical Conference And Exhibition PROCEEDINGS, v. Omega, Formation Evaluation And Reservoir Geology*. Society of Petroleum Engineers, 187–199.

——, BORNEMANN, E. T., EZELL, J. R. & SCHWALBACH, J. R. 1997. Real-time formation dip from a logging-while-drilling tool, SPE-38657. *In*: *Annual Technical Conference And Exhibition Proceedings*, omega, *Formation Evaluation And Reservoir Geology, Part 1*. Society of Petroleum Engineers, 41–54. Also published in 1997 in brief form. *Journal of Petroleum Technology*, **49**(9), 969–970.

ROUILLAC, D. 1994. *Cement Evaluation Logging Handbook*. Editions Technip, Paris.

RUBEL, H. J., SCHEPERS, R. & SCHMITZ, D. 1986. High resolution televiewer logs from sedimentary formations, paper GG. *In*: *10th European Formation Evaluation Symposium Transactions*. Society of Professional Well Log Analysts, Aberdeen Chapter. Later reprinted in 1990. *In*: PAILLET, F. L., BARTON, C., LUTHI, S., RAMBOW, F. & ZEMANEK, J. (eds) *Borehole Imaging*. Society of Professional Well Log Analysts Reprint Volume, 291–305.

RUZYLA, K. 1992. Quantitative image analysis of reservoir rocks – pitfalls, limitations, and suggested procedures. *Geobyte*, **7**(6), 7–20.

SAFINYA, K. A., LE LAN, P., VILLEGAS, M. & CHEUNG, P. S. 1991. Improved formation imaging with extended microelectrical arrays, SPE-22726. *In*: *SPE Annual Technical Conference And Exhibition Proceedings, v. Omega, Formation Evaluation And Reservoir Geology*. Society of Petroleum Engineers, 653–664.

SALIMULLAH, A. R. M. & STOW, D. A. V. 1992*a*. Application of FMS images in poorly recoverd coring intervals–examples from ODP Leg 129. *In*: HURST, A., GRIFFITHS, C. M. & WORTHINGTON, P. F. (eds) *Geological Applications Of Wireline Logs II*. Geological Society, London, Special Publications No. **65**, 71–86.

—— & STOW, D. A. V. 1992*b*. Wireline log signatures of resedimented volcaniclastic facies, ODP Leg 129, west central Pacific. *In*: HURST, A., GRIFFITHS, C. M. & WORTHINGTON, P. F. (eds) *Geological Applications Of Wireline Logs II*. Geological Society, London, Special Publications No. **65**, 87–97.

SANDERS, L. & BURKHART, C. J. 1996. The downhole video camera – open cased hole applications in the Permian Basin. *In*: *43rd Annual Short Course Proceedings*. Southwestern Petroleum Short Course Association, Lubbock, Texas, 280–285.

—— & FUCHS, K. 1996. Borehole imaging – the future of formation evaluation, SPE-35159. *In*: *SPE Permian Basin Oil And Gas Recovery Conference Proceedings*. Society of Petroleum Engineers, 131–138.

SCHAAR, R. G. 1992. The borehole televiewer and the Formation Microscanner; two innovative ways to detect and evaluate subsurface fractures. *In*: *6th National Outdoor Action Conference On Aquifer Restoration, Ground Water Monitoring, And Geophysical Methods. Ground Water Management*, **11**, 17–30.

SCHALLER, H. B., KILPATRICK, ROBERT, & STRATTON, ROBERT, 1972. The acoustic televiewer – A new method for inspection of down-hole tabular goods, SPE-4000. Society of Petroleum Engineers, presented at 47th annual fall meeting.

SCHEPERS, R. 1991. Development of a new acoustic borehole imaging tool. *Scientific Drilling*, **2**(4), 203–214.

SCHLUMBERGER, 1993. *Ultrasonic Imaging – USI, Ultrasonic Imager; UBI, Ultrasonic Borehole Imager*. Schlumberger Wireline and Testing, Houston, Document No. SMP-9230.

SCHMITT, D. R. 1991. A field based system for the digitization of ultrasonic borehole televiewer data in real time, paper M. *In*: *13th Formation Evaluation Symposium Transactions*. Canadian Well Logging Society. Also published in 1991 SCHMITT, D. R., NEIMAN, H., HOLM, L. & MACKINNON, J. 1991. Ultrasonic borehole televiewer logging – real time digitization during logging on a PC based system. *In*: *4th International Symposium On Borehole Geophysics For Minerals, Geotechnical And Groundwater Applications Proceedings*. Society of Professional Well Log Analysts, Minerals and Geotechnical Logging Society Chapter-at-Large, 339–346.

——1993. Fracture statistics derived from digital ultrasonic televiewer logging. *Journal of Canadian Petroleum Technology*, **32**(2), 34–43.

SEILER, D. 1995. Borehole visualization in 3-D, paper UUU. *In*: *36th Annual Logging Symposium Transactions*. Society of Professional Well Log Analysts.

——, EDMISTON, C., TORRES, D. & GOETZ, J. 1990. Field performance of a new borehole televiewer tool and associated image processing techniques, paper H. *In*: *31st Annual Logging Symposium Transactions*. Society of Professional Well Log Analysts.

——, KING, G. & EUBANKS, D. 1994. Field test results of a six arm microresistivity borehole imaging tool, paper W. *In*: *35th Annual Logging Symposium Transactions*. Society of Professional Well Log Analysts.

SERRA, O. 1989. *Formation Microscanner Image Interpretation*. Schlumberger Educational Service, Houston, SMP-7028.

SLATT, R. M., JORDAN, D. W. & DAVIS, R. J. 1994. Interpreting Formation MicroScanner log images of Gulf of Mexico Pliocene trubidites by comparision with Pennsylvanian turbidite outcrops, Arkansas. *In*: Weimer, P., BOUMA, A. H. & PERKINS, B. F. (eds) *Submarine Fans And Turbidite Systems; Sequence Stratigraphy, Reservoir Architecture, And Production Characteristics, Gulf Of Mexico And International* (*15th GCS SEPM annual research conference proceedings*). Society of Sedimetary Geology Gulf Coast Section, 335–348.

SMITH, M. B., ROSENBERG, R. J. & BOWEN, J. F. 1982. *Fracture width-design vs. measurement [TV], SPE-10965*. Society of Petroleum Engineers, 57th Annual Fall technical Conference and Exhibition, preprint.

SMITS, J. W., BENIMELI, D., DUBOURG, I., FAIVRE, O., HOYLE, D., TOURILLON, V., TROULLIER, J-C. & ANDERSON, B. I. 1995. High resolution from a new laterolog with azimuthal imaging, SPE-30584. *In*: *SPE Annual Technical Conference And Exhibition PROCEEDINGS, v. Omega, Formation Evaluation And Reservoir Geology*. Society of Petroleum Engineers, 563–576.

SOVICH, J., KLEIN, J. & GAYNOR, N. 1996. A thin bed model for the Kuparuk A sand, Kuparuk River field, North Slope, Alaska, paper D. *In*: *37th Annual Logging Symposium Transactions*. Society of Professional Well Log Analysts. Also published in 1996, An integrated petrophysical study of the thin beds in the Kuparuk A sand, Kuparuk River field, north slope, Alaska. *In*: PACHT, J. A., SHERIFF, R. E. & PERKINS, B. F. (eds) *Stratigraphic Analysis Utilizing Advanced Geophysical, Wireline, And Borehole Technology For Petroleum Exploration And Production* [*17th annual research conference proceedings*]. SEPM Society of Sedimentary Geology Foundation, Gulf Coast Section, 295–303

SOVICH, J. P. & NEWBERRY, B. 1993. Quantitative applications of borehole imaging, paper FFF. *In*: *34th Annual Logging Symposium Transactions*. Society of Professional Well Log Analysts.

SPANG, R. J., SLATT, R. M., BROWNE, G. H., HURLEY, N. F., WILLIAMS, E. T., DAVIS, R. J., KEAR, G. R. & FOULK, L. S. 1997. Fullbore Formation Micro Imager logs for evaluating stratigraphic features and key surfaces in thin-bedded turbidite successions. *Gulf Coast Association of Geological Societies Transactions*, **47**, 643–645.

STANDEN, E., NURMI, R., EL-WAZEER, F. & OZKANLI, M. 1993. Quantitative applications of wellbore images to reservoir analysis, paper EEE. *In*: *34th Annual Logging Symposium Transactions*. Society of Professional Well Log Analysts.

STANG, C. W. 1989. Alternative electronic logging technique locates fractures in Austin chalk horizontal well. *Oil and Gas Journal*, **87**(45), 42–45.

STARCHER, M. G., MURPHY, J. R., ALEXANDER, P. D. & WHITTAKER, J. L. 1995. Video camera log used for water isolation in the Main Body 'B' pool, Elk Hills field, Kern County, California – water and oil, SPE-29654. *In*: *Western Regional Meeting Proceedings*. Society of Petroleum Engineers, 383–394.

STOCK, J. M., HEALY, J. H., HICKMAN, S. H. & ZOBACK, M. D. 1985. Hydraulic fracturing stress measurements at Yucca Mountain, Nevada and relationship to the regional stress field. *Journal of Geophysical Research*, **90**(B10), 8691–8706. Later reprinted in 1990. *In*: PAILLET, F. L., BARTON, C., LUTHI, S., RAMBOW, F. & ZEMANEK, J. (eds) *Borehole Imaging*. Society of Professional Well Log Analysts Reprint Volume, 355–370.

STRAUB, A., LEBERT, F. & VALLA, P. 1989. ELIAS: a slimhole electrical imaging probe, paper K. *In*: *3rd International Symposium On Borehole Geophysics For Minerals, Geotechnical, And Groundwater Applications Proceedings*. Society of Professional Well Log Analysts, Minerals and Geotechnical Logging Society, Chapter-at-Large, 161–181.

——, GROS, Y. & KRUCKEL, U. 1990. Structural description by electrical imaging of a borehole in granite, paper LL. *In*: *13th European Formation Evaluation Symposium Transactions*. Society of Professional Well Log Analysts, Budapest Chapter. Later published in 1991 as, STRAUB, A., KRUCKEL, U. & GROS, Y. 1991. Borehole electrical imaging and structural analysis in a granitic environment. *Geophysical Journal International*, **106**(3), 635–646.

——, SANO, K., IMAMURA, S. & OHHASHI, T. 1995. Development of a borehole imaging system, the Elias probe, paper L. *In*: *1st Annual Well Logging Symposium Of Japan Proceedings*. Society of Professional Well Log Analysts, Japan Chapter.

STUART-BRUGES, W. P. 1984. A dipmeter for use in oil based muds, paper 32. *In*: *9th International Formation Evaluation Symposium Transactions*. Society of Professional Well Log Analysts, Paris Chapter (SAID).

SULLIVAN, K. B. 1994. Formation evaluation of a thinly bedded and fractured reservoir using high-resolution logs and borehole images, Santa Ynez unit, offshore California. *In*: *2nd International Symposium On Well Logging Transactions*. Society of Professional Well Log Analysts, Beijing Chapter, 296–303.

—— & SCHEPEL, K. J. 1995. Borehole image logs – applications in fractured and thinly bedded reservoirs, paper T. *In*: *36th Annual Logging Symposium Transactions*. Society of Professional Well Log Analysts.

SUZUKI, K., KAJIHARA, K., OTSUKA, Y., TASAKI, S. & KAWAMOTO, T. 1996. Estimation of rock quality using RQD observed by BHTV. *In*: *Korea–Japan Joint Symposium On Rock Engineering*, 237–242.

SVOR, T. R. & MEEHAN, D. N. 1991. Quantifying horizontal well logs in naturally fractured reservoirs, part I SPE-22704. *In*: *SPE Annual Technical Conference And Exhibition Proceedings, v. Omega, Formation Evaluation And Reservoir Geology*. Society of Petroleum Engineers, 469–480. Part II, published as SPE-22932. *In*: *SPE Annual Technical Conference And Exhibition, v. Sigma, Reservoir Engineering*. Society of Petroleum Engineers, 471–480.

TAHA, M. 1997. *Borehole electrical imagery – a synergetic approach to delineating reservoir characteristics*. The Search for Oil and Gas in Latin America and the Carribean, Schlumberger Surenco, no. **5**, 37–51.

TANG, X. M. & MARTIN, R. J. 1994. *High Resolution Evaluation Of Formation Flow Properties From A Borehole Acoustic Imaging Tool*, final report. Gas Research Institute Report GRI-94/0521.

TANIGUCHI, S., KAMEWADA, S., YONEDA, H. & KOKUBU, H. 1989. Development of inspection method for diagnosing aging tunnels [in Japanese]. *In*: *21st Symposium On Rock Mechanics Proceedings*. Committee on Rock Mechanics, Japanese Society of Civil Engineers, 191–195.

——, YASUDA, T. & KAMWADA, S. 1993. Investigation method for oldaged tunnels. *In*: PASAMEHMETOGLU, A. G., KAWAMOTO, T., WHITTAKER, B. N. & AYDAN, O. (eds) *Assessment And Prevention of Failure Phenomenon in Rock Engineering*. A. A. Balkema, Rotterdam, The Netherlands, 810–806.

TANIMOTO, C., KISHIDA, K., ANDO, T., MURAI, S. & MATSUMOTO, T. 1995. Immediate image and its analysis of fractured/jointed rock mass through the borehole scanner. *In*: MYER, L. R., TSANG, C.-F., COOK, N. G. W. & GOODMAN, R. E. (eds) *Fractured And Jointed Rock Masses*. A. A. Balkema, Rotterdam, The Netherlands, 219–227.

TAYLOR, T. J. 1983. Interpretation and application of borehole televiewer surveys, paper QQ. *In*: *24th Annual Logging Symposium Transactions*. Society of Professional Well Log Analysts. Reprinted in 1990. *In*: FOSTER, N. H. & BEAUMONT, E. A. (eds) *Formation Evaluation I – Log Evaluation*. AAPG Treatise of Petroleum Geology Reprint Series No. 16, 591–609.

——1991. A method for identifying fault related fracture systems using the borehole televiewer, paper JJ. *In*: *32nd Annual Logging Symposium Transactions*. Society of Professional Well Log Analysts.

TENZER, H. 1994, Fracture detection in hot dry rock drillholes at Soultz sous Forets and Bad Urach by borehole measurements. *In*: *Geothermics 94 In Europe, Symposium Communications*. Editions BRGM, Orleans, France, BRGM document no. **230**, 133–137.

TENZER, H., BUDEUS, P. & SCHELLSCHMIDT, R. 1992. Fracture analyses in Hot Dry Rock drillholes at Soultz and Urach by borehole televiewer measurements. *Geothermal Resources Council Transactions*, **19**, 317–321.

TETZLAFF, D. & PAAUWE, E. 1997. Combined formation imaging provides more than the sum of its parts. *In*: *Depth*. Western Atlas International, Houston, Texas, **1**, 47–49. Reprinted in condensed form in 1998, Formation imaging aided by combining acoustics, resistivity. *Offshore*, **58**(5), 96, 98–99, 166.

THAPA, B. B. 1994. *Analysis of in-situ joint strength using digital borehole scanner images*. PhD thesis, University of California Berkeley.

THAPA, B. B., HUGHETT, P. & KARASAKI, K. 1997. Semi-automatic analysis of rock fracture orientations from borehole wall images. *Geophysics*, **62**(1), 129–137.

THOMAS, S., CORBETT, P. & JENSEN, J. 1996. Permeability and permeability anisotropy characterization in the near wellbore – a numerical model using the probe permeameter and micro-resistivity image data, paper JJJ. *In*: *37th Annual Logging Symposium Transactions*. Society of Professional Well Log Analysts.

THOMPSON, L. B. & SNEDDEN, J. W. 1996. Geology and reservoir description of 1Y1 reservoir, Oso field, Nigeria, using FMS and dipmeter. *In*: PACHT, J. A., SHERIFF, R. E. & PERKINS, B. F. (eds) *Stratigraphic Analysis Utilizing Advanced Geophysical, Wireline, And Borehole Technology For Petroleum Exploration And Production*. 17th annual research conference proceedings. Society of Sedimentary Geology Foundation, Gulf Coast Section, 315–328.

TORRES, D., STRICKLAND, R. & GIANZERO, M. 1990. A new approach to determining dip and strike using borehole images, paper K *In*: *31st Annual Logging Symposium Transactions*. Society of Professional Well Log Analysts, 20. Also published in 1990 as paper JJ. *In*: *13th European Formation Evaluation Symposium Transactions*. Society of Professional Well Log Analysts, Budapest Chapter.

TROUILLER, J-C., DELHOMME, J-P., CARLINE, S. & ANXIONNAZ, H. 1989. Thin-bed reservoir analysis from borehole electrical images, SPE-19578. *In*: *SPE Annual Technical Conference And Exhibition Proceedings, v. Omega, Formation Evaluation And Reservoir Geology*. Society of Petroleum Engineers, 61–72. Later reprinted in 1990. *In*: PAILLET, F. L., BARTON, C., LUTHI, S., RAMBOW, F. & ZEMANEK, J. (eds) *Borehole Imaging*. Society of Professional Well Log Analysts Reprint Volume, 217–228.

USWAK, G. & MCLAFFERTY, N. 1993. Cement imaging through ultrasonics, paper CIM 93-30. *In*: *44th Annual Technical Meeting Preprints*, **2**. Canadian Institute of Mining and Metallurgy. Petroleum Society. Also published in 1993 as paper 93-302. *In*: *CADE/CAODC Spring Drilling Conference Proceedings*. Canadian Association of Drilling Engineers, Calgary, Alberta.

VERDUR, H., STINCO, L. & NAIDES, C. 1991. Sedimentological analysis utilizing the circumferential borehole acoustic image, paper II. *In*: *32nd Annual Logging Symposium Transactions*. Society of Professional Well Log Analysts.

VERHOEFF, E. K. & CHELINI, V. 1993. Radial anisotropy of petrophysical characteristics of reservoir rocks using borehole image analysis, paper V. *In*: *15th European Formation Evaluation Symposium Transactions*. Society of Professional Well Log Analysts, Norwegian Chapter.

—— & FROST, E., JR. 1995. Lithofacies characterization using statistical properties of borehole images, paper I. *In*: *1st Annual Well Logging Symposium Of Japan Proceedings*. Society of Professional Well Log Analysts, Japan Chapter.

VOIGHT, E. & HABERLAND, J. 1996. Application of an acoustic image device to obtain full structural information in halite, paper KKK. *In*: *37th Annual Logging Symposium Transactions*. Society of Professional Well Log Analysts.

WALBE, K. 1986. Utilization of the borehole television camera in conjunction with regular open and cased-hole logging in the Devonian shale of the Appalachian Basin, SPE-15610. Society of Petroleum Engineers, 61st annual meeting preprint. A similar paper, Utilization of the borehole television camera in open and cased holes, published in 1989, paper 89-DT-92. *In*: *Distribution/Transmission Conference Proceedings*. American Gas Association Operation Section, 605–606.

WALBE, K. A. 1992. Computer enhancement of borehole television imaging, SPE-24449. *In*: *7th SPE Petroleum Computer Conference Proceedings*. Society of Petroleum Engineers, 247–257.

WALBE, K. & COLLART, D. 1991. Use of the borehole television camera and the low-volume flowmeter to identify and measure gas flow in low-permeability formations, SPE-21835. *In*: *SPE Joint Rocky Mountain Regional Meeting And Low-Permeability Reservoirs Symposium Proceedings*. Society of Petroleum Engineers, 299–306.

WALTHAN, R. R., HUNTORO, T., HENDROWIBOWO, K., SUMATRI, H. & LLOYD, P. M. 1996. Formation imaging in geothermal wells – a key to improved reservoir characterization in the Kamojang field of west Java, paper IPA96-2.6-072. *In*: *25th Annual Convention Proceedings*, **2**. Indonesian Petroleum Association, Jakarta, 440–452.

WARNER, D. 1996. Delineation of high permeability zones using borehole geophysics and borehole video data in the karst upper Floridan Aquifer, Albany Georgia [abstract]. *Geological Society of America, Abstracts with Programs*, **28**(2), Southeastern Section, 48.

WHITTAKER, J. L. & LINVILLE, G. D. 1996. Well preparation –essential to successful video logging, SPE-35680. *In*: *66th SPE Western Region Meeting Proceedings*. Society of Petroleum Engineers, 297–308.

——, GOLICH, G. M. & SMOLEN, J. J. 1997. Diagnosing horizontal well production in the Belridge field with downhole video and production logs, SPE-38295. *In*: *Annual Technical Conference and Exhibition Proceedings, v. Omega, Formation Evaluation and Reservoir Geology*. Society of Petroleum Engineers, Part 2, 369–382.

WILEY, R. 1980. Borehole televiewer – revisited, paper HH. *In*: *21st Annual Logging Symposium Transactions*. Society of Professional Well Log Analysts.

WILLIAMS, C., JACKSON, P., LOVELL, M., HARVEY, P. & REECE, G. 1994. Numerical simulation of downhole electrical conductance imaging, paper O. *In*: *16th European Formation Evaluation Symposium Transactions*. Society of Professional Well Log Analysts, Aberdeen Chapter. Later published in 1995 as, Numerical simulation of downhole electrical images. *Scientific Drilling*, **5**(2), 93.

WILLIAMS, C. G., JACKSON, P. D., LOVELL, M. A. & HARVEY, P. K. 1997. Assessment and interpretation of electrical borehole images using numerical simulations. *The Log Analyst*, **38**(6), 34–44.

WILLIAMS, G. R. 1991. *The Formation Microscanner As A Measure Of Heterogeneity In Core-Log Studies*. MSc thesis, University of Texas, Department of Petroleum Engineering.

—— & SHARMA, M. M. 1991. *The Formation Microscanner As A Measure Of Heterogeneity In Core-Log Studies*. Gas Research Institute Report No. GRI-91/0239. A paper based on this report also published in 1991 as, Quantification of log-core correlation using the formation microscope [sic, should be microscanner], SPE-23524. *In*: *SPE Annual Technical Conference And Exhibition Proceedings, v. Omega, Formation Evaluation And Reservoir Geology*. Society of Petroleum Engineers, 985–994.

WONG, S. A., STARTZMAN, R. A. & KUO, T-B. 1989*a*. A new approach to the interpretation of wellbore images, SPE-19579. *In*: *SPE Annual Technical Conference And Exhibition Proceedings, v. Omega, Formation Evaluation And Reservoir Geology*. Society of Petroleum Engineers, 73–82.

——, —— & ——1989*b*. Enhancing borehole image data on a high-resolution PC, SPE-19124. *In*: *SPE Petroleum Computer Conference Proceedings*. Society of Petroleum Engineers, 37–48. Later reprinted in 1990. *In*: PAILLET, F. L., BARTON, C., LUTHI, S., RAMBOW, F. & ZEMANEK, J. (eds) *Borehole Imaging*. Society of Professional Well Log Analysts Reprint Volume, 443–455.

WOODS, T. 1995. Ultrasonic logging in cased boreholes for corrosion evaluation. *Geothermal Resources Council Transactions*, **19**, 129–135.

XIE, G. 1990. Digital borehole televiewer, paper EE. *In*: *International Well Logging Symposium Transactions*. Society of Professional Well Log Analysts, Beijing Chapter.

XU, C. & YUN, H. 1994. Fullbore Formation Micro-Image logging aids reservoir evaluation. *In*: *2nd International Symposium On Well Logging* Transactions. Society of Professional Well Log Analysts, Beijing Chapter, 104–110.

XU, J. & JACOBI, R. D. 1997. Fracture analysis techniques for data from cores and downhole video images [abstract]. *Geological Society of America, Abstracts with Programs*, **29**(1), Northeast Section, 91.

YE, L. & KERR, D. R. 1996. Use of microresistivity image logs in detailed reservoir architecture reconstruction of Glenn Sandstone, Glenn Pool field, northeastern Oklahoma. *In*: *AAPG Mid-Continent Section Meeting Transactions*. Tulsa Geological Society, 203–213.

YE, S. J., RABILLER, P. & KESKES, N. 1997. Automatic high resolution sedimentary dip detection on borehole imagery, paper O. *In*: *38th Annual Logging Symposium Transactions*. Society of Professional Well Log Analysts.

——, —— & ——1998, Automatic high resolution texture analysis on borehole imagery, paper M. *In*: *39th Annual Logging Symposium Transactions*. Society of Profesional Well Log Analysts.

YU, G., CIVAROLO, M., PAINCHAUD, S., STRACK, K.-M. & TETZLAFF, D. 1997. Analog modeling for acoustic imaging tools, paper CCC. *In*: *38th Annual Logging Symposium Transactions*. Society of Professional Well Log Analysts.

——, PAINCHAUD, S. & STRACK, K.-M. 1998. Analog and numerical modeling for borehole resistivity- and acoustic-imaging tools, paper KKK. *In*: *39th Annual Logging Symposium Transactions*. Society of Profesional Well Log Analysts.

ZEMANEK, J., CALDWELL, R. L., GLENN, E. E., JR., HOLCOMB, S. W., NORTON, L. J. & STRAUS, A. J. D. 1968. *The Borehole Televiewer – A new logging concept for fracture location and other types of borehole inspection*, SPE-2402. Society of Petroleum Engineers, 43rd Annual meeting preprint. Later published in 1969. *Journal of Petroleum Technology*, **21**(6), 762–774.

——, STOZESKI, B. & WANG, Z. 1990. The operational characteristics of a 250 kHz focused borehole imaging device, paper I. *In*: *31st Annual Logging Symposium Transactions*. Society of Professional Well Log Analysts.

ZOBACK, M. D., MOOS, D., MASTIN, L. & ANDERSON, R. N. 1985. Well bore breakouts and in situ stress. *Journal of Geophysical Research*, **90**(B7), 5523–5530. Discussion and reply in 1986, *Journal of Geophysical Research*, **91**(B14), 14 161–14 164. Later reprinted in 1990. *In*: PAILLET, F. L., BARTON, C., LUTHI, S., RAMBOW, F. & ZEMANEK, J. (eds) *Borehole Imaging*. Society of Professional Well Log Analysts Reprint Volume, 399–406.

Bibliography

AL-BUSAIDI, R. 1997. The use of borehole imaging logs to optimize horizontal well completions in fractured water-flooded carbonate rocks. *Geoarabia*, **2**, 19–34.

ALLAN, G. W. & MANLY, G. L. 1991. Porosity determination from acoustic borehole imaging logs, paper X. *In*: *13th Formation Evaluation Symposium Transactions*. Canadian Well Logging Society.

ANDERSON, R. N. & ZOBACK, M. D. 1983. The implications of fracture and void distribution from borehole televiewer imagery for the seismic velocity of the upper oceanic crust at Deep Sea Drilling Project Holes 501 and 504B. *In*: CANN, J. R., LANGSETH, M. G., HONNOREZ, J., VON HARZEN, R. P., WHITE, S. M. *et al.* (eds) *Deep Sea Drilling Project, Initial Reports*, **69**. US Government Printing Office, Washington, DC.

ARIZA, M., STARTZMAN, R. & KHACHATRYAN, A. 1991. Automatic fracture detection from high resolution wellbore images using artificial neural networks. *In*: *Conference On Artificial Intelligence In Petroleum Exploration And Production Proceedings*. Texas A&M University, Department of Petroleum Engineering, College, Station, Texas, 185–192.

BATEMAN, R. M. 1986. Determination of high angle fracture plane orientations from SHDT dipmeter recordings using a microcomputer. *Canadian Well Logging Society Journal,* **15**(1), 85–100.

BELFIELD, W. C. & SOVICH, J. P. 1994. Fracture statistics from horizontal wellbores, paper HWC94-37. *In*: *SPE/CIM/CANMET International Conference On Recent Advances In Horizontal Well Applications, Preprints*. CIM Petroleum Society, Calgary, Alberta.

BELL, J. S. & GOUGH, D. I. 1982. The use of borehole breakouts in the study of crustal stress. *In*: M. D. ZOBACK & B. C. HAIMSON (eds) *Workshop On Hydraulic Fracturing Stress Measurements Proceedings*. US Geological Survey Open-file Report 82-1075, 539–557.

BELL, J. & OSORIO, M. 1997. *Hi-res logs as reservoir characterization tools*. The Search for Oil and Gas in Latin America and the Carribean, Schlumberger Surenco, **5**, 42–59.

BIGELOW, E. L. 1990. *Cement Evaluation*. Atlas Wireline Services, Houston, TX.

BONNER, S., FREDETTE, M., LOVELL, J., MONTARON, B., ROSTHAL, R., TABANOU, J., WU, P., CLARK, B., MILLS, R. & WILLIAMS, R. 1996. Resistivity while drilling – images from the string. *Schlumberger Oilfield Review*, **8**(1), p. 4–19.

BOURKE, L. T., Corbin, N., BUCK, S. G. & HUDSON, G. 1993. Permeability images – a new technique for enhanced reservoir characterization. *In*: ASHTON, M. (ed.) *Advances In Reservoir Geology*. Geological Society, London, Special Publications No. **6**, 219–232.

BOUROZ, C., ANGELIER, J. & BERGERAT, F. 1990. Sancerre-Couy drilling project – a comparison between FMS images and cores. *Bulletin de la Societe Geologique de France*, **6**(5), 779–787.

BUCZAK, J. M. J. 1989. FMS, dipmeter, and petrophysics, paper 10. *In*: *Log Analysis Software Evaluation And Review (LASER) Symposium Transactions*. Society of Professional Well Log Analysts, London Chapter.

BUSCHER, M. 1989. Formation MicroScanner – a direct comparison of 2-pad versus 4-pad tool, paper O. *In*: *12th International Formation Evaluation Symposium Transactions*. Society of Professional Well Log Analysts, Paris Chapter (SAID).

CALDER, P. N., BAUER, A. & TOMICA, J. 1972. Surveying rock structure using a borehole television probe. *In*: *Annual meeting proceedings. Canadian Institute of Surveying*, 253–264.

CHABERNAUD, T. J. 1994. High-resolution electrical imaging in the New Hebrides Island arc – structural analysis and stress studies, chapter 34. *In*: GREENE, H. G., COLLOT, J-Y., TOKKING, L. B. *et al.* (eds) *Proceedings Of The Ocean Drilling Program, Scientific Results*, Ocean Drilling Program, Texas A&M University, College Station, **134**, 591–606.

CHARDAC, J-L. & EL-SAYED, H. 1995. Applications of a new multi-resistivity and imaging tool in the Middle East, SPE-29838. *In*: *SPE Middle East Oil Show Proceedings*, **1**. Society of Petroleum Engineers, 589–596.

CHITALE, V. & SHETTY, R. 1995. State-of-the-art electrical imaging of boreholes. *In*: *1st India International Petroleum Conference Proceedings*, **2**. Oil and Natural Gas Corporation, 433–434.

CLADER, P. N., BAUER, A. & TOMICA, J. 1972. Surveying rock structure using a borehole television probe. *In*: *Annual Meeting Proceedings*. Canadian Institute of Surveying, Quebec City.

COLL, C., CORTIULA, B. & RONDON, L. 1996. Accurate reservoir evaluation from borehole imaging techniques and thin bed analysis – case studies in shaly sands and complex lithologies in Lower Eocene sands, block III, Lake Maracaibo, Venezuela, SPE-36150. *In*: *4th Latin American And Carribean Petroleum Engineering Conference Proceedings*, Society of Petroleum Engineers, **3**, 697–708.

COMMITTEE ON FRACTURE CHARACTERIZATION AND FLUID FLOW 1996. Fracture detection methods, chapter 4. *In*: *Rock Fractures and Fluid Flow; Contemporary Understanding and Applications*. National Academy Press, Washington, DC., 167–242.

COOPER, M. 1992. The analysis of fracture systems in the subsurface thrust structures from the foothills of the Canadian Rockies. *In*: MCCLAY, K. R. (ed.) *Thrust Tectonics*. Chapman & Hall, London, 391–405.

COX, J. W. 1982. Long-axis orientation in elongated boreholes. *The Technical Review*, **30**(3), 15–25. Later published in 1983 as, Long axis orientation in elongated boreholes and its correlation with rock stress data, paper J. *In*: *24th Annual Logging Symposium Transactions*. Society of Professional Well Log Analysts.

CRUDEN, D. M., LAU, J. S. O., AUGER, L. F. & BISSON, J. G. 1987. Subsurface fracture surveys using a borehole television camera and acoustic televiewer. *Canadian Geotechnical Journal*, **24**, 499–508.

CULL, R. 1988. Formation Microscanner tool imaging and laterolog resistivity measurements in interpretation of crystalline terrains. *In*: A. BODEN & K. G. ERIKSSON (eds) *Deep Drilling in Crystalline Bedrock*, Volume **2**. Springer-Verlag, Berlin, 315–327.

DANGERFIELD, J., KNIGHT, I. & FARRELL, H. 1992. Characterization of faulting and fracturing in Ekofisk field from seismic, core, and log data. *In*: LARSEN, R. M., BREKKE, H., LARSEN, B. T. & TALLERAAS, E. (eds) *Structural And Tectonic Modelling And Its Application To Petroleum Geology*. Elsevier, Amsterdam, Norwegian Petroleum Society Special Publication No. 1, 397–407.

DAVEY, E. C. & SHOREY, W. R. J. 1984. *An Assessment Of The Simplec Borehole Acoustic Televiewer*. Atomic Energy of Canada Limited, Report CRNL-2400.

DAVISON, C. C., KEYS, W. S. & PAILLET, F. L. 1982. *Use Of Borehole-Geophysical Logs And Hydrologic Tests To Characterize Crystalline Rocks For Nuclear-Waste Storage*, Whiteshell nuclear research establishment, Manitoba and Chalk River nuclear laboratory, Ontario, Canada. Batelle National Laboratory, Project Management Division, Office of Nuclear Waste Isolation, Technical Report 418.

DELHOMME, J. P. 1992. A quantitative characterization of formation heterogeneities based on borehole image analysis, paper T. *In*: *33rd Annual Logging Symposium Transactions*. Society of Professional Well Log Analysts.

DEZAYES, C., VILLEMIN, T., GENTER, A., TRAINEAU, H. & ANGELIER, J. 1993. Paleostress analysis from fault and fracture geometry in the Hot Dry Rock boreholes at Soutz-sous-forets (Rhinegraben). *Terra Nova*, **5**, 217.

DOLFUS, D., CAMBRAY, H. & PEZARD, P. 1997. From FMS images to formation density – an artificial neutral network approach. *Scientific Drilling*, **6**(2), 59–66.

DRAXLER, J. K., LINGNAU, R. & WOHRL, T. 1990 Formation Microscanner application in crystalline rocks, paper KK. *In*: *13th European Formation Evaluation Symposium Transactions*. Society of Professional Well Log Analysts, Budapest Chapter.

EHLEN, J., HEVENOR, R. A., KEMENY, J. M. & GIRDNER, K. 1995. Fracture recognition in digital imagery. *In*: DAEMEN, J. J. K. & SCHULTZ, R. A. (eds) *Rock Mechanics; Proceedings Of The 35th, US Symposium*. A. A. Balkema, Rotterdam, 141–146.

EUBANKS, D. 1994. Application of borehole imaging in geologic studies. *In*: GIBBS, J. F. (ed.) *Synergy, Equals Energy – Teams, Tools, And Techniques*. West Texas Geological Society, Fall Symposium, 111–118.

FAM, M. 1997. *EMI And CAST Processing And Interpretation Workshop Notes*. Halliburton Energy Services, Houston, variously paginated.

FENS, T. W. & EPPING, W. J. M. 1996. Calibrated core-image digitisation, a prerequisite to make real quantitative use of coreslab images, paper SCA-9642. *In*: *International SCA Symposium Proceedings*. Society of Professional Well Log Analysts, Society of Core Analysts Chapter-at-Large.

FINKBEINER, T., BARTON, C. A. & ZOBACK, M. D. 1997. Relationships among in-situ stress, fractures, and faults, and fluid flow – Monterey Formation, Santa Maria Basin, California. *AAPG Bulletin*, **81**(12), 1975–1999.

FISCHER, J. A., FISCHER, J. J. & BULLWINKEL, R. J. 1994. Results of a downhole Formation Microscanner study in a Juro-Triassic-aged sedimentary deposit (Passaic Formation). *In*: LINK, P. K. (ed.) *Hydrogeology, Waste Disposal, Science And Politics*, 30th symposium on engineering geology and geotechnical engineering proceedings. Idaho State University, Pocatello, College of Engineering, 359–370.

FOCKE, J., HEINE, C. J. & NURMI, R. 1996. The outcrop on your desktop. *Middle East Well Evaluation Review*, **17**, 21–37.

FOLLOWS, E. 1997. Integration of inclined pilot hole core with horizontal image logs to appraise an aeolian reservoir, Auk field, central North Sea. *Petroleum Geoscience*, **3**(1), 43–56.

FORBES, D. & USWAK, G. 1992. Detection of gas migration behind casing using ultrasonic imaging methods. *Journal of Canadian Petroleum Technology*, **31**(6), 18–25.

FOWLER, M. L. 1996. The role of borehole-imaging technologies in improved recovery (examples from the Department of Energy's Class Program), SPE/DOE-35460. *In*: *SPE/DOE 10th Improved Oil Recovery Symposium Proceedings*, **2**. Society of Petroleum Engineers, 603–610.

FRISINGER, M. R. & GYLLENSTEN, A. 1986. Fracture detection in North Sea reservoirs, paper Q. *In*: *10th European Formation Evaluation Symposium Transactions*. Society of Professional Well Log Analysts, Aberdeen Chapter.

GARCIA-CARBALLIDO, C. 1997. Reservoir characterization using borehole images. Aberdeen University, UK, unpublished MSc thesis. An expanded abstract published in 1998. *In*: *AAPG Annual Convention Abstracts*.

GOETZ, J. F., SEILER, D. D. & EDMISTON, C. S. 1991, Fracture description in the Austin Chalk using CAST images. *In*: CHUBER, S. (ed.) *Austin Chalk Exploration Symposium, Geology, Geophysics, And Formation Evaluation, Abstracts And Short Papers*. South Texas Geological Society, 51–63.

GOLKE, M. & BRUDY, M. 1996. Orientation of crustal stresses in the North Sea and Barents Sea inferred from borehole breakouts. *Tectonophysics*, **266**, 25–52.

GOUGH, D. I. & GOUGH, W. I. 1987. Stress near the surface of the earth. *In*: G. W. WETHERILL, A. L. ALBEE & F. G. STEHLI (eds) *Annual Review Of Earth And Planetary Sciences*, volume 15. Annual Reviews Inc., Palo Alto, 545–66.

GRAHAM, R. L., WALBE, K. & MCBANE, R. A. 1986. Application of the borehole TV camera. *Eastern Devonian Gas Shales Technology Review*, **3**(3), December, 19–23.

GRAHAM, W. L., SILVA, C. I., LEIMKUHLER, J. M. & DE KOCK, A. J. 1997. Cement evaluation and casing inspection with advanced ultrasonic scanning methods, SPE-38651. *In*: *Annual Technical Conference And Exhibition Proceedings*, omega, *Formation Evaluation And Reservoir Geology*, Part 1. Society of Petroleum Engineers, 79–89.

GRENOILLEAU, J.-E. 1994. Breakthroughs in downhole video services. *In*: *14th World Petroleum Congress Proceedings*, **2**. John Wiley and Sons, New York, 243–244.

HAIMSON, B. C. & SONG, I. 1995. A new borehole failure criterion for estimating in situ stress from breakout span. *In*: FUJII (ed.) *8th International*

Congress On Rock Mechanics, **2**. International Society for Rock Mechanics, Tokyo, Japan, 341–346.

——, LEE, M. Y., BAUMGARTNER & RUMMEL, F. 1990. Plate tectonics and structure inferences from in situ stress measurements and fracture logging in drillhole CY-4, Troodos ophiolite, Cyprus. *In*: MALPAS, J., MOORES, E. M., PANAYIOTOU, A. & XENOPHONTOS, C. (eds) *Ophiollites; Oceanic Crustal Analogues* [proceedings of the symposium 'Troodos 1987']. Geological Survey Department, Cyprus Ministry of Agriculture and Natural Resources, Nicosia, 131–138.

HANSEN, S. M. 1988. Application of the Formation MicroScanner imaging tool in the Permian basin. *In*: *35th Annual Southwestern Petroleum Short Course Proceedings*. Southwestern Petroleum Short Course Association, 143–154.

HARVEY, P. K., JACKSON, P. D., LOVELL, M. A., WILLIAMSON, G., BALL, J. K., ASHU, P., SMITH, A. S. & FLINT, R. 1994. Structural implications from fluid flow and electrical resistivity images in aeolian sandstones, paper LL. *In*: *35th Annual Logging Symposium Transactions*. Society of Professional Well Log Analysts.

——, LOVELL, M. A., JACKSON, P. D., ASHU, A. P., WILLIAMSON, G., & SMITH, A. S., BALL, J. K. & FLINT, R. F. 1995. Electrical resistivity core imaging III – characterization of an aeolian sandstone. *Scientific Drilling*, **5**(4), 165–176.

HARVEY, P. K. H., PEZARD, P., ITURRINO, G. J., BOLDREEL, L. O. & LOVELL, M. A. 1995. The sheeted dike complex in hole 504B–observations from the integration of core and log data, chapter 26. *In*: ERZINGER, J., BECKER, K., DICK, H. J. B. & STOKKING, L. B. (eds) *Proceedings Of The Ocean Drilling Program, Scientific Results*, Texas A&M University, Ocean Drilling Program, College Station, Texas, **137/140**, 305–311.

HEALY, J. H., HICKMAN, S. H., ZOBACK, M. D. & ELLIS, W. L. 1984. *Report On Televiewer Log And Stress Measurements In Core Hole USW-G1, Nevada Test Site, December 13–22, 1981*. US Geological Survey Open-File Report 84-15.

HICKMAN, S. H., SVITEK, J. F. & LANGSETH, M. G. 1984. Borehole televiewer log of 395A. *In*: HYNDMAN, R. D. *et al.* (eds) *Deep Sea Drilling Project, Initial Reports*, US Government Printing Office, Washington, DC, **78B**, 709–715.

HILL, R. E. 1992 *Analysis Of Natural And Induced Fractures In The Barnett Shale–Mitchell Energy, T.P. Sims No. 2.* Gas Research Institute Report No. GRI-92/0094.

HILLIS, R. R., MIDLREN, S. C., PIGRAM, C. J. & WILLOUGHBY, D. R. 1997. Rotation of horizontal stresses in the Australian North West continental shelf due to the collision of the Indo-Australian and Eurasian plates. *Tectonics*, **16**, 323–335.

HOWARD, M. 1989. Advances in borehole imaging with second generation CBIL borehole televiewer instrumentation, paper CC. *In*: *12th Formation Evaluation Symposium Transactions*. Canadian Well Logging Society.

HUANG, L., RICHERS, D. & ROBINSON, J. E. 1993. Well-log imaging and its appliction to geologic interpretation. *In*: *Computerized Basin Analysis – The Prognosis Of Energy And Mineral Resources*. Plenum Press, 163–184.

JACKSON, P. D. & LOVELL, M. A. 1991. The correspondence of electrical current and fluid flow in rocks – the impact of electrical resistivity core imaging, paper J. *In*: *14th European Formation Evaluation Symposium Transactions [THAMES Symposium]*. Society of Professional Well Log Analysts, London Chapter.

—— & SHIPBOARD SCIENTIFIC PARTY, ODP LEG 133 1991. Electrical resistivity core scanning – a new aid to the evaluation of fine-scale structure in sedimentary cores. *Scientific Drilling*, **2**, 41–54.

——, GUNN, D. G., FLINT, R. C., *et al.* 1997. A noncontacting resistivity imaging method for characterizing whole round core while in its liner. *In*: LOVELL, M. A. & HARVEY, P. K. (eds) *Developments In Petrophysics*. Geological Society, London, Special Publications No. **122**, 1–10.

——, LOVELL, M. A., HARVEY, P. K., BALL, J. K., *et al.* 1992. Electrical resistivity core imaging – theoretical and practical experiments as an aid to reservoir characterization, paper VV. *In*: *33rd Annual Logging Symposium Transactions*. Society of Professional Well Log Analysts.

——, ——, ——, *et al.* 1994. Advances in resistivity core imaging, paper GG. *In*: *35th Annual Logging Symposium Transactions*. Society of Professional Well Log Analysts.

——, ——, ——, *et al.* 1994. Electrical resistivity core imaging – a new technology for high resolution investigation of petrophysical properties, paper I. *In*: *16th European Formation Evaluation Symposium Transactions*. Society of Professional Well Log Analysts, Aberdeen Chapter, 14. Later published in 1995. *Scientific Drilling*, **5**(4), 139–151.

JORGENSEN, S. D. & JAMES, S. W. 1988. Integration of Stratigraphic High Resolution Dipmeter data into the development of the Minnelusa 'B' sand reservoir in Hawk Point field, Campbell County, Wyoming. *In*: DIEDRICH, R. P., DYKA, M. A. K. & MILLER, W. R. (eds) *Eastern Powder River Basin-Black Hills*. Wyoming Geological Association, 39th Field Conference Guidebook, 105–116.

JUTTEN, J. & MORRISS, S. L. 1990. Cement job evaluation, chapter 16. *In*: NELSON, E. B. (ed.) *Well Cementing*. Elsevier, Amsterdam, Developments in Petroleum Science No. 28, 16-1–16-44.

KAUBACH, S. E., HAMLIN, H. S., BUEHRING, R., BAUMGARNER, R. W., JR. & MONSON, E. R. 1990. *Application Of Borehole Imaging Logs To Geologic Analysis, Cotton Valley Group And Travis Peak Formation, GRI Staged Field Experiment Well, East Texas*. Gas Research Institute Report No. GRI-90/0222.

KAWAMOTO, T., WHITTAKER, B. N. & AYDAN, O. 1993. Assessment and prevention of failure phenomena in rock engineering. *In*: PASAMEHMETOGLU, A. G. (ed.) *International Symposium On Assessment And Prevention Of Failure Phenomena*

In Rock Engineering Proceedings., A. A. Balkema, Rotterdam, The Netherlands, 801–806.

KERZNER, M. G. 1986. *Image Processing In Well Log Analysis.* IHRDC Press, Boston.

KESSELS, W. & KUCK, J. 1992. Computer-aided matching of plane core structures with borehole measurements for core orientation. *Scientific Drilling*, **3**(3), 225–238.

KEYS, W. S. 1980. The application of the acoustic televiewer to the characterization of hydraulic fractures in geothermal wells. *In*: *Raft River Well Stimulation Experiments, Geothermal Reservoir Well Stimulation Program.* US Department of Energy Report DOE/AL/10563-T7, A1–A11. Also published in 1980. *In*: *Geothermal Reservoir Stimulation Symposium Proceedings*, 176–202.

——1997. Borehole imaging logs. *In*: *A Practical Guide To Borehole Geophysics In Environmental Investigations.* CRC Lewis Publishers, Boca Raton, Florida, 97–102.

KIERSTEIN, R. A. 1984. *True Location And Orientation Of Fractures Logged With The Acoustic Televiewer (Including Programs To Correct Fracture Identification).* US Geological Survey Water-Resources Investigations Report 83-4275.

KIRCHOFF-STEIN, K. S., SILVER, E. A., MOOS, D., BARTON, C. A. & JARRARD, R. 1991. Results of borehole televiewer observations in the Celebes and Sulu Seas. *In*: SILVER, E. A., RANGIN, C., VON BREYMANN, M. T. *et al.* (eds) *Proceedings Of The Ocean Drilling Program, Scientific Results*, Texas A&M University, Ocean Drilling Program, College Station, Texas, **124**, 105–118.

KUBIK, W. T. 1993. *Characterization Of Natural Fractures In The Devonian Shale – GRI Experimental Development Research Area, Pike County, Kentucky, topical report.* Gas Research Institute Report No. GRI-93/0157.

LANDRY, P. G. 1992. The applications of acoustic cement bond logging to well casing evaluation and remediation programs. *In*: *6th National Outdoor Action Conference On Aquifer Restoration, Ground Water Monitoring, and Geophysical Methods.* Ground Water Management, **11**, 717–728.

LANYON, G. W. 1986. Interactive computer analysis of borehole televiewer data, paper S. *In*: *10th European Formation Evaluation Symposium Transactions.* Society of Professional Well Log Analysts, Aberdeen Chapter.

LAU, J. S. O. 1980. Borehole television survey. *In*: *13th Canadian Rock Mechanics Symposium Transactions*: Canadian Institute of Mining and Metallurgy, Toronto, Ontario, 204–210.

——, BISSON, J. G., ELLIOT, H. M., MARMENT, R. M., MCEWEN, J. H. & STONE, D. C. 1981. *A Progress Report On Borehole Television Surveys And The GSC Data File 1979/80.* Atomic energy of Canada Limited, Technical Record TR-135.

——, AUGER, L. F. & BISSON, J. G. 1982. *Preliminary Reports On Borehole Acoustic Televiewer Logging.* Atomic Energy of Canada Limited, Technical Record TR-215.

LAUBACH, S. E., BAUMGARDNER, R. W., Jr. & MEADOR, K. J. 1987. *Analysis Of Natural Fractures And Borehole Ellipticity, Travis Peak Formation, East Texas, topical report.* Gas Research Institute Report GRI 86/0211.

LOGAN, M. H., HEGARTY, D. M., SHIDELER, W. M. & BURTON, L. R. 1968. *Computer Determination Of True Dip And Strike For Planar Structures Intersected By An Inclined Drill Hole Based On Drill Hole TV Camera Observations.* US Bureau of Reclamation, Engineering Geology Division, Denver, Colorado.

LORENZ, J. C., ARGUELLO, J. G., STONE, C. M., HARSTAD, H., TEUFEL, L. W. & BROWN, S. R. 1995. *Prediction Of Fracture And Stress Orientations – Subsurface Frontier Formation, Green River Basin, topical report.* Gas Research Institute Report GRI-95/0151.

LOVELL, M., PEZARD, P. & HARVEY, P. 1990. Ultrasonic borehole images of the subsurface crust. *Geology Today*, **6**(5), 154–156.

——, HARVEY, P., JACKSON, P., FLINT, R. & GUNN, D. 1994. Characterisation of thinly-bedded reservoirs – a role for electrical resistivity core imaging? paper H. *In*: *16th European Formation Evaluation Symposium Transactions.* Society of Professional Well Log Analysts, Aberdeen Chapter.

LOVELL, M. A. & JACKSON, P. D. 1991. Electrical flow in rocks; the application of high resolution electrical resistivity core measurements, paper H. *In*: *32nd Annual Logging Symposium Transactions.* Society of Professional Well Log Analysts.

——, HARVEY, P. K., JACKSON, P. D., BALL, J. K., ASHU, P., FLINT, R. & GUNN, D 1994. Electrical resistivity core imaging – towards a 3-dimensional solution, paper JJ. *In*: *35th Annual Logging Symposium Transactions.* Society of Professional Well Log Analysts.

——, ——, ——, *et al.* 1995. Electrical resistivity core imaging II – investigation of fabric and fluid flow characteristics. *Scientific Drilling*, **5**(4), 153–164.

LUCAS, P. 1993. Reservoir characterisation using imaging logs – a future trend in reservoir development? *In*: *Advances In Reservoir Technology Proceedings.* The Petroleum Science and Technology Institute, London.

LUTHI, S. M. 1992. Part 4, wireline methods – borehole imaging devices. *In*: MORTON-THOMPSON, D. & WOODS, A. M. (eds) *Development Geology Reference Manual.* AAPG Methods in Exploration Series, No. **10**, 163–166.

MADDOX, S. 1996*a*. Downhole video is more than a repair tool. *Petroleum Engineer International*, **69**(5), 47–48.

MARTIN, P. & BERGERAT, F. 1996. Palaeo-stresses inferred from macro- and microfractures in the Balazuc-1 borehole (GPF programme); contribution to the tectonic evolution of the Cevennes border of the SE basin of France. *Marine And Petroleum Geology*, **13**(6), 671–684.

MASTIN, L. & HEINEMANN, B. 1989. First evaluation of four arm caliper and borehole televiewer measurements in the KTB pilot hole below 500 m depth. *In*: *SPE/ISRM Symposium Rock At Great Depth Proceedings.* Society of Petroleum Engineers.

MASTIN, L. G., HEINEMANN, B., KRAMMER, A., FUCHS, K. & ZOBACK, M. D. 1991. Stress orientation in the KTB pilot hole determined from wellbore breakouts,. *Scientific Drilling*, **2**(1), 1–12.

MCCALLUM, R. E. & BELL, J. S. 1995. Diagnosing natural and drilling-induced fractures from borehole images of western Canadian wells, BELL, J. S. & BIRD, T. D. (eds) *Energy From Sediments; Oil And Gas Forum '95 Proceedings*. Geological Survey of Canada Open File Report 3058, 79–82.

MEEHAN, D. M. 1994. Rock mechanics issues in petroleum engineering. *In*: NELSON, P. P. & LAUBACH, S. E. *Rock Mechanics Models And Measurements Challenges From Industry*. A. A. Balkema, Rotterdam, 3–17.

MERCADIER, C. G. L. & LIVERA, S. E. 1993. Application of the formation micro-scanner to modelling of Paleozoic reservoirs in Oman. *In*: FLINT, S. S. & BRYANT, I. D. (eds) *The Geological Modelling Of Hydrocarbon Reservoirs And Outcrop Analogues*. Blackwell Scientific Publications, Oxford, International Association of Sedimentologists Special Publication No. **15**, 125–142.

MEREDITH, J. A. & TADA, R. 1992. Evidence for Late Miocene cyclicity and broad-scale uniformity of sedimentation in the Yamato Basin, Sea of Japan, from Formation Microscanner data. *In*: TAMAKI, K., SUYEHIRO, K., ALLAN, J., MCWILLIAMS, M. *et al.* (eds) *Proceedings Of The Ocean Drilling Program, Scientific Results*, Texas A&M University, Ocean Drilling Program, College Station, Texas, part 2, **127/128**, 1037–1046.

MILDREN, S. D., HILLIS, R. R., FETT, T. & ROBINSON, P. H. 1994. Contemporary stresses in the Timor Sea–implications for fault-trap integrity. *In*: *West Australia basins symposium proceedings*. 291–300.

MOORE, S. D. 1991. Video logging – your downhole eye. *Petroleum Engineer International*, **63**(10), 22–23.

MOORE, T. K. 1988. *Initial Borehole Acoustic Televiewer Data Processing Algorithms*. Los Alamos National Laboratory Report No. LA-011327-MS, 178.

MOOS, D. & ZOBACK, M. D. 1990. Utilization of observations of well bore failure to constrain the orientation and magnitude of crustal stresses – Application to continental, Deep Sea Drilling Project, and Ocean Drilling Program boreholes. *Journal of Geophysical Research*, **95**(B6), 9305–9325.

MORIN, R. H. & FLAMAND, R. 1996. Quantifying lithospheric stress from the shape of oval fractures in a borehole. *In*: AUABERTIN, M., HASSANI, F. & MITRI, H. (eds) *Rock Mechanics Tools And Techniques*. A. A. Balkema, Rotterdam, The Netherlands, 883–889.

—— & PAILLET, F. L. 1996. Analysis of fractures intersecting Kahi Puka Well 1 and its relation to the growth of the island of Hawaii. *Journal of Geophysical Research*, **101**(B5), 11 695–11 699.

——, ANDERSON, R. N. & BARTON, C. 1989, Analysis and interpretation of the borehole televiewer log; information on the state of stress and the lithostratigraphy at hole 504B, chapter 10. *In*: Becker, K., SAKAI, H. *et al.* (eds.) *Proceedings Of The Ocean Drilling Program, Scientific Results*, Texas A&M University, Ocean Drilling Program, **111**, 109–118.

——, NEWMARK, R. L., BARTON, C. A. & ANDERSON, R. N. 1990. State of lithospheric stress and borehole stability at Deep Sea Drilling Project site 504B, eastern equatorial Pacific. *Journal Of Geophysical Research*, **95**(B6), 9293–9303.

——, MOOS, D. & HESS, A. E. 1992. Analysis of the borehole televiewer log from DSDP hole 395A – results from the Dianut Program. *Geophysical Research Letters*, **19**(5), 501–504.

MOSNIER, J. 1992. Detection of fractures in a borehole by electrical imaging [in French]. *Memoires de la Societe Geologique de France*, **161**, 25–30.

NEWMARK, R. L., ZOBACK, M. D. & ANDERSON, R. N. 1984. Orientation of in situ stresses in the oceanic crust. *Nature*, **311**, 424–429. Later published in 1985 as, Orientation of the in situ stresses near the Coast Rica rift and Peru-Chile trench – Deep Sea Drilling Project Hole 504B. *In*: *Initial Reports Of The Deep Sea Drilling Project*, US Government Printing Office, **83**, 511–515.

——, ANDERSON, R. N., MOOS, D. & ZOBACK, M. D. 1985. Sonic and ultrasonic logging of hole 504B and its implications for the structure, porosity, and stress regime of the upper 1 km of the oceanic crust. *In*: Initial Reports Of The Deep Sea Drilling Project, **83**. US Government Printing Office, 479–510. Also published in 1985 as, Structure, porosity and stress regime of the upper oceanic crust – Sonic and ultrasonic logging of DSDP hole 504B. *Tectonophysics*, **118**, 1–42.

NIELSEN, D. L., THORN, D. & LUTZ, S. J. 1993. Application of borehole imagery for mapping reservoir heterogeneity. *In*: THOMPSON, H. A. (ed.) *1992 International Gas Research Conference Proceedings*, **1**. Government Institutes, Inc., Rockville, MD, 23–31.

NOBLETT, B. R. & HOWARD, L. P. 1991. Using the Circumferential Borehole Imaging Log to identify and orient vertical fractures in the Austin Chalk. *In*: CHUBER, S. (ed.) *Austin Chalk Exploration Symposium, Geology, Geophysics, And Formation Evaluation, Abstracts And Short Papers*. South Texas Geological Society, 105–109.

NURMI, R. D. 1984. Geological evaluation of high resolution dipmeter data, paper YY. *In*: *25th Annual Logging Symposium Transactions*. Society of Professional Well Log Analysts.

NURMI, R., STANDEN, E. & HUSSEINEI, M. I. 1991. Old sandstone, new horizons. *Schlumberger Middle East Well Evaluation Review*, no. **11**, 10–31.

ORTENZI, A. & FIORANI, M. 1993. Petrographic applications of image analysis. *In*: SPENCER, A. M. (ed.) Generation, Accumulations And Production Of Europe's Hydrocarbons III. Springer-Verlag, Berlin, European Association of Petroleum Geologists Special Publication No. **3**, 369–375.

OZKANLI, M. & STANDEN, E. J. W. 1993. Fracture morphology in Cretaceous carbonate reservoirs from southeast Turkey. *First Break*, **11**(8), 323–333.

PAAUWE, E. 1996. Borehole imaging in the 21st century, paper 10. *In*: *35th Annual Conference Proceedings*. Ontario Petroleum Institute, London, Ontario, Canada,.

PAILLET, F. L. & HESS, A. E. 1986. *Geophysical Well-Log Analysis Of Fractured Crystalline Rocks At East Bull Lake, Ontario Canada*. US Geological Survey Open-File Report, 86-452.

—— & ——1987. *Geophysical Well Log Analysis Of Fractured Granitic Rocks At Atikokan, Ontario, Canada*. US Geological Survey Water Resources Investigations Report No. 87-4154.

—— & OLLILA, P. 1994. *Identification, Characterization, And Analysis Of Hydraulically Conductive Fractures In Granitic Basement Rocks, Millville, Massachusetts*. US Geological Survey Water-Resources Investigations Report 94-4185.

—— & OLSON, J. D. 1994. *Analysis Of The Results Of Hydraulic-Fracture Stimulation Of Two Crystalline Bedrock Boreholes, Grand Portage, Minnesota*. US Geological Survey Water-Resources Investigations Report 94-4044.

PETERS, C. A., SCHULTZ, P. K. & COBB, C. C. 1994. Development of an electro-optical logging cable and video system for offshore and other downhole applications, OTC-7608. *In*: *26th Annual Offshore Technology Conference Proceedings*, **4**. Society of Petroleum Engineers, 885–892. Also published in 1994 as, Downhole video system for harshest environment. *Sea Technology*, **35**(7), 12–18, 21.

PEZARD, P. A. & LUTHI, S. M. 1988. Borehole electrical images in the basement of the Cajon Pass scientific drillhole, California; fracture identification and tectonic implications. *Geophysical Research Letters*, **15**(9) (supplement), 1017–1020.

—— & LOVELL, M. 1990. Downhole images; electrical scanning reveals the nature of subsurface oceanic crust. *EOS Transactions*, **71**(20), 709, 718.

—— & ANDERSON, R. N. 1990. In situ measurements of electrical resistivity, formation anisotropy, and tectonic context, paper M. *In*: *31st Annual Logging Symposium Transactions*. Society of Professional Well Log Analysts.

——, ——, HOWARD, J. J. & LUTHI, S. M. 1988. Fracture distribution and basement structure from measurements of electrical resistivity in the basement of the Cajon Pass scientific drillhole, California. *Geophysical Research Letters*, **15**(9), (supplement), 1021–1024.

PILKINGTON, P. E. 1992. Cement evaluation–past, present, and future, SPE-20314. *Journal of Petroleum Technology*, **44**(2), 132–140.

POULTON, H. N. & STROZESKI, B. B. 1989. Ultrasonic Diplog; a system for measuring dips in wells drilled with oil-base mud, paper GG. *In*: *12th International Formation Evaluation Symposium Transactions*. Society of Professional Well Log Analysts, Paris Chapter.

RAMBOW, F. H. K. 1988. Cement evaluation in fiberglass casing; a case for pulse-echo tools, paper VV. *In*: *29th Annual Symposium Transactions*. Society of Professional Well Log Analysts.

RAMONES, M., CARMONA, R., MARCOS, J. & QUINN, T. 1997. Applications of acoustic image logs, SPE-39013. *In*: *5th SPE Latin America And Caribbean Petroleum Engineering Conference Proceedings*. Society of Petroleum Engineers.

RICHARDS, S., WOOD, J., USWAK, G. & DREBIT, G. 1991. The application of multiple corrosion measurements, paper U. *In*: *13th Formation Evaluation Symposium Transactions*. Canadian Well Logging Society.

RINK, M. & SCHOPPER, J. R. 1977. On the application of image analysis to formation evaluation, paper 17. *In*: *5th European Logging Symposium Transactions*. Society of Professional Well Log Analysts, Paris Chapter (SAID). Later published in 1978. *The Log Analyst*, **19**(1), 12–22.

ROBERTS, D. L. & RICHARDS, J. W. 1987. Casing corrosion evaluation using ultrasonic techniques; a unique approach for west Texas wells. *In*: *34th Annual Short Course Proceedings*. Southwestern Petroleum Short Course Association, Lubbock, Texas, 20–33.

—— & WALTER, J. L. 1985. Cement and casing evaluation using sonic and ultrasonic techniques. *In*: *32nd Annual Short Course Proceedings*. Southwestern Petroleum Short Course Association, Lubbock, Texas, 67–95.

ROESTENBURG, J. 1988. Subsurface structural and stratigraphic analysis using in situ wellbore formation images and core comparisons, paper IPA 88-11.06. *In*: *17th Annual Convention Proceedings*. Indonesian Petroleum Association, Jakarta, **1**, 307–321.

ROSENFELD, A. 1984. Image analysis. *In*: EKSTROM, M. P. (ed.) *Digital Image Processing Techniques*. Academic Press, New York, Series on Computational Techniques, **2**, 257–288.

SALIMULLAH, A. R. M. 1993. Volcaniclastic facies and sequence, Leg 129, chapter 6. *In*: LARSON, R. L. & LANCELEOT, Y. *et al.* (eds) *Proceedings Of The Ocean Drilling Program, Scientific Results*. Ocean Drilling Program, Texas A&M University, College Station, Texas, **129**, 507–527.

—— & STOW, D. A. V. 1995. Ichnofacies recognition in turbidites/hemiturbidites using enhanced FMS images – examples from ODP Leg 129. *The Log Analyst*, **36**(4), 38–49.

SANDERS, L. 1992. Fracture identification using contemporary acoustic tools in the Permian Basin. *In*: *39th Annual Short Course Proceedings*. Southwestern Petroleum Short Course Association, Lubbock, Texas, 237–252.

SAPRU, A., NURMI, R. & CHAKRAVORTY, S. 1995. Geological, petrophysical, and reservoir applications of borehole imagery in India and the Middle East. *In*: *Petrotech 95, 1st Petroleum Confrence Proceedings*. India Oil and natural Gas Corporation, New Delhi, India, **2**, 435–448.

SAUCIER, A., HUSEBY, O. K. & MULLER, J. 1997. Electrical texture characterization of dipmeter microresistivity signals using multifractal analysis. *Journal of Geophysical Research*, **102**(B5), 10 327–10 337.

SCHIUMA, M. F. 1993. Description and classification of fractures from acoustic imaging logs [in Spanish]. *In*: *12th Argentine Geologic Congress Transactions*, **1**. Geological Association of Argentina/ Argentina Petroleum Institute, Buenos Aires, 360–374.

SCHLUMBERGER, 1992. *Corrosion Evaluation*. Schlumberger Wireline & Testing, Houston, Texas, Document No. SMP-9110.

——1992. *FMI Fullbore Formation MicroImager*. Schlumberger Educational Services, Houston, TX, Document No. SMP-9210.

SCHMIDT, M. G., LI, X., STEINSIEK, R. R. & MA, T. A. 1992. Numerical model for predicting ultrasonic transducer pulse-echo response to borehole fractures, paper U. *In*: *33rd Annual Logging Symposium Transactions*. Society of Professional Well Log Analysts.

SCHULTZ, P. K. 1995. DHV systems allow downhole vision. *American Oil and Gas Reporter*, **38**(9), 60–65.

SHEIVES, T. C., TELLO, L. N., MAKI, V. E., JR., STANDLEY, T. E. & BLANKINSHIP, T. J. 1986. *A comparison of new ultrasonic cement and casing evaluation logs with standard cement bond logs*, SPE-15436. Society of Petroleum Engineers, presented at 61st Annual Technical Conference and Exhibition Preprints.

SNOW, D. M., LITTLEFIELD, B. A., MILLER, D., MUELLER, F. A. & WAITE, T. 1997. A case history – good foam cement jobs can be done and effectively evaluated by using a USI log, SPE-37536. *In*: *SPE International Thermal Operations And Heavy Oil Symposium Proceedings*. Society of Petroleum Engineers.

SPRINGER, J. E. & THORPE, R. K. 1981. *Borehole Elongation Versus In Situ Stress Orientation*. Lawrence Livermore National Laboratory Report UCRL-7018.

STANDEN, E. 1989. The analysis of fracture anomalies on electrical wellbore images, paper G. *In*: *12th Formation Evaluation Symposium Transactions*. Canadian Well Logging Society.

——1991. Egypt's resolution revolution [FMI]. *Schlumberger Middle East Well Evaluation Review*, no. **11**, 27–31.

——1991. Tips for analyzing fractures on electrical wellbore images. *World Oil*, **212**, no. **4**, 99–117.

STARKEY, J. & SAMANTARAY, A. K. 1994. A microcomputer-based system for quantitative petrographic analysis. *Computers and Geosciences*, **20**(9), 1285–1296.

STARTZMAN, R. A. & KUO, T. B. 1989. A knowledge-based approach to the interpretation of well-bore images. *In*: *Artificial Intelligence In Petroleum Exploration And Production Conference Proceedings*. Texas A&M University, College Station, 35–42.

STEELE, W. M., ATKINS, R. L., BRACKETT, D. A. & SCHMITT, T. J. 1988. Orientations of fractures measured from sonic televiewer logs of selected crystalline-rock wells in the Piedmont and Blue Ridge physiographic provinces of Georgia. *In*: *International Conference On Fluid Flow In Fractured Rocks Proceedings*. Georgia State University, Department of Geology, 162–187.

STOCK, J. M., HEALY, J. H. & HICKMAN, S. H. 1984. *Report On Televiewer Log And Stress Measurements In Core Hole USW G-2, Nevada Test Site*. US Geological Survey Open File Report OF 84–172.

STORM, E. 1979. Fisheye camera for surveying NTS boreholes. *In*: *Energy and Technology Review*. Lawrence Livermore National Laboratory Report UCRL-52000-79-10, 25–27.

STROZESKI, B. B., HILLIKER, D. J. & OLIVER, D. W. 1989. Theoretical and experimental development of the ultrasonic DIPLOG System, paper DD. *In*: *30th Annual Logging Symposium Transactions*. Society of Professional Well Log Analysts.

TANDIRCIOGLU, A. & KUMSAL, K. 1993. Fracture evaluation using FMI log data in the Upper Cretaceous carbonate reservoirs, southeastern Turkey. *In*: *4th Annual Archie Conference Proceedings*. Society of Exploration Geophysicists, 61–66.

TANDOM, P. M., NHUN, T. X., TJIA, H. D. & SPAGNUOLO, S. A. 1997. Fractured basement reservoir characterisation, offshore southern Vietnam, paper D. *In*: *3rd Well Logging Symposium Of Japan Proceedings*. Society of Professional Well Log Analysts, Japan Chapter.

THOMPSON, L. B., DOBLIN, M., SABIN, B. & SYMMONS, D. 1993. Deriving 3-D fracture networks from borehole images, wirelin data, and 3-D computer visualization modeling. *In*: *4th Annual Archie Conference Proceedings*. Society of Exploration Geophysicists, 57–60.

THORN, D. R. 1993. Evaluation techniques and applications of subsurface image logs. *In*: *New Technology For Independent Producers, Conference Proceedings*. Rocky Mountain Association of Geologists, Denver, section **6**, 53–61.

TOMICA, J. 1971, *Slope Stability Studies Using A Borehole Television Probe*. MSc thesis, Queen's University, Kingston, Ontario, Canada.

UDA, S. 1994. Image processing of borehole wall. *In*: *7th International Congress Proceedings*, **2**. International Association of Engineering Geology, 1041–1047.

WALLS, J. D., DVORKIN, J. & MAVKO, G. 1996. *In-Situ Stress Magnitude And Azimuth From Well Log Measurements*, final report. Gas Research Institute Report GRI-95/0356.

WILLIS-RICHARDS, J. & JUPE, A. 1995. Evaluation of fracture orientation statistics from borehole imaging. *Scientific Drilling*, **5**(2), 55–60

WOLFSBURGER, R. T., WELYCHKA, E. P. & STANDEN, E. J. W. 1988. Formation MicroScanner applications in southwestern Ontario. *In*: *27th Annual Conference Transactions*. Ontario Petroleum Institute, London, Ontario, Canada. Later published in 1990 as, Formation MicroScanner applications in the study of subsurface Paleozoic carbonate rocks; southwestern Ontario, Canada. *Carbonates and Evaporites*, **5**(2), 123–140.

WONG, S. A. 1989. *Automatic Feature Detection From Wellbore Images*. PhD thesis, Texas A&M University.

WONN, J. W. 1979. *Development Of An Acoustic Sensor For A Geothermal Borehole Televiewer*. US Department of Energy Report ALO-5391-T2, 264.

YASSIR, N. A. & ZERWER, A. 1997. Stress regimes in the Gulf Coast, offshore Louisiana – data from well-bore breakout analysis. *AAPG Bulletin*, **81**(2), 293–307.

ZEMANEK, J., GLENN, E. E., NORTON, L. J. & CALDWELL, R. L. 1970. Formation evaluation by inspection with the borehole televiewer. *Geophysics*, **35**, 254–269. Later reprinted in 1990. *In*: PAILLET, F. L., BARTON, C., LUTHI, S., RAMBOW, F. & ZEMANEK, J. (eds) *Borehole Imaging*. Society of Professional Well Log Analysts Reprint Volume, 55–70.

ZEWER, A. & YASSIR, N. A. 1994. Borehole breakout interpretation in the Gulf Coast, offshore Louisiana. *In*: NELSON, P. P. & LAUBACH, S. E. (eds) *Rock Mechanics Models And Measurements Challenges From Industry*. A. A. Balkema, Rotterdam, 225–232.

ZOBACK, M. D. & ANDERSON, R. N. 1982. Ultrasonic borehole televiewer investigation of oceanic crustal layer 2a, Costa Rica rift. *Nature*, **295**, 375–379.

Microresistivity and ultrasonic imagers: tool operations and processing principles with reference to commonly encountered image artefacts

PHILIP S. CHEUNG

Schlumberger Ribaud Product Center, 26 Rue de la Cavée, B.P. 202, 92142 Clamart, France

Abstract: Borehole images obtained from microresistivity and ultrasonic imager tools form an important source of geological information. The interpretation of these images is often hampered by the presence of artefacts, arising from peculiarities of the logging tools, and/or unexpected borehole conditions. In this paper, the operations of microresistivity and ultrasonic imager tools are reviewed, with a special emphasis on the tool features and borehole conditions that may give rise to artefacts in the resultant borehole images. The origin and characteristics of some commonly encountered artefacts are discussed, pointing out how they can be identified. Processing methods that may help to remove or reduce the effects of the artefacts are described. This information should help an interpreter to initially better understand the origin and significance of commonly encountered artefacts in terms of tool response, and secondly ensure that undesirable effects of such artefacts are minimised by using available data processing methods.

The author was invited to write this article to complement the paper by Lofts & Bourke (1999) *this volume*, titled 'The recognition of artefacts from acoustic and resisitivity borehole imaging devices'. As there was little time to gather publishable examples, permission was kindly given to use the examples presented by Lofts & Bourke. These examples from Lofts & Bourke will be referred to in short, for example, L&B 'Mud Smear', where the description in quotes is the label used by Lofts & Bourke in their paper.

The purpose of this paper is to summarize the information concerning the operation and data processing of microresistivity and ultrasonic imaging tools which is essential to the correct interpretation of the borehole images. In particular, it is intended to help an interpreter to initially better understand the origin and significance of commonly encountered artefacts in terms of tool response, and secondly ensure that undesirable effects of such artefacts are minimized by using available data processing methods. The following approach will be adopted in the paper. For each of the two types of imaging tools, the tool operations and sensor response will be outlined. Special features that may lead to artefacts are then discussed. The discussion will include pointers as to how artefacts may be recognized, and if processing methods are available, their theory and limitations. The physical basis for the appearance of various artefacts will be described in some detail, to enable an interpreter to assess their significance. In contrast, the various processing methods are only briefly mentioned, since details of algorithms would be beyond the scope of this article. Finally, the two tool types are discussed together in a summary section.

Microresistivity imagers

Tool operation

The following is a brief and general description of the operations of microresistivity imaging tools. Details of the actual tools available today are given by Ekstrom *et al.* (1987), Safinya *et al.* (1991), Seiler *et al.* (1994) and Lacazette (1996). In the complex logging environment, an understanding of the tool response can usually be achieved only by comparing logged data, cores and results from mathematical modelling. Trouiller *et al.* (1989) and Williams *et al.* (1997) provide further details of this process.

Microresistivity imagers are equipped with a number of pads with conducting surfaces, on which arrays of button electrodes are placed. When logging, the pads are pressed against the borehole wall and a voltage is applied between the lower section of the sonde, this includes the pads, and its upper section, which is separated from the lower section by an insulating section (Fig. 1). In consequence, currents will flow from the lower to the upper section of the sonde. Some currents will flow directly up the borehole

From: LOVELL, M. A., WILLIAMSON, G. & HARVEY, P. K. (eds) 1999. Borehole Imaging: applications and case histories. Geological Society, London, Special Publications, **159**, 45–57. 1-86239-043-6/99/$15.00. © The Geological Society of London 1999.

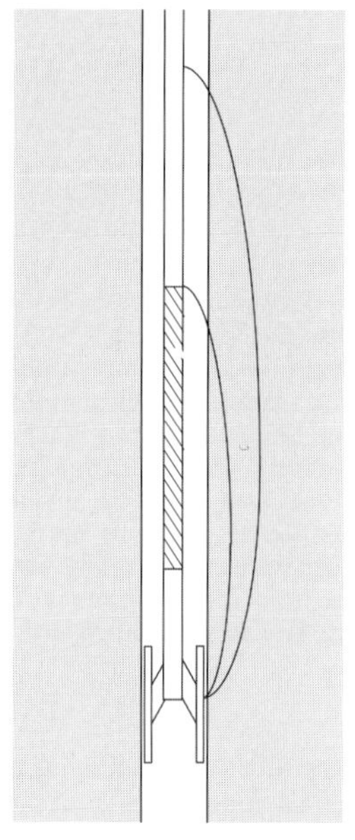

Fig. 1. Schematic showing a microresistivity tool and the path of the current emerging from a tool button.

through the mud. Currents emerging from the buttons will be focused by the conducting pad face into the formation, before returning eventually to the upper section of the sonde. The button currents are modulated by the conductivity of the formation directly in front of them, thereby allowing a qualitative conductivity image of the borehole wall to be constructed.

Figure 2 shows the button array typical of microresistivity tools currently in use. While the number of buttons may vary, their diameter is of the order of 5 mm, and they are usually arranged in two rows with a lateral offset to obtain a lateral sampling of half the button diameter, thus avoiding aliasing. The buttons on each pad image a strip of the borehole wall of fixed width. Different tools carry a different number of pads, on a different number of arms to provide different degrees of lateral coverage. In all cases, percentage cover of the borehole wall decreases as hole size increases.

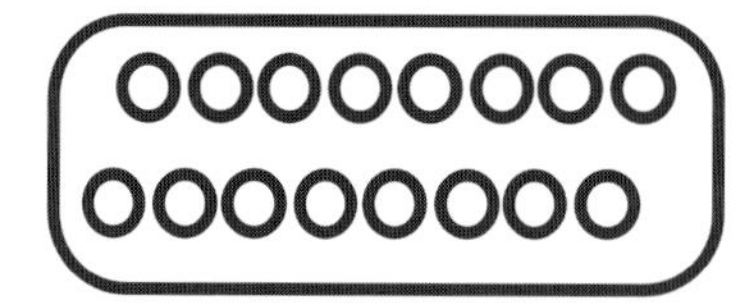

Fig. 2. Typical button arrangement on the pad of a microresistivity tool. Note that the image is formed by interlacing data from two rows of buttons which are offset in depth.

In order that the pads pack into a compact space when the tool is closed, the pads from most tools are located at different depths. For four-arm tools carrying two pads per arm, the second pad (referred to as a flap) is attached alongside and below the first pad, fixed directly on the arm. For six-arm tools, the odd and even numbered pads are displaced in depth. This depth offset of the pads, and the offset of the buttons on each pad, imply that the data acquired must be correctly aligned in depth to produce a correct image. Unless corrected, irregular tool speed during logging will give rise to certain artefacts in the image, this is discussed in more detail later.

Button sensor response

Since the pad face is a conductor forming an equipotential surface, the currents leaving it travel initially in the direction normal to the pad. In particular, the current flowing out of a button will be focused initially into a cylinder with a cross-sectional area equal to the button area. Almost immediately, the current diverges as the focusing influence of the pad face diminishes. Since the return electrode is effectively the entire upper section of the sonde, the current diffuses into a vast volume before entering the return electrode. This is shown schematically in Fig. 1.

The magnitude of the button current is equal to the applied voltage divided by the resistance of the current path. This resistance can be thought of qualitatively as consisting of two resistances in series. The first corresponds to the small volume facing the button where the current is focused, and the second, to the volume where the current is diffused. In each case:

$$\text{resistance} = \text{formation resistivity} \times (\text{path length}/\text{path cross-sectional area})$$

where the formation resistivity is some weighted average value for the volume concerned. The two resistances are in general comparable in magnitude: the path length and cross-sectional area being both small for the focused path, and large for the diffused path. By contrast, the focused resistance, pertaining to small volume, gives a high resolution measure of the formation resistivity, while the diffused resistance is a low resolution measure.

The button current as a function of depth will therefore consist of two distinct components: a high resolution component corresponding to the focused volume, with a resolution as high as the button size (5 mm); and a low resolution background corresponding to the diffused volume with a resolution as low as the sonde length

(≈10 m). On an image, one observes essentially the variations of high resolution component.

The button current is a qualitative conductivity measurement. While the concept of the focused and diffused components is useful in helping to understand the response of the tool, it is in practice difficult to separate the two components to give a quantitative conductivity.

Depth of investigation. When discussing the depth of investigation of a microresistivity device, it is clear that one refers to the high resolution component, since the information conveyed in the image is contained essentially in this component. From the above discussion, we expect the high resolution component to have a shallow depth of investigation. Indeed, if the depth of investigation were much deeper than one or two centimetres, the image will be a juxtaposition of features at different radial depths, which is not commonly observed.

Williams *et al.* (1997) investigated in detail the button response with respect to small inhomogeneities by mathematical modelling. Leaving aside subtle differences between the response to small conductive and resistive objects placed at different distances from the borehole wall, they arrive at a depth of investigation of 1–2 cm. Bourke *et al.*(1989) quoted a value of ≈2.5 cm. The electrical penetration parameter used to correct dips measured by microresistivity devices is another measure of depth of investigation. Using modelling results as well as the comparisons of dip values obtained from logs (dipmeter and image data) and core measurements, values between 1–2 cm are normally used. This evidence points to a very shallow depth of investigation.

A special circumstance should be noted concerning thin conductive beds or conductive fractures, sandwiched between two resistive layers. In such cases the depth of investigation is increased. An explanation for this phenomenon is given in the section *Distortions by Strong Contrast*. When computing dips for conductive features (for example, when the resistivity contrast between adjacent beds exceeds a factor of 10), it will be necessary to take into account a larger than normal electrical penetration. In theory, if rough values of the resistivity contrast and dip magnitude are known, then it is possible to compute a suitable electrical penetration parameter. In practice, this approach is not always feasible. It would be safer in such cases to compare dips from microresistivity devices with cores or other measurements. Hansen & Parkinson (1999) report a depth of investigation of 5–7 cm in a study comparing images from microresistivity and ultrasonic devices.

Trouiller *et al.* (1989) mention that in thick homogeneous or finely laminated beds, the depth of investigation of a button is similar to that of a shallow laterolog, which is roughly 25 cm. This comment has caused some confusion in the past. In fact, it applies only to the low resolution component, or the background value of the image (Trouiller, pers. comm. 1989). This does not mean that small inhomogeneities more than about 2 cm from the borehole wall would be observable.

Resolution. When the pad makes direct contact with a formation with no strong resistivity contrasts (e.g. <10), the theoretical resolution of the button approaches the value of its diameter (5 mm). In practice, the resolution can be much poorer (≈1 cm). Apart from errors arising from acquisition, such as irregular tool speed, there are two other factors which affect resolution: distortion by strong contrast and standoff. These will be discussed later under the heading, '*Special features related to artefacts*'.

Response to fractures. Luthi & Souhaité (1990) have shown by simulation that there will be an increase of button current in the neighbourhood of a fracture filled with conductive fluid. More precisely, the integral of the increase in the button current across the fracture trace is proportional to the product of the fracture aperture and the conductivity of the fluid in the fracture. By contrast, the apparent width of a fracture is unrelated to its aperture. In the majority of cases, the apparent width would be of the order of the button resolution of ≈5 mm, while the aperture would rarely exceed this value.

Given the conductivities of the fluid and the surrounding rock, an effective aperture can be obtained from the microresistivity button current response. Provided there is a strong resistivity contrast between the fluid in the fracture and the surrounding rock, apertures down to 10 μm are quantifiable. This technique has been applied to logged data by Hornby *et al.* (1992).

The work of Luthi & Souhaité (1990) is limited to fracture dip angles below 40°. To apply the technique successfully, the interpreter must take care to select only open fractures (since the method will give an 'answer' for any conductive feature), and have an accurate knowledge of the conductivity of the in-filling fluid.

Special features related to artefacts

Tool speed variations. Most logging systems record data as a function of cable depth, i.e. a frame of data is recorded each time the cable at wellhead is detected as having moved a fixed distance, this being $\approx$2.5 mm in the case of a microresistivity imaging tool. Since the cable is elastic, the movement of the cable at wellhead does not always correspond to the movement of the tool at a given instant of time. A log will be stretched/compressed locally when the tool speed is below/above the cable speed i.e. the data attributed to a given depth interval is in fact acquired over a smaller/larger distance. In the extreme case, the tool may become stuck i.e. the tool remains stationary while the cable continues to wind in. In this case, the same data will be recorded over several cable depths. This results in pronounced stretching of the image (L&B 'Tool Speed Irregularities – Stuck zones'). As the tool eventually frees itself, it accelerates due to the buildup of tension in the cable, causing the image to be severely compressed. When a tool is conveyed by tubing or casing, stick-and-slip can also occur. Since these are less elastic than a cable, the motion is more jerky, in a series of short stick-and-slips. All logs and images suffer from alternating compression and stretching to some degree unless corrected. Quite naturally, the effects on images are particularly dramatic. The distortions would lead to errors in bed thickness and dip estimation.

Stretching and compressing have another effect unique to microresistivity imagers. Since the sensors are not at the same depth, they will pass a given depth at different times. At these different times, the tool speed will not in general be the same. Thus at a given depth interval, the degree of stretching or compression will be different for the sensors offset in depth. The resultant image obtained by assembling the data from the different sensors will therefore contain depth mismatches. When each pad contains two rows of buttons interlaced laterally but separated in depth, the depth mismatch of the two rows will give rise to saw-teeth distortions across the pad image (L&B 'Tool speed irregularities – sticky zones'). Similarly, when adjacent pads suffer from depth mismatch, geological features will become discontinuous going from one pad to the next (L&B 'Pad-flap or button offsets').

Correction for tool speed is done using an accelerometer located along the tool axis. Chan (1984) describes how the acceleration can be integrated, using the cable depth as a constraint by means of a Kalman filter, to give estimates of the tool velocity and depth. This is an important advance on earlier efforts, because the Kalman filter allowed the constraint to be applied in a continuous fashion. Chapellat *et al.* (1996) have proposed a second method which uses the correlations between button values measured on the two rows to estimate the degree of compression and stretching, and hence the tool velocity relative to cable speed. This latter method is usually applied after the accelerometer based method although it can also be used on its own, as when the accelerometer malfunctions. It is most obviously effective in correcting 'saw-teeth' but leads to significant improvements even when 'saw-teeth' are not apparent.

With these methods, and their refinements over the years, speed correction problems have become less frequent. There are nevertheless cases, especially in highly deviated wells, when speed correction is unsatisfactory. It is worth pointing out that speed correction relies on measurements which themselves suffer from error. For example, the accelerometer measurement has at least three limitations:

- imperfect separation of tool acceleration and the gravitational acceleration component which are measured simultaneously by the accelerometer
- precision and frequency (transient) response of the accelerometer
- synchronisation of the accelerometer and the imager sensors

Furthermore, compression implies undersampling, so that even if the tool speed is calculated accurately, recovery of the image may be imperfect.

Image orientation/Tool rotation. As the logging cable is wound around its drum, it is twisted round its axis. This in turn causes the tool to rotate as the twist unwinds. The rotation is usually quite slow. Fast rotation occurs when the tool is prevented from rotating, for example by pad friction, so that a large torque builds up before finally, the tool is free to rotate. Fast rotation also occurs frequently at the start of a log, when during the downward trip, the tool touches the bottom and stops rotating before all the twist in the cable has been completely unwound.

Inclinometry data acquired with an imager tool allow the image to be correctly oriented, even when the tool rotates. One rotation every ten metres, is often quoted as the limit above which inclinometry measurements and dip computations would begin to lose accuracy. Modern instrumentation can in fact cope with rotation rates up to one rotation every two metres. Correctly oriented images have been obtained for up to one rotation per metre. It is possible that some dip computation programs, based on button correlation methods, may run into difficulty if the tool rotates significantly (e.g. $>90°$) within the length of the correlation window, which is usually of the order one metre. Dips picked interactively on a workstation would be unaffected by rotation, as long as the image itself is correctly oriented.

It must be pointed out that *apparent* excessive rotation can be caused by (a) incorrect speed correction, when a log is compressed/stretched erroneously, and (b) when the inclinometry measurements are in error. Apart from outright failure of the instruments, magnetometer measurements are notably prone to error in the proximity of casing. These possibilities should always be investigated if the image orientation appears to be abnormal. In the early days of horizontal wells, images were often oriented incorrectly, using 'North' as the reference, i.e. orienting the image such that 'North' is the left edge of the image. In horizontal wells, the definition of 'North' in the plane perpendicular to the tool axis can be ambiguous, and the top-of-hole direction should be used as reference instead.

Signal strength (applied voltage and gain). During acquisition, the voltage applied between the pad and the current return may be adjusted using a feedback loop, so as to obtain a button current level above noise, and below saturation for the electronics. In addition, on some tools, a gain can be applied to low signal levels before digitization, so as to avoid the introduction of quantization noise. Despite these mechanisms, in adverse conditions, such as when the ratio of formation resistivity to mud resistivity is very high, the signal level will be low and affected/dominated by noise. In such cases, the image will have a speckled aspect as illustrated by L&B 'Extremes of condition'.

If a variable voltage and/or gain have been applied, the measured currents must be corrected in order that they reflect the formation conductivity. Such a correction is usually taken care of automatically in the processing. If the correction is omitted by mistake, or if it is impossible to make the correction because the voltage and gains data are lost, anomalies in image intensity will occur. Without the corrections, the contrast between different beds will be reduced or even reversed for widely separated beds. However, these errors are not often easy to notice. A much clearer indication of this omission is a mismatch in intensity between the images from pads/flaps in the neighbourhood of a boundary of sharp contrast, where a gain or voltage change has taken place. Because the gains and voltages are applied as functions of time, images from pads/flaps which are offset in depth will show the effects of the gains and voltage changes at different depths (equal to the depth offset), leading to an intensity mismatch which is easily recognizable.

Button equalization, faulty buttons. The response of individual buttons are seldom identical because of limitations in manufacturing, and/or accidental factors during logging. Among the latter are uneven pad contact, mechanical wear, and deposition of mud/oil on the pad face. These lead to stripes in the image when buttons on the same pad have a markedly different response, or unevenness in the intensity of the different pads. In addition, certain buttons may fail to work at all, when the electronics fail. Such failures may be permanent or intermittent. e.g. L&B 'Button or Flap Death'.

To remove this unequal response and to detect faulty buttons, the following post-acquisition processing may be carried out. The mean and variance of the buttons are computed over a long window (≈ 5 m). If the mean and variance for a given button are close enough to the median values for all buttons, then the button response is 'equalized' by a 'gain and offset' operation, such that mean and variance become equal to their respective median values without disturbing the detailed variations. Buttons which are defective are detected as having abnormal mean and variance values. Values of faulty buttons occurring in isolation are replaced by interpolation using values from adjacent working buttons. Equalization works if it can be assumed that over an interval of ≈ 5 m all buttons functioning normally would have the same average response.

L&B 'Drilling induced breakout' shows an interval where the images from different pads are not correctly equalized. In this interval, the buttons evidently do not have the same average response, since the pads are affected differently by the breakout. So the equalization correction breaks down.

Striping (mud smear) problems are observed in L&B 'Mud smears', L&B 'Drilling Induced Fractures' and L&B 'Spiral Hole'. Either the equalization correction has not been applied, or it has broken down because the buttons are affected differently by fractures, or irregular contact in the spiral hole.

L&B 'Faulty button correction' is an example where a circuit failure has caused every third button to fail. (In the tool concerned, the data from every third button is processed by the same electronic circuit.) This in turn causes the speed correction using button correlations to work incorrectly. The result is vertical striping throughout the image (faulty buttons uncorrected), and discontinuous bedding features and 'saw-teeth' (incorrect speed correction) in the left hand image. In the corrected right hand image, the faulty buttons are correctly detected and excluded from the speed correction calculation. The image is therefore improved in two ways: by a better speed correction and by the replacement of faulty button values by interpolation.

Distortion by strong contrast. Currents tend to follow a path of minimum resistance, directing themselves towards conductive regions and away from resistive ones. When imaging small objects such as vugs and nodules, the conductive ones will appear larger, as currents from nearby buttons are drawn towards them (i.e. the object is 'seen' by more buttons than it should), while resistive ones appear smaller. Similarly, conductive beds will appear thicker while resistive ones appear thinner (Trouiller *et al.* 1989). These distortions are proportional to the conductivity contrast across the object boundary. When the contrast is high, halo effects are observed, as shown in L&B 'Halo effects conductive/resistive' and 'Fracture aureoles'. These observed effects are confirmed by modelling (Trouiller, *pers. com.* 1989) and a simple explanation based on the modelling results is given below.

The response of the button current to inhomogeneities is most readily understood in terms of current lines which indicate the *direction* followed by a current originating from a point on the pad surface. Figures 3a–d show the current lines (shown as arrows) emerging from the pad face and entering the formation in front of four different conductivity inhomogeneities:

(a) conductive nodule
(b) resistive nodule
(c) dipping resistive plane or fracture
(d) dipping conductive plane or fracture

The diagrams are schematics and do not correspond to precise computations. The areas shown correspond to the first few centimetres of formation facing the pad. The resistivity contrast between the conductive areas (dark), and resistive one (light) would be about 1:100.

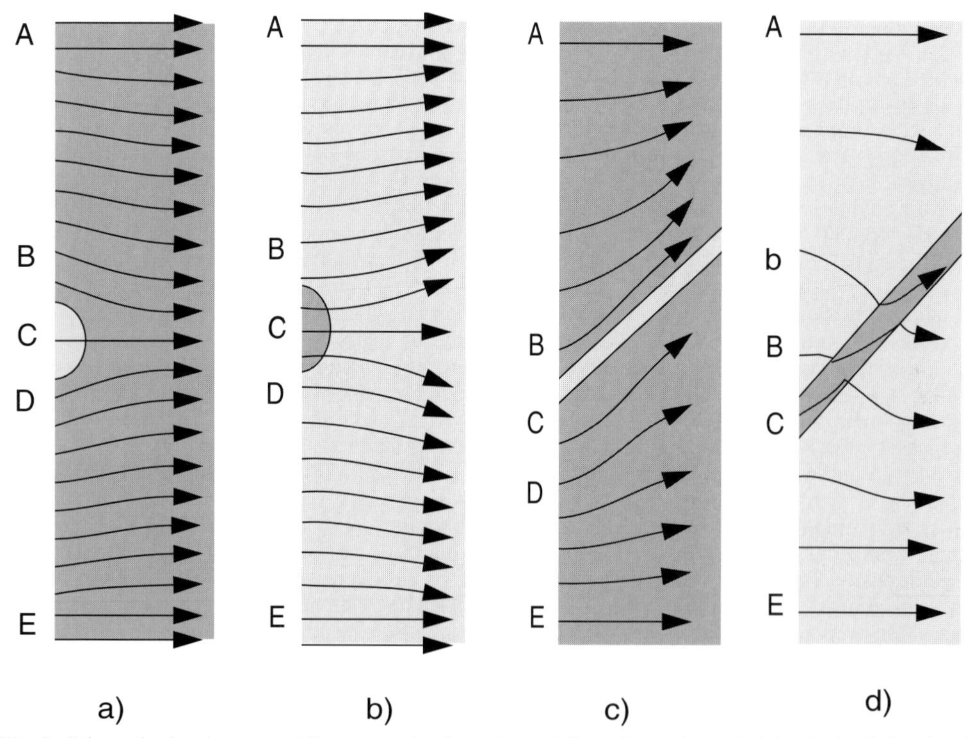

Fig. 3. Schematic showing current lines emerging from the pad face of an microresistivity device (left side of diagram) into the formation in the neighbourhood of: (**a**) a small resistive nodule, (**b**) a small conductive nodule or vug, (**c**) a thin resistive bed or fracture, and (**d**) a thin conductive bed or fracture. The shading of the formation is light for resistive and dark for conductive.

As the conductive pad face is an equipotential surface, the current lines are initially perpendicular to the pad. Subsequently, the lines are directed in space according to the following 'tendencies':

- aligned perpendicular to the surface of conductive bodies, since these tend to be equipotential surfaces
- aligned parallel to the surface of highly resistive bodies, since the current tries to avoid such regions
- avoid each other since current lines cannot intersect

The current emerging from a button will follow a path prescribed by a tube containing all the current lines originating from the button surface. For a given potential applied between the pad and the return electrode, the current will be proportional to the conductance of the path. The conductance of a given length of path will be proportional to the product of the conductivity of the formation along the path, and the cross-sectional area of the tube of current lines. Referring to the diagrams, we can say that the button current will be relatively strong if firstly the current lines pass through a conductive part of the formation, and secondly if the current lines are divergent, indicating a large cross-sectional area. Conversely, the current will be relatively weak if the path passes through a resistive region or if the current lines are squeezed together.

In Fig. 3a, if a button was placed at position A or E far away from the resistive nodule, the homogeneous background formation will read a normal current: the tube of current lines emerging from such a button will be entirely in the background formation, and the cross-sectional area of the tube will be 'average' since the current lines are neither divergent nor convergent. At position C facing the nodule, the current is low because the path passes through the resistivity nodule. These positions correspond to the bright zone in L&B 'Halo effects – resistive'. At positions B and D immediately on either side of the nodule, the current lines are entirely in the background formation, but spread out round the back of the nodule, implying an increased cross-sectional area. The current is therefore higher than average. At these positions, we observe the apparent conductive (dark) rim surrounding the resistive nodule.

In Fig. 3b, buttons at positions A or E far away from the conductive nodule, will again read a normal current for the homogeneous background formation. At position C facing the nodule, the conductance along the current path is high: firstly because the tubes of current lines pass through the conductive nodule, and secondly, because the current lines are divergent. Thus the current for a button at C will be high, corresponding to the black areas in L&B 'Halo effects – conductive'. At positions B and D immediately on either side of the nodule, the current lines are entirely in the background formation, but they are squeezed together, implying a decreased cross-sectional area. The current is therefore lower than average. At these positions, we observe the apparent resistive (bright) halo surrounding the conductive nodule.

Figure 3c represents the case of a dipping resistive bed or fracture, at the up-dip azimuth. The current lines are squeezed together at position B, so the button current at B will be lower than that at A. By contrast, at position D the current lines spread out, so the button current from D will read higher than that at A. This is observed in L&B 'Proximity effects – Fracture aureoles'. Above the crest of the fracture sinusoid, the image is white (low current) and below the crest, it is black (high current). Turning Fig. 3c upside down gives the situation at the down-dip azimuth. Below the lowest point of the sinusoid, the image is white (low current) and above it, the image is black (high current). L&B 'Cement mottling' is a more complicated example of the same effect.

Figure 3d, which represents the case of a dipping conductive bed or fracture in the up-dip azimuth, illustrates the increased depth of investigation mentioned earlier. High currents will be registered by a button placed at C facing the conductive layer, but also at B and to a lesser extent b where the current path is drawn into the conductive layer deeper into the formation. In other words, at B and b the depth of investigation of the button is increased. On the image, the apparent position of the conductive feature will therefore be above C. Again, Fig. 3d can be turned upside down to see that the apparent position of the conductive bed will be below its true position in the down-dip azimuth. If a dip is computed from the button response, the dip magnitude will be too high. Note that unlike the three earlier cases, there are no obvious anomalies in the image to indicate that the tool response is distorted.

In general, it is difficult to correct for these geometrical distortions because the resistivities and geometry of the heterogeneities are unknown. Even if these are known, the only way forward would be extensive modelling which is impractical on a routine basis. The interpreter should be cautious when using images to obtain bedding thickness, sand count, and dips, in the presence of strong resistivity contrasts.

In L&B 'Proximity features', dark patchy features are interpreted as 'Benthic shells in a mudstone' close to but not intersecting the borehole wall. While it is possible for heterogeneities with strong contrast situated away from the borehole wall to be observed in a microresistivity image, it would be difficult in general to distinguish them from weakly contrasted ones exposed on the borehole wall. The fact that microresistivity devices have a finite depth of investigation can pose ambiguities in the interpretation.

Standoff/mudcake. Since the resistivity contrast between mud and mudcake is small, mudcake and standoff have similar effects on microresistivity devices. If there is standoff between pad and formation, due to mudcake, rugosity, washouts, breakouts, drilling damage or well deviation, the focusing effect of the pad face in the formation will diminish, leading to a loss in resolution initially, and eventually, to total loss of the image. Generally speaking, small standoffs (<5 mm) are barely noticeable while standoffs greater than 2 cm will render an image too blurred to be of value. In addition, opposite a large standoff, the button responds to the mud conductivity.

L&B 'Spiral Hole', 'Drilling induced breakout', 'Rugose Hole' and 'Mudcake Buildup' are examples where standoffs have caused blurring or loss of parts of the image. The caliper logs should be used for confirmation if standoff appears to be a problem. Again, standoff effects cannot be corrected in practice.

Ultrasonic imagers

Tool operation

The borehole televiewer tool was invented by Zemanek *et al.* (1969). Modern versions commercially available include the CBIL (Faraguna *et al.* 1989), CAST (Seiler *et al.* 1990) and UBI (Hayman *et al.* 1994). The above references give a detailed description of the characteristics of these tools.

Ultrasonic tools work on the pulse-echo principle. Ultrasonic pulses at a frequency between 250 kHz and 500 kHz are emitted from a rotating transducer towards the formation, at up to about 250 times per revolution. The same transducer captures the reflected pulses, from each of which an amplitude and a transit time are extracted and recorded. An image of the borehole wall can be reconstructed with either the amplitude or transit time data. To make an image, the logging speed must match the rotation speed of the transducer i.e. one rotational scan is made per vertical sampling interval. Focused transducers are used in modern tools to improve signal strength and decrease sensitivity to rugosity.

The commercial tools differ in the way the transit time and amplitude are measured. Some use the method of the original televiewer. The transit time is measured as the time when the signal first exceeds a threshold, in a time window where the reflected signal is expected to arrive. The amplitude is measured as the largest amplitude value measured in the window. The transit time and amplitude so measured correspond to the same arrival if the first arrival is the strongest. This is most often but not always the case. In later tools, the reflected waveform in the window is digitized. The amplitude and transit time are then obtained by digital processing methods. For example, the waveform envelope is computed to obtain the amplitude and transit time corresponding to its peak. In either case, the detection window moveout is regulated by a feedback loop based on the previous detection.

Tool response

Transit time measurement. The transit time gives the distance between the transducer and the borehole wall once the mud velocity is known. The mud velocity is often logged by the ultrasonic imager, so that the transit time image is transformed into an image of the borehole radius. The radius image is above all a survey of the hole shape and is the preferred method to study breakouts (Barton *et al.* 1997). Bed boundaries, fractures and small homogeneities will be visible in the transit time/radius image if they cause the borehole radius to change. Sand and shale beds are often distinguishable on the radius image because of differences in the borehole radius (Hayman *et al.* 1994). Only large fractures and vugs are normally visible on the transit time image.

Amplitude measurement. Considerably more details are observed on the amplitude image than the radius image. The amplitude is influenced by the following factors:

- surface roughness/continuity
- acoustic impedance contrast between mud/mudcake and formation

- angle of incidence
- length of path traversed and attenuation in mud

The first effect is usually the most important from the interpretation point of view, allowing fractures, vugs, beds and other inhomogeneities to be observed. Changes in acoustic impedance contrast allows beds to be distinguished, if a sufficiently large acoustic impedance contrast exists. The last two effects do not convey any information which is not already in the radius image. One of the aims of processing is normally to remove these latter two effects.

Depth of investigation and resolution. The depth of investigation of a reflection measurement is effectively zero. The beam width is maintained at ≈4 mm at 500 kHz and ≈8 mm at 250 kHz for different hole sizes by using transducers with either different focal lengths or mounted with a different radial offset. The beam width doubles for a given transducer if the transducer-reflector distance is increased to about 5 cm above the nominal hole radius. While beam size controls resolution, objects smaller than beam size can be detected.

Although the theoretical resolution of ultrasonic imagers is similar to that of a microresistivity imager, comparison of the two types of images shows that more small features are visible in the latter. This is because the range of resistivity values occurring in rocks (two or three decades) is much larger than the range of acoustic impedance values (less than a decade).

Response to fractures. Schmidt *et al.* (1992) give a detailed account of the pulse-echo response to fractures. This shows that apertures below 0.1 mm can be detected. Like the microresistivity measurement, the integrated amplitude response can be related to fracture aperture. However, since the amplitude is also sensitive to the roughness at the intersection of the fracture with the borehole wall, this method gives only a rough indication of aperture.

Comparison with microresistivity images show that the ultrasonic devices detect only the larger, open fractures. This should not necessarily be viewed as a disadvantage, especially when both types of devices are used in synergy.

Special features related to artefacts

Speed correction. Ultrasonic images can be subject to irregular logging speed, much as other logs (L&B 'Tool Speed Irregularities – Stuck zones, Stick zones'). This results in stretching and compressing of the image as explained earlier. Unlike the microresistivity imagers, the image is not derived from interlacing sensors which are offset in depth, so there are no 'saw-teeth' mismatches. In one sense this is a benefit while in another sense, it is a disadvantage: (1) just because saw-teeth artefacts cannot occur does not mean that image compression has not occurred, and, (2) the off-depth microresistivity sensors do allow an 'image-based' speed correction to be applied.

A problem relating to tool speed is the requirement that the rate of rotation of the transducer matches the vertical sampling. If the tool is rotating too fast or too slow relative to the logging speed, periodically, there will be a surplus or deficit of one scan. The surplus is usually discarded and the deficit made up by repeating a previous scan. In either case, there will be a discontinuity in the image. If this occurs frequently, the image will have a 'blocky' character. This requirement means that the logging speed for ultrasonic imagers is two or three times lower than that for microresistivity imagers. Experience shows that low logging speed often results in an increase in tool speed variations.

Tool eccentering. Eccentering causes the amplitude to increase at those azimuths where the tool is closer to the borehole wall and vice versa. Small to moderate eccentering (<1 cm say) induces a variation in signal strength which may obscure finer details on the borehole wall (L&B 'Eccentralisation/Standoff'). This is essentially a display problem: when changes in radius dominate the changes in amplitude, the smaller amplitude variations due to fine details translate into one or two adjacent colours on the colour scale, and become difficult to visualize. Data are not lost. On the other hand, with severe eccentering, the signal may be lost altogether from certain azimuths where the reflected pulse no longer returns to the transducer.

Provided there is only a small loss of data, the effects of eccentering can be readily removed. From the radius data, the actual location of the tool centre relative to the borehole centre can be determined by solving a relatively simple geometrical problem, and the effects of eccentering on the radii values can be corrected. To correct the amplitude, Menger (1994) describes a method using the known tool position in the borehole and known characteristics of the tool. However, this method requires that the tool is calibrated uphole in a fluid similar to the mud used. More generally, a variety of image filters can be applied to remove the most obvious effects of eccentering. For example, the effects of eccentering in the form of the first and second harmonic variation of the amplitude can be removed by a suitably designed filter.

Irregular hole shape. Irregular hole shapes such as washouts (L&B 'Washout') and spiral holes caused by drilling (L&B 'Tool Orbiting', L&B 'Spiral Hole') pose serious problems for the ultrasonic scanners. Hole shape irregularities produce very much the same effects as eccentering, except that the changes in signal are much less gradual and less predictable. When there is severe signal loss, then as with severe eccentering, there is nothing to be done. When signal loss is limited, the eccentering correction can help to improve the image in some cases.

Certain deformations on the borehole wall caused by the drilling process give rise to 'vertical stripes' e.g. L&B 'Keyseat Furrow', L&B 'Stabilizer grooving'. These effects can often be diminished using image filters, or by 'equalizing' the image along vertical stripes as described under 'Button Equalization' for microresistivity imagers.

Signal loss. As mentioned earlier in connection with eccentering and irregular hole shape, signal loss can result when the reflected pulse fails to return to the transducer, due to large angles of incidence. In addition, signal loss can also result from attenuation of the signal in heavy muds, or when a sudden increase in travel time, due to an abrupt change in hole shape, causes the return pulse to fall outside the detection time window.

Signal loss is usually indicated by white spots on an ultrasonic image. L&B 'Signal Loss' shows an example of signal loss due to an irregular/enlarged hole.

Mudcake. Information concerning *in situ* mudcake properties is difficult to come by, so it is impossible to be quantitative concerning details of the interaction between the ultrasonic pulse and the mudcake. Qualitatively, mudcake *in situ* should behave much like mud, except that the velocity and attenuation of sound are both increased. The acoustic contrast between mud and mudcake is weak as compared to that between mudcake and formation, so the pulse is effectively reflected at the formation surface. Thus the presence of mudcake usually causes a drop in amplitude and a decrease in transit time. Normally, it would be difficult to detect mudcake since many lithology changes are also associated with the same changes. However, since the mudcake is fragile, the passage of centraliser/stabilizer arms will remove some mudcake along their tracks. This gives rise to artefacts on the ultrasonic images (amplitude and radius) which are difficult to mistake. See L&B 'Stabilizer grooving', 'Noise: 60 Hertz' and 'Mudcake Build-up'. (Note that the tracks of the arms are not strictly vertical, but should be parallel to a curve such as relative bearing (RB), giving tool rotation.)

In all the three examples cited above, it is evident that bedding features can be observed even in the presence of mudcake. It is unclear if these are observable because the change of lithology has led to changes in reflectivity and rugosity at the formation surface, or to the thickness and consistency of the mudcake itself.

60 Hz noise. L&B '60 Hz Noise' shows an artefact which is particular to the UBI tool. It appears only on the amplitude image.

If the 60 Hz variation of the power supply is not effectively screened out, the pulse emitted by the transducer, and hence the received amplitude, will be modulated by a 60 Hz variation. Since the transducer rotates about seven times a second, and the rotation is coherent with the 60 Hz voltage variation, there will be eight periods of image amplitude variations per rotation, with a small phase shift between rotations, resulting in the observed eight-fold spiral or diagonal stripes. The 60 Hz noise can be readily removed by a filter.

Wood grain pattern. The L&B 'Wood Grain' examples for the UBI and CBIL tools have quite unrelated causes. As reported in L&B, the patterns for the CBIL/STAR tool are caused by interference of reflections associated with the housing surrounding the transducer. The pattern for the UBI is caused by a bias in the interpolation algorithm. The artefact appears on both the amplitude and the radius images of the UBI, with an exact correspondence between the patterns. An explanation for this bias effect is as follows.

During acquisition, the reflected waveform is sampled by the UBI at discrete time intervals and digitized. The transit time is obtained as the time corresponding to the peak in the waveform envelope. At the first instance, the transit time obtained would be an exact multiple of the sampling interval Δ. An interpolation formula is then applied to obtain more accurate values of the peak location and amplitude. This yields a transit time which can be expressed as an exact multiple of Δ plus a fractional part. The three-point interpolation formula used introduces a bias if the waveform envelope near the peak is non-parabolic.

Figure 4 shows the values of transit time in a small part of a scan. The continuous curve shows the slowly varying transit time values that the tool should have measured. The dashed line shows the discretized values $(n-1)\Delta$, $n\Delta$, $(n+1)\Delta$ measured before interpolation. The dotted lines and arrows show the sampled values obtained by interpolation. The interpolated values are observed to show a preponderance of fractional values centred at say 0.5Δ. This gives rise directly to the wood grain pattern in the transit time (or radius) image; contour lines are compressed at values close to integer values of Δ, and spaced out whenever the fractional values are $\approx 0.5\Delta$.

Note that the wood grain is observable only when the transit time changes by amounts much smaller than Δ between many successive samples. Δ is a small quantity ($0.1\,\mu s$) which corresponds typically to a radius change of <0.08 mm in water. So the borehole must be smooth and the tool well centred to observe the bias effect. A corresponding pattern appears on the amplitude image also, since the amplitude varies approximately linearly with transit time when the transit time varies within close limits.

The bias can be removed only on a statistical basis since it depends on the shape of the waveform envelope, which is variable. If we take many transit times measured over a large depth interval, then the fractional part should be uniformly distributed between zero and one since all fractional values are equally likely to occur. If however, the pattern shown in Fig. 4 is repeated systematically, then a histogram of counts versus fractional values would have a peak centred at ≈ 0.5. From such a histogram, one can derive a correction to the fractional part of the transit time that will yield a uniform distribution, and hence remove the wood grain pattern from the transit time image. From a crossplot of amplitude and transit time, the amplitude can be modified in parallel with the transit time, and the wood grain in the amplitude image is reduced. Since the processing for the amplitude involves this additional correlation, removal of the wood grain pattern is less effective.

Gain calibration. In the UBI tool, a variable gain is applied to the emitted pulse according to the strength of the previously received signal. Since the received amplitude is divided by the gain before recording, the amplitude is nominally corrected for the applied gain. However, the gain applied electronically may not lead to a proportional gain in acoustic amplitude. It is therefore necessary to calibrate the gains, and re-correct the amplitudes. If the gains are not calibrated, the image will appear spotty as shown

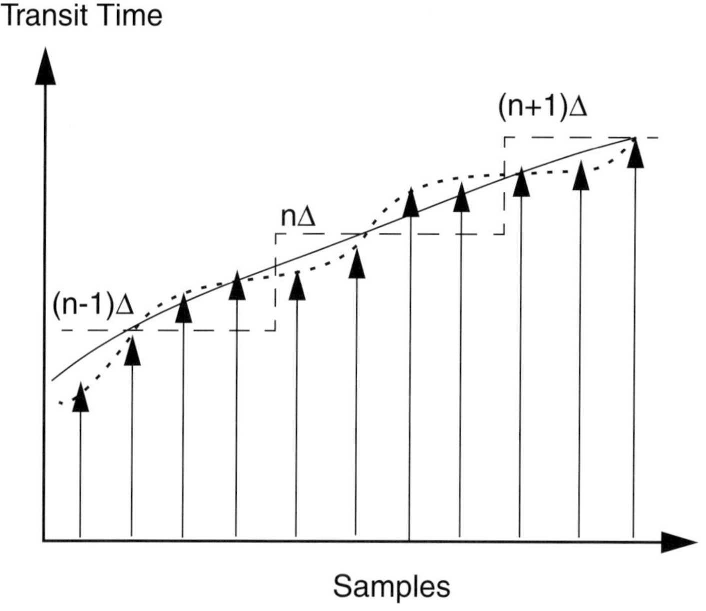

Fig. 4. Schematic showing how wood grain patterns arise as a result of bias in the interpolation of transit times. The full line shows the values that should have been measured. The dashed line shows the discretized values measured before interpolation. The dotted line and arrows show the values after interpolation.

in the L&B 'Honey-comb' example. This artefact is found only on the amplitude image of the UBI.

The calibration is done statistically. The gain applied may take one of 16 levels according to the strength of the previously received signal. To allow the calibration to work, this level is then *randomly* increased or decreased by one level. Now, if we take from the entire image, a large number of pairs of successive amplitudes (nominally corrected) where the gain has changed from level n to $n+1$, and compute the ratio of these pairs of amplitudes, we would expect the average value to be one. If the measured ratio is in fact $1:x$, then the gain of levels n and $n+1$ must be modified in the ratio $x:1$.

Discussion and summary

In this paper, the essential aspects of tool operation and processing of microresistivity and ultrasonic imaging tools have been described with reference to the artefacts described by Lofts & Bourke. From the measurement point of view, many features which are actually on the borehole wall are non-artefacts, and these have been excluded from the discussion. It is perhaps worthwhile pointing out that breakouts and induced fractures which are classed as artefacts in the sense of Lofts & Bourke, provide important data for the study of in-situ stress and well stability (Barton *et al.* 1997). Similarly, marks left by packers, sidewall core samplers, fluids samplers etc. provide important information when analysing the corresponding test/data. Mud smears and mudcake effects also make a positive contribution to interpretation, by indicating or confirming oil-bearing and permeable zones respectively.

An understanding of the similarities and differences between the microresistivity and ultrasonic images is important in the choice of tool for a given environment, and can be helpful even when interpreting the images separately. Some of the more important points concerning the use and interpretation of the two types of images are summarized below. It is noted that the two types of images are complimentary. Hayman *et al.* (1994) and Hansen & Parkinson (1999) show how interpretation can indeed be enhanced when both types of images are available.

Tool response. Microresistivity devices appear to 'see more'. This is mainly because many lithologies and geological features such as fractures, are more readily distinguishable on the basis of resistivity contrast as opposed to acoustic impedance contrast and other properties on which the ultrasonic response depends. The slightly higher resolution of the microresistivity devices also contribute by making small heterogeneities and thin beds distinct.

Microresistivity buttons have a finite but variable depth of investigation, and the response is distorted in the presence of strong resistivity contrast. These effects can give rise to ambiguities concerning dip values and the exact location of small features in certain cases. Ultrasonic devices respond only to features which are on the surface of the borehole wall so there is no ambiguity concerning their location.

Microresistivity imagers see large as well as relatively small fractures, whether they are open or in-filled with conductive or resistive material. Moreover, if the in-filling material is conductive, and its conductivity known, an estimate can be made of the aperture, even for values much below tool resolution. Ultrasonic imagers see mainly large, open fractures. There are no reliable methods to estimate aperture.

As the microresistivity physics is somewhat complex, it is worthwhile mentioning that mathematical modelling is indispensable for clarifying the underlying causes of anomalies observed.

Hole condition. Both tool types are affected by irregular hole shape, but especially the ultrasonic tool. Spiral holes associated with turbo-drilling appear to be a particularly serious problem. Since bad holes are also the underlying cause of speed correction problems, a good borehole is a crucial factor for good images.

Mud, mudcake. Microresistivity devices only work in conductive mud. Ultrasonic imagers work in all types of mud in principle, provided attenuation is not too severe. The presence of mudcake is an important indication of permeability. We mentioned earlier how the ultrasonic imager provides a robust way to detect mudcake. Microresistivity imagers are insensitive to mudcake by comparison.

Hole size and coverage. Microresistivity tools will operate in hole diameters up to about 40 cm, which correspond to the maximum opening of the arms. Coverage of the borehole wall, which is never 100% except in very small holes, will decrease linearly with hole size. For a hole diameter of 20 cm, the coverage is typically 10% per pad (i.e. 80% for a eight-pad tool, 60% for a six-pad tool). Resolution will be unchanged with hole size for the same pad contact conditions.

Ultrasonic tools gives 100% coverage, unless signal is lost due to very large tool eccentering.

The maximum hole size for ultrasonic imagers is limited by the attenuation of sound in the mud, and the focal point of the transducer. In a light mud, the maximum hole diameter would be about 30 cm. At the largest hole sizes, there is a decrease of resolution, due to defocusing, signal attenuation and increase in the lateral distance between samples on the borehole wall.

Conclusion

In this paper, the essential aspects of tool operation and processing of microresistivity and ultrasonic imaging tools are described. A particular reference has been made to the artefacts described by Lofts & Bourke, with the intention of giving a fuller explanation of the origins of the anomalies, and indicating whether the distortions can be remedied by processing. With the few exceptions stated explicitly, the discussion should be valid for all the commercially available tools since they operate in a similar way.

The author thanks the two referees for many useful suggestions, and J.-C. Trouiller for his help to clarify various issues concerning the microresistivity tool.

References

BARTON, C. A., MOOS, D., PESKA, P. & ZOBACK, M. D. 1997. Utilizing wellbore image data to determine the complete stress tensor: application to permeability anisotropy and wellbore stability. *The Log Analyst*, **38**(6), 21–33.

BOURKE, L., DELFINER, P., TROUILLER, J.-C., FETT, T., GRACE, M., LUTHI, S. M., SERRA, O. & STANDEN, E. 1989. Using Formation MicroScanner images. *The Technical Review*, **37**(1), 16–40.

CHAN, D. S. K. 1984. Accurate depth determination in well logging. *IEEE Transactions on Acoustics, Speech and Signal Processing*, **32**(1), 42–48.

CHAPELLAT, H., BERARD, M. V. & CHEUNG, P. S. 1996. *Method and Apparatus for Determining a Depth Correction for a Logging Tool in a Well*. US Patent No. 5522260.

EKSTROM, M. P., DAHAN, C. A., CHEN, M. Y., LLOYD, P. M. & ROSSI, D. J. 1987, Formation imaging with microelectrical scanning arrays: *The Log Analyst*, **28**(3), 294–306.

FARAGUNA, J. K., CHACE, D. M. & SCHMIDT, M. G. 1989. An improved borehole televiewer system: image acquisition, analysis and integration. *Transactions of the 30th Annual Logging Symposium of the Society of Professional Well Log Analysts*, paper UU.

HAYMAN, A. J., PARENT, P., CHEUNG, P. S. & VERGES, P. 1998. *Society of Petroleum Engineers Production and Facilities*, **13**, 5–13.

HANSEN, T. & PARKINSON, D. N. 1999. Insights from simultaneous acoustic and resistivity imaging: *This volume*.

HORNBY, B. E., LUTHI, S. M. & PLUMB, R. A. 1992, Comparison of fracture apertures computed from electrical borehole scans and reflected Stoneley waves: an integrated interpretation. *The Log Analyst*, **33**(1), 50–66.

LACAZETTE, A. 1996. *The STAR (SimulTaneous Acoustic and Resistivity) Imager*. London Petrophysical Society (SPWLA Chapter) presented at the LPS Borehole Imaging Seminar.

LOFTS, J. C. & BOURKE, L. B. 1999. The recognition of artefact images from acoustic and resistivity devices: *This volume*.

LUTHI, S. M. & SOUHAITÉ, P. 1990. Fracture apertures from electrical borehole scans. *Geophysics*, **55**(7), 821–833.

MENGER, S. 1994. New Aspects of the Borehole Televiewer Decentralization Correction. *The Log Analyst*, **35**, 14–20.

SAFINYA, K., LE LAN, P., VILLEGAS, M. & CHEUNG, P. S. 1991. Improved formation imaging with extended micro-electrical arrays. *Society of Petroleum Engineers*, **SPE-22726**, presented at the 66th SPE Annual Technical Conference and Exhibition, Dallas, Texas 1991.

SCHMIDT, M. G., LI, X. STEINSIEK, R. R. & MA, T. A. 1992. Numerical model for predicting ultrasonic transducer pulse-echo response to borehole fractures. *Transactions of the 33rd Annual Logging Symposium of the Society of Professional Well Log Analysts*, paper **U**.

SEILER, D., EDMISTON, C., TORRES, D. & GOETZ, J. 1990. Field performance of a new borehole televiewer tool and associated image processing techniques. *Transactions of the 31st Annual Logging Symposium of the Society of Professional Well Log Analysts*, paper **H**.

——, KING, G. & EUBANKS, D. 1994. Field test results of a six-arm microresistivity borehole imaging tool. *Transactions of the 35th Annual Logging Symposium of the Society of Professional Well Log Analysts*, paper **W**.

TROUILLER, J-C., DELHOMME, J-P., CARLIN, S. & ANXIONNAZ, H. 1989. *Thin-bed reservoir analysis from borehole electrical images. Society of Petroleum Engineers*, **SPE-19578**, presented at the 64th SPE Annual Technical Conference and Exhibition, San Antonio, Texas 1989.

ZEMANEK, J., GLENN, E. E., HOLCOMB, S. V., NORTON, L. J. & STRAUS, A. J. D. 1969. The borehole televiewer – A new logging concept for fracture location and other types of borehole inspection: *Journal of Petroleum Technology*, **21**, 762–774.

WILLIAMS, C. G., JACKSON, P. D., LOVELL, M. A. & HARVEY, P. K. 1997. Assessment and interpretation of electrical borehole images using numerical simulations: *The Log Analyst*, **38**(6), 34–43.

The recognition of artefacts from acoustic and resistivity borehole imaging devices

J. C. LOFTS & L. T. BOURKE

Baker Atlas Geoscience, Kettock Lodge, Campus 2 Aberdeen Science and Technology Park, Bridge of Don, Aberdeen, AB9 8GU, Scotland, UK

Abstract: Image artefacts, that is, non-geological features present on acoustic or micro-resistivity borehole image logs are a fact of life. In poor quality image logs, they can constitute the bulk of the data-set although fortunately, such features generally constitute the minority of a data-set. The recognition of artefact features must form part of the log quality control process and should also form part of the interpretation process. By applying sound log quality control principles, these artefacts can be identified systematically and the remaining image data will be geologically related. This paper provides an up-to-date summary of borehole image artefacts, their cause and recognition. It also, for the first time, considers artefacts arising from acoustic (ultrasonic) as well as micro-resistivity images. Four main artefact image categories related to data acquisition, borehole wall condition, post processing of data and measurement physics form the basis for the systematic identification procedure. These are discussed and developed herein.

Acoustic and micro-resistivity borehole image logs have become an invaluable and widely accepted source of information for sub-surface reservoir characterization over the last decade. Interpreters are frequently presented with images that consist of confusing non-geological shapes that reduce the confidence of an interpretation. These non-geological features are commonly referred to as image artefacts. They are features that do not arise from *in-situ* geological lithology or which depart significantly from their physical nature on the borehole wall or which are not visible in core. In a poor quality image log they may represent the vast majority of the data present. Fortunately, artefacts generally represent a small proportion of the data-set although in general terms all data sets contain some artefacts. Figure 1 illustrates a small section of micro-resistivity log showing no fewer than five artefact features present together.

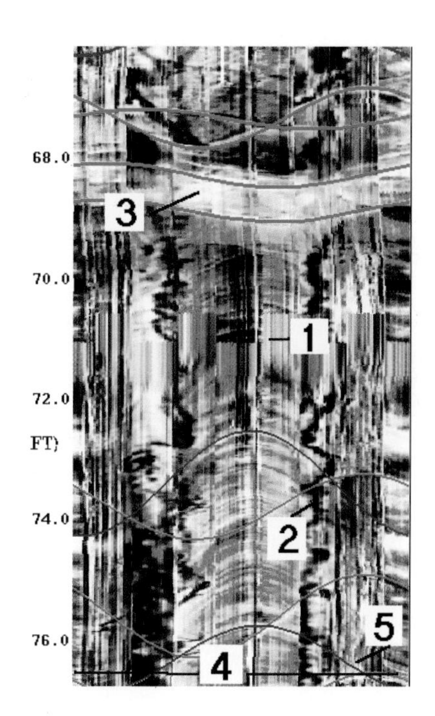

Fig. 1. A small section of micro-resistivity log showing no fewer than five artefact features (labelled by arrows) 1. tool speed irregularities, 2. drilling induced fracturing, 3. cementation mottling, 4. incorrect image display (no pad-flap offset) 5. vertical stripes indicate mud smear on button electrodes.

The ability to recognize genuine formation features is an essential part of image log interpretation and therefore implies relevant geological experience and use of localized geological knowledge (Harker *et al.* 1989). The need to recognize artefacts within an image data-set and to assess their effect on image quality is therefore key to any subsequent image interpretation. The recognition of these non-geological features must be an inherent part of the log quality control (LQC) procedure. By applying sound log

From: LOVELL, M. A., WILLIAMSON, G. & HARVEY, P. K. (eds) 1999. Borehole Imaging: applications and case histories. Geological Society, London, Special Publications, **159**, 59–76. 1-86239-043-6/99/$15.00.

quality control principles, artefacts can be recognized and identified systematically. This must be the first stage of any interpretation and should form part of an interpreters thought process during all stages of interpretation. The process of examining an image log for artefacts requires the interpreter to think of an image log in terms of the tool measurement principles, the nature of the borehole and how the images were generated.

Previous independent work by both authors has attempted to classify artefact images. Bourke (1989) classifies artefact features observed in early micro-resistivity images whilst Lofts *et al.* (1997) classify artefacts common in horizontal or highly deviated well bores. Bourke (1989) provides a comprehensive guide to log quality control.

The previous classification of artefacts by Bourke (1989) and adopted by Lofts *et al.* (1997), forms the basis of this contribution although it is extended to account for modern image logs. This paper considers artefacts arising from both acoustic (ultrasonic) as well as micro-resistivity image logs and provides an up-to-date catalogue of artefacts and an indication of their cause in the context of the adopted systematic approach. Examples are drawn from image logs acquired with currently available image-logging tools.

Micro-resistivity image technology

The most commonly observed artefacts of micro-resistivity images are attributed to the pad technology of the tool and to the complex pathway taken by the passive electrical current of these devices. See Ekstrom *et al.* (1987), Bourke *et al.* (1989) and Safinya *et al.* (1991) for detailed accounts of the technology. Micro-resistivity devices, with various arrangements of pads, commonly suffer from *tool sticking*, differences in friction between pad and the borehole wall (*mud smear*, *washout*, *tool stand-off*), and *offsets* between pads. These produce the most common micro-resistivity artefacts. In addition, micro-resistivity tools are generally conductivity 'seeking' devices where electrical current will tend to flow towards the most conductive lithology/fluid present, Williams (1996). This produces very complex current pathways and also accounts for a number of common artefacts such as current 'halo' effects and proximity effects (described later).

Acoustic (ultrasonic) image log technology

Ultrasonic imaging was primarily introduced for use with oil-based drilling fluids. This involves firing a high frequency (250 or 500 kHz) ultrasonic pulse from a rotating transducer towards the borehole wall and simultaneously measuring the amplitude and transit time of the returned pulse (Faraguna *et al.* 1989; Hayman *et al.* 1994). The pulse is fired between 140 and 250 times per revolution of the transducer depending on the tool design. Consequently, this device is particularly sensitive to the condition of the borehole wall and the acoustic impedance of the fluid in the borehole, resulting in artefacts such as hole spiralling, and stabilizer grooving and rugosity effects which are described in Tables 1 and 2.

Artefact classification

Four main artefact image types have previously been recognized, these include:

Acquisition artefacts

These are features arising from the acquisition of the image data. These can be subdivided into artefacts associated with the *drilling* process and artefacts associated with the *logging* process.

Drilling related features. These result from perturbations that accompany the drilling of a borehole such as tool orbiting and drill-bit slide, or during maintenance of the borehole such as wiper trips and reeming, or features that merely represent manifestations of the drilling process such as bit-at-rest and side-track windows. Separate features are described in Tables 1 and 2.

Logging related features. These can be attributed to physical problems associated with tool malfunction, such as tool eccentralization, mud smearing, or 'dead' button or pads, excessive tool rotation, decreased signal-noise ratio. They may also result from changes in formation properties that cause tool speed irregularities, or from extremes-of-condition, such as very resistive formations like evaporites.

Borehole wall artefacts

These are non-geological features developed on the borehole wall through drilling, well clean-up or logging operations. Whilst relating to acquisition they are not exclusive to it. They commonly include washouts, rugosity, key seating, mud-cake effects, tools marks and drilling induced fracturing. Borehole wall conditions are the greatest cause of artefacts in acoustic images due to the sensitivity of the measurement.

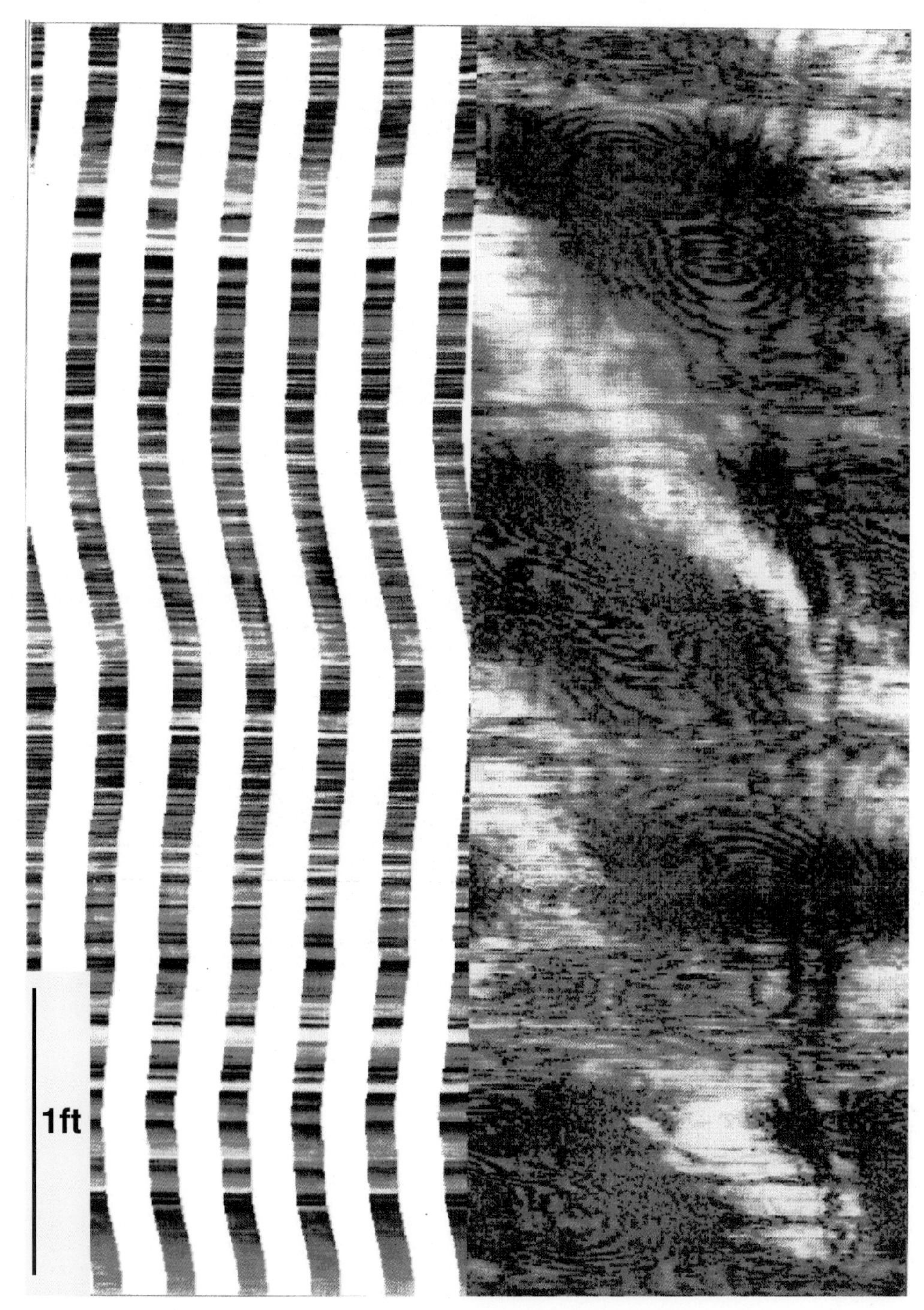

Fig. 2. A 5 ft vertical section of ultrasonic (right) and micro-resistivity images (left). This indicates the sensitivity of the acoustic image to the borehole wall condition which in this case is hole spiralled, whereas the micro-resistivity image shows no adverse effect or artefact. The acoustic image itself demonstrates two of the most common artefact features seen an acoustic image, hole spiralling and 'woodgrain' artefacts.

Table 1. *Summary of the most common micro-resistivity and acoustic artefacts observed on borehole image logs.*

Artefact Category	Artefact	Description	Micro-resistivity – Occurrence and severity	Acoustic-Occurrence and severity	Comments
1. Acquisition-Drilling	Tool Orbiting	Borehole is 'roller coasting' (bit whirl)	Common/mild	Common/severe	Affects acoustic images most
	Drill bit slide	Driller allows drill bit to slide, not rotate	Common in RAB tools only	N/A	Only relevant to the LWD RAB tool
	Side-track window	Window left when side tracking	Rare/Severe	Rare/Severe	Will effect all images
	'Bit/stabilizer at rest'	Bit/stabilizer marks on borehole wall	Common/Mild	Common/Mild	Common when pipe conveyed
	Wiper trip/reeming	Cleaning hole allows scratching	Rare/No effect	Common/Mild to severe	Affects acoustic images only
Acquisition-Logging	Eccentralization/ stand-off	Tool not centred in borehole, dark/light images	Common/no real effect	Common/Severe	Severe when pipe conveyed
	Mud smear	Mud buildup on electrodes, missing button data	Common/Mild-severe	N/A	Affects electrical images only
	Tool Speed irregularities	Yo-yo effect of logging tool due to speed variations	Common/Mild to severe	Common/Mild to severe	Worse when pipe conveyed
	Sticky zones	Logging tool catches on a ledge, dark image patches	Common /Severe	Common/Severe	Interpretation is often impossible
	Button or flap 'death'	Electrical fault on button/flap, missing button data	Possible/Mild to severe	N/A	Affects electrical images only
	Excessive tool rotation	Tool rotates >1 rotation per 30 ft.	Common/Mild to severe	Common/Mild to severe	Accuracy of dip estimation reduced
	Noise: '60 Hertz'	Acquisition current interference, diagonal stripes	N/A	Common/Mild to severe	Common in acoustic images
	'Woodgrain 1'	Sampling inaccuracy resembling woodgrain	N/A	Can be common/Severe	Common in acoustic images
	'Honeycomb'	Gain correction inaccuracy – honeycomb pattern	N/A	Common/Mild-severe	Common in acoustic images
	Signal Loss – Spotty image	Degradation in signal/noise – spotty image	When mud lubricants added (up to 8%) /Mild to severe	Common in washouts or heavy muds/Severe	Variety of causes. Common in acoustic images with heavy muds
	Specking – Extremes of condition	Dynamic range of tool exceeded (very high/low acoustic impedance/resistivity lithologies.	Possible in some high resistivity lithologies as speckling/Severe	Rare/Severe	Variety of causes. Effect is generally severe
2. Borehole Wall Artefacts	Washout	Enlargement of hole diameter	De-focusing of image/ Severe	Dark image/Severe	Will effect all images
	Rugose hole	Pitted borehole wall	Rare/Mild	Common/Mild to severe	Generally affects acoustic images only
	Key-seat Furrow	Ovalisation due to bit wear on underside of wellbore	Noticeable in deviated wells/Mild	Common in deviated wells/ Mild to severe	Will effect all images, can be filtered out
	Spiral Hole	Borehole wall is grooved due to bit	Common/Mild	Common/severe	Effects acoustic tools most
	Mudcake	Build up of mudcake generally over permeable lithologies	Common/Mild to severe	Common/Can be severe	Obliterates acoustic image interpretation
	Tool Marks: Stabilizer grooving	Logging stabilizers scratch borehole wall	Not noticeable	Common without care/ Mild severe	Acoustic tools most sensitive

	Sampling tool marks	Round marks made by probe & packer sampling tools	Rare but possible/Mild	Rare but possible/Mild	Easily identified at regular spacings
	Sidewall core samples	Marks made by sidewall corer	Rare but possible/Mild	Rare but possible/Mild	Easily identified as round
	Cable 'slap'	Marking of wireline cable on borehole wall	Uncommon/No affect	Uncommon/Mild	Effects acoustic tools most
	Drilling induced fractures	Zigzag fractures parallel to well bore axis	Common/Mild	Common/Mild	Useful for in *situ-stress* determination
	Drilling induced breakout	Broad parallel grooves separated by 180 degrees	Common but rarely identified/Mild	Easily detected/Mild	Useful for in *situ-stress* determination
3. Processing Artefacts	Incorrect hole diameter	Wrong assignment of diameter produces erroneous dips	Uncommon/Severe	Uncommon/Severe	Wrong apparent dip is calculated
	Incorrect colour assignment	Saturated images – concealment of detail in images. Apportioning the data into unequal bins for display	Possible/Locally severe	Possible/Locally severe	Will effect all images
	Incorrect gain correction	Under saturated/over saturated images	Possible/mild-severe	Possible/mild	Will effect all images
	Faulty Emex correction	'Tiger striped' images	Possible/Severe	N/A	Relevant to Schlumberger tools only
	Pad-flap or button offsets	Incorrect speed correction 'sawtooth' patterns on images	Common/Mild to severe	N/A	Common in lectrical tools
	Window length	Normalization window produces multiples of high contrast layers	Rare/Mild to severe	Rare/Mild to severe	Refer to Bourke (1989)
	Multiple-pass image offsets	Depth mismatch between logging runs. Correct with depth matching	Rare/Mild	Rare/Mild	Refer to Bourke (1989)
4. Measurement Derived Artefacts	Current Gather Effects:				
	Halo effects – Conductive	Light rim of a conductive feature due to current draw	Common in heterogeneous rocks	N/A	Common in electrical tools
	Halo effects – Resistive	Dark rim of a resistive feature due to current drain	Common in heterogeneous rocks	N/A	Common in electrical tools
	Cement mottling	Mottling fabric caused by highly resistive cements	Common in fractured rocks	N/A	Common in electrical tools
	Proximity effects:				
	Fracture aureoles	Light patches at the apex of fracture sinusoids	Common in fractured rocks	N/A	Common in electrical tools
	Proximal features	Imaging of features close to the borehole wall	Common in heterogeneous rocks	N/A	Common in electrical tools
	'Woodgrain 2'	Multiples, resembling woodgrain 1	Common on early CBIL images	Common in heavy mud/ Severe	Common in acoustic images
	Residual Hydrocarbon Effects	Blotchy image caused by gas/oil saturation	Rare/ Severe	Common – Distorts resistivity map	Care when using thresholding techniques

A brief description and indication of the effect and severity of each artefact on both micro-resistivity and acoustic images. An explanation of tool mnemonics is given in the text.

Table 2. *Description of the most common micro-resistivity and acoustic artefacts observed on borehole image logs.*

Artefact	Cause – description	Micro-resistivity	Acoustic
Acquisition – Drilling Tool Orbiting	Borehole is 'roller coasting' (effect caused by certain bit-stabilizer configurations). Affects acoustic tools most severely, often obscuring interpretation. Generally, effect is minimal on resistivity images where it rarely completely impairs interpretation. Left: OBDT false images showing diagonal tool orbiting. Calipers in blue/cyan. (ft). Right: UBI amplitude image showing severe orbiting of borehole, eradicating any useful information. Calipers in blue/cyan. (ft). Box below right: hole orbiting stops with coring @xx42 ft.	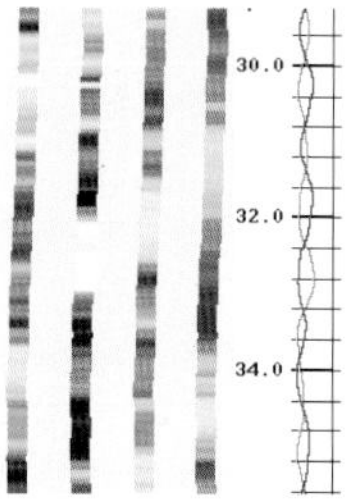	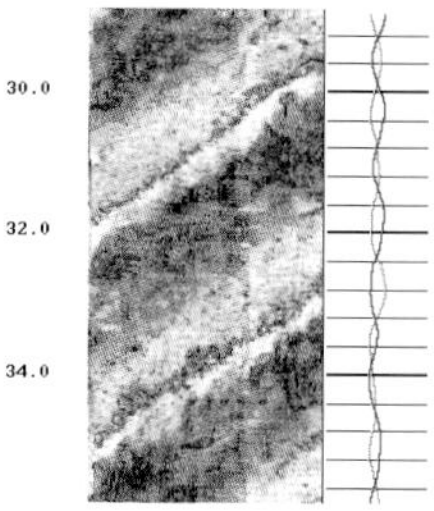
Drill bit slide	Driller allows drill bit to slide, not rotate. Only relevant to LWD (logging while drilling) tools. Left: RAB tool showing a severe slide interval. This results in data under-sampling and a smear of the button data in the top half of the image (above xx01). This occurs when the drill bit slides rather than rotates during drilling. Only relevant to LWD tools. (scale ft, horizontal). Right: Described in 'Tool orbiting' (above).	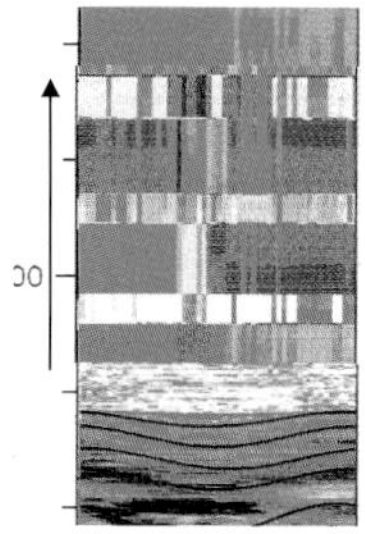	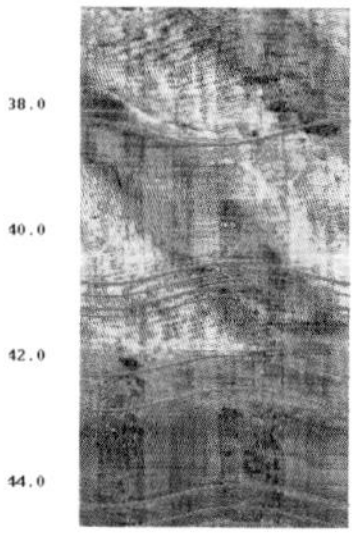
Side-track window	Hole in borehole wall produced when side-tracking. Severe in all imaging tools as this is a physical gap in the borehole wall. Often seen as a conical shape in images up to 20 ft in length. Left: See 'Bit-at-rest' artefact (below). Right: Horizontal well (20 ft section) showing UBI acoustic image over a side-track window. Note, interpretation is still possible on the right side. (ft).	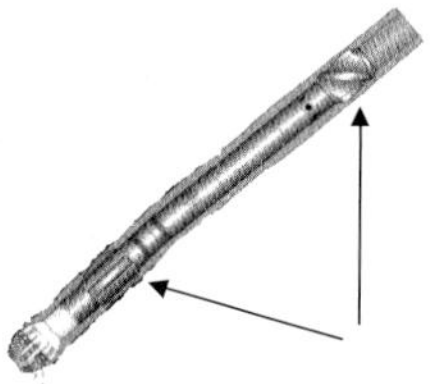	
'Bit-at-rest'	Bit/stabilizer imprints produced during drilling. Effect is mild in all images. Consists of a horizontal mark across the image (impression of the bit, left @xx96 ft) and a set of diagonal marks immediately above (impression of the near-bit-stabiliser, (left at xx95 ft). Right: Artefact in an acoustic image. Bit mark at xx52 ft, stabilizer at xx51 ft. (ft, both deviated). Above Left: Bit-stabilizer configuration that can be responsible for 'Bit-at-rest' artefact. Note, the near-bit-stabiliser that produces the diagonal mark.	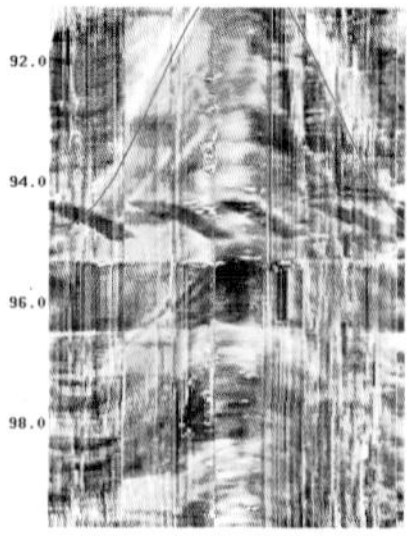	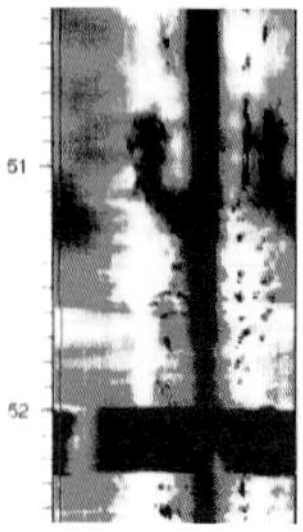

Artefact	Cause – description	Micro-resistivity	Acoustic
Wiper-trip & Reeming	Processes used to clean the borehole or during coring produce a scratching of the borehole wall. Mild effect in micro-resistivity images and it rarely impairs interpretation. Mild to severe in acoustic tools because these respond much more to the borehole wall. Right: Semi-horizontal to diagonal regular scratch marks caused by the reeming of the borehole wall on a microresistivity device. (one pad is also malfunctioning as shown by the black areas to the right of the image. (ft, deviated, bit size – 6 inches).	Mild effect on micro-resistivity images.	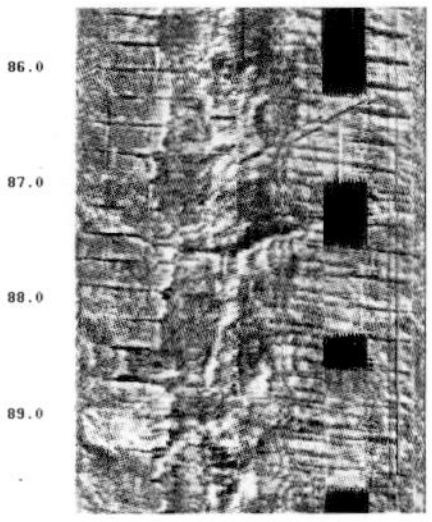
Debris material material in borehole	Material from drilling that is lodged in borehole wall. Right: Acoustic image showing a disused drill-bit that is lodged in the borehole wall shown by density spike (red curve) and a high amplitude acoustic (light area) image half way down the diagram surrounded by a low amplitude area (dark image)		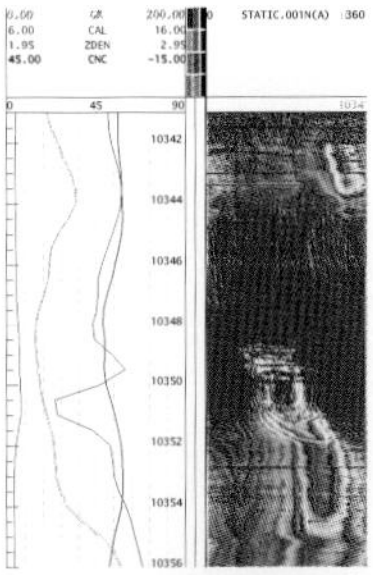
Acquisition – Logging Eccentralization/ stand-off	Logging tool is not centred in the borehole. This causes one side of the image to be too near the borehole wall, and one side too far away. Effect is mild on micro-resistivity images because these are pad devices that touch the borehole wall, unless the hole diameter exceeds the operational diameter of the tool. On acoustic images however, it is often severe (especially the transit-time image) as these are centred devices (held centrally by stabilizers/stand-offs). Common with pipe-conveyed logging. Right: Acoustic amplitude image showing a broad vertical darkening down the left side of the image and a light strip down the far-right side. (NB there is also a very dark vertical strip, this is a stabilizer mark). This is caused by the eccentralization in this horizontal borehole. Tool is falling to the low-side of the hole which is the lighter side of this image, to the far-right and is confirmed by a broad 'key-seat' to the right. (ft, bit size 8 inches, referenced to north).		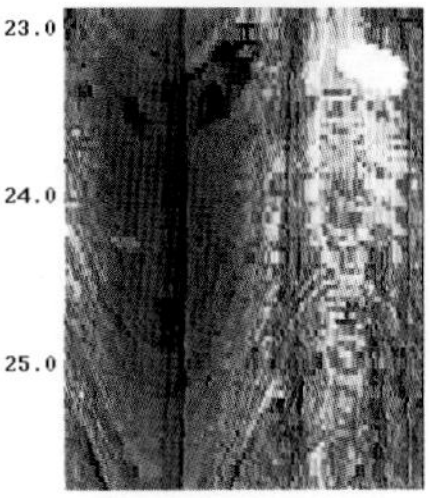
Mud smear	Mud build-up on electrode buttons, missing button data causes a vertical striping. Effect is mild to severe on micro-resistivity images depending on mud used, tool speed and tool type. Left: Mud smear on separate buttons of a FMI micro-resistivity device causing vertical striping. Buttons become smeared then clean by themselves due to pad friction with the borehole wall. The effect does not severely inhibit interpretation in this example. (m, deviated, bit size 6 inches).	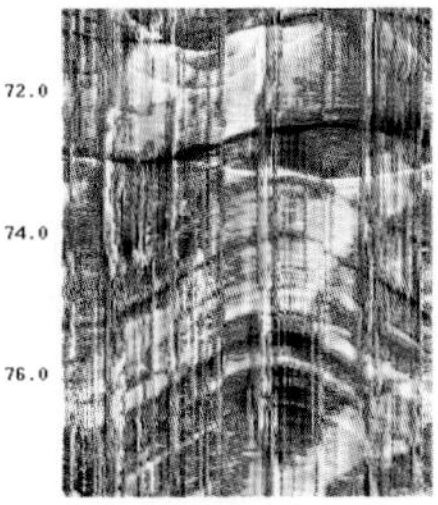	Not applicable to acoustic tools as they do not have electrodes.

Table 2. (*continued*).

Artefact	Cause – description	Micro-resistivity	Acoustic
Tool Speed Irregularities 1) Stuck Zones	Erratic breaks in an image resulting in a compression/expansion of the image over a small interval. Caused by sudden differences in tool speed and the yo-yo effect of wireline cable. Interpretation is often impaired. Refer also to Bourke (1989), Fig. 3a. Left: 4 pad micro-resistivity device with a tool stick and resulting data expansion over the base of the image (below xx75 ft) and compression just above at xx74 ft. Right: Acoustic image with expansion at the base (xx8–10 ft) and compression at xx7 ft.	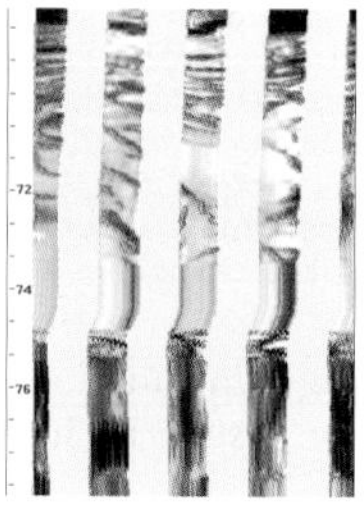	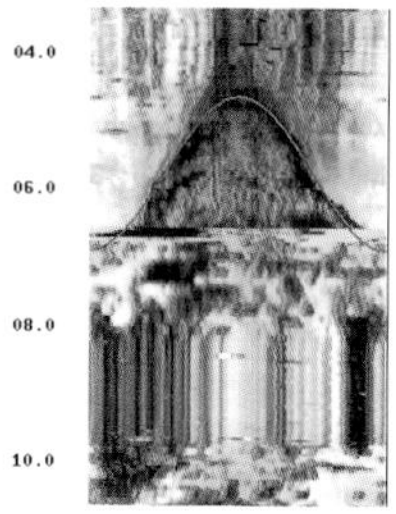
Tool Speed Irregularities 2) Sticky zones	Logging tool has suffered minor sticking possibly visualized as a minor tool juddering. Often the appearance of 'saw tooth' on micro-resistivity images. Interpretation is not often impaired. Left: single pad of a micro-resistivity device with a saw-tooth appearance, common within sticky zones. (vertical extent 200 mm). Right: Acoustic image over a 'sticky' interval where the tool motion is erratic or 'jerky' but has not caused a severe stick. The results are blocky patches across the image at xx57.4 ft, xx58.3 ft and xx59.4–59.9 ft.	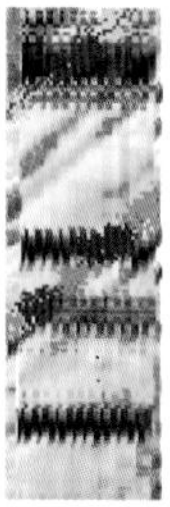	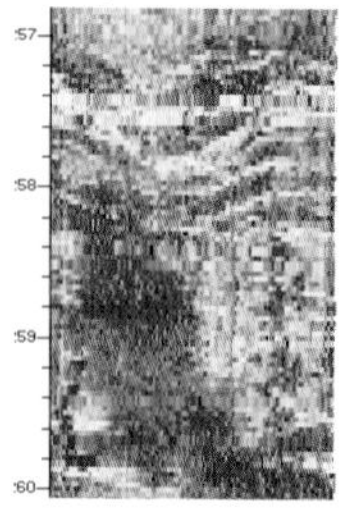
Button or flap 'death'	Electrical fault on micro-resistivity devices usually a single button or complete pad, resulting in missing button data and interrupted image. Mild-severe to severe affect on interpretation. Right: One single pad of an 8 pad-flap FMI micro-resistivity device is malfunctioning as shown by the interrupted image on the right of the image above xx00.0 ft. Interpretation is not inhibited in this example. (ft, vertical, bit size 6 inches).	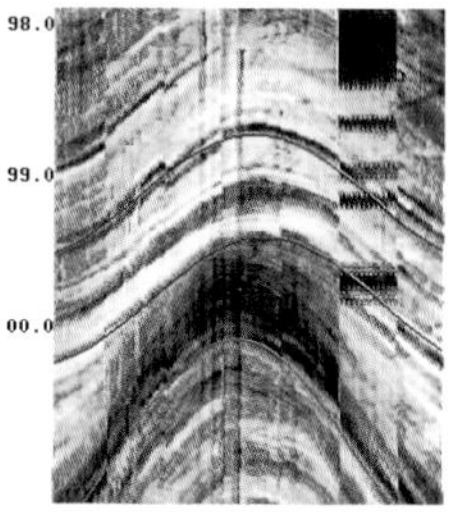	Not applicable to acoustic devices as these are centred devices.
Faulty inclinometry data	Either one or all of the 3 accelerometer or magnetometers are faulty which give an incorrect orientation of the images. Resulting dip magnitude and azimuth will therefore be in error. Effect is severe on subsequent interpretations and thorough checks of inclinometry data should always be made before interpretation. The artefact is common at the top of logging runs where the inclinometer enters the pipe ahead of the image tool (which causes the magnetometers to malfunction due to the magnetic effect of the casing). The result is incorrectly displayed image data over an interval generally related to the length of the tool. Right: Acoustic amplitude image showing zigzag patterns especially at xx5.0 m and just below xx6.0 m although other minor rotation disruptions are present in between. These are caused by magnetometers that are malfunctioning. The subsequent dips azimuth and magnitude are likely to be affected.		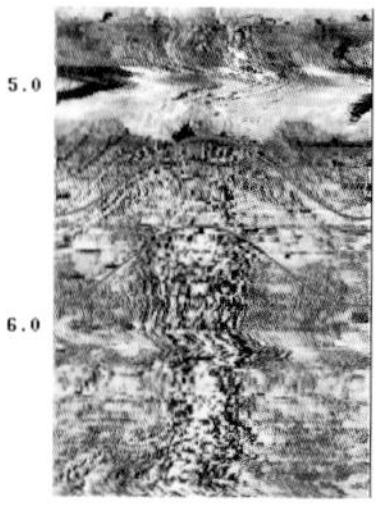

Artefact	Cause – description	Micro-resistivity	Acoustic
Excessive tool rotation	Logging tool rotates at a speed greater than the acquisition system can sample data. Results in an under-sampling of the inclinometry data. Generally a problem when the tool rotates >1 rotation per 30 ft (10 m). The result is commonly reduction of the precision and accuracy of the dip magnitude. This artefact is common with a mild to severe effect on interpretation in both micro-resistivity and acoustic images.		Notable by a fast rotation of the orientation Pad 1 Azimuth and Relative baring curves. Not often noticeable on images at high display scales (<1:20)
Noise: '60 Hertz'	Acquisition current noise – operating frequency interference, resulting in diagonal stripes across images. Can be common with a mild-severe effect in all tools. Right: acoustic image (UBI) with clear right-to-left diagonal stripes. Generally visible at low EMEX voltage. Can be successfully filtered. Interpretation is still possible although there is also a vertical stripping attributed to stabilizer grooves scarring the borehole.		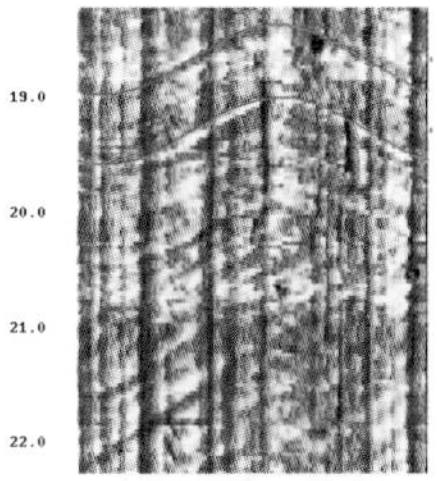
Processing Noise: 'Woodgrain'	Processing noise resembling woodgrain texture. Only applicable to acoustic images. This artefact can be common with severe results often inhibiting interpretation (right). Caused by systematic errors in acquisition peak interpolation. This can be ameliorated by filtering. A similar artefact is caused by an interference multiple effect (similar to a fresnel effect), see derived image artefacts. Left: 'Woodgrain' effect occurring within a hole spiral from a STAR2 tool. Common in heavy muds. (4 ft, vertical). Right: woodgrain texture clear to the left of the UBI image, caused by peak interpolation errors. (ft, deviated, 12.25 inch bit size).		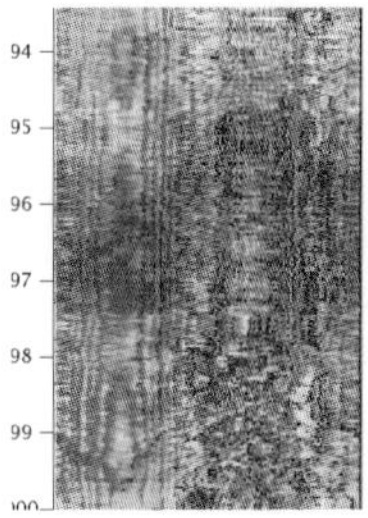
Processing Noise 'Honey-comb'	Regular horizontal comb pattern on the amplitude of acoustic images. Caused by errors in calculation of the amplitude gain during acquisition. The tool changes its gain on each sample, each gain is corrected based on statistics from the previous sample if correction is miscalculated it results in honeycomb pattern. Can be common with mild-severe effect. N/A in micro-resistivity devices Left: amplitude image with honeycomb texture noticeable to right of image. Interpretation is impaired (ft). Right: honeycomb texture stops below xx79.5 ft.	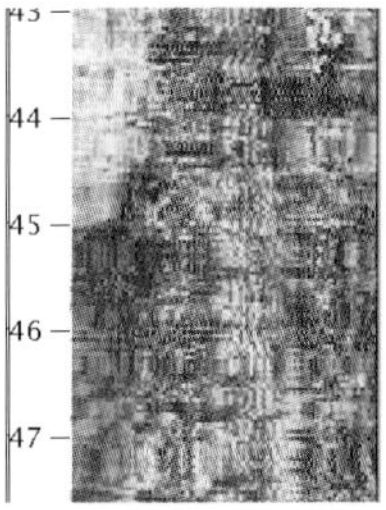	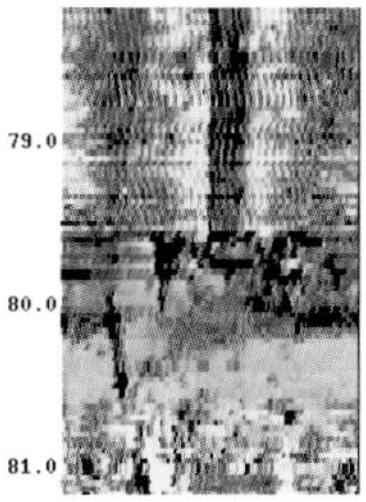

Table 2. (*continued*).

Artefact	Cause – description	Micro-resistivity	Acoustic
Signal Loss	Degradation in signal/noise ratio. Seen in all devices for various reasons. Rare but can be mild to severe in effect. Left: 8 pad-flap micro-resistivity device showing a spotted texture attributable to oil based mud lubricant additives (up to 8%) in water based mud. The effect does not inhibit interpretation in this example. (ft). Right: More common in acoustic tools are the white patches on travel time image representing areas where returned travel time (TT) falls outside of the measurement window. See dispersion (right side) of TT at xx65 ft. Common in washouts. (ft).	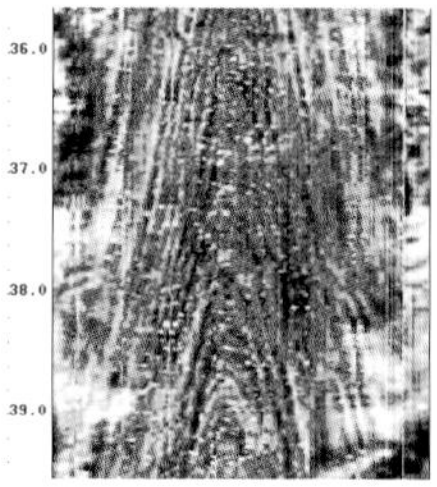	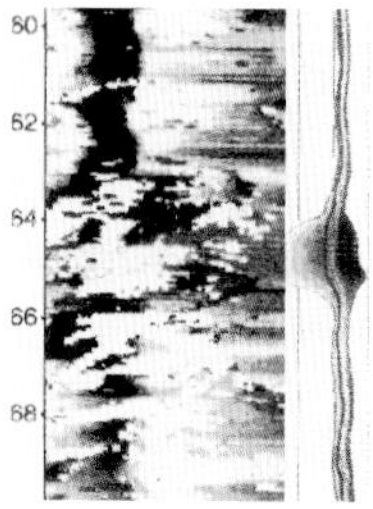
Extremes of condition	Dynamic range of tool exceeded. Common in high resistive lithologies as a crystalline like speckling on an resistivity image. Left: 'Crystalline speckling' in a very highly resistive lithology (evaporite) where the micro-resistivity tool has reached its maximum dynamic range. Lithology at top left is a shale. (ft).	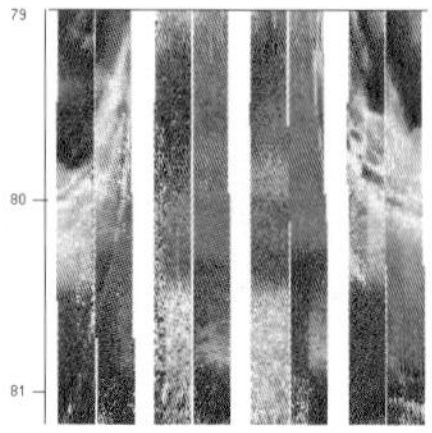	Not applicable to acoustic images

Artefact	Cause – description	Micro-resistivity	Acoustic
2. Borehole Wall Artefacts Washout	Enlargement of the borehole hole diameter and corresponding de-focusing of image and giving a patchy image. Severe effect on all image logs. Left: Micro-resistivity image and calipers (left in blue/red increasing to right) showing de-focusing of the image as a washout is encountered above 4170 m. Interpretation is inhibited above. Right: dark area on an acoustic log and associated caliper deflection (left) indicate a large washout at xx153 ft.	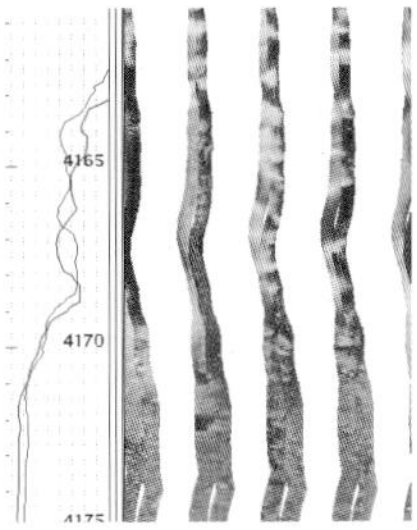	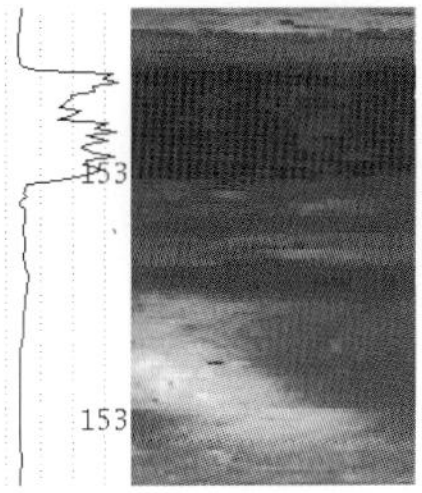
Rugose hole	Pitted borehole wall. Common on acoustic logs with a mild to severe effect on interpretation. Rarely noticeable on micro-resistivity images. Right: Acoustic image showing dark and light patches with a corresponding erratic caliper (black curve on left is the caliper, increasing to the right). This corresponds to rugosity of the borehole wall. (ft, 8 inch bit size).	Rare in micro-resistivity	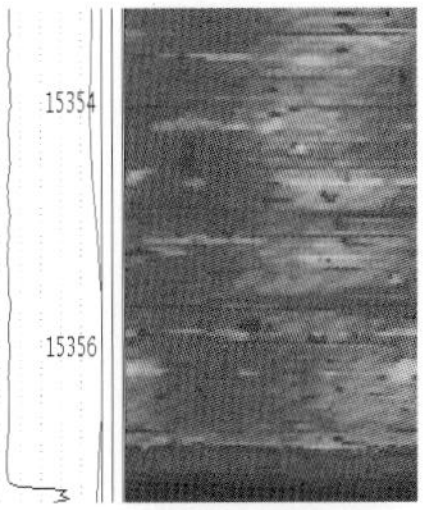
'Key-seat Furrow	Ovalization due to bit wear on underside of the wellbore. Noticeable in deviated wells with often only mild effects as a broad vertical stripe. Left: micro-resistivity FMI image showing a broad vertical key-seat furrow towards the right hand side. (ft, 6.25 inch bit size), Right: Vertical key-seat furrow in an acoustic CBIL image. More visible just right of centre (see arrows). (ft, 8.5 inch bit size).	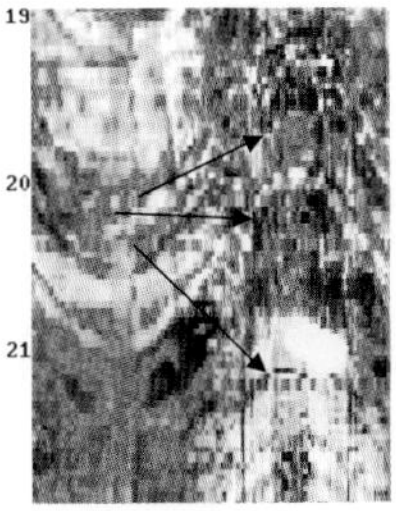	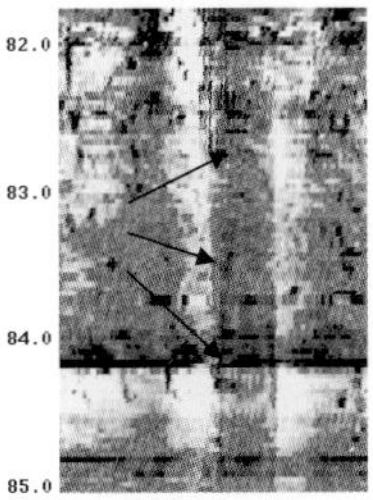
Mudcake Build-up	Build up of mudcake generally over permeable lithologies. Effect can be mild to severe, often inhibiting interpretation. Left: Micro-resistivity FMS image showing a transition from a shale above 11 935 ft to a permeable sand below. A mud cake has formed over the sand giving a blur to the image and masking bedding. Gamma-ray to left (increasing to the right). Right: Acoustic UBI image showing mudcake buildup and latter stabilizer scratching over sand layers in a sand-shale sequence. These are indicative of mudcake build-up in sands. Gamma-ray to right. (ft)	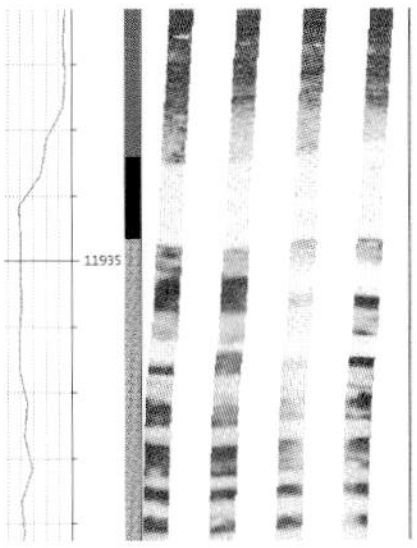	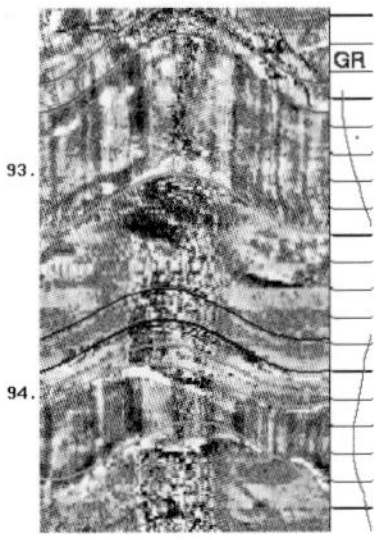

Table 2. (*continued*).

Artefact	Cause – description	Micro-resistivity	Acoustic
Spiral Hole	Borehole wall is grooved due to scratching by the bit. Often leaves a diagonal mark similar to hole orbiting. Common effect in all logs, although more severe on acoustic images. Left: Horizontal FMI showing regular diagonal marks representing catching of the bit which is scratching the borehole wall. Button smear is also seen as vertical marks on this image. (ft, 6 inch bit size). Right: Black diagonal marks on a acoustic tool are scratch marks that are filled with drilling mud appear as low acoustic impedance. (ft, 6 inch bit size). From Lofts *et al.* (1997).	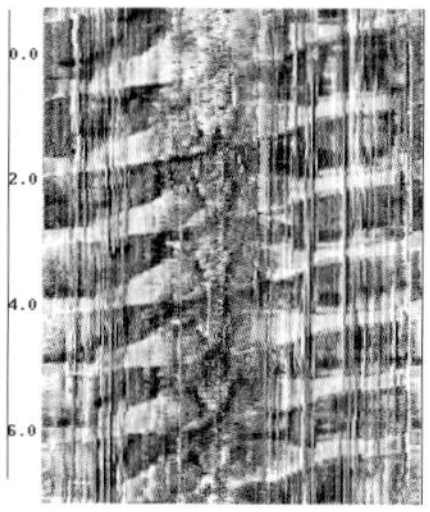	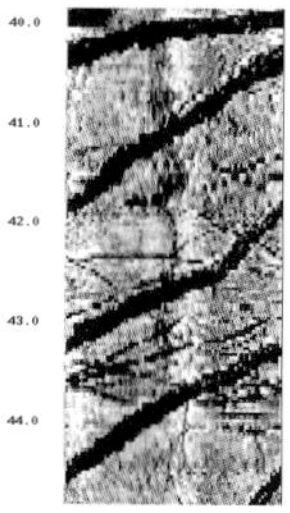
Tool Marks: A. Stabiliser grooving	Vertical striping on the borehole wall caused by scratching of the logging tool stabilizers. Acoustic tools are highly sensitive to the borehole wall condition as is seen here. Right: Light and dark vertical stripes in an acoustic CBIL image. Often only partially obscure interpretation. Bedding can in fact be identified below xx46 ft shown as a green sinusoid.	Rarely noticeable in micro-resistivity images.	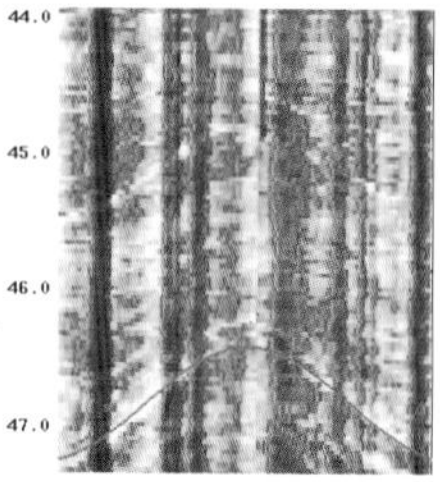
B. Sampling tool probe & packer marks	Marks made by probe/packer sampling tools. Seen in all tools as distinctive rounded marks left in mudcake or soft lithologies like the massive sand on the right. Commonly mistaken for cementation nodules. Effects are rarely serious on interpretation. Left: Resistivity image of two pads of an FMS tool showing an RFT probe/packer on mudcake. (ft, unknown bit size). Right: Two probe/packer marks on a micro-resistivity FMI image in a soft massive sand lithology. (ft, 8.5 inch bit size).	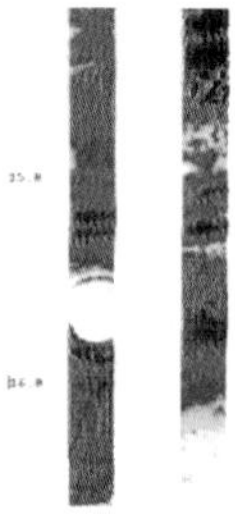	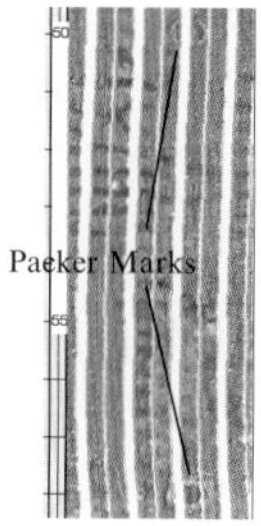 Packer Marks
C. Sidewall core sample points	Marks made by a mechanical sidewall corer device. Usually seen as dark spots where mud fluid replaces the sample that has been extracted. Left: Spilt image showing two sidewall core samples (left) and a zoom showing a single core sample and resistivity profile over that core sample (right). These are easily identified as they are symmetrically round shapes. (ft) unknown bit size.	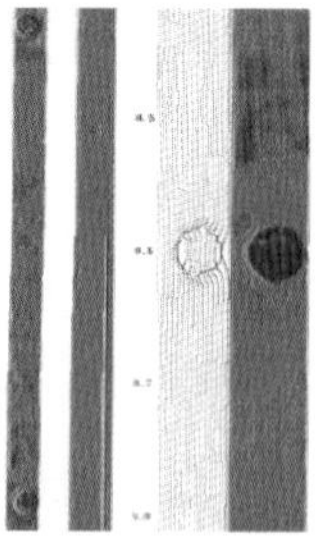	Possible in acoustic images, similar in appearance to micro-resistivity images

Artefact	Cause – description	Micro-resistivity	Acoustic
Cable 'slap'	Marking of wireline cable on borehole wall. A very subtle artefact and relatively uncommon. Most noticeable in acoustic images. Right: acoustic CBIL image showing two marks made by the cable 'slapping' the borehole wall, one almost vertical and one slightly diagonal in the centre of the image (see arrows). (ft, 8.5 inch bit size).	Not common in micro-resistivity images.	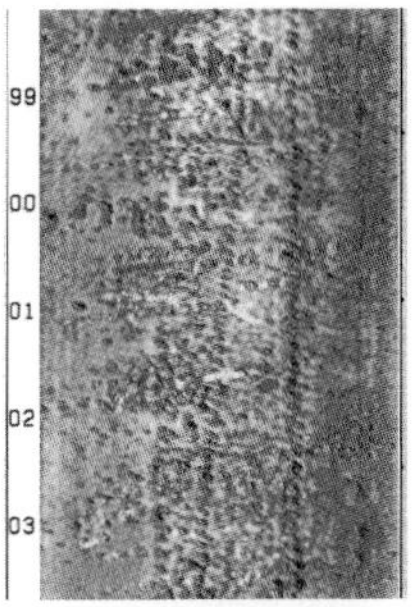
Drilling induced fractures	In general, zigzag fractures parallel to well bore axis. Caused by drilling process and indicates the predominant in-situ stress direction (sigma 1). Easily identified and useful in situ-stress determination. Left: Micro-resistivity FMI image showing two parallel zigzagging vertical fractures in a deviated well bore. Image also suffers from some mud smear on various buttons especially to the left. (ft 6.25 inch bit size). Right: Hairline zigzag fractures very clearly visible in an acoustic image parallel to the borehole axis. (ft, 8.5 inch bit size).	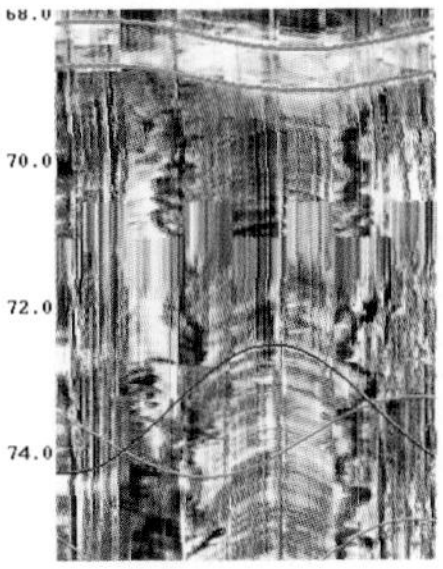	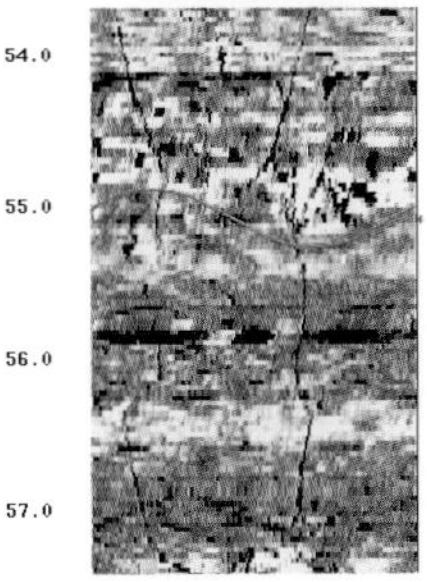
Drilling induced breakout	Broad parallel grooves separated by 180° and parallel to borehole axis. Caused by shear failure fractures orthogonal to the main in-situ stress direction and therefore useful in situ-stress determination. Generally, they do not fully inhibit interpretation. Left: Micro-resistivity image showing the broad breakout features which are dark as they are filled with conductive drilling mud. (ft, 6.25 inch bit size). Right: Acoustic image showing breakouts as wide dark area which abruptly terminate at xx02 ft as the lithology changes.	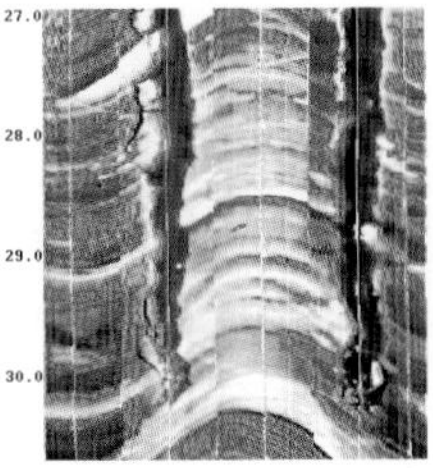	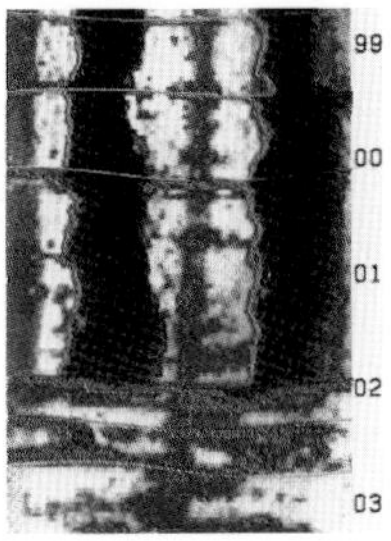

Table 2. (*continued*).

Artefact	Cause – description	Micro-resistivity	Acoustic
3. Processing Artefacts Incorrect hole diameter	Wrong assignment of borehole diameter in processing produces an incorrect calculation of apparent dip. The resulting true dip is therefore in error.	Effect not noticeable but can be severe.	Effect not noticeable but can be severe.
Normalization window length artefacts	Distinctive colour tones caused by an extremely high/low resistivity feature which skews the histogram such that the background colour is modified for a distance corresponding to the window length for normalization	Not common in modern images. Refer to Bourke (1989), Fig. 10a, b and c.	Not common in modern images
Multiple pass image offsets	Depth mismatch between two logging passes over the same interval. Easily corrected with modern depth matching.	Not common in modern images Refer to Bourke (1989), Fig. 12.	Not common in modern images Refer to Bourke (1989).
Incorrect colour assignment	Results from the skew of the image histogram by an anomalous high or low resistivity/amplitude image. Saturated looking images concealment of detail.	Not common in modern images	Not common in modern images
Incorrect gain correction	Wrong gain correction is applied during processing of images. Resulting in under saturated/over saturated images. Can appear in all logs. Left/right: Static acoustic images showing a raw uncorrected image that is saturated (left image) and one that has had a proper gain correction applied (right image). More detail is visible and the image is not so washed out. (ft, 8.5 inch bit size).	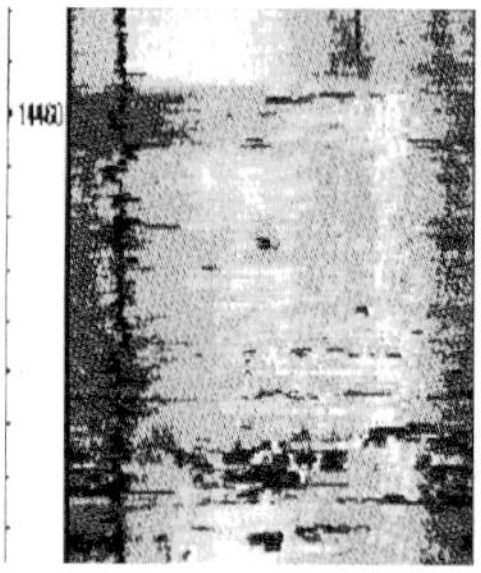	
Faulty button correction	'Tiger striped' images that appear fuzzy. This results from the incorrect processing and correction for the faulty electrodes. Generally only occurs if there is a faulty acquisition circuitry. Left: FMI tool with an inferior correction applied. Right: Same image with correct processing applied.	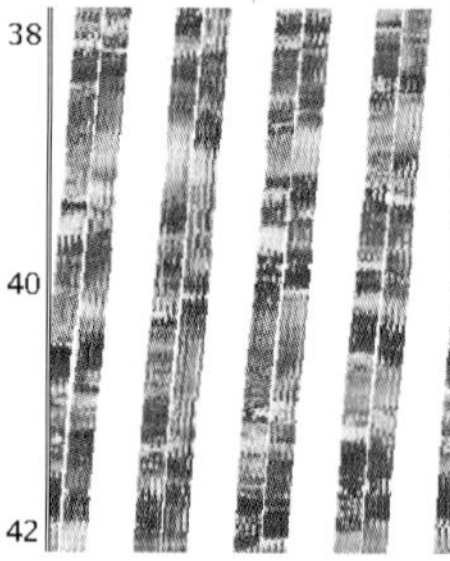	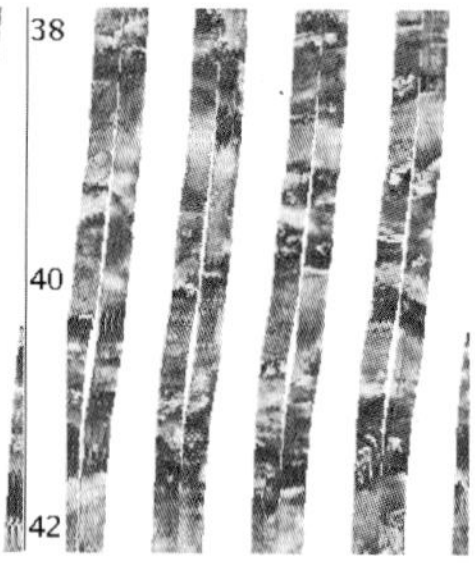
Pad-flap or button offsets	Incorrect speed correction applied. Offset of the pad to the flap in pad-flap micro-resistivity tools or buttons on different rows of a pad device. Refer also to Bourke 1989. Left: Large pad-flap mismatch on an FMI tool as a result of incorrect speed correction. (ft, bit size 8.5 inch)	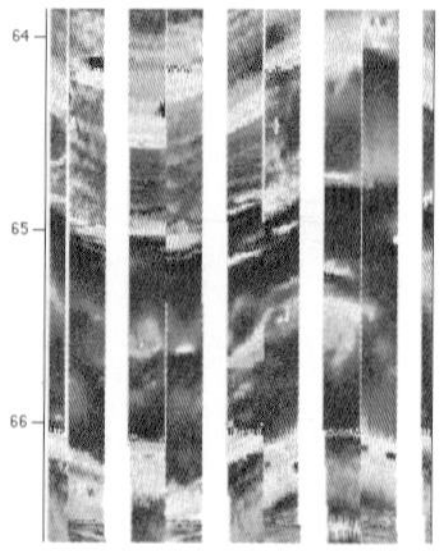	Not applicable to acoustic tools.

Artefact	Cause – description	Micro-resistivity	Acoustic
4. Derived Artefacts Current Gather Effects: Halo effects – Conductive	Light rim or 'halo' surrounding a conductive feature. This results from the flow of current toward the most conductive pathway. This 'current gather' leaves a resistive rim where current is lacking around such a feature. Left: Current gather seen around pyrite nodules, which are extremely good conductors of electrical current. The best example is a dark nodule at xx06.5 ft. (ft, bit size 8.5 inch).	306 307 308	Not applicable to acoustic tools.
Halo effects – Resistive	Dark rim or 'halo' surrounding a resistive feature (light). This results from the drain of current away from the most resistive areas to the more conductive pathway. This 'current gather' leaves a conductive rim where current is gathering around such a feature. Left: Large resistive cemented nodule (light colour) surrounded immediately by a dark halo indicating current gather.	12 13 14	Not applicable to acoustic tools.
Cement mottling	Mottling fabric caused by highly resistive cements. Again this is an effect caused by current gather. In the case of cements, they are generally resistive and result in a mottle fabric as current drains to a more conductive area. Left: White 'mottled' area following the outside of the red sinusoid is to current gather over a resistive bedding horizon. (Horizontal well with shallow bedding makes bedding surfaces here very steep sinusoids). (ft, bit size 8.5 in.).	66.0 68.0 70.0	Not applicable to acoustic tools.
Proximity effects. Fracture aureoles	Light patches at the apex of fracture sinusoids. Due to the depth of investigation (approx. 0.5–1 inch) the fracture is detected by the tool before the fracture plane is encountered in the borehole wall. Refer to Bourke (1989), Fig 5b. Left: Resistive fracture (shallow sinusoid) with a light aureole at the apex of the sinusoid.	30.0 32.0 34.0	Not applicable to acoustic tools.

Table 2. (*continued*).

Artefact	Cause – description	Micro-resistivity	Acoustic
Proximity features	Patchy texture caused by current gather/ around resistive and conductive features (such as nodules) proximal to the borehole wall but not actually touching. Refer to Williams (1996). Not applicable to acoustic tools.Left: Benthic shells in a mudstone showing dark blotchy texture which is attributed to proximal features not actually touching the borehole wall and not seen in core at this depth.		Not applicable to acoustic tools.
Woodgrain 2	Processing noise resembling woodgrain texture. Explained as a multiple or interference effect which occurs in heavy muds and within heavy mudcake intervals. With reflections from tool housing. Only applicable to acoustic images. This artefact can be common with severe results often inhibiting interpretation (right). Left: Woodgrain effect in a heavy mudcake on a CBIL image. (ft, 8.5 inch bit size) Right: Woodgrain effect commences where heavy mudcake starts in a permeable zone above xx32 ft. (ft, 8.5 inch bit size)	52.5 53.0 53.5 54.0	32.0

Micro-resistivity images are illustrated on the left hand side, acoustic images to the right. An indication of the depth units and bit size is given in brackets after the main text in each box. An explanation of tool mnemonics is given in the text.

Processing artefacts

These arise as a consequence of subsequent data processing and image generation post acquisition. Examples of this class of artefact include the incorrect use of normalization parameters, the incorrect distribution of colour scales in the process of displaying the images, or incorrect speed correction.

Measurement 'derived' artefacts

These are features which arise from the interaction of the physics of the measuring device and the actual geological feature. They have a different appearance on images than they would visually appear if seen on the outside of a core or in the borehole wall. These effects are predominantly seen on micro-resistivity images because of the complex nature of the flow of electrical current throughout a rock formation. This gives rise to effects such as halos, aureoles, cementation mottling and effects close to the borehole wall although not actually touching (termed here 'proximal' features).

Additional classification

In addition to using artefact categories for systematic identification, artefacts can also be graded on a scale of severity. Such a scale would range from an artefact which masks a small part of an image (e.g. drilling induced fracturing), where interpretation is mildly impaired, to extreme cases where the image is totally obscured by the artefact and geological data exists (e.g. woodgrain texture or hole-orbiting). Extreme disruption of the image occurs more commonly in acoustic images (Bourke 1999).

An interval may contain more than one artefact image. Figure 1 illustrates a small section of micro-resistivity log displaying no less than five artefact features. This is also seen in the acoustic image in Fig. 2 where two artefacts occur together (right image), diagonal hole spiralling and an artefact resembling 'woodgrain' (see Table 2 for a description). The presence of multiple artefacts is usually additive and may reduce the quality of the image and confidence of the subsequent interpretation.

Due to the inherent differences in the acquisition technology between micro-resistivity and acoustic devices, different artefacts may be exclusive to one image technology. This is illustrated in Fig. 2 which shows an acoustic image (right) and micro-resistivity image (left) over the same section of lithology. Hole spiralling has severely reduced the quality of the acoustic image log whereas the micro-resistivity image shows no adverse effect or artefact. The acoustic image demonstrates two of the most common acoustic image artefact features, hole spiralling (due to borehole wall morphology) and 'woodgrain' (due to excessive mud weight, see Table 2).

A summary of the artefacts recognized in these four categories is summarised in Table 1.

Table 2 illustrates the most common artefacts and includes a brief description and their effect and severity on micro-resistivity and acoustic image interpretation.

Conclusions and recommendations

- Systematic recognition of artefacts is the first stage in log quality control prior to any interpretation. Artefacts should be related to the acquisition practice so that they can be effectively filtered out with greater confidence during interpretation. The ability of the interpreter to recognize and disregard artefacts (effectively blank them out) should significantly enhance confidence and accuracy of the geological interpretation.
- When two or more artefacts are present they are generally additive in lowering the confidence of an image interpretation.
- Micro-resistivity images are most affected by either pad contact problems or the complexities of current flow within rocks.
- Acoustic (ultrasonic) image logs are highly sensitive to borehole conditions and drilling mud properties. These are the most common causes of artefacts observed on acoustic image logs.
- As tool design and processing improve further, some artefact types will doubtless disappear but new ones may be created.
- As interpreters become more familiar with the causes this information can be related back to the acquisition companies for modification of tool design and acquisition practices.

Acknowledgements are given to those oil companies who have contributed artefacts in the preparation of this paper but who wish to remain anonymous.

Glossary of tool mnemonics and terms

CBIL – Circumferential Borehole Imaging Log (Western Atlas)

FMS – Formation MicroScanner (Schlumberger)
FMI – Fullbore Formation MicroImager (Schlumberger)
OBDT – Oil Based Dipmeter Tool (Schlumberger)
RAB – Resistivity at Bit Tool (Schlumberger)
RFT– Repeat Formation Tester (Schlumberger)
STAR2 – Simultaneous Acoustic & Resistivity Borehole Imager (Western Atlas)
EMEX – variable emitter-exciter current/voltage used by Schlumberger's passive imaging devices (**FMS** and **FMI**) to maximise signal-noise.

FMI, **FMS**, **UBI**, **RAB**, **RFT** and **OBDT** are all marks of Schlumberger. CBIL and STAR2 are marks of Western Atlas Inc.

References

BOURKE, L. T. 1989. Recognizing artefact images of the Formation MicroScanner. *SPWLA 13th Annual Logging Symposium Transactions, June, Paper WW.*

——, DELFINER, P., TROUILLER, J. C., FETT, T., GRACE, M., LUTHI, S., SERRA, O., STANDEN, E. 1989. Using Formation MicroScanner Images. *The Technical Review*, **37**(1), 14–16.

——1999. A summary of best acquisition practices for acoustic image logs. *This volume.*

EKSTROM, M. P., DAHAN, C. A., CHEN, M. Y., LLOYD, P. M. & ROSSI, D. J. 1987. Formation Imaging with Microelectrical Scanning Arrays. *The Log Analyst*, **28**(3).

FARAGUNA, J. K., CHACE, D. M. & SCHMIDT, M. G. 1989. An improved borehole televiewer system: image acquisition, analysis and integration. *SPWLA Thirtieth Annual Logging Symposium Transactions.* June 11–14, Denver.

HAYMAN, A. J., PARENT, P., CHEUNG, P., VERGES, P. 1994. Improved borehole imaging by ultrasonics. Society of Petroleum Engineers. Paper. SPE 28440.

HARKER, S. D., MCGANNN, G. J., BOURKE, L. T. & ADAMS, J. T. 1989. Methodology of Formation MicroScanner Image interpretation in Claymore and Scapa Fields (North Sea) *In*: HURST, A., LOVELL, M. A., MORTON, A. C. (eds) *Geological applications of wireline logs.* Geological Society of London, Special Publication, **48**.

LOFTS, J. C., BEDFORD, J., BOULTON, H., VAN DOORN, J. A. & JEFFREYS, P. 1997. Feature recognition and the interpretation of images acquired from horizontal wellbores. *In*: LOVELL, M. A & HARVEY, P. K. (eds) *Developments in Petrophysics*, Geological Society Special Publication, **122,** 345–365.

SAFYINA, K. A., LE LAN, P., VILLEGES, M. & CHEUNG, P. S. 1991. *Improved formation imaging with extended micro-electrical arrays.* Society of Petroleum Engineers, Paper SPE 22726.

WILLIAMS, C. G. 1996. *Assessment of electrical resistivity profiles through development of three-dimensional numerical models.* Unpublished PhD thesis. University of Leicester, February.

A methodology for applying a non unique, morphological classification to sine wave events picked from borehole image log data

ROBERT TRICE

Enterprise Oil Italiana S.p.A., Via dei due Macelli 66, 00187 Roma, Italy
(e-mail: robert.trice@rome.entoil.com)

Abstract: Sine wave events are interactively picked as a representation of planar and pseudo-planar events within a borehole. This paper presents a logic and terminology by which sine wave features can be assigned a classification based on their morphological character. This morphological classification is non unique with respect to geological interpretation but can act as a template from which geological interpretations can be applied. The methodology presented has been written as a result of working on problematic image interpretations from a range of sedimentological, structural and petrophysical case studies. Application of the technique is considered of specific benefit to geoscientists with limited practical experience in image log analysis, to practising image log specialists when dealing with problematic data sets and as a methodology by which a base case interpretation can be derived in equity or shared data sets. An application of the technique is provided as a short case study.

This paper presents a logic and method by which planar events can be classified from borehole image log data. The methodology is applicable to resistivity and acoustic images and is designed to provide a preliminary, non unique, interpretation which can be viewed as a base case from which other more refined interpretations evolve. The paper is not intended to be a definitive article on the subject of geological classification but is intended to provoke thought on how sine wave features are classified or could be classified. Defining a transferable method of classification is a reasonable goal for the image analysis community as there is an apparent increasing trend in procuring and interpreting image data which necessitates the sharing of more interpretations and ideas between companies and or institutions. In addition, the increased application of image data to quantitative analysis necessitates an equitable methodology by which differing interpretations can be measured. The method is considered of particular benefit for: (a) geoscientists new to practical borehole image interpretation (b) describing unfamiliar formations or facies types (c) problematic images (d) problematic classifications and (e) providing a frame of reference for discussion where several interpretations are being debated.

Geological classification of planar features in outcrop, core and image log data

The purpose of borehole image log surveys is to provide a remote sensed scan of the borehole wall resulting in data that can be processed to form a visually interpretable image. The resolution and the colour contrast seen in these images is sufficient that detailed geological information such as rock texture and planar features can be defined, at the 'core scale'. Planar features (sine wave features) are routinely defined and catalogued by an image analyst by means of interactive software and a high quality graphics terminal. This is normally achieved through the identification of a minimum three equidistant points spaced along a given observed feature which results in a computer fitted sine wave. This sine represents the depth and attitude of the feature, comparison of the sine wave to the feature gives an indication of the events planarity. Such information can be applied quantitatively in structural, stratigraphic, sedimentological petrophysical, and rock mechanical studies (Rider 1996). As the medium for input to the various geological applications is primarily an interpreted sine wave, (reflecting an interpreted planar feature) it is important

From: LOVELL, M. A., WILLIAMSON, G. & HARVEY, P. K. (eds) 1999. Borehole Imaging: applications and case histories. Geological Society, London, Special Publications, **159**, 77–90. 1-86239-043-6/99/$15.00.

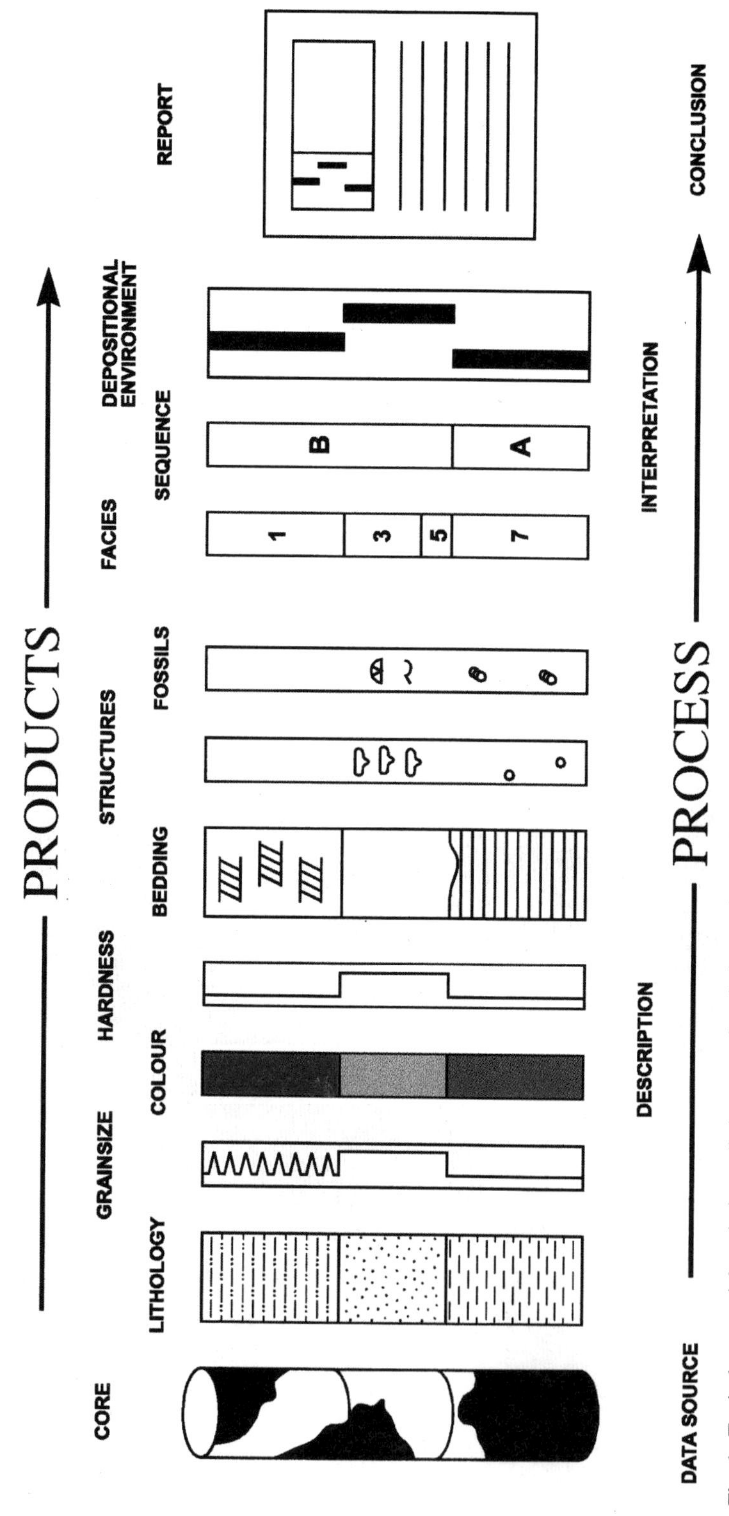

Fig. 1. Typical sequential analysis applied to core description and reporting.

that classification methods limit the opportunity for mis-classification.

Before considering the interpretation of image logs it is worth considering outcrop and core descriptions. In the field it is common to find that significant geological events can be readily observed and classified by the relationship between a given planar or non-planar feature and that of the host rocks bulk texture. This situation is typically afforded by good exposure and the ability for the observer to assess the characteristics of the host rock including scale, colour, grain size, structural attitude, apparent chemical composition and apparent mechanical state. This situation is also the case for planar features, which can be interpreted as simply an abrupt change (interface) in grainsize or rock colour as in some bedding planes and laminations. In contrast, planar features may be of a different mineralogy as in mineralized veins or void as in open joints. For many individual geological features a geoscientist will make a field interpretation without any conscious thought or measurement. This is based on an individual's ability to draw on a broad base of experience, as in the case of recognizing (classifying) bedding planes, faults and fractures. In these instances the geoscientist is drawing on a subconscious store of previous experience combined with the ability to rapidly absorb and compare a mass of related geological information in the surrounding outcrop. It is only when there is an absence of a good scale comparison or no clear structural trend, typically resulting from poor exposure, that basic classification becomes problematic.

For core descriptions the situation is much the same but the geoscientist observing and recording features of the core, is limited to a highly reduced volume of rock and is therefore reliant on more assumptions in formulating an interpretation. Despite this limitation a skilled core logger will reduce his reliance on assumptions and approach the core with a fixed process. This logical sequential chain results in an interpretation. An example of such an approach is depicted in Fig. 1 and portrays the relationship between data and products. As the data is progressively interpreted a series of products is achieved. The absolute approach and classification used in the core description will depend on the application that the description is to be put too and the type of material that is being described.

Interpretation of image logs requires a slightly different approach to that of core in that the data is representative of a one dimensional sample and planar events of less than 1 mm in aperture (Hornby *et al.* 1990) can be detected. However the luxury of lithological colour contrast and physical sampling is not available, as the formation is remotely sensed. The only advantage the image analyst has, is the common situation that there is usually a greater 'vertical' extent of image data on which to base an interpretation. By panning up and down the image and looking for continuity and deviations in a given data trend the interpreter can apply the stratigraphic principle of 'what goes up, must go sideways' to make inferences on bedding continuity away from the borehole. Assumptions can be reduced rapidly in instances where side wall cores and wholecores have been correctly depth correlated to borehole image data. In these cases the 'hard rock' can provide reliable textural information and close approximations to planar information that are detected on the image. However, 'hard rock' information is rarely available for the whole image logged interval and for this reason it is necessary for the image analyst to make interpretations based on assumption, previous experience and analytical 'common sense'. The following sections on sine wave classification, image fabric assessment, morphological characterisation and geological characterization describe the sequential processes that an image analyst can proceed through to create a preliminary geological interpretation. Final interpretations occur after numerous iterations, integrating dip picks with other relevant data and the grouping of dips through graphical techniques such as histograms and stereonets.

Classification of sine waves picked from borehole images

When an image is being interpreted the analyst makes a series of judgements similar to those a geologist would make looking at outcrop or core. These judgements fall into a series of subconscious and conscious decisions. The subconscious include assessing the image fabric and the morphology of an apparent sine wave whilst the conscious is concerned with the cataloguing of the sine wave into a pigeonhole, based on local or general experience. The degree to which the subconscious effort dominates the conscious will depend on a range of factors based on experience, data quality and the scale of the feature relative to the borehole environment, and the amount of logged section available. Typically the analyst will scan the image for a confident bedding plane pick before interpreting other features. As in field geology the reference to consistent bedding trend is essential before more detailed sedimentological or structural interpretations can be made. To make this bedding pick the analyst must first assess both the image fabric and any sine wave features that are present.

Assessment of image fabric

The borehole image, after appropriate processing, consists of two basic components that require interpretation namely a fabric and a sine wave feature. The fabric is the image response to the bulk lithology making up the formation. In electrical resistivity imaging logs, the response is a direct effect of the rocks porosity and permeability whilst in the case of the borehole televiewer there is a strong overprint of the borehole wall texture which may or may not reflect a specific lithological characteristic. For a given fabric type it is commonly a qualitative decision which defines at what scale the fabric is described This decision is usually based on depth boundaries defined by geological inferences or is simply a function of the interactive depth window that is being used. A less qualitative approach relies on computer derived assessments based on fuzzy logic or frequency analysis The method by which the fabric is picked is not of significance in this paper. What is significant, however, is the relationship of the sine wave to the fabric and how this relationship can be used to classify a given sine wave feature from a morphological and geological perspective. Excellent examples of image fabric correlated to core logs can be seen in Pilenko (1988).

Sine wave classification

A sine wave has a morphological character which leads the image analyst to apply a geological classification such as bedding plane, fracture or truncation surface. The characteristics that can be used to describe the sine wave in a morphological sense and then to apply the morphology to a basic geological classification are described in the following sections.

The morphological definition of sine waves is based on the following characteristics

(i) response of formation to sensing medium, i.e. whether it is transmissive or reflective
(ii) angular attitude to other sine waves
(iii) sine wave attitude to fabric
(iv) trace length continuity.

(i) The difference of transmissive or reflective as used in this terminology is whether the sine wave is receptive to the physical process that is sensing it. An analogy to this would be that of a black and white painted strip which is being scanned remotely by a human eye (Fig. 2). The eye will be responding to the visible part of the electromagnetic spectrum that is being reflected back from the strip. In this instance the black lines would be transmissive as black is a light absorber whereas white would be reflective as it is a light reflector. For imaging logs that are based on the response of the formation to local electrical fields transmissive refers to events that are relatively electrically conductive and reflective events are those that are relatively resistive to electrical current. Acoustic imaging logs, borehole televiewers, rely on the time duration and amplitude of a reflected acoustic pulse. Transmissive in this instance is relatively low amplitude and relatively low transit time whereas reflective is the converse.

(ii) Angular attitude is used to describe the angular attitude (relative to horizontal) of one sine wave relative to another sine wave when one or more sine waves share the same depth and can have parts of their trace length present in the same depth and borehole azimuthal space (Fig. 3). Under such conditions one sine wave appears to be overlain or cut by another. Within the classification used in this paper a sine wave is deemed to overlay another sine wave if there is a depth and azimuthal coincidence to part of the sine wave and if the sine wave in question has a higher relative angle. If the angle is less, then the sine wave is deemed to be overlain.

(iii) Sine wave attitude falls into four categories: (a) interface (b) displacive (c) fabric bound and (d) fabric cutting (Fig. 4). Interface sine waves mark the boundary between two distinct image fabrics, such events being interfaces, do not have a specific image resistivity or conductivity character. A version of the interface where there is an angular component that overlays one fabric over another is termed displacive. Fabric bound events are sine waves that occur solely within a given fabric. This does not mean that they are specific to a type of fabric but that the trace length of the sine wave is confined to a given fabric. Fabric bound events do not overlay interface sine waves but can overlay or be overlain by fabric cutting sine waves. Fabric cutting sine waves are effectively non fabric bound events and therefore have their trace length within one or more fabrics. Fabric cutting sine waves by definition overlay interface sine waves and can overlay fabric bound sine waves or be overlain by fabric bound sine waves.

(iv) Trace length continuity is basically a description of the relationship of the fitted

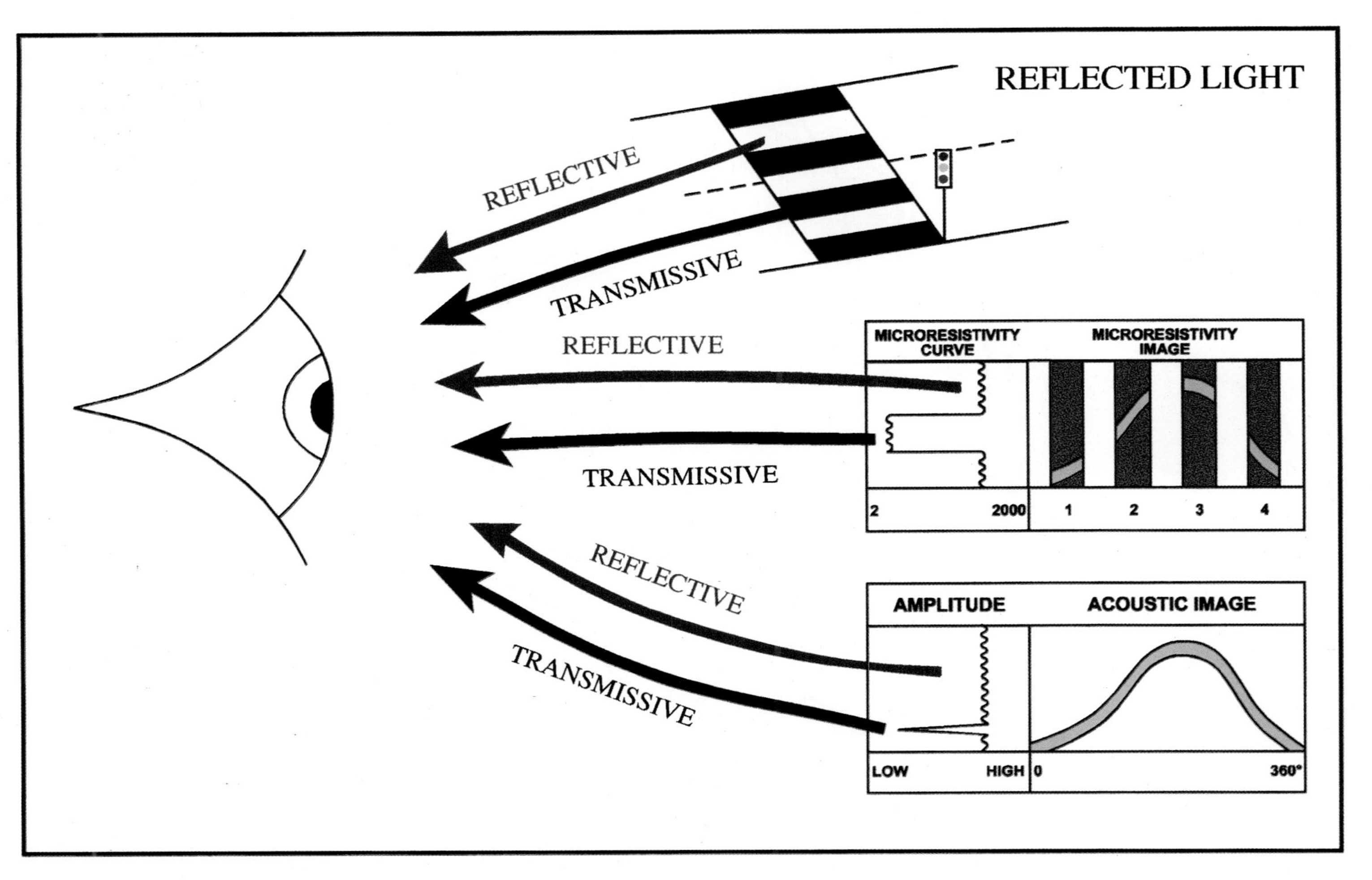

Fig. 2. Schematic diagram portraying the concept of resistive and transmissive sine wave features.

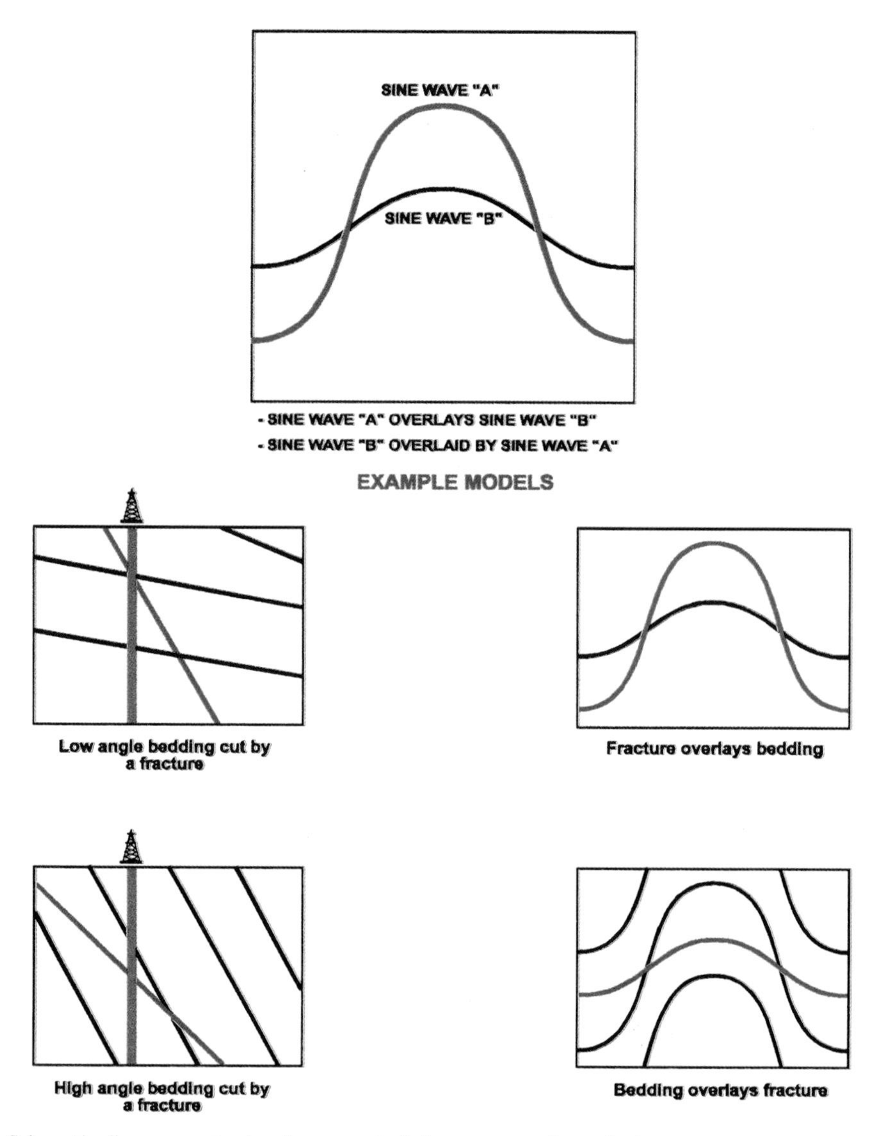

Fig. 3. Schematic diagram portraying the concept of sine wave angular attitude.

sine wave length and the apparent length of the observed image feature. In the instance of acoustic images it is possible for the trace length and the observed image feature to have equivalent lengths, i.e. to be continuous, as the borehole televiewer image is a 360° representation of the borehole wall. For electrical imaging logs the overall image is commonly less than 360° and under these conditions it is not possible for an absolutely continuous sine wave to be present. In the case of electrical image logs it is therefore necessary to make some allowance in use of the continuous descriptor. Continuous is used in this latter case to describe sine waves that are associated with an image feature that is present on all of the electrical pads. Partial sine waves are where an overlaying sine wave appears truncated by the overlain sine wave. This occurs in cases such as current laminations where the lamination abuts, or is truncated by the upper bounding surface. Discontinuous sine waves are image events that have apparent lengths less than the describing sine wave (Fig. 5). Discontinuity may be caused by the event present in the image being non planar or because the event itself is discontinuous. Causes for discontinuity

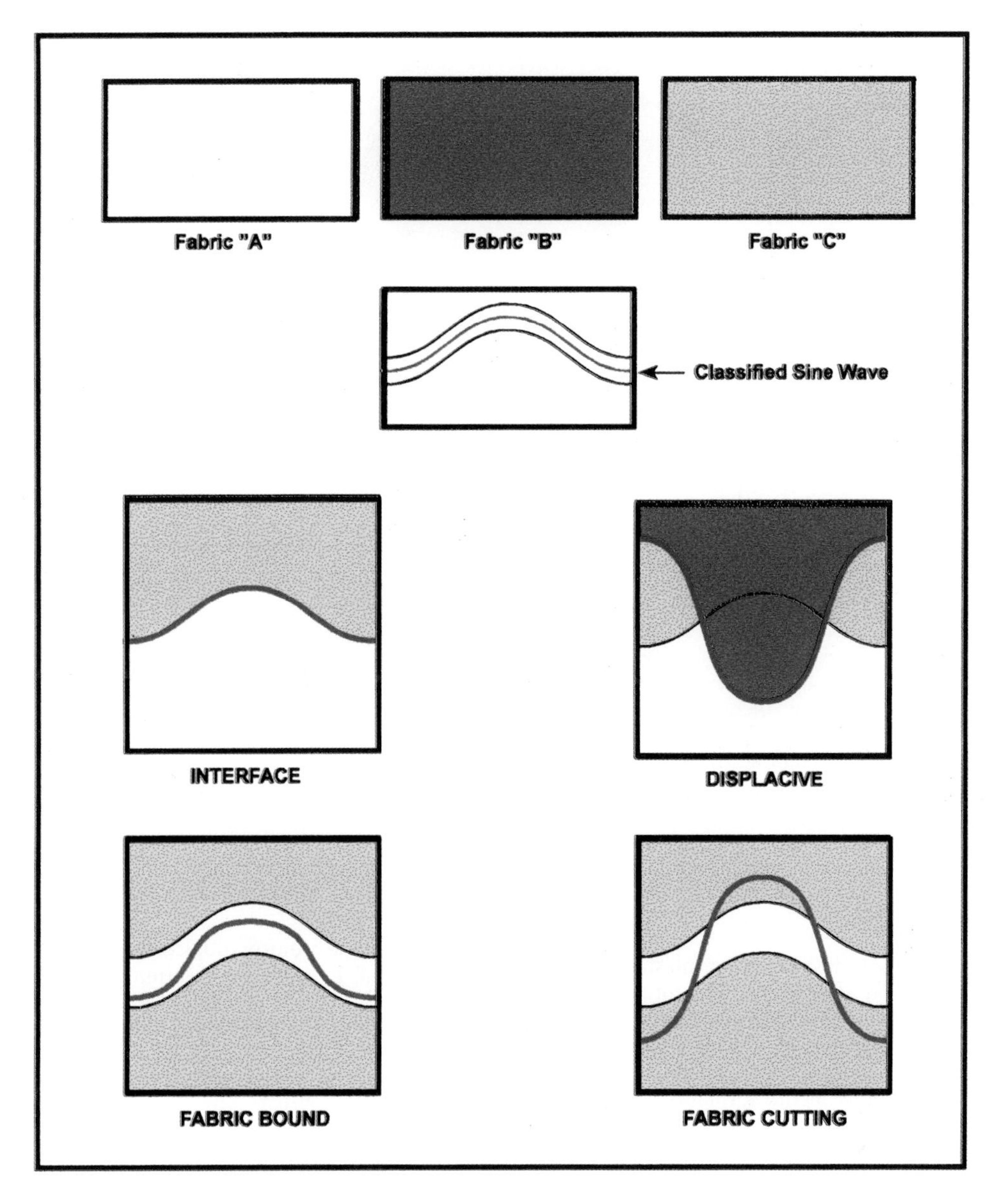

Fig. 4. Schematic diagram demonstrating the four attitudes of sine waves with respect to image fabric.

include hole shape, borehole condition, poor signal and non planar geological events.

The above descriptive terminologies can be used in a sequence, the end product of which is morphological characteristics which can be used in defining a basic geological classification (Fig. 6). In this figure the descriptive process of the sine waves morphological characteristics starts with the identification of fabric types. Once the sine waves have been 'classified' the geological classification is initiated with an assigned geological label, if appropriate. After numerous iterations and graphical/statistical processing a definitive geological classification can be applied.

Geological classifications

Using the above morphological nomenclature sine wave types can be placed into geological meaningful categories that can be used as

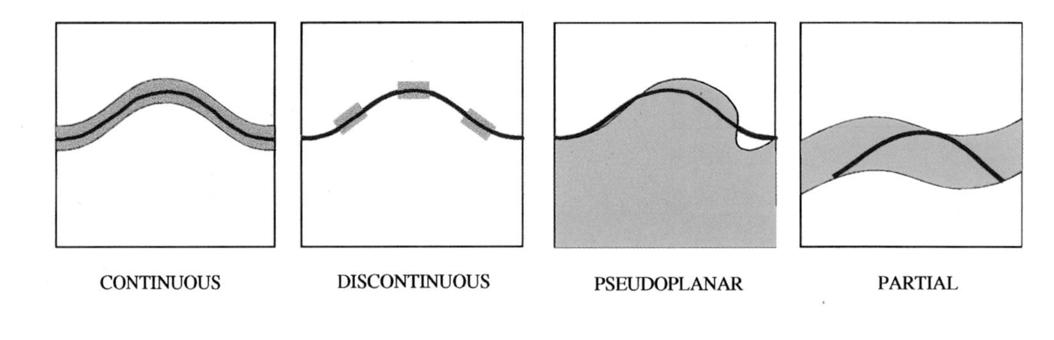

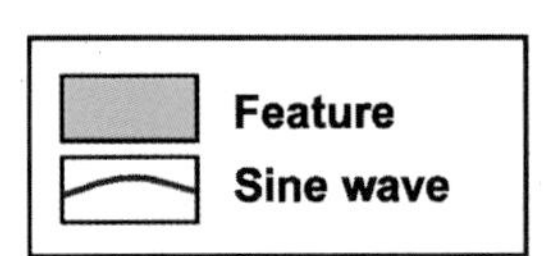

Fig. 5. Schematic diagram portraying the concept of sine wave continuity with respect to a given image feature.

part of a preliminary interpretation. Natural geological features within this classification are those common enough to be present in routine analysis and do not include specialist terms that may be applied to a specific depositional or structural setting. Geological feature classification falls into four basic categories, interface events, laminations, fractures, displacive and pseudoplanar features.

Interface features

Bedding is the most common interface event as the boundary between two different rock types is a horizon. Other interfaces include variations in the vertical profile of a bed which could be attributed to grainsize, permeability or compositional variations in the formation. In addition interface features can clearly mark fluid contacts in permeable formations.

Laminations

Laminations are conductive or resistive fabric bound events and can be partial or complete in trace length. When laminations have a sedimentary origin they are either parallel or inclined to interface features. If the lamination is of a structural or diagenetic alteration to an original sedimentary component the lamination will also tend to be parallel to the interface. Examples of laminations include current bedding, shale partings, thin interbeds, strata bound fractures, bedding laminations and stylolites.

Fractures and veins

Fractures. Fractures are geologically diverse and fall into two distinct categories with respect to image analysis, the fabric bound fracture and the non fabric bound fracture. In the case of the former, the fracture is morphologically identical to a lamination, in the case of the latter the morphology of a fracture is one that overlays interfaces and or laminations. Fractures are interpreted as being transmissive features.

Veins. Veins are morphologically identical to fractures but in this classification are reflective features. The classification of vein and fracture is from the morphological perspective of the image and is not intended to reflect the potential fluid flow properties of a given sine wave feature.

Microfractures. These events are partial sine waves and have the appearance of multi-oriented laminations. The frequency and azimuthal variation of the lamination is commonly such that the microfractures appear as an image fabric. Microveins are reflective versions of microfractures.

Drilling induced fractures. Drilling induced sine wave features, are often ignored by image analysts as background noise. However, drilling induced fractures, tensile wall failure fractures, 'apparently natural fractures' are important measurements for understanding the stress state of the formation and can lead to a mechanical stratigraphy of the borehole. Stress analysis is

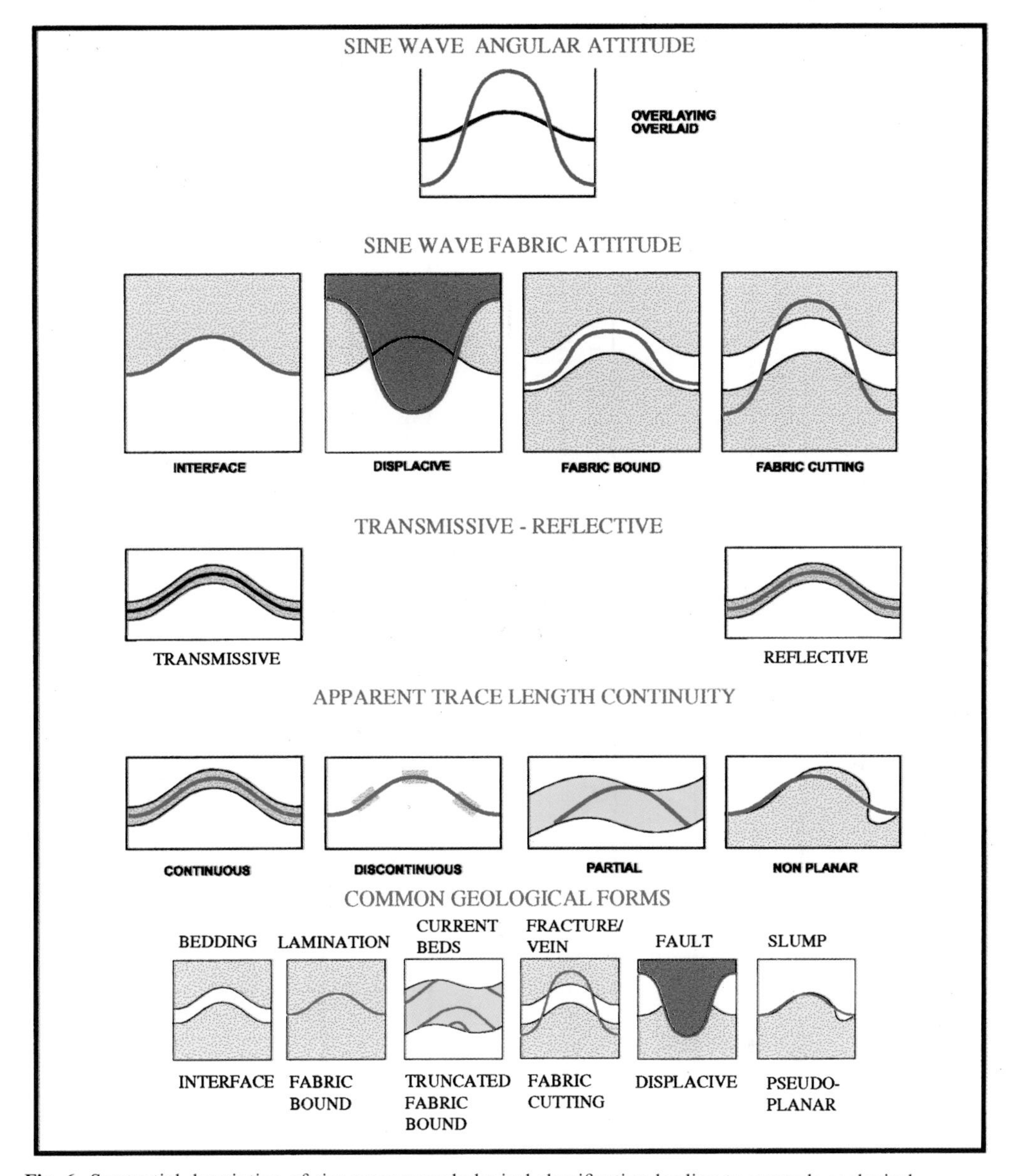

Fig. 6. Sequential description of sine wave morphological classification leading to example geological classification.

increasingly becoming a routine operation and is essential to the correct classification of natural fracture in fractured reservoir studies (Peska & Zoback 1997).

Induced fractures. These are the classic borehole parallel fracture that are transmissive and follow the borehole as single conductive pseudoplanar features, that often terminate or anastomise at interface events.

Tension fracture. Tension fractures or tensile wall failure fractures are partial, transmissive sine wave features. On electrical imaging logs they tend to occur on two opposing calliper pads

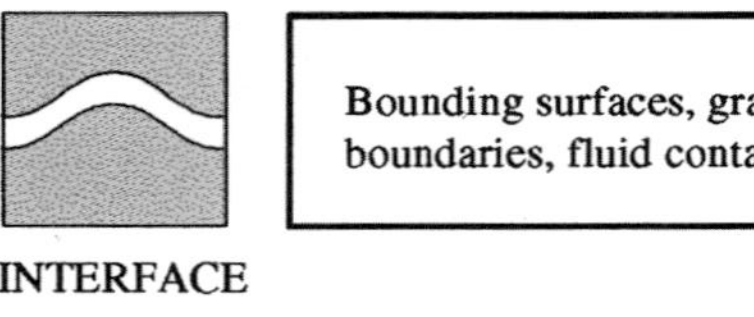
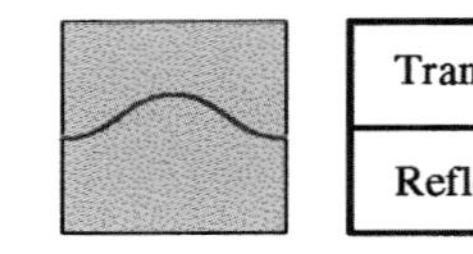

Fig. 7. Summary of basic morphological types and examples of geological interpretation.

and may be linked by a sine fit. Tension fractures form at 90° to the azimuth of minimum horizontal stress.

Shear fractures. These are the most problematic of all drilling induced fractures and are hybrid events grading from partial to full sine wave fits (Trice 1999). Shear fractures appear identical to natural fractures i.e. they are transmissive laminations formed as either fabric bound or fabric cutting features and are developed as full or partial sine waves. Shear fractures strike parallel to maximum horizontal local stress or regional (far field) stress field. These fractures can only be confidently identified after extensive experience in a given area and through detailed integration of core and other wireline data. Shear fractures may result from further displacement of an original fracture or be associated with apparently competent lithologies. Acoustic imaging logs provide the best technique for establishing the type of shear damage present (Cesaro *et al.* 1996).

Displacive features

Displacive features are numerically rare in a given population of sine wave picks however they often prove essential in any structural or stratigraphic interpretation providing key information with respect to palaeo-transport directions and fault orientations.

Fault. Fault planes fall into a range of morphological categories from the simple to the complex. A single fracture observed in the borehole with no apparent associated change in lithology or structural dip has the potential to be a fault. Fault planes can be confidently defined by the image analyst when present as an interface with a displacive attitude, associated with a change in structural dip trend. However this is not a unique definition as faults can often be associated with no or negligible change in structural trends. In addition fault planes are rarely simple either in outcrop or from an image analyst's perspective. In the borehole setting the fault plane can consist of a variety of natural and drilling induced sine wave features and image textural components. The actual fault plane may not be visible to the image log being masked by poor hole conditions or a complex of microfractured rock, brecciation and or high intensity of fractures. A single sine wave that classifies a fault's orientation is therefore rare, however in the event that bedding drag or rollover trends are present any fractures or veins interpreted as being parallel to the axis of the deformation trend can be used to infer the orientation of a fault. In such instances as the fault plane is assumed rather than picked this author applies the term an *apparent fault plane*.

Sedimentary features. Cross stratification, channel bases and slump planes are all examples of features that can be classified as displacive features. The orientation of the displacive feature is likely to be perpendicular or parallel to palaeoflow and is therefore a key feature to identify.

Unconformity. Defining an unconformity is perhaps the most elusive feature from an image analyst's perspective and can be only identified confidently when there is a structural element to the feature. In fact unconformities can have the morphological characteristics of interfaces and pseudoplanar features. Unconformities are perhaps best classified when regional stratigraphic information is available, under these circumstances the image log becomes a very powerful tool in characterizing the unconformity surface.

Pseudoplanar. The definition of a confident sine wave fit to a geological event can be compromised by the orientation and diameter of the borehole to the geological feature, or by the geological feature having a character that is not strictly planar. Morphologically the fitted sine wave is partial however the feature visible on the image is complete. Such sine waves have a pseudoplanar morphology. Examples of the latter include lenses, flaser bedding, soft sedimentary deformation, induced fractures and high amplitude stylolites. Within carbonate sequences pseudoplanar features are important considerations as they can mark exposure surfaces, caliche horizons and karst events.

Example case study

The above approach has been used in a wide range of sedimentological, structural and fracture case studies. One example presented here is from an offshore exploration well penetrating a platform carbonate reservoir section. The example is chosen as it is both simple and it achieved significant net results.

The penetrated section consists of a sequence of low porosity carbonate interbedded with highly porous limestone layers comprising reworked rudists. The image data was good as verified by correlation to core photographs. Correlation vugs, and individual rudists, could all be seen clearly. For the purpose of this paper the relatively simple problem of bedding will be considered. Figure 8 is a representative example

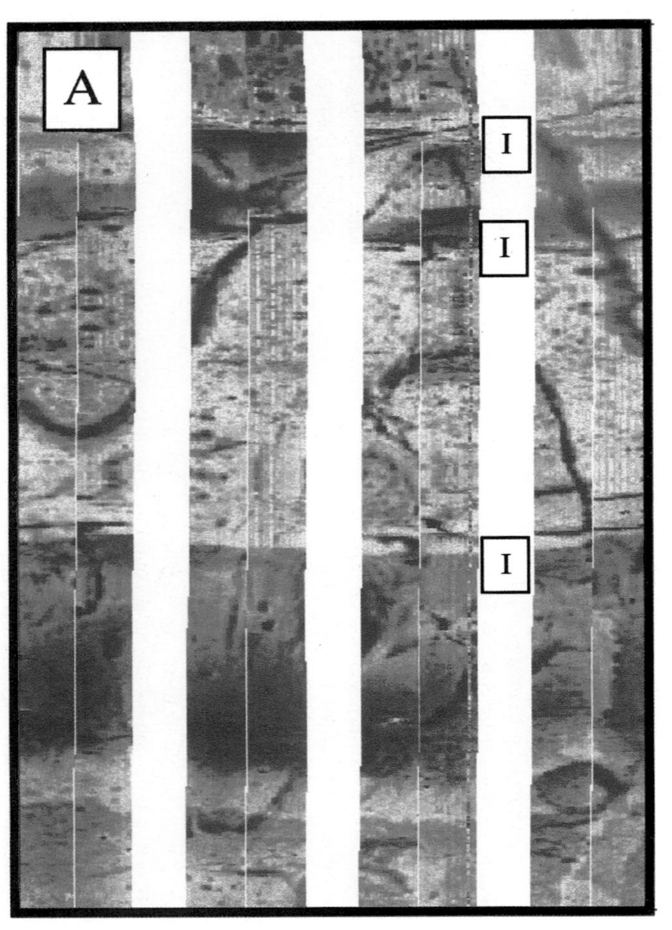

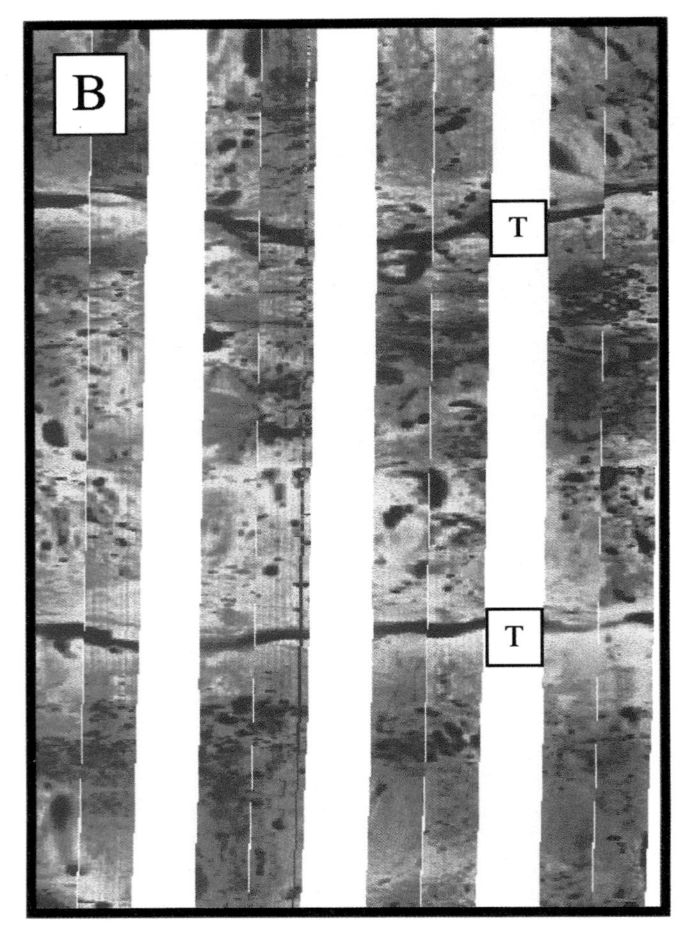

Fig. 8. Two metre dynamically normalized FMI images with sine wave features interpreted as I, 'bedding' and T, 'transmissive laminations'. Alternate interpretations place T and I as 'bedding'.

of a borehole image through the sequence where intervals of discrete fabric are separated by interface events pseudoplanar interfaces, transmissive laminations and transmissive pseudo planar laminations. A previous worker took the standard approach to the data set which was to classify the interfaces and laminations as bedding. This is not unreasonable as the laminations and interfaces are parallel and are clearly related from the perspective of geometric attitude. By doing so the interpretation is set early in the analysis and as such may be overlooked in later iterations.

The morphological approach taken was to classify the interface events as bedding (see Fig. 8a I) whether the interfaces were planar or non planar in their morphological characteristics. For the transmissive fabric bound sine waves a planar (Fig. 8b uppermost T) and pseudoplanar sine wave fit were noted (Fig. 8b lowermost T). In both these instances the features were described as transmissive laminations. The two sine wave types of 'bedding' and 'transmissive lamination', were then compared on stereonets and tadpole plots. From this graphical analysis it was clear that the two events where parallel, existing in the same stereo net space. However two questions remained (a) what were the transmissive laminations? and (b) were they significant? Core and field analogues indicated that these features where unlikely to be shale or electrically conductive marls. As they had been classified as discreet entities, further comparisons were possible using other graphical techniques, but no meaningful relationships were noted with other sine wave features or image based facies. A final comparison was made using linear plots comparing all interpreted sine waves with routine wireline logs. In this comparison it was clear that there was a positive relationship between the 'conductive laminations' and conductive blips, seen on the shallow and deep lateral logs. Supportive data from the literature indicated that low angle open fractures flushed with saline fluid could be expected to have an identical lateralog response to that of the 'transmissive laminations'. From this analysis it was clear that the 'transmissive laminations' were behaving as open fractures and not as simple bedding planes. Further geological and petrophysical analysis led to the conclusion that the 'transmissive laminations' were probably the result of a dissolution processes, acting on bedding planes similar to that described from karst reservoirs. The fact that the 'transmissive laminations' had been described as discrete entities, meant that these low angle 'fractures' could be analysed from a frequency perspective and output as a fracture class to fracture modelling software. This new interpretation was considered significant as it contributed to a more detailed geological and petrophysical analysis, providing potentially increased connectivity to high angle open fractures, and also contributed to a major revision of the existing exploration model.

Discussion

Classifications are descriptive tools used as an aid in describing, sorting and conveying information about the characteristics of various phenomena. Classification is an essential part in the process of evaluating a potential reservoir however classifications are never perfect, being a convenient compromise, often open to misinterpretation. Defining a classification system for use in image log analysis is no exception, as the measuring medium, the location and the environment of measurement is not ideal. The classification methodology presented here has been used with some success by thinking about a morphological classification before assigning a geological classification. This sequential approach proposed is not intended as a mechanical recipe for every sine wave requiring classification. Such a route would be tedious and time consuming. The method is made available as a mental template which can be used when appropriate and as a disciplinary reminder that sine waves are picked from images not from real rock. The morphological approach can be readily appreciated by geoscientists that have no previous image log interpretation, or who are unfamiliar with basic field geological terminology. The morphological approach allows for a catalogue of sine wave features to be determined upon which later more geological classifications can be based. The morphological classification can therefore act as a baseline population from which problematic intervals can be discussed in geologically complex situations, in cases where accurate numerical counts of a given event are necessary and in cases where groups of companies may wish to share a common data as in equity discussions.

Conclusion

The methodology discussed in this paper allows for sine wave features to be classified in a non unique way from which a more refined geological interpretation can be derived. It also allows for a ready reinterpretation at a later date if required. The application of a first phase of morphological classification gives the analyst an opportunity to reduce the number of assumptions that are applied in the first pass analysis. The technique is relatively simple to comprehend and has the advantage of being accessible to geoscientists unfamiliar with image analysis or basic field geology techniques. This approach provides a series of logical steps that break down the often intuitive dominated approach that is applied by the image log analyst. Such a breakdown becomes useful in (a) explaining problematic image features where a specific geological interpretation may not be possible (b) in training applications and (c) in introducing geoscientists unfamiliar with field geology or core descriptive techniques to image log analysis. This paper is not intended to be a definitive catalogue of geological features identifiable from borehole images but is offered as a template from which definitive sine wave classifications can be achieved.

I would like to gratefully acknowledge my colleagues Carrado Belfiore and Gloria Verdecchi for their skill and patience in draughting the figures in this publication.

References

CESARO, M. GONFALINI, M. CHEUNG, P. & ETCHECOPAR 1996. Shaping up to stress in the Apennines. *In: Italy 2000 Value added reservoir characterisation.* Schlumberger, Milan, 65–73.

HORNBY, B. E., LUTHI, S. M. & PLUMB, R. A. 1990. Comparison of Fracture Apertures Computed from Electrical Borehole Scans and Reflected Stoneley Waves: An Integrated Interpretation. *31st Annual Well Logging Symposium S.P.W.L.A.*, paper L, Lafayette, Lousiana.

RIDER, M. 1996. *The Geological Interpretation of Well Logs.* Whittles Publishing, Scotland, UK.

PESKA, P. & ZOBACK, M. D. 1997. *Constraining Complete Stress Tensor Using Drilling Induced Tensile Fractures in Inclined Boreholes.* Stanford Rock Physics & Borehole Geophysics Project (**63**). Department of Geophysics School of Earth Sciences Stanford, California.

PILENKO, T. 1988. *Formation Microscanner Applications in Italy.* Schlumberger Italiana S.p.A.

TRICE, R. C. 1999. An example of the application of borehole image logs in constructing three dimensional static models of producing fracture networks in the Apulian Platform, Southern Apennines. *This volume.*

Methodologies for multi-well sequence analysis using borehole image and dipmeter data

JEREMY PROSSER,[1] STUART BUCK,[2] SHAUN SADDLER[2] & VINCE HILTON[2]

[1] *Baker Atlas Geoscience, 2nd Lvl, Sheraton Court, 207 Adelaide Terrace, East Perth, WA 6004, Australia*

[2] *Baker Atlas Geoscience, Campus 2 Balgownie Science and Technology Park, Bridge of Don, Aberdeen.*

Abstract: Borehole image logs provide high-resolution directional data sets, in addition to a wealth of information concerning sedimentary fabric and texture. The data derived from these images are typically used for lithofacies characterization, palaeotransport analyses, and fault and unconformity recognition. These data are ideal input variables to the development of correlative models, particularly within a sequence stratigraphical context. However, in field settings characterized by low structural dip (less than 15°), unconformity recognition can be very difficult within shallow marine sediments, which would have had very low depositional or sedimentary dip. In the study described, integration of borehole image data with core sedimentological and ichnofabric analyses has enabled calibration of images for lithofacies recognition, identification of bioturbation fabrics, and the degree of bioturbation present. In particular, comparison of the results of dip trend analyses with the results of ichnofabric analyses has indicated that subtle, though consistent variations in shale bed dip azimuth correspond closely with the locations of sequence boundaries. This has provided increased confidence in unconformity recognition in un-cored well sections, and enabled correlations to be developed. Lessons learned from interpretation and calibration of high sample density borehole image logs (FMI) within cored well sections has enabled interpretive methodologies to be transferred to lower density FMS and SHDT data sets. Similarly, examination of core-photographs as a continuous 'photographic image log' in the workstation environment has revealed that when displayed at appropriate scales, even vintage core photographic plates contain a phenomenal amount of data. This is obviously useful for calibration of dipmeter data where, due to climate etc., cores are poorly if at all preserved. Perhaps more surprisingly, comparison of core-photographic data with existing cores has revealed that when displayed on a workstation, core photographs are suitable for detailed ichnofabric analysis. This has allowed reinterpretation of effectively 'lost' data within the context of recent advances in sedimentology. The procedures described constitute an innovative approach to the integration of different data types gathered using existing, proven techniques.

This study aims to present a methodology for the integrated interpretation of borehole image and dipmeter data within shallow marine settings. The data described are from a South-east Asian oilfield. The hydrocarbons occur within northwesterly dipping sediments, on the limbs of a large, complexly faulted anticlinal structure. Reservoirs occur within a variety of sandstone facies, including those of marine shelf, shallow marine (shoreface) and paralic to deltaic environments (tidal distributary channels and tidal flats). The studied successions are characterized by low structural dip (less than 15°), and the identification of unconformities or correlative surfaces can be very difficult. Seismic data provide some insight to identification of key surfaces at large scale, but high-resolution correlations are difficult.

This study will illustrate a system used for the characterization of lithofacies from FMI data, and how interpretations were extended to lower sample density FMS and SHDT data. The lithological interpretations were used to determine lithofacies stacking patterns that were integrated with dipmeter data to provide interpretations of depositional environment. Core data from key wells were fully integrated with borehole image log data in order to calibrate borehole image log interpretations. Where core was no longer available, photographic data were used to provide calibration input. In particular, ichnofabric analyses performed either upon core

From: LOVELL, M. A., WILLIAMSON, G. & HARVEY, P. K. (eds) 1999. Borehole Imaging: applications and case histories. Geological Society, London, Special Publications, **159**, 91–121. 1-86239-043-6/99/$15.00.

or photographic material provide invaluable data for the recognition of key stratal surfaces required for sequence analysis.

Borehole image log and dipmeter data used

The dipmeter data used during this study was acquired using the Schlumberger Stratigraphic High Resolution Dipmeter Tool (SHDT), Formation MicroScanner (FMS) and MicroImager (FMI) tools.

The FMI is a high-resolution microresistivity imaging device based on the FMS tool (Lloyd *et al.* 1986, Serra 1989). The FMI device has an array of 24 measuring electrodes on each of the four orthogonal pads, with similar arrays on laterally and vertically offset 'flaps' attached to each pad. The FMI tool has 5 mm (0.2 inch) diameter electrodes arranged in two rows 4 mm (0.15 inch) apart, providing eight strips of microresistivity data with a very high lateral and vertical sampling around the borehole. These data are processed to provide eight *c.* 7 cm-wide strips of microresistivity image, spaced in pairs at 90° around the borehole. This gives approximately 75% coverage of the borehole wall in the 8.5 inch hole section. The data also provide eight SHDT curves for standard dipmeter processing. The FMS tool has a very similar set-up, but comprises only four orthogonal pads, and lacks the 'flaps' which provide increased borehole coverage in the FMI tool.

The SHDT comprises two electrodes on each of four orthogonal pads, and records resistivity data at a 0.1 inch sampling rate. The resolution of the SHDT tool is much less than FMI or FMS, due to the smaller number of samples acquired.

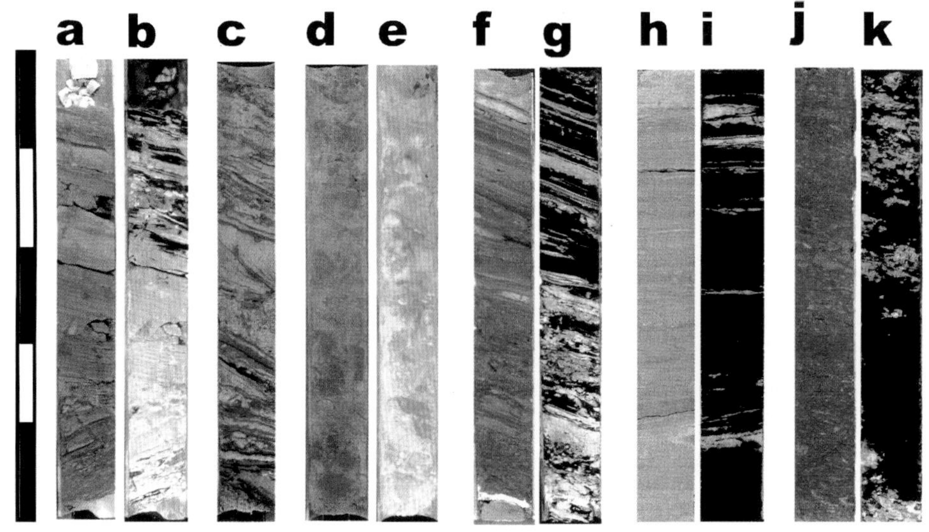

Fig. 1. Scale bar is 1 m.

a–b Plane light (a) and UV light photographs (b) illustrating laminated, cross stratified sandstone lithofacies Sl, with mud-draped forsets (Sd) in upper part of core interval. Minor bioturbation is evident in lower part of core interval, Sl(Sb).

c Plane light photograph illustrating sandstone with mud-draped inclined lamination Sd some intervals of which are partly bioturbated to give lithofacies Sd(Sb).

d–e Plane light (d) and UV light photographs (e) illustrating intensely bioturbated sandstone lithofacies Sb, with *Ophiomorpha* burrow evident in center of core interval.

f–g Plane light (f) and UV light photographs (g) illustrating, laminated heterolithic lithofacies Hl with minimal bioturbation (upper part of core interval). The lowermost half of the cored interval comprises intensely bioturbated lithofacies Hb at its base, overlain by lithofacies Hb(Hl) which contains some relict lamination.

h–i Plane light (h) and UV light photographs (i) illustrating laminated, laminated heterolithic lithofacies Hl within uppermost part of cored interval. The remainder of the core is dominated by laminated mudstone lithofacies Ml.

j–k Plane light (j) and UV light photographs (k) illustrating bioturbated mudstones (Mb) dominating lower part of illustrated core. Bioturbated mudstones in lower part of core pass upwards into mudstones with minor sand-filled burrows, and finally intensely bioturbated hereolithics (Hb) in the uppermost 30 cm.

Lithofacies characterization

Core observations revealed three broad lithofacies were present within the studied sections, i.e. sandstones, mudstones, and heterolithic successions comprising centimetre-decimetre scale interbedded sandstone and mudstone beds. Examples of some of these lithofacies are illustrated in Fig. 1. Initially, lithofacies identification was carried out independently of core using FMI images in conjunction with openhole log suites. The FMI interpretations were then calibrated against core data in order to quantify the accuracy of the technique. When the limitations of the technique were identified, and a scheme for lithofacies identification formalized, the technique was then applied to lower sample density FMS and SHDT data. During core examination, analysis of ichnofabrics visible in the core provided important information for interpretation of depositional environment.

Core ichnofabric analyses

The ichnology of sedimentary successions describes the texture and internal structures arising from bioturbation and bio-erosion (Ekdale & Bromley 1983). Ichnofabric analysis thus involves recognition of burrow types, their organization within sediments, and their temporal heirarchy (described as teiring). It provides detailed information concerning original depositional conditions, the composition and diversity of epi- and infaunal communities responsible for the bioturbation, and the changing nature of the substrate through the history of bioturbation (i.e. soft substrate versus firmground or lithified hardground communities).

In the marine and marginal marine setting, the ichnofabrics of different depositional environments are often distinctive. Figure 2 illustrates the ichnofabric characteristics of the different depositional settings observed within cores from the studied succession. Gradual changes in burrow communities may reflect transition through successive, spatially adjacent sedimentary environments, as is common within individual parasequences. Abrupt changes in ichnofabric may reflect the presence of a key stratal surface or sequence boundary, about which a landward or basinward shift in facies occurs.

Where core data are available, ichnofabric analysis can clearly provide detailed information of importance for the construction of correlative reservoir models. However although ichnofabric studies have been commonplace since the pioneering work of Frey (1975), their importance and routine application with the context of sequence stratigraphic reservoir correlation for the oil industry has become popular only during the last decade. In tropical climates, this often poses a problem for the reinterpretation of vintage cores, since lithologies are often poorly consolidated, and preservation of core is often very poor. The reader is directed to Pemberton (1992) for further references concerning ichnofabric analyses and their application to stratigraphical correlation.

The examination of standard A4 format core-photographs as a continuous 'photographic image log' in the workstation environment has revealed that when displayed at appropriate scales, even vintage core photographic plates contain sufficient data for detailed ichnofabric analysis. This has allowed reinterpretation of effectively 'lost' data within the context of recent advances in sedimentology. It has also been useful for calibration of fabrics visible in borehole image logs with those evident in core. Generally, bioturbation within borehole image logs is characterized by intense mottling at a variety of scales. Individual burrows are often evident within image logs, and in some cases distinctive burrow types, such as mud-lined *Ophiomorpha* burrows, can be identified.

The procedure for creating continuous photographic image logs involved scanning of A4 size presentation core photographs at a resolution of 200DPI using a desktop scanner. The 200DPI resolution approximates a true, 'core scale' resolution of approximately 50DPI. Scanned images were saved in TIFF format, loaded into the Z&S RECALL software as RGB colour data files, and one metre core sticks were interactively spliced into continuous depth-based core photographic image logs, and depth-matched to dipmeter and wireline data.

Once in the workstation environment, the core photographic image log can be manipulated by the application of image filters, or by dynamic normalization in order to enhance features present within the scanned cores. These can then be displayed at high magnification (<1:3 scale) on a workstation, where they can be used for ichnofabric analysis. The limit to the magnification that can be applied to the scanned images is dependent upon the magnification scale and image quality of the original core photographs. Poor quality core photographs will produce degraded ('pixely') images when enlarged. An example of a scanned core image used for ichnofabric analysis is provided in Fig. 3. The photographic image log may also be used for net sand determination by interactive thresholding of images in the workstation environment (Sovich *et al.* 1996).

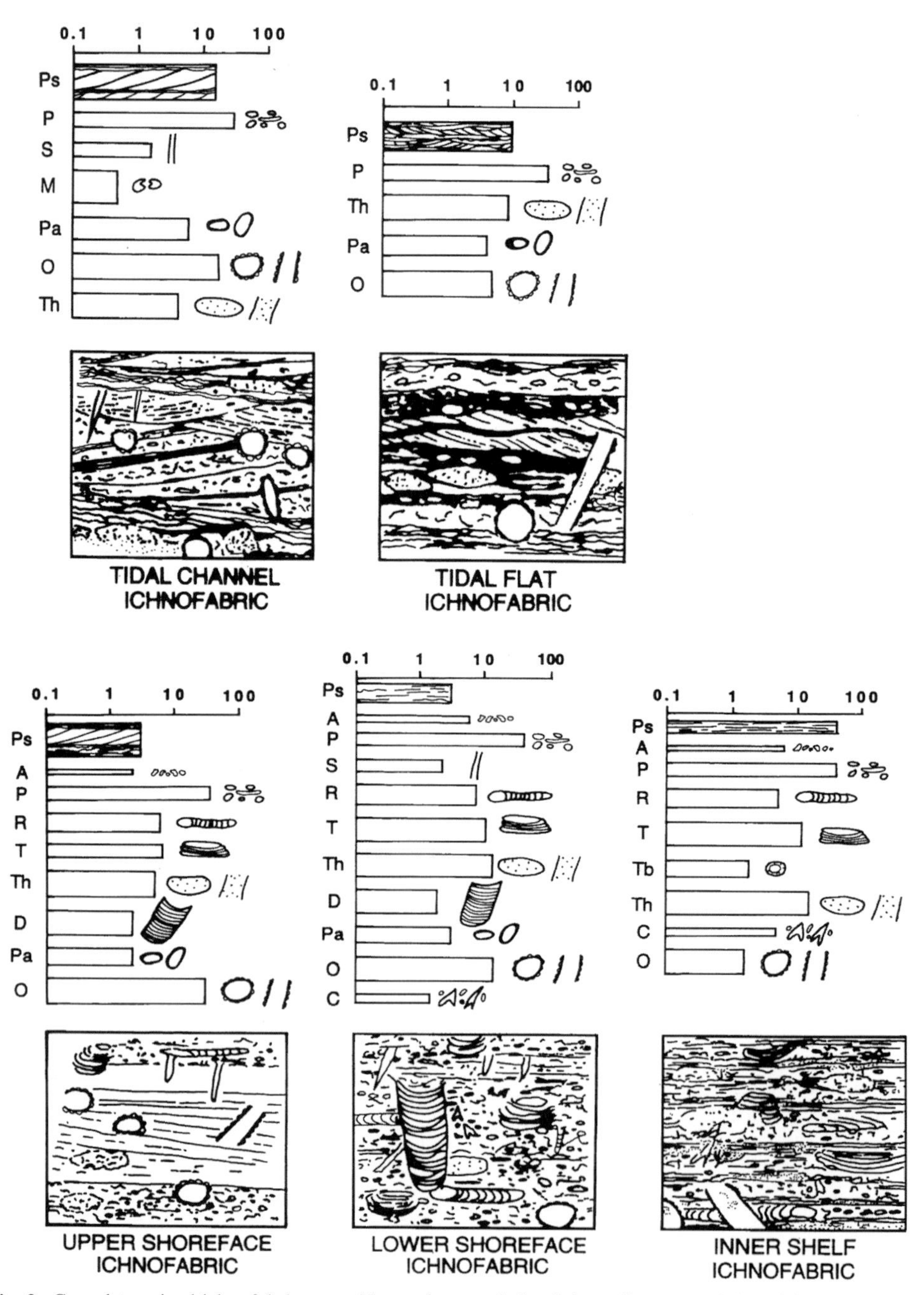

Fig. 2. Core determined ichnofabric assemblages characteristic of the sediments and depositional environments studied. Ps = Primary sedimentary structure, P = *Planolites*, S = *Skolithos*, M = *Monocraterion*, Pa = *Palaeophycus*, O = *Ophiomorpha*, Th = *Thallasandoides*, A = *Arenicolites*, R = *Rhyzocorralium*, T = *Teichichnus*, D = *Diplocraterion*, C = *Chondrites*, Tb = *Terebelina*. The scale indicates the percentage of different ichnofabrics present and percentage of primary structure present.

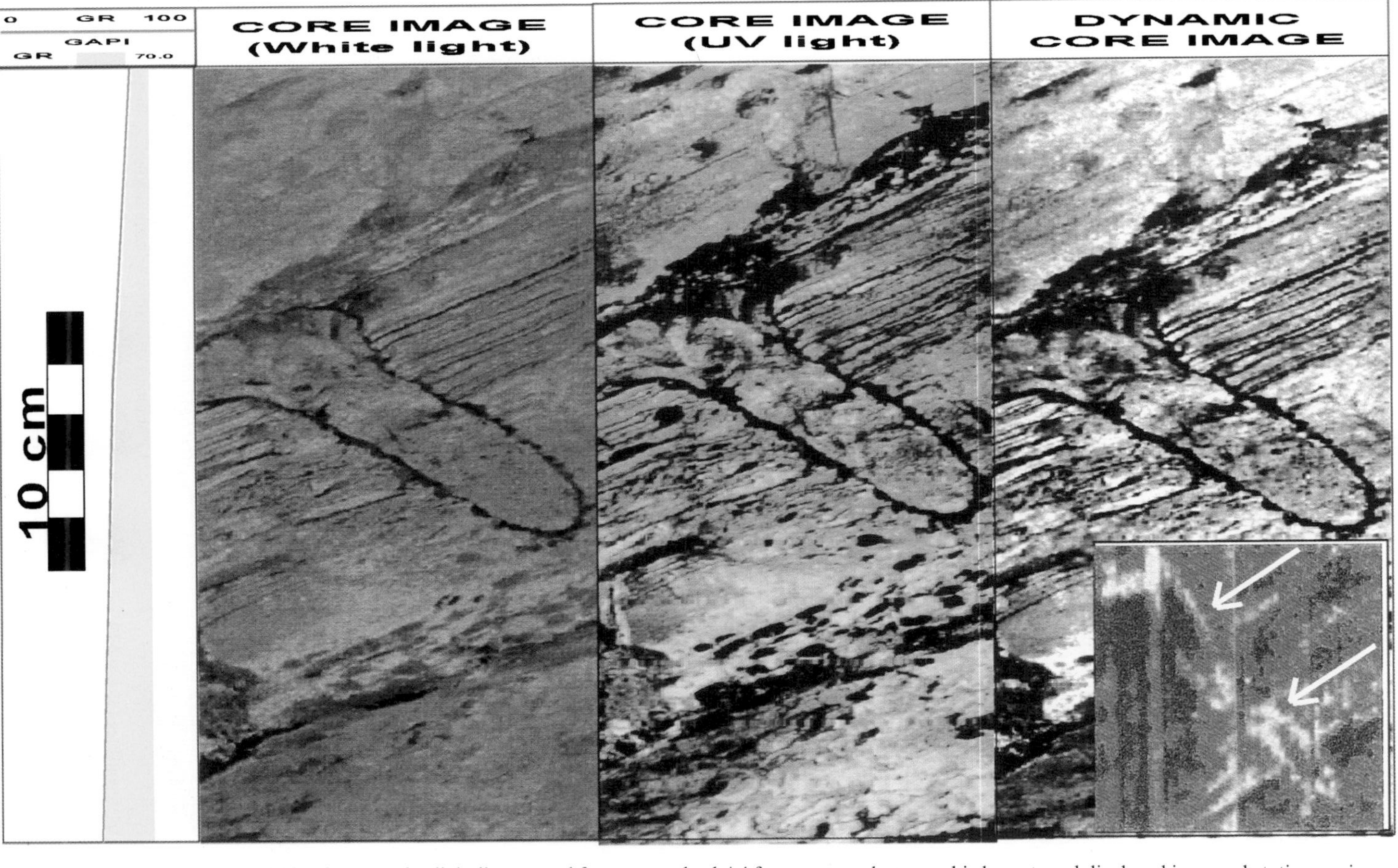

Fig. 3. Example of detail present within photographs digitally scanned from a standard A4 format core photographic layout, and displayed in a workstation environment. The three images presented are RGB colour, greyscale, and a dynamically normalized grey-scale image using a 20 cm normalization window. A large, distinctive *Ophiomorpha* burrow is clearly illustrated. The large scale of these burrows coupled with their distinctive pelleted clay margins makes them readily identifiable within borehole image logs (inset).

Lithofacies analysis from image log and wireline data

Sedimentological interpretation of FMI images and dipmeter data were carried out with the aid of gamma ray, density, neutron porosity, sonic and resistivity logs. The gamma ray log response was found to be one of the most useful aids to lithofacies interpretation when used in conjunction with borehole image logs. Where core calibration was possible, a series of log cut-offs were identified, which were used to help guide interpretation of lithologies. Identification of centimetre-decimetre scale bedded heterolithic sandstone-mudstone successions proved the greatest challenge, and confusion with mudstones or argillaceous sandstones was common. However, heterolithic lithofacies tended to be characterized by a high frequency of large resistivity contrasts at centimetre-decimetre scale within static image logs, and a log response intermediate between that of sandstones and mudstones. They were most readily identified where sandstone-mudstone inter-bedding occurred at scale of *c.* 20 centimetres, this often being sufficient to result in a clearly serrated gamma ray log profile. Typical examples of these log cut-offs are indicated in Table 1.

Lithofacies types were classified according to a simple scheme using mnemonics based upon interpreted lithology and contained fabric, the latter being determined from borehole image log and associated dip data. Examples of identified lithofacies are summarized in Table 2 and illustrated in Fig. 1.

The hierarchical combination of different litho-facies mnemonics was used to provide detailed descriptions of lithofacies types. In these descriptions, the enclosure of lithofacies mnemonics in parenthesis was used to denote the minor presence of a lithofacies type, or poor development or preservation of a sedimentary structure, e.g.

Table 2. *Lithofacies types used to describe the sediments within studied successions*

Inferred lithology/ grain size	Image log fabric	Lithofacies mnemonic	Interpreted sand content
Sandstone	laminated	Sl	>60%
	cemented	Sc	>60%
	cross bedded	Sx	>60%
	mud draped	Sd	>60%
	bioturbated	Sb	>60%
Heterolithics	laminated	Hl	10–59%
	bioturbated	Hb	10–59%
Mudstone	laminated	Ml	<10%
	bioturbated	Mb	<10%

Hb (Hl) bioturbated heterolithics with relict lamination or minor laminated intervals.

Sb (Sl) bioturbated sandstone with relict lamination.

Sl (Sb) laminated sandstone with minor bioturbation.

The hierarchical lithofacies nomenclature scheme applied to a description of lithofacies from borehole image logs was also used to provide a simple bioturbation index as illustrated for sandstones in Table 3.

Although the lithofacies nomenclature and bioturbation index described in Table 3, is based upon qualitative visual estimation, it was found that lithofacies could be assigned with a high degree of consistency, even in examples where several workers were integrating data from different wells. Figure 4 illustrates examples of different lithofacies types for sandstone and mudstone lithologies, together with their bioturbation index.

Having characterized lithofacies types using high resolution FMI and FMS borehole image logs, the technique was extended to lower

Table 1. *GR cut-off values for different lithofacies for a typical study well*

Typical study well
Sandstone (< 60 API)
Heterolithics (50–70 API)
Mudstone (65–75 API)
60 API 65 API 70 API

Table 3. *Simple bioturbation index (applied to sandstone lithofacies) based upon mnemonics scheme used for FMI interpretation of*

Lithofacies	Approximate degree of bioturbation	Bioturbation index
Sl	Minimal <10%	1
Sl (Sb)	approximately 25%	2
Sl/Sb	50%	3
Sb (Sl)	approximately 75%	4
Sb	near total 100 %	5

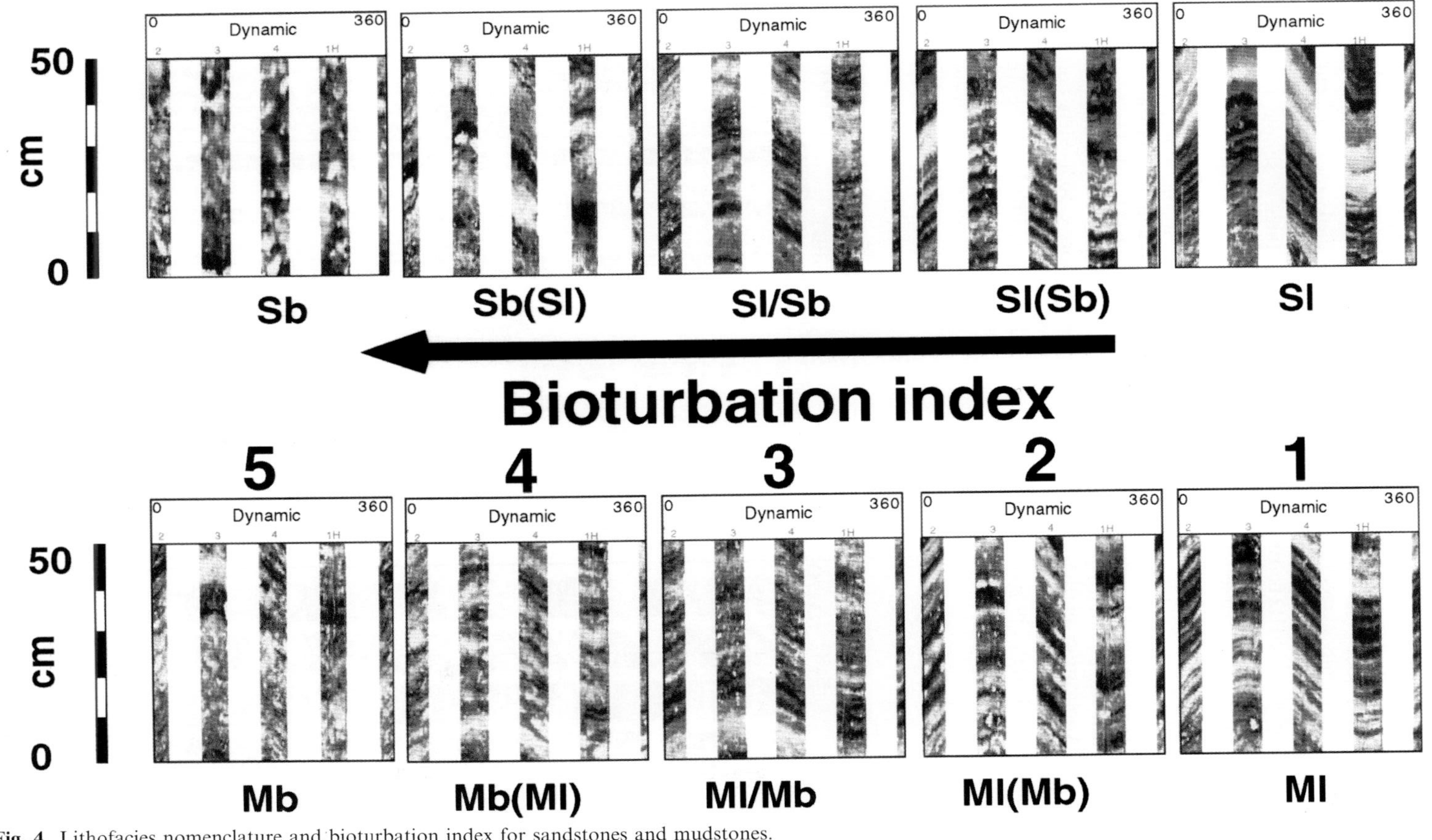

Fig. 4. Lithofacies nomenclature and bioturbation index for sandstones and mudstones.

resolution SHDT data. To develop confidence in SHDT interpretations and an understanding of their limitations, high resolution FMI data were resampled to create lower sample density 'pseudo-SHDT' data sets. The two data sets were then used to critically assess the degree to which sedimentary fabric and bioturbation could be evaluated from SHDT. Examples of these data sets are illustrated in Figs 5–7. It was found possible to classify sedimentary fabric within the SHDT data by evaluating curve activity, the spatial scale at which resistivity variations occur, the degree of correlation of resistivity elements around the borehole, and the orientation of surfaces manually picked from SHDT images. Application of the above outlined lithofacies scheme and five-fold bioturbation index to SHDT data proved successful, but interpretations sometimes appeared slightly biased towards end-member lithofacies, simply because of the lower sampling densities. e.g. it is relatively easy to identify a well laminated lithology, but minor amounts of bioturbation within a decimetre scale laminated bed may be difficult to detect in SHDT data.

Having identified the different lithofacies types present within borehole and dipmeter images, it was then a simple process to use these data to quantify net sand contents for studied sections.

Results have indicated that in terms of an overall comparison, considering both the lithology and contained fabric, an accurate match between FMI lithofacies classification and core lithofacies classification, can typically be achieved for *c.* 50%–90% of sandstone lithologies and *c.* 30%–50% of mudstones and heterolithics. The differences between the two techniques should not be considered in terms of an inadequacy of any particular technique

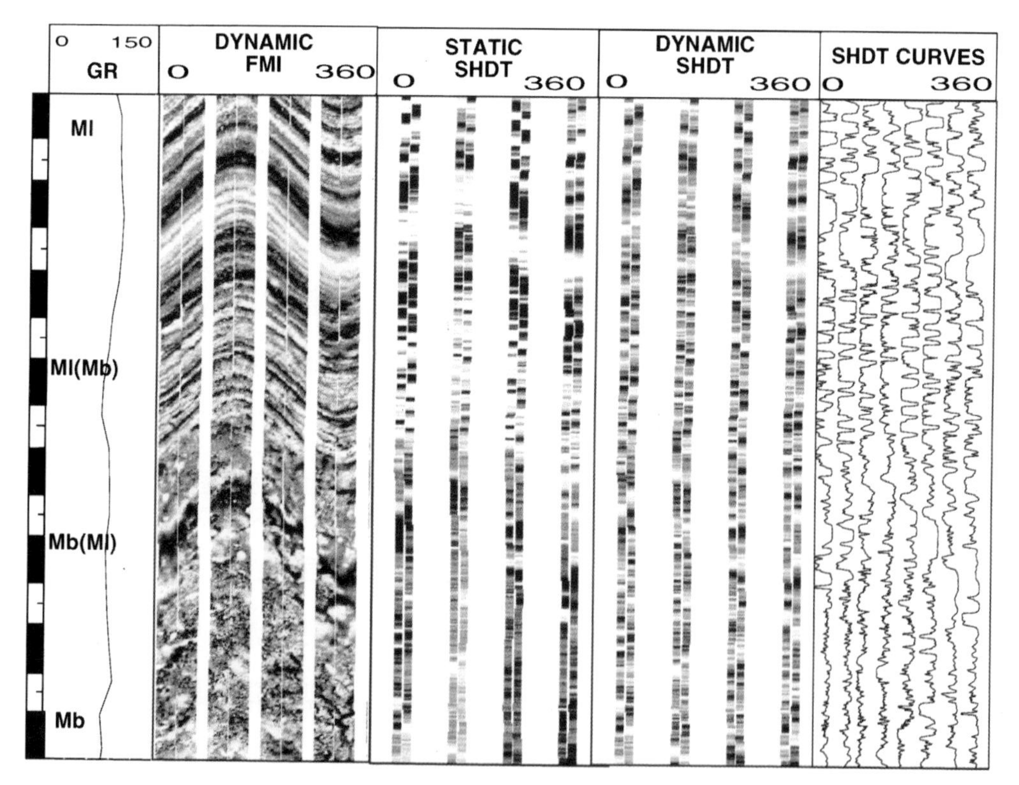

Fig. 5. FMI log and SHDT image strips for a succession of mudstones containing a variety of differing degrees of bioturbation. Well laminated mudstones (Ml) are clearly evident in the FMI images. The SHDT image strips and resistivity curves also reveal a high degree of correlation of features around the borehole within this interval. Bioturbated mudstones (Mb) are characterized by highly mottled fabric and a lack of features that are correlated around the borehole. The images displayed are characterized by an overall increase in bioturbation downward through the succession. Each division of the scale bar is 0.2 m.

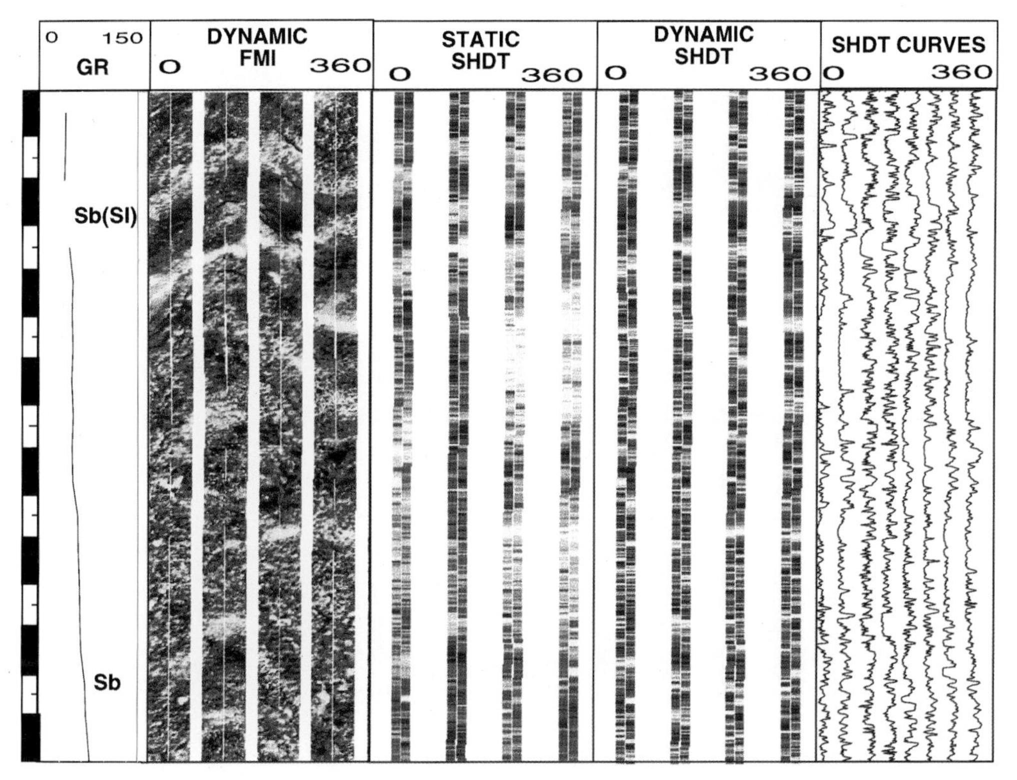

Fig. 6. FMI log and SHDT image strips for a bioturbated sandstone succession. Bioturbated sandstones with relict decimetre scale bedding fabric Sb(Sl), are clearly evident in the FMI images. A decimetre scale fabric traceable around the borehole is also interpretable from SHDT image strips for this interval, and some correlations are clearly evident within otherwise highly variable SHDT curves. SHDT image strips and resistivity curves within highly bioturbated sandstones (Sb), are characterized by a chaotic fabric of non correlated resistivity elements. Each division of the scale bar is 0.2 m.

to characterize the studied successions. Some differences clearly arise as a result of a fundamental conflict existing between features detectable during analysis of core within the visible light spectrum, and those features detected during microresistivity analysis of the borehole wall. The FMI log may identify resistivity contrasts which represent responses to features that are not apparent in the visible light spectrum. For example, where the visual appearance of the core is that of a massive (i.e. structureless sandstone), the FMI tool may detect subtle resistivity contrasts indicative of relict lamination or a bioturbation fabric. Similarly, core description may allow direct observation of complex fabrics and lithological variations (e.g. grain-size related) beyond the resolution of the FMI log. The image log and core-based techniques are complimentary, and should be used in conjunction with one another wherever possible. Calibration of lithofacies interpretations with core data reveals that interpretation conflicts may also arise as a result of:

- Poor borehole quality, and hence image log quality.
- Confusion of mudstones and heterolithics, which may have very similar wireline log response. This is a particular problem where sandstone beds are less than 20 cm in thickness. In this case, they are not always readily identified using the gamma ray log (or neutron porosity and density logs), and often there is no serrated profile within the gamma ray log to indicate the presence of sands interbedded with mudstones.
- Differences in the resolution scales of the FMI tool (mm–cm scale) and parameters which may be visually determined from core data (mm scale).

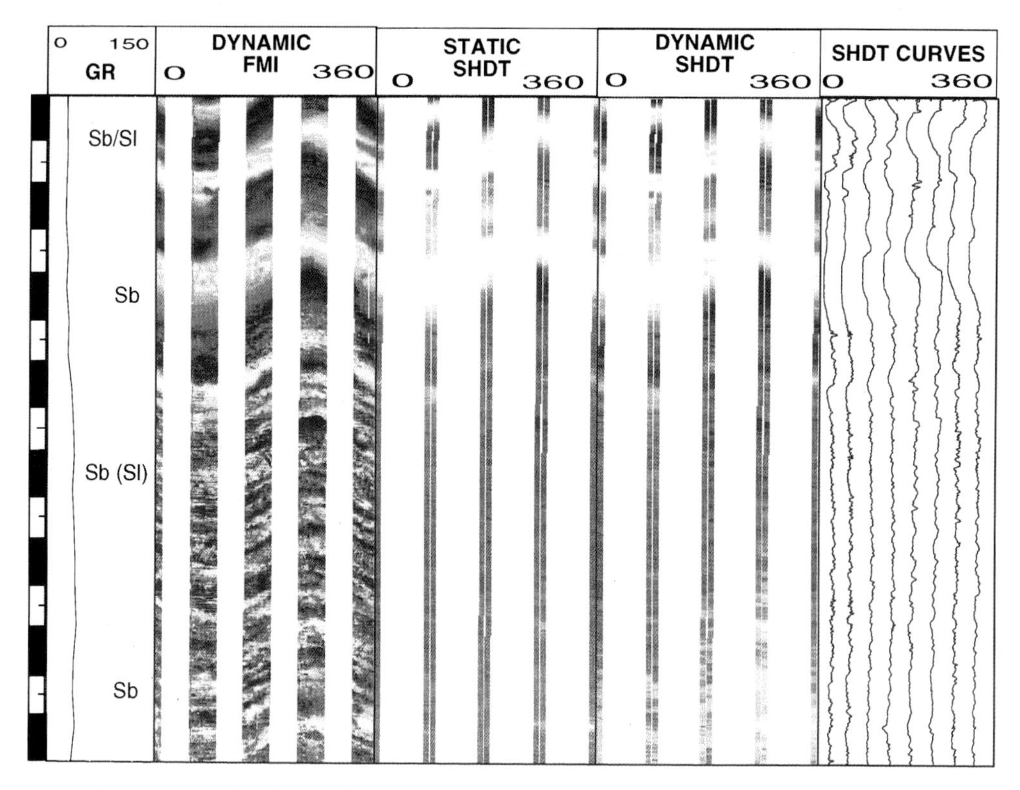

Fig. 7. FMI log and SHDT images through sandstones displaying varying degrees of bioturbation. The FMI images indicate sandstones to be highly bioturbated and contain fabrics at different scales. The uppermost sandstone is bioturbated, and clearly displays decimetre scale surfaces which are not sharp and are often disrupted. These fabrics are interpreted as decimetre scale bed boundaries, defining bioturbated sandstone beds in which little or no internal lamination is preserved. The lower half of the interval displayed also illustrates highly bioturbated sandstones which are characterized by an intensely 'mottled' image character. However, in this case a centimetre scale well defined fabric is clearly evident within both the FMI log and SHDT image strips. A slightly higher gamma ray log response within this interval suggests the presence of slightly more argillaceous lithologies. Each division of the scale bar is 0.2 m.

- The larger surface area surveyed by the borehole image log compared to the slabbed core surface may in some instances provide greater insight into the presence or relative abundance of certain fabrics such as bioturbation, cementation mottling etc.

Dip computations

Following identification of the different lithofacies types present within the studied strata, it was then necessary to develop a classification scheme to categorize dip features observed within the borehole image logs and dipmeters. Two types of dip computation were carried out on the available data.

'Manual dips' were computed by interactive dip picking of fully processed (speed corrected, depth-shifted, etc.) borehole images, or 'image strips' which were generated from dipmeter data. Manual dip picking was carried out using RECALL software, which provides an environment in which borehole image and dipmeter data may be processed, displayed on a workstation and interpreted. Manual dip picking involved the 'fitting' of sinusoids to surfaces evident in image logs. Due to the variable data quality within image log and dipmeter data, only a limited number of dip types were selected to categorize these surfaces, based upon comparison of FMI data with accompanying cores. Core calibration revealed that the studied successions were highly bioturbated. The bioturbation is typically evidenced as a pervasive, mottled fabric through the borehole image logs. The intensity of mottling increases with increasing bioturbation. Lithofacies were identified and categorized by using wireline log data to guide lithology interpretation, and borehole image or dipmeter logs to characterize sedimentary fabric. In order to obtain a shale bedding data set of suitable size for characterization of tectonic tilt, it was necessary to restrict the bioturbated bedding dip category to sandstone and heterolithic beds, i.e., all shales were characterized simply as shale, regardless of the extent of their contained bioturbation fabrics. Structural dip was removed from data sets to provide 'rotated' or 'sedimentary dips' for features identified using stereonet based techniques in a workstation environment. Dip feature categories used are summarized in Table 4.

In addition to manually picked dips, a pad-to-pad (PTP) interval correlation was performed upon SHDT curves from dipmeter and borehole image logs, using a 1 m correlation interval, 0.5 m step distance and a 60° search angle (referenced to borehole axis). The results of the automatic dip correlation typically compare only moderately well with the manually picked dips, from the borehole images or dipmeter images (Fig. 8). This is largely due to the high degree of bioturbation present within the studied sections. A wide variety of different 'quality' dip surfaces are typically evident.

In some cases, bioturbated intervals in image logs are characterized by highly discontinuous

Table 4. *Dip feature categories*

Lithological, structural or diagenetic type	Dip feature category used	Description	Colour coding
Mudstones	Shale Bedding	Confident bedding features, with consistent magnitudes, and variable azimuths, which generally conform to a known structural attitude. Note, all mudstone beds were characterized as shales regardless of degree of bioturbation.	Black
Sandstones	Parallel Lamination	Low true dip angle (typically much less than 10°), parallel – slightly discordant with respect to shale bedding. Moderate – good confidence features.	Blue
	Cross Bedding	High true dip angle (typically ≥15°), discordant to bedding. High confidence, variable dip azimuth.	Red
Heterolithics	Heterolithic Lamination	Low true dip angle (typically much less than 10°), typically parallel – slightly discordant with respect to shale bedding. Moderate – good confidence features within laminated lithologies of high resistivity contrast.	Cyan
Bioturbated Sandstones and Heterolithics	Bioturbated Bedding Surfaces	Highly variable true dip angle, since this category may have originated as either parallel lamination, heterolithic lamination or cross-bedding fabrics. They may be parallel – highly discordant to bed boundaries, and are characterized by zones of resistivity contrast several centimeters thick, rather than distinct, sharp surfaces. Moderate confidence only, with variable dip azimuth trends. Used for bioturbated surfaces within lithologies other than shales.	Green
Diagenetic Features	Cemented Horizons	High resistivity, variable dip magnitude and azimuth features, which may be concordant or highly discordant to bedding.	Magenta
Structural Features	Conductive Fractures	Typically high true dip angle, conductive features, discordant to bed boundaries etc. No bedding offset visible.	Yellow
	Resistive Fractures	Typically high true dip angle, resistive features, discordant to bed boundaries etc. No bedding offset visible.	Orange
	Fault	High true dip angle features, discordant to bed boundaries, visible bedding offset.	Purple

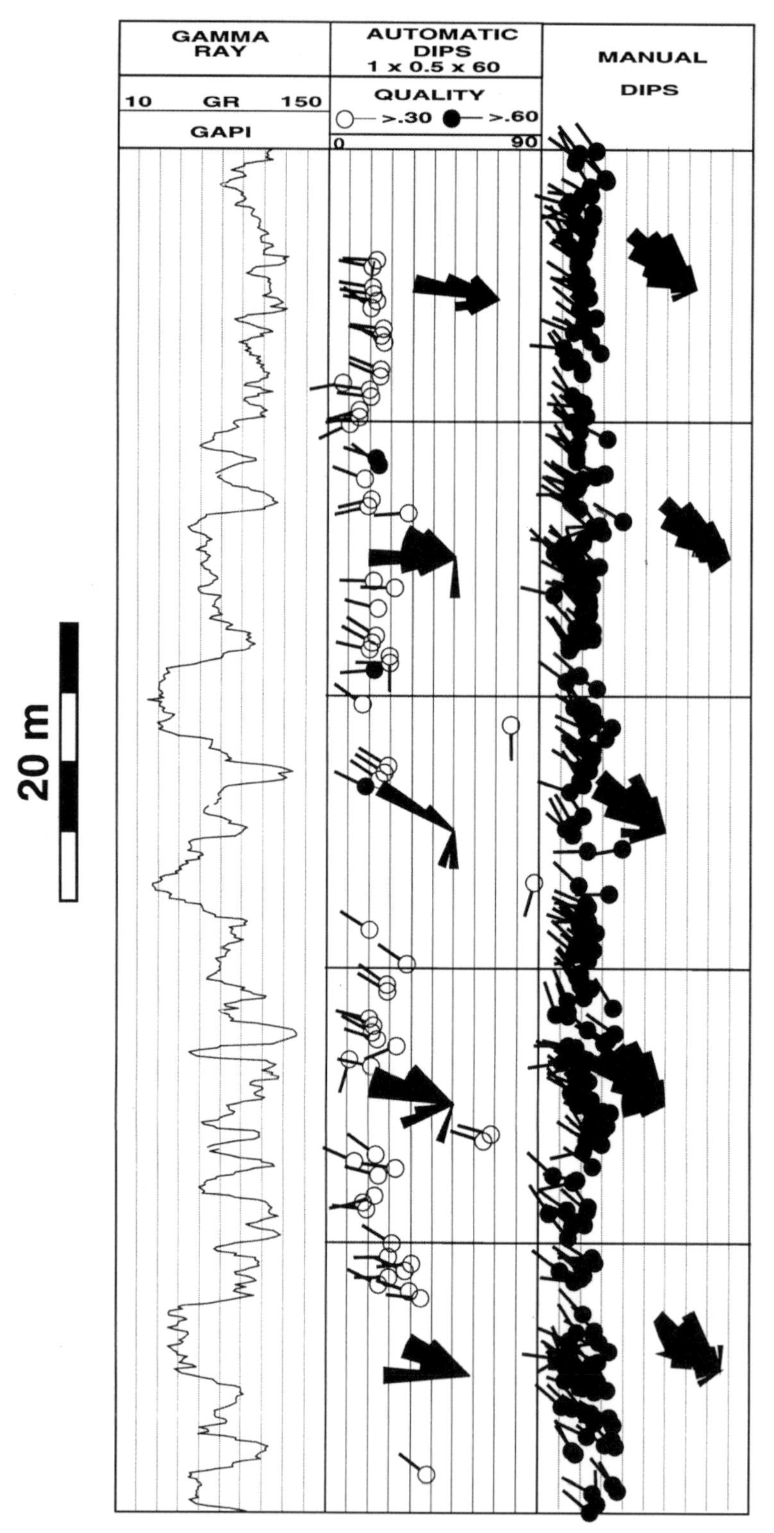

Fig. 8. Comparison of manual and automatic dips through a typical study section. Automatic dips calculated using 1 m window × 0.5 m step size × 60° search angle. Low quality automatic dips correspond to the presence of bioturbated lithofacies.

resistivity elements. However, in other intensely bioturbated intervals, primary bedding surfaces are clearly still present (inclined at a variety of dip angles), but are characterized by resistivity contrasts which occur over zones several centimetres thick, rather than by sharp, distinct contrasts. Manual interpreted dips from such intervals are thus of moderate confidence only and mathematically resolved 'automatic dips' of even lower confidence.

All orientation data cited is referenced using the standard convention of dip/dip azimuth. For example 10°/100° indicates a dip of 10° (measured from the horizontal) towards 100° (referenced clockwise from north). Borehole orientation data follows a similar convention of deviation/deviation azimuth. For example, 88°/045° indicates a 88° deviation from the vertical towards 045° (NE).

Lithofacies associations

Observed vertical transitions in lithofacies types identified in borehole images enabled lithofacies to be grouped into genetically related successions of strata or *Lithofacies associations*, which have some environmental significance (Collinson 1969, Walker 1992). Five Lithofacies Associations were identified within the studied data set, and reflect deposition within shelf, lower shoreface, upper shoreface, tidal channel and tidal flat environments. The deposits are characteristic of those typically found along shallow marine siliciclastic shorelines.

Lithofacies Association I. Lithofacies Association I is argillaceous, mainly comprising Ml (Mb) and Ml/Mb and Mb(Ml) with lesser Ml (FMI bioturbation index 1–4). Rare, thin (decimetre scale) units of Sb or Hb may also be present. The mudstones occur at the base of successions displaying overall upward cleaning (and coarsening) gamma ray log trend. A lower degree of bioturbation is evident within the basal part of this Lithofacies Association. In borehole image logs, bioturbation fabrics are represented by resistive mottling, or by distinctive horizontal burrows. Individual burrows are resistive relative to surrounding mudstone lithologies, suggesting the burrows are probably sand filled. Calibration with core indicates that larger horizontal burrows are most likely to represent *Thalassanoides* or *Teichichnus*.

Generally, Lithofacies Association I forms relatively thick deposits (up to 8 m, average *c.* 3 m), and is characterized by blocky to serrate, overall high gamma-ray log response (>70 API), reflecting the presence of a predominantly argillaceous succession of lithofacies types. Gamma ray log response (typically >70 API) may decrease slightly upward. Lithofacies Association I typically passes upward into heterolithic deposits of Lithofacies Association II. (see Fig. 9)

The mudstones of Lithofacies Association I display low sedimentary dip (typically <5°), with wide ranging dip azimuths (covering 360° spread) indicative of their original deposition as parallel stratified sediments upon a flat lying substrate.

Comparison with cored sections reveals Lithofacies Association I to comprise the deposits of marine shelf environments. Figure 9 provides an example of lithofacies and dip-data through Lithofacies Association I. The presence of argillaceous lithofacies with lamination or relict lamination fabrics reflects deposition within low energy muddy environments. Sedimentation was dominated by suspension fallout of argillaceous material, resulting in the accumulation of laminated mudstone lithofacies (Ml, etc.). Very fine-grained sediments accumulate on the shelf and typically grade into the sands and silts of the lower shoreface. A high proportion of bioturbated lithofacies within parts of the interpreted shelf deposits indicate extensive colonization of the sediment substrate was possible, perhaps close to fair-weather wave base. The preserved relict lamination may reflect slightly deeper shelf conditions below storm wave base.

Lithofacies Association II. Lithofacies Association II mainly comprises heterolithic lithologies, predominantly Hb and Hb (Hl), with minor Mb, Ml, Hl and rare Sb or Sb (Sl). Heterolithic lithologies comprise centimetre-decimetre scale interbedded sandstones and mudstones with high degree of bioturbation, and rare thin sands (decimetre scale). These typically form units 3–10 m thick but may reach thickness of 15–20 m. They occur towards the base of facies successions which display overall upward cleaning (and coarsening) gamma ray log trend. Fine-grained lithofacies predominate within the lower part of this lithofacies association (Mb and Hb with minor preservation of laminated lithofacies Ml and Hl), and beds thicken- and coarsen-upward into interbedded fine-grained, bioturbated, heterolithic lithofacies Hb. Minor beds of Sb may be present in the upper 'coarser' parts of the successions.

Beds are typically characterized by an abundance of bioturbated bedding fabrics, and removal of tectonic tilt reveals these to have

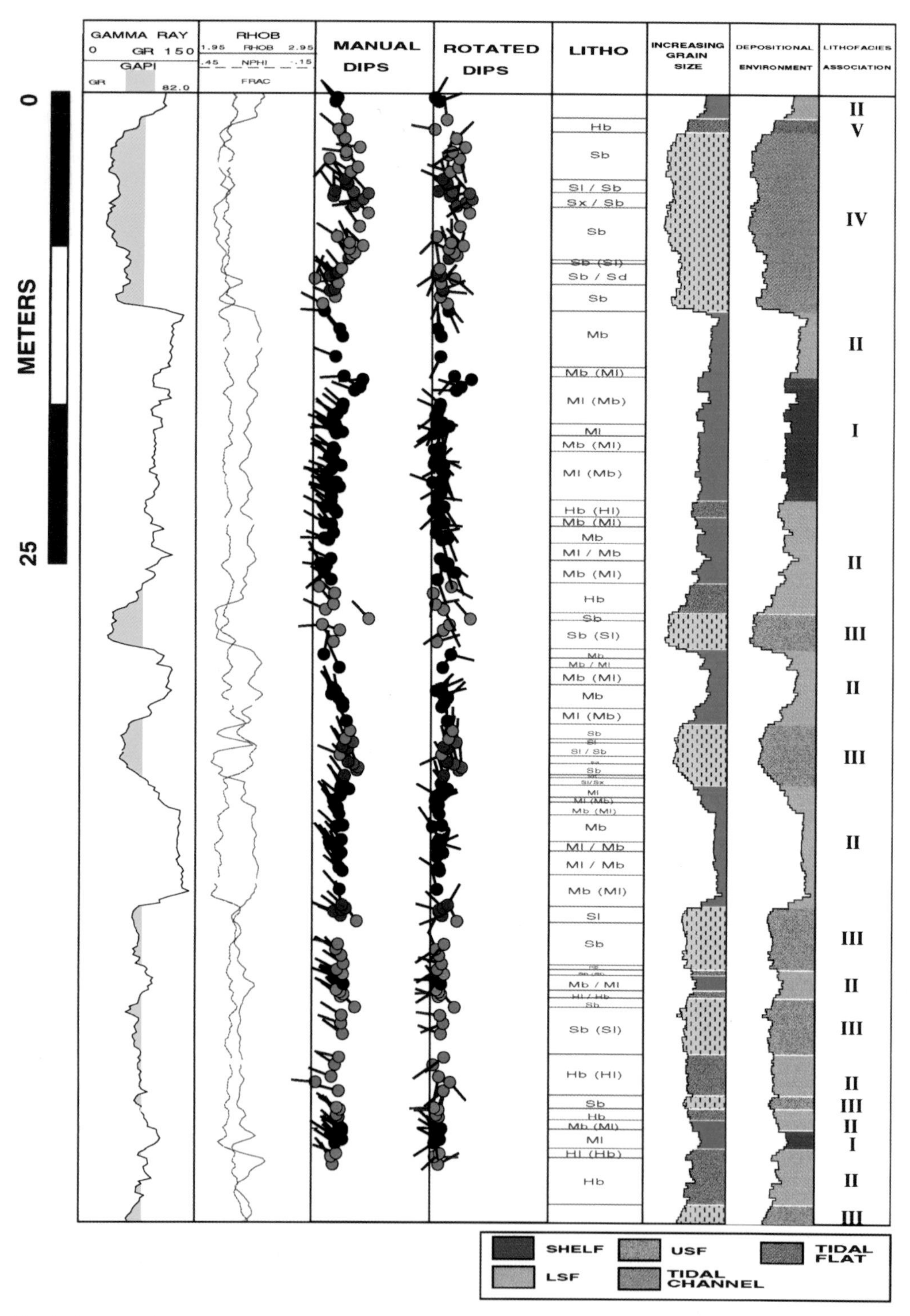

Fig. 9. An example of dip data through Lithofacies Associations I–V representing deposits of shelf, lower shoreface (LSF), upper shoreface (USF), tidal channel and tidal flat environments. Note elevated dips within tidal channel deposits. Relative grain size variations are inferred from gamma ray log trends following core observations.

had <10° sedimentary dip. Removal of structural dip reveals interbedded mudstone bedding fabrics to be characterized by packages of gently discordant bedding features typically with wide ranging (up to 360° spread) dip azimuths. Comparison with cored sections reveals Lithofacies Association II comprises the deposits of lower shoreface environments. Examples of dip-data through Lithofacies Association II are illustrated in Figs 9–10. A high proportion of bioturbated lithofacies within parts of the interpreted shoreface deposits indicate extensive colonization of the sediment substrate. The intense bioturbation of heterolithic sediments may increase vertical communication (i.e. vertical permeability) through this lithofacies. In borehole image logs, the main fabric is one of intense mottling, and often the images are characterized by a patchy fabric of centimetre-decimetres scale resistivity variations. Individual decimetre scale sand beds often contain distinctive burrows with serrated conductive margins indicative of the pelleted form *Ophiomorpha*. Core calibration also indicates that where larger burrows are evident in image logs they are likely to represent *Thalassanoides*, *Diplocraterion, Rhyzocorralium* or *Teichichnus*. However, these individual burrow types cannot be distinguished using image logs alone.

Lithofacies Association III. Lithofacies Association III mainly comprises sandstone lithologies Sb with minor Sb (Sl). Rarely lamination fabrics are preserved within lithofacies Sl (Sb) or Hl (Hb). The lithofacies association is typically 5–10 m in thickness, but may form successions up to 20 m thick. The sandstones rest gradationally upon Lithofacies Association II. FMI derived lithofacies types comprise less argillaceous, perhaps coarser grained lithologies than Lithofacies Association II.

Comparison with cored sections reveals Lithofacies Association III to comprise the deposits of upper shoreface environments. Good examples of dip-data through Lithofacies Association III are illustrated in Fig. 9. A low gamma ray log response within upper shoreface deposits of Lithofacies Association III indicates that they contain a significant proportion of clean, potentially high reservoir quality sandstones. A relatively homogeneous fabric is predicted to have been generated by intense bioturbation, and the sand-body geometry would be sheet-like and continuous. Bioturbation is characterized by intense mottling and disruption of bedding fabrics. Occasionally large individual burrows can be recognized in FMI images, some of which display the pelleted margins characteristic of *Ophiomorpha*. Core calibration indicates that detailed characterization of larger, non-pelleted, sand filled horizontal and vertical burrow types visible in images is not possible. However, very generally, larger sand filled burrows observed in FMI typically correspond to *Thallasanoides* and *Diplocraterion* traces within the same cored interval.

The upper shoreface lithofacies association is characterized by moderate angle sedimentary dips (typically <10° but rarely up to 15°). Dip categories present are dominated by bioturbated bedding, with lesser amounts of parallel lamination, and rare heterolithic bedding fabrics. Cross-bedding is rarely preserved. Following removal of structural dip, bioturbated bedding surfaces typically show highly variable dip azimuths, reflecting deposition as low amplitude, strongly three dimensional bedforms. However, sometimes northwesterly and northerly trends were found to predominate within these distributions. This may reflect sediment transport in a direction oblique to the palaeocoastline, which has been inferred, from previous regional scale studies of sedimentary facies variation and structural development, to have had broadly NNE–SSW orientation. The poorly preserved parallel laminations within sandstone, heterolithic bedding and cross-bedding, also typically display highly variable dip azimuths.

Lithofacies Association IV. FMI derived lithofacies are sand dominated, mainly comprising Sb, Sl/Sx, Sb(Sl) and minor Sd. Lithofacies Association IV occurs as units <3 m to >15 m in thickness. Lithologies containing some preserved sedimentary structure e.g. Sb(Sl), Sl/Sx, Sl(Sb), dominate successions, often comprising over 80% of their total. Lithofacies with relict lamination tend to pass upward into more intensely bioturbated sandstone lithofacies (Sb).

FMI interpretation of manually picked dips indicates the presence of cross stratification within Lithofacies Association IV. Bedding surfaces separating cosets of strata typically occur at decimetre to metre scale, whereas inclined 'cross-bedding' surfaces occur at a decimetre scale.

Dip data sets for Lithofacies Association IV are typically small, and they are dominated by bioturbated bedding surfaces. Clearly defined cross bed surfaces, parallel lamination fabrics, and minor heterolithic bedding fabrics may also be present. Few distinct dip patterns (e.g. upward increasing or decreasing bedding dips) can be identified within Lithofacies Association IV, but lower angle dips are common within its upper parts. In some cases Lithofacies Association IV is characterized by a marked erosive base.

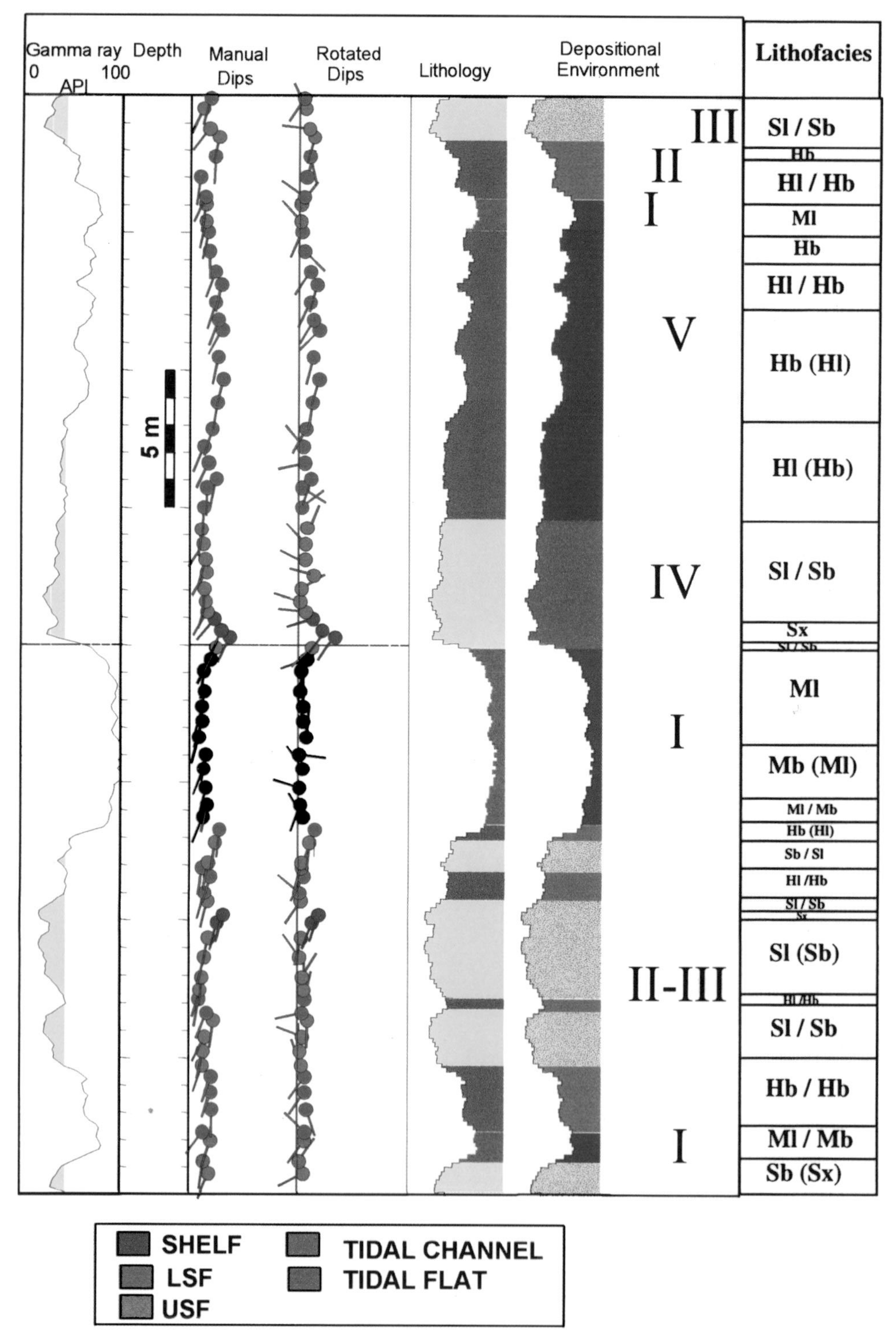

Fig. 10. An example of dip data through Lithofacies Associations I–V representing deposits of shelf, lower shoreface (LSF) upper shoreface (USF), tidal channel and tidal flat environments.

Sediment dispersal patterns interpreted for Lithofacies Association IV are variable, and may comprise unimodal, strongly bimodal or polymodal azimuthal distributions. However, within thicker deposits, unimodal azimuths tend to predominate. Comparison with cored sections reveals Lithofacies Association IV comprises the deposits of tidal channel environments. The bioturbated sandstone lithologies often contain distinctive, large, *Ophiomorpha* burrows, which are distinguished from other burrows in FMI images by their conductive (i.e. clay rich), serrated margins. Calibration with core indicates that large burrows without clay margins are most likely to represent *Thalassanoides*. An example of dip-data through a sharp-based example of Lithofacies Association IV is illustrated in Fig. 9.

The term 'tidal channel' may encompass a range of features and sub-environments characteristic of the near-shore zone. The discrete nature of the sandstone bodies assigned to this lithofacies association, and their sequential relationships with sediments of lithofacies associations representative of lower shoreface, marine shelf and tidal flat deposits, suggest that they are most likely to represent tidally-influenced estuarine or distributary channels, rather than tidal inlets dissecting barrier island deposits. Tidal forces would have dictated flow within channels and flow directions would therefore have reversed at the transitions between the ebb and flood tidal stages. The impact of flow reversal on channel morphology, sediment accumulation and orientation of preserved sedimentary fabrics may have been limited if channel activity was restricted to a particular tidal flow stage (Dalrymple *et al.* 1992). In this case, azimuthal distributions with a strongly dominant mode might be expected for dip data. Within the studied data sets, sedimentary fabrics within tidal channel deposits may display a variety of azimuthal distributions, ranging from polymodal to unimodal. Tidally influenced estuarine and distributary channels are also frequently characterized by progressive lateral migration across low relief inter-tidal and shallow sub-tidal flats. This could account for polymodal azimuthal distributions observed within some channel fill successions.

Evidence of primary stratification within tidal channel deposits testifies to the development and migration of bedforms, with the locally abundant cross-bedding indicating dunes and sand waves. The common presence of lithofacies with relict cross stratification Sb(Sl) within the lower parts of some channel fills reflects the development of tractional bedforms upon a moving substrate during high energy flows. A higher proportion of bioturbated lithofacies (Sb) within tidal distributary channel successions infers that conditions within the tidal channels were suitable for extensive frequent faunal colonization of substrates, presumably during periods of low energy discharge, or temporary channel abandonment. Well developed (i.e. thoroughly churned) lithofacies Sb also commonly occurs near the top of distributary channel fills where it may reflect faunal colonization during shoaling of channel systems prior to their complete abandonment.

The low gamma ray log response, typical of Lithofacies Association IV, indicates that channel fill successions contain a significant proportion of clean, potentially high reservoir quality sandstones. In this respect, the orientation of this lithofacies association has important implications with respect to development of water flood strategies. Channels having their long axis oriented perpendicular to the advancing water flood front may act as thief zones, resulting in by-passing of oil within adjacent lithofacies associations. Channels oriented with their long axis parallel to the advancing water flood front may be subject to more efficient sweep. Dipmeter data obviously provide important insight as to bedform orientation and hence flow within channels. However, interpretation of channel drainage directions is often hampered by small dip data sets from intensely bioturbated channel fill successions.

The tidal channel deposits of Lithofacies Association IV most commonly rest upon the shoreface deposits of Lithofacies Association III, or occur interbedded with tidal deposits of Lithofacies Association V.

Lithofacies Association V. Lithofacies Association V comprises heterolithic successions of strata dominated by Hl, HL(Hb), and Hl/Hb, with lesser Sb, Sb (Sl), Mb and Mb(Ml). The lithofacies are thinly interbedded (decimetre-metre scale), and are characterized by overall high, but variable and distinctly serrate gamma ray log response. Lithofacies Association V is characterized by low angle sedimentary dip ($<10^{\circ}$ dip), with wide ranging azimuth indicative of their original deposition as low angle-parallel stratified sediments upon a flat lying substrate.

The deposits of Lithofacies Association V are distinguished from those of Lithofacies Association I and II by:

- Increased interpreted sandstone / heterolithic ratio

- Frequent preservation of lamination fabrics (as seen in FMI), especially within heterolithic strata
- Association with sandstone lithofacies of interpreted tidal channel origin

Comparison with core data indicates that Lithofacies Association V comprises the deposits of tidal flats. Examples of dip-data through Lithofacies Association V are illustrated in Figs 9–10.

Tidal flats are low relief environments which may be developed within a variety of marginal to shallow marine settings, including estuaries, the margins of back-barrier lagoons and on the open coast. They are commonly subdivided into intertidal and subtidal flats, with the transition (occurring around the mean low water mark) reflected by a progressive increase in the influence of tidal currents on sedimentation in an offshore direction (Weimer *et al.* 1982). The intertidal zone has been further subdivided on the basis of sedimentation style into proximal mud flats, which pass gradationally into mixed mud flats and sand flats in a seaward direction (Dalrymple 1992). Heterolithic sediments are the main product of tidal flat processes, with the fine interbedding of sand and mud reflecting the alternation of traction currents and phases of suspension deposition characteristic of the tidally-influenced flow regime. The interbedded sand and mudstone successions characteristic of Lithofacies Association V reflect the presence of thinly interbedded sandflats and mud flats. Calibration with core indicates that larger burrows visible within FMI images through Lithofacies Association V are most likely to represent *Ophiomorpha* and *Thalassanoides*. A background mottling within FMI images through Lithofacies Association V is often generated by smaller burrow forms including *Planolites* and *Palaeophycus*. With the exception of the distinctive *Ophiomorpha*, identification of specific burrow types is not feasible from FMI images alone.

Limitations to the identification of tidal facies. The identification of tidal deposits within uncored intervals is commonly problematical. Tidal channel or tidal flat deposits cannot be readily distinguished from shoreface/shelf facies successions on the basis of log profile alone. For example, serrate overall high gamma ray log profiles may be developed in both tidal flat and lower shoreface/shelf deposits. Recent studies (Prave *et al.* 1996) have also demonstrated how 'upward coarsening' log profiles may be developed in both environmental settings, further complicating interpretation. Although log motif is equivocal in terms of interpretation, many 'upward cleaning' gamma ray profiles within the studied sediments also appear to display a succession of lithofacies types and dips which core calibration reveals are typical and shelf-shoreface lithofacies associations. These intervals were thus most readily interpreted as shelf-shoreface lithofacies successions.

Sedimentary tidal indicators can include centimetre-scale mud drapes, frequent reactivation surfaces and tidal bundles. Mud draped sandstones are evident near the base of some tidal channel deposits identified from both core and image logs, and some intervals may comprise inclined heterolithic strata. The presence of cross-bedding fabrics alone is not sufficient to assign a tidal channel interpretation to a lithofacies association. Cross bedding fabrics may be generated directly within the upper shoreface environment or may reflect the incorporation of minor tidal channel deposits into shoreface sandstone bodies. Channelized tidal deposits of Lithofacies Association IV may be distinguished from the upper shoreface deposits of Lithofacies Association II by somewhat higher dip magnitudes. However, since bioturbation is abundant within both these Lithofacies Associations, the amount of dip data upon which interpretations may be made is often quite small, reducing interpretation confidence.

More significantly, the identification of cross-bedding may allow depositional environments to be distinguished on the basis of palaeocurrent distributions. Tidal deposits are often characterized by bimodal palaeocurrent orientations reflecting the systematic reversal of tidal currents and the associated migration of sandy bedforms in opposing directions. However, as the flow regime within a particular tidal channel may be dominated by a single phase of the tidal cycle (e.g. 'ebb-dominated'), indicators of variable or reversing flows may prove to be small scale and features recording them therefore beyond the resolution of the FMI images.

Confident identification of tidal flat deposits posed the greatest problem, with these sediments containing very similar heterolithic lithofacies to those which dominate lower shoreface deposits. Tidal flat interpretations were often influenced by an association with sandstone bodies interpreted as the deposits of tidal channels, and by a high degree of preserved lamination fabrics within heterolithic lithologies. In the absence of core, the interpreted distribution of tidal flat facies was thus partly a function of interpreted sequential patterns.

Table 5. *Tidal facies interpretation confidence scale*

Confidence	Criteria
Grade 1 (very high)	Tidal interpretations from image log and dipmeter data can be directly corroborated by core recovered from the same interval in the well.
Grade 2 (high)	No core is available for the study well. Units interpreted as of tidal origin are clearly correlatable in terms of a suite of open-hole log responses with cored intervals interpreted as tidal deposits in adjacent wells.
Grade 3 (moderate)	No core is available. A combination of log response, contained fabric, dip data and association with other lithofacies suggests a tidal origin. The interval is broadly correlateable with units in other wells known from core data to contain tidal deposits. However, no direct well log correlation can be established.
Grade 4 (low)	No core is available. A combination of log response, contained fabric, dip data and association with other lithofacies suggests a tidal origin. No direct log correlation can be established with tidal facies interpreted from a cored well. The interval may correlate with other low confidence tidal facies interpretations.

In order to document the interpretation confidence, and hence 'risk' associated with interpretations, it was found necessary to rank interpretation confidence on the basis of gamma-ray log response, sequential lithofacies successions, dip variations and core information. The ranked scheme applied could be equally useful in any depositional setting where an interpretation conflict arises from the use of dipmeter or borehole image logs. The ranked scheme applied is summarized in Table 5.

Environmental synthesis

Borehole images and dipmeter data together with wireline log character indicate that the studied sediments comprise five lithofacies associations, indicative of deposition within a shallow marine setting dominated by shoreface and related environments (Fig. 11). Marine shelf sediments (Lithofacies Association I) are almost exclusively mudstones deposited from suspension. Lower shoreface deposits (Lithofacies Association II) are predominantly a heterolithic assemblage of thinly bedded sandstones and mudstones that were probably deposited by a variety of fair-weather and storm-generated processes. Upper shoreface deposits (Lithofacies Association III) are predominantly sand-prone and reflect reworking, mainly by near-shore wave- and storm-dominated processes. However, the shoreface setting also appears to have been subject to significant tidal influence, with evidence of tidally-influenced estuarine or distributary channels (Lithofacies Association IV) and the local development of tidal flats (Lithofacies Association V). Sediments of all settings may be overprinted be varying degrees of bioturbation which reflect the relative intensity of infaunal reworking.

The lithofacies and lithofacies associations identified from borehole image logs and dipmeters form the basic building blocks required for sequence analysis. The repeated transitions from shelf or lower shoreface depositional environments into those of upper shoreface deposits have enabled subdivision of the studied strata into *lithofacies successions* or parasequences. A parasequence is defined as a relatively conformable succession of genetically related beds or bedsets bounded by marine-flooding surfaces or their correlative conformities (Van Wagoner *et al.* 1990; Swift *et al.* 1991). An idealized facies relationship within a shallow marine upward-coarsening parasequence is summarized in Fig. 12 (from Prave *et al.* 1996). Parasequences can be grouped into parasequence sets which are defined as genetically related groups of parasequences forming distinctive stacking patterns and bounded by major marine-flooding surfaces. Progradational facies stacking patterns within parasequence sets have been identified where the thickness of facies associations forming the upper part of the shelf-shoreface facies succession (lower shoreface-upper shoreface) increase progressively upward through the stacked parasequences. i.e. the net sand content increases upward through the successive parasequences within a parasequence set. Retrogradational facies stacking patterns within parasequence sets have been identified where the thickness of shelf mudstones increases progressively upward through the stacked parasequences. That is, the net sand content decreases upward through the successive parasequences within a parasequence

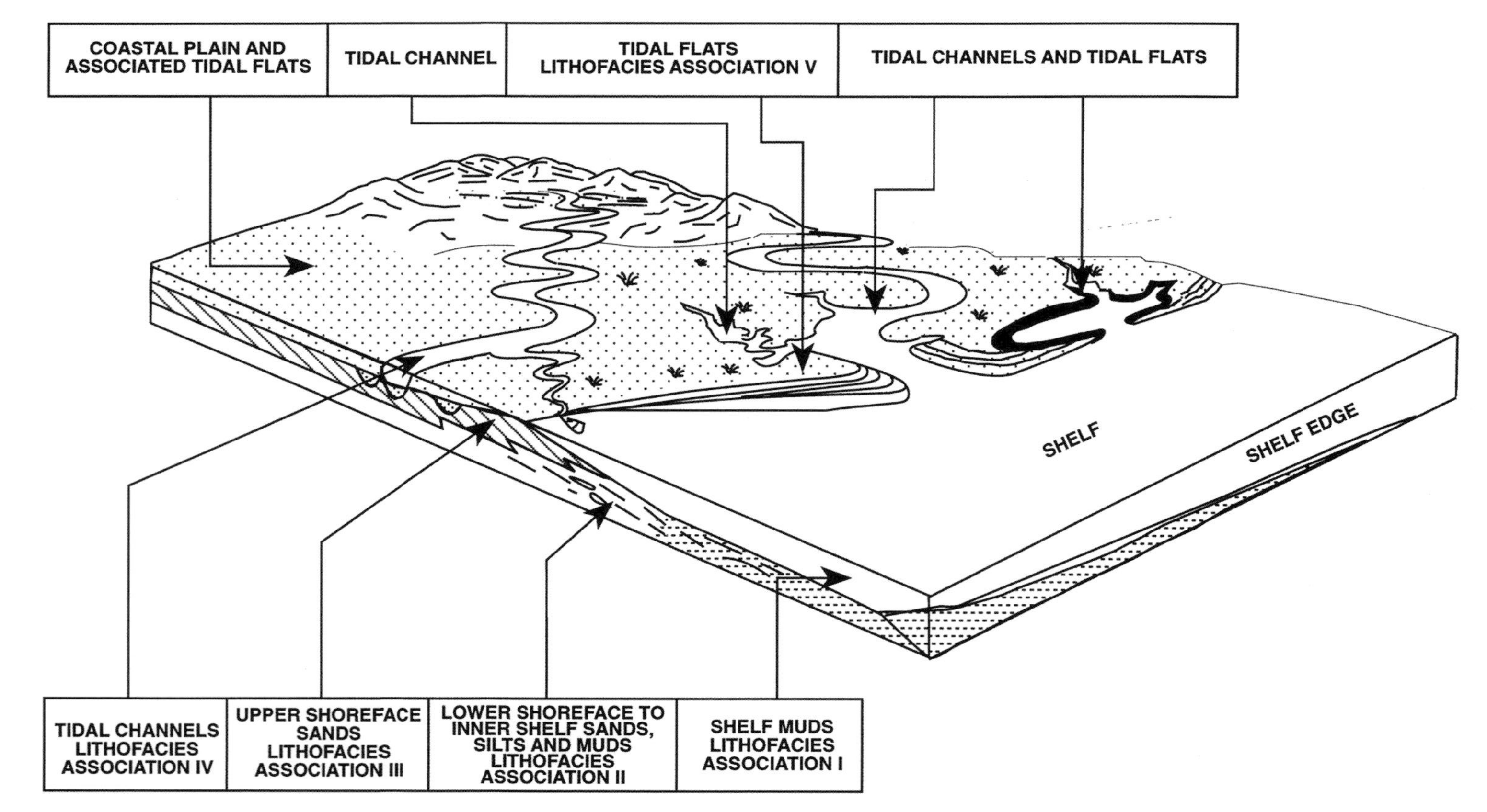

Fig. 11. Sedimentological model applicable to the studied succession.

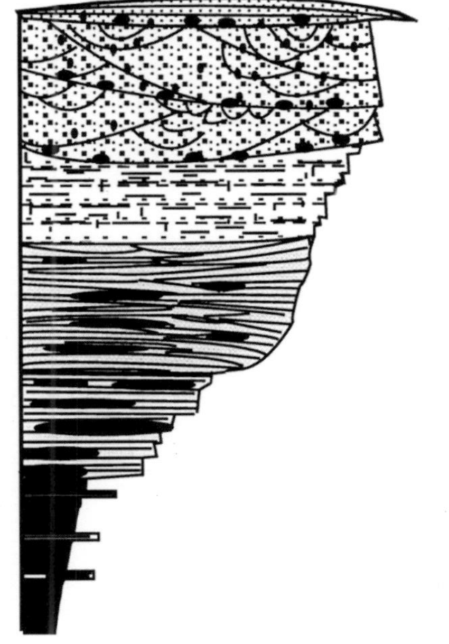

Fig. 12. Idealized facies relationships within a shallow marine upward coarsening sequence (after Prave *et al.* 1996).

set. Within the studied Miocene succession, complete parasequence sets range from approximately 10 m to over 40 m in thickness (average thickness = *c*. 15–20 m).

Transitions between the lithofacies associations identified using FMI images were further evaluated using embedded Markov chain analysis to identify those which occur with a potentially significant probability. Figure 13 summarizes the results of Markov chain analysis for a typical well. Figure 13 indicates that shelf deposits mainly translate into lower shoreface deposits which in turn mainly translate into upper shoreface deposits. Upper shoreface deposits mainly translate into lower shoreface deposits. Neither shelf, lower shoreface or upper shoreface deposits display a strong tendency to translate into tidal lithofacies. Transitions between tidal and fully marine lithofacies occur between tidal channels/tidal sandflats and shelf deposits.

When the abundance of lithofacies associations and depositional environments are expressed as a percentage of each parasequence set and plotted versus depth, the studied successions become further divisible into a series of large-scale sequence stratigraphic units, and a complex series of transgressive and highstand systems tracts is typically evident.

An example of a sequence stratigraphical model interpreted from FMI data is illustrated in Fig. 14. The fundamental building blocks used in construction of the model illustrated were the lithofacies identified by integration of wireline log data with fabrics observed in borehole image logs. Variation in shale bedding orientation data (described in later sections) were also of key importance for interpretation of observed variations in lithofacies stacking patterns within the context of sequence development and the identification of unconformities. For example, in the model illustrated (Fig. 14), a sequence boundary located at the base of tidal deposits within the lower part of lithostratigraphic Unit B (parasequence sets PSS3-PSS4) is characterized by a subtle variation in tectonic tilt, interpreted as indicating the presence of an unconformity. A similar change in tectonic tilt occurs about a sequence boundary at the base of parasequence set PSS16, within lithostratigraphic Unit I (see well III in Fig. 18).

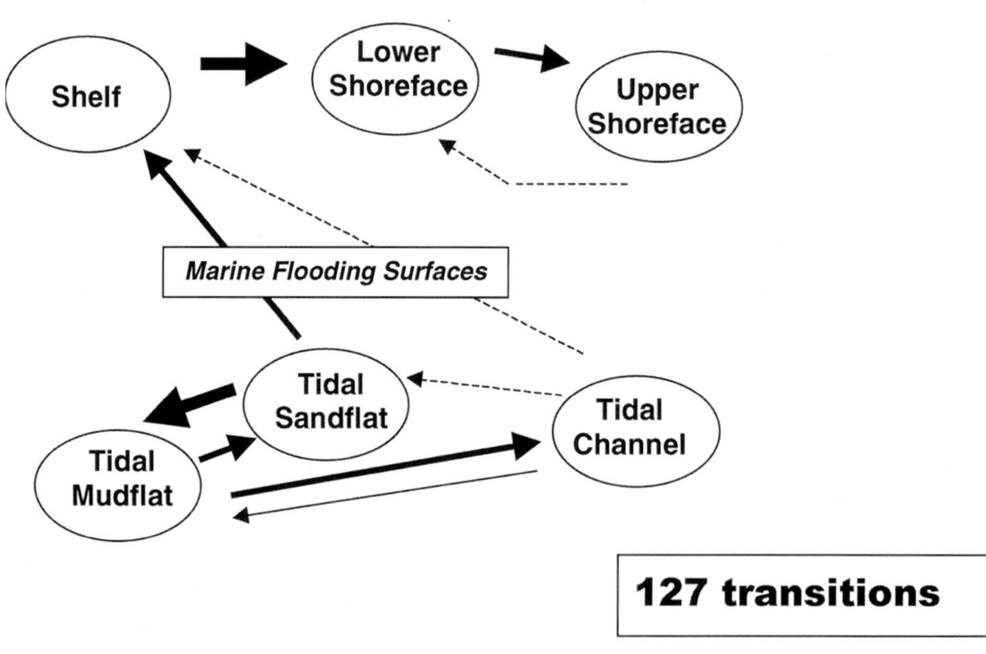

Fig. 13. Facies relationship diagram illustrating the results of Markov chain analysis of transitions for lithofacies associations within a typical study well. The results are based upon analysis of 127 transitions. For these analyses, Lithofacies Association V has been sub-divided into tidal sandflats versus tidal mudflats, depending upon argillaceous content. The relative probability of a facies transition is indicated by the thickness of the connecting line (heaviest line = highest probability, dashed line = lowest probability). Only transitions with values greater than 0.1 difference between transition probability and independent trials probability, are plotted.

Approximate Reservoir	Parasequence Set*	Stacking Pattern	SEA LEVEL RISING ←	System Tract
R-S	PSS22	Progradational	Flooding Surface	HST
P-Q	PSS21	Progradational	Flooding Surface	
N-O	PSS20	Progradational	Maximum Flooding Surface	
N	PSS19	Retrogradational	Sequence Boundary	TST
M-N	PSS18	Progradational	Flooding Surface	HST
L-K	PSS17	Progradational	Maximum Flooding Surface	
I-K	PSS16	Retrogradational	Sequence Boundary	TST
H-I	PSS15	Progradational	Flooding Surface	HST
G-H	PSS14	Progradational	Flooding Surface	
G	PSS13	Progradational	Flooding Surface	
F-G	PSS12	Progradational	Maximum Flooding Surface	
F	PSS11	Retrogradational ?	Sequence Boundary	TST
F	PSS10	Progradational	Flooding Surface	HST
F	PSS9	Progradational	Flooding Surface	
E-F	PSS8	Progradational	Flooding Surface	
D-E	PSS7	Progradational	Maximum Flooding Surface	
D	PSS6	TIDAL		TST
C-D	PSS5	Retrogradational	Sequence Boundary Maximum Flooding Surface Sequence Boundary	TST
B-C	PSS4	Progradational	Flooding Surface	HST
A-B	PSS3	Progradational	Flooding Surface	
A	PSS2	TIDAL		TST
	PSS1	Progradational	Sequence Boundary	HST ?

LOW CONFIDENCE (GRADE 4) POSSIBLE TIDAL FACIES

Fig. 14. Example of a sequence stratigraphic style interpretation based upon borehole image and dipmeter interpretation. Note how the presence of tidal facies could dramatically change interpretation.

The repeated alternation of lower and upper shoreface environments reflects the rise and fall of relative sea level, the controls on which are likely to have been climate, eustasy and tectonism. Syndepositional growth faulting and associated differential subsidence have been shown to have controlled the thickness of shallow marine clastic successions in the study area, and hence also control their degree of amalgamation and preservation potential (Edwards 1995).

Climatically controlled eustatic sea-level fluctuations are a possible explanation for the high-frequency of depositional cycles observed within some Miocene shallow marine clastic sequences (Galloway 1989; Edwards 1995). This type of eustatic component could certainly account for the development of the parasequences observed, but the development of the parasequence sets and their stacking patterns are likely to reflect the interaction of eustatic fluctuations with a variable, fault-related subsidence regime. The development of correlative sequence boundaries and associated tidal deposits within this succession appears to have occurred episodically rather than as part of long-term eustatic cycles. The large-scale sequential characteristics are therefore suggestive of the superposition of syndepositional tectonic forcing of environmental conditions on longer term eustatic trends.

The importance of the sequence stratigraphical sub-division of reservoirs further becomes evident when summations of net sandstone content are compared for a conventional lithostratigraphical reservoir subdivision, and a reservoir sub-divison based upon the recognition of parasequence sets (Fig. 15). The two scenario's presented in Fig. 15 both display sand contents determined from analysis of wireline log data in conjunction with borehole images. However, the interpreted vertical distribution of sand will clearly vary depending upon the nature of the framework (i.e. layer boundary model) within which it is analysed. For example, in Fig. 15, the lithostratigraphical reservoir subdivision 'F' contains approximately 33% sandstone. However, the same interval can be sub-divided into four sequence stratigraphical units (parasequence sets) with net sand contents ranging from 15%–80%. On the scales illustrated, reservoir models constructed using purely lithostratigraphically driven log correlations would thus result in different predicted fluid distribution and possibly flow unit definition to those constructed using a sequence stratigraphic architecture, delineated using borehole image log interpretations.

Figure 14 also illustrates a very important aspect with regard to quantifying the confidence of correlative reservoir models developed for production purposes. The pattern of highstand and transgressive systems tracts developed was based upon interpretation of FMI data for a well from which no core material were available. Tidal channel and tidal flat deposits are known to occur at similar levels within wells of only a few kilometers spacing, but little evidence for extensive development of tidal facies is present within these intervals in the well illustrated (Fig. 14). The FMI data suggest that most lithofacies observed are best interpreted as forming shelf-shoreface facies successions, and little correlation is possible with known tidal facies within adjacent wells on the basis of wireline log response. However, because tidal facies are abundant within adjacent wells and it is known that interpretation conflicts arise in recognition of tidal lithofacies, stratigraphical intervals which could possibly constitute a grade four tidal interpretation (Table 5) are also illustrated in the model (Fig. 14). It is important to note that if valid, these tidal facies interpretations could significantly alter any correlative models developed. Thus, at all times during study of this type of succession, it is recommended to try to quantify the validity of environmental and sequence stratigraphic interpretations. That a

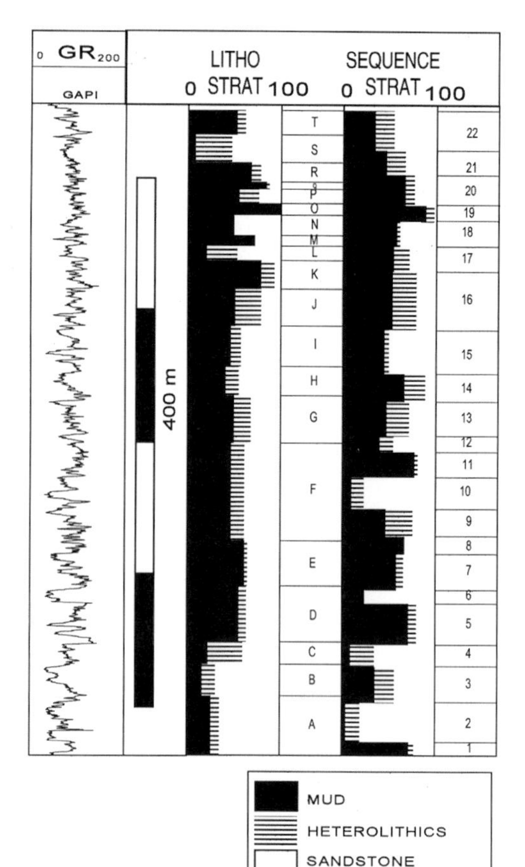

Fig. 15. Net sand content determined for stratigraphical sub-divisions based upon lithostratographical reservoir units compared with net sand calculated for parasequence sets.

model may be either applicable or inappropriate is obvious. However, an understanding of how inappropriate a model may be, and the probable consequences of its wrongful application, are vital for risk assessment. These confidence criteria should be routinely built into the interpretation of multi-well borehole image log data-sets.

Structural characterization

Structural dip determination

Structural dip is best identified from well-bedded intervals which are assumed to have been originally horizontally-stratified (e.g. mudstones or perhaps parallel lamination within sandstone lithologies). Structural dip interpretation was an iterative process requiring both initial evaluation of automatic computed dips to identify general data trends, and also manual picking of shale bed dips to evaluate detailed structural dip magnitude and azimuth variation. Shale beds were identified as having gamma ray log response in excess of 70 API.

Two different types of zonation were used to provide a framework within which to evaluate structural data:

Structural zonation. Studied intervals were subdivided into structural zones using dip trend analysis, together with analysis of cumulative shale bedding dip azimuth vector plots. These were identified essentially independent of traditional lithostratigraphical driven reservoir subdivisions or sequence stratigraphical reservoir subdivisions.

Parasequence set zonation. Structural dips were computed and compared for each parasequence set identified on the basis of lithofacies stacking patterns interpreted from image logs and dipmeter data. This aided the identification of key stratal surfaces (unconformities) related to the reservoir. Sequence stratigraphic surfaces identified invariably approximated with some of the lithostratigraphical reservoir zone boundaries defined previously using conventional well-log correlations. This is perhaps predictable, since conventional correlations in marine settings are frequently based upon matching of spikes in gamma-ray profiles, which in the studied example effectively correspond to flooding surfaces near the base of parasequences.

The averaged results from manually picked shale bedding throughout the studied intervals indicated mainly NW dipping successions, characterized by low moderate (7°–15°) structural dips, with often wide azimuthal spread. A combination of low structural dip and wide ranging azimuths within individual stratigraphical or sequence stratigraphic defined units is not ideal for identification of unconformities or correlative surfaces. The scatter observed in dip data reduced precision in structural dip estimates, and limited interpretation of dip trends (Fig. 16). However, in spite of these constraints, subdivision of the reservoirs using dip data was found to be possible. The often wide azimuthal dispersion (20°–90° scatter) of bedding plane poles from the individual parasequence sets was predictable for succession of strata largely deposited as essentially flat-lying, three dimensional, undulatory bedforms with polymodal dip azimuth in a shallow marine settings.

Cumulative shale bedding azimuth vector plots (Hurley 1994) were found to be of key importance in gross subdivision of the studied successions. Figure 17 illustrates a cumulative shale bedding azimuth vector plot together with azimuth histograms of shale bedding data. Figure 18 illustrates cumulative shale bedding azimuth vector plots and correlated surfaces for four studied wells. Deviations in trends in cumulative shale bedding dip azimuth vector plots are of variable magnitude. Some are sufficiently small that their identification as unconformities on the basis of dipmeter data alone would normally be a gross over-interpretation. However, where cores are available, these deviations can clearly be shown to coincide with the locations of sequence boundaries identified by the superposition of different sedimentary environments, and by their ichnofabrics. This enabled subtle shale bedding azimuth variations within non-cored wells to be interpreted as of sequence stratigraphic significance with a confidence otherwise not possible. In interpreting these subtle variations in bedding azimuth, it is necessary to rigorously locate their stratigraphical location, the lithofacies association in which they occur, and to critically assess their validity in terms of data sample size, overall variation and hence significance.

In terms of sequence stratigraphical significance, these subtle variations in structural dip azimuth commonly correspond to marine flooding surfaces or sequence boundaries, and may be correlated through several wells. However, because they are subtle features, they many not be identified in all wells. Their absence does not necessarily indicate that successions are strictly 'conformable', or lack surfaces of sequence stratigraphical significance. Generally, the less subtle a variation in structural dip azimuth then the more widespread is its correlation.

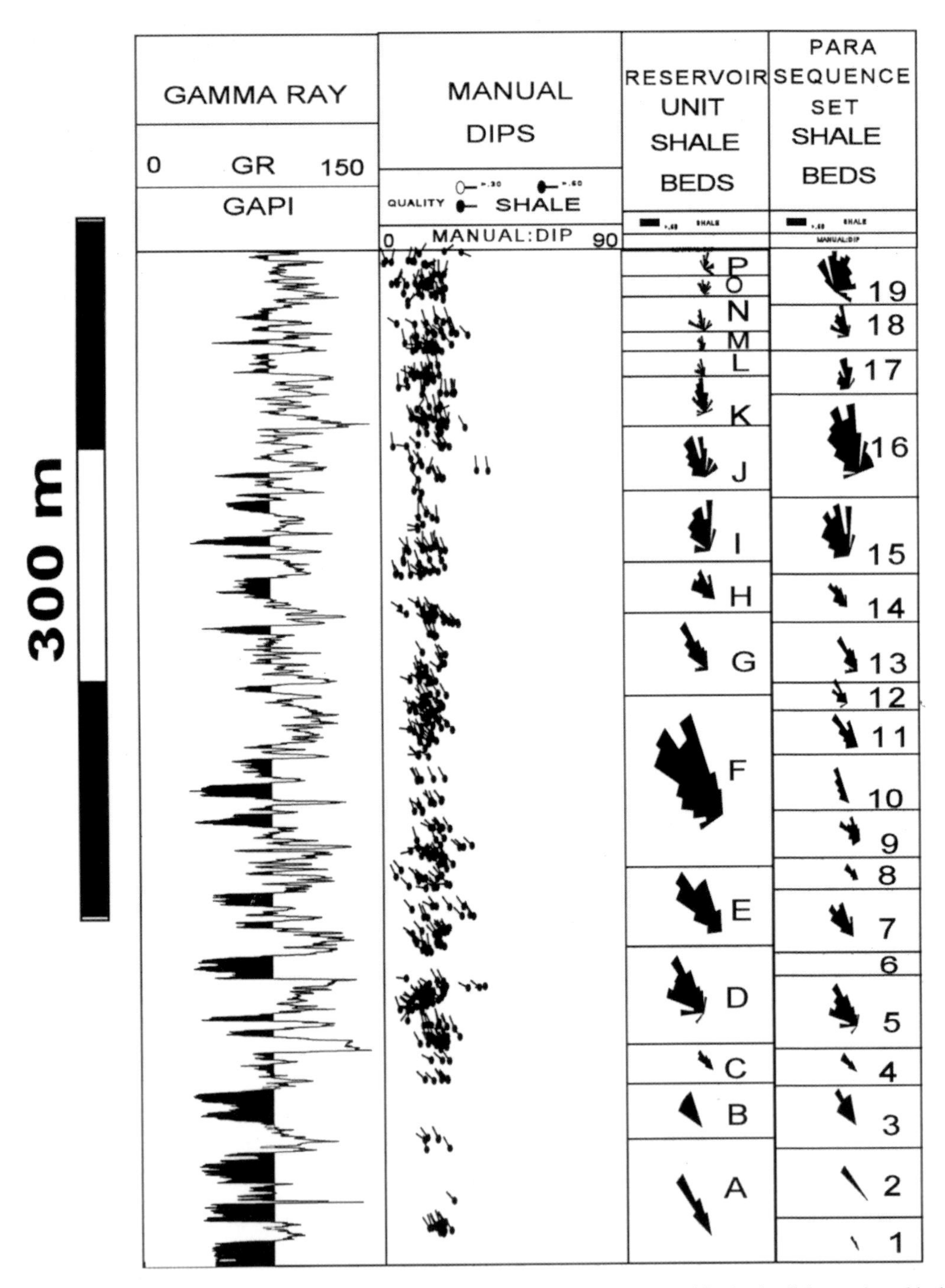

Fig. 16. An example of shale bedding dip data. Azimuth histograms are illustrated for both a lithostratigraphical determined reservoir correlation based upon matching of wireline log trends, and a sequence stratigraphical reservoir subdivision based upon intergration of dipmeter and wireline log data.

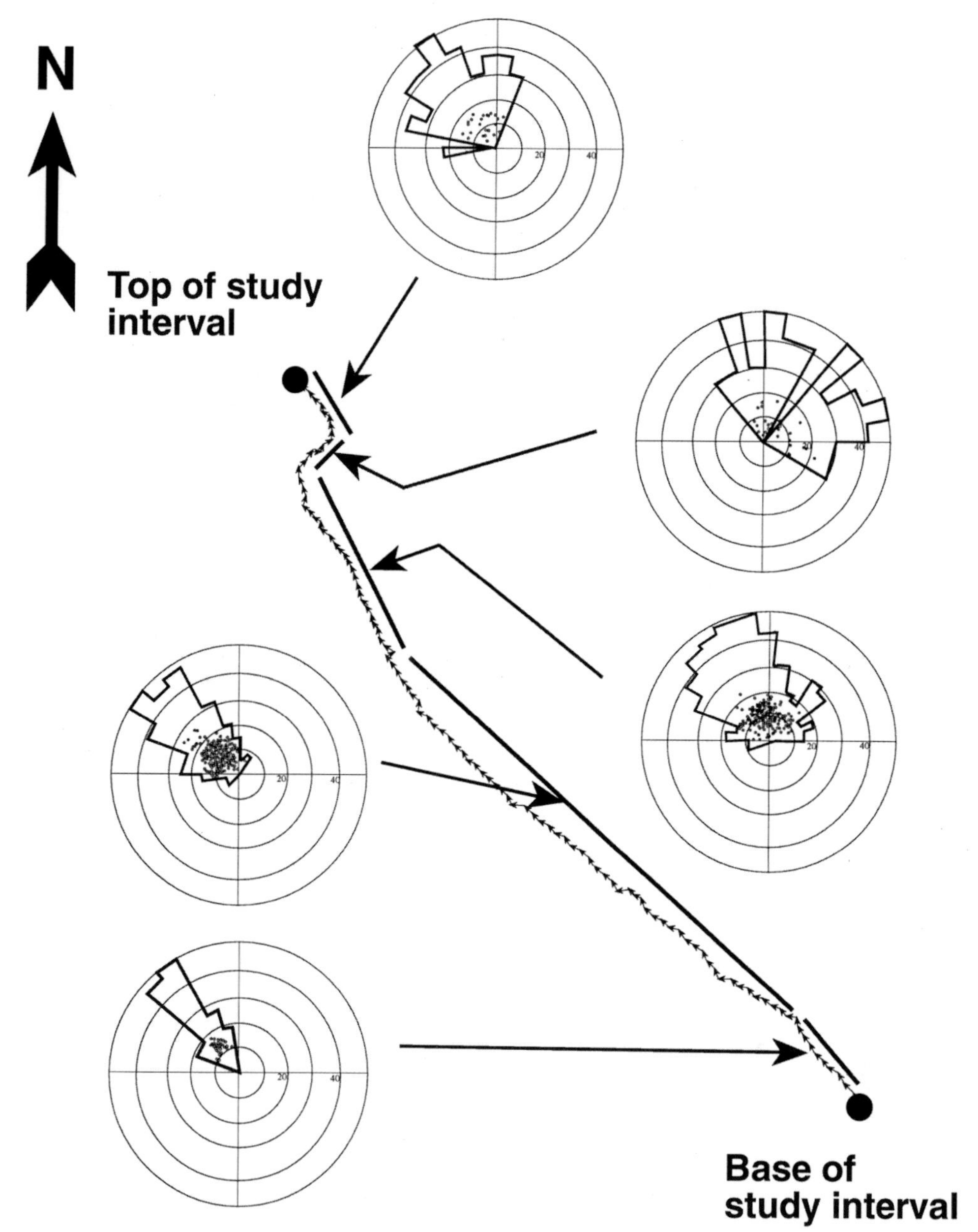

Fig. 17. Example of a cumulative shale bedding dip azimuth vector plot together with azimuth histograms of shale bedding data. The cumulative dip azimuth vector plots were found to be of key importance in gross subdivision of the studied successions, and were especially useful in recognition of unconformities characterized by only subtle variations in dip azimuth. The cumulative dip azimuth vector plot shown is from the same well as that illustrated in the sequence stratigraphic model shown in Fig. 14.

Dip azimuth variations due to faulting were also identified in cumulative shale bedding azimuth plots. They were distinguished from unconformities due to evidence of fault associated roll-over in adjacent sediments, and their association with visible fault and fracture zones within image logs.

Cumulative azimuth vector plots were also constructed for the different dip categories following removal of tectonic tilt (i.e. for

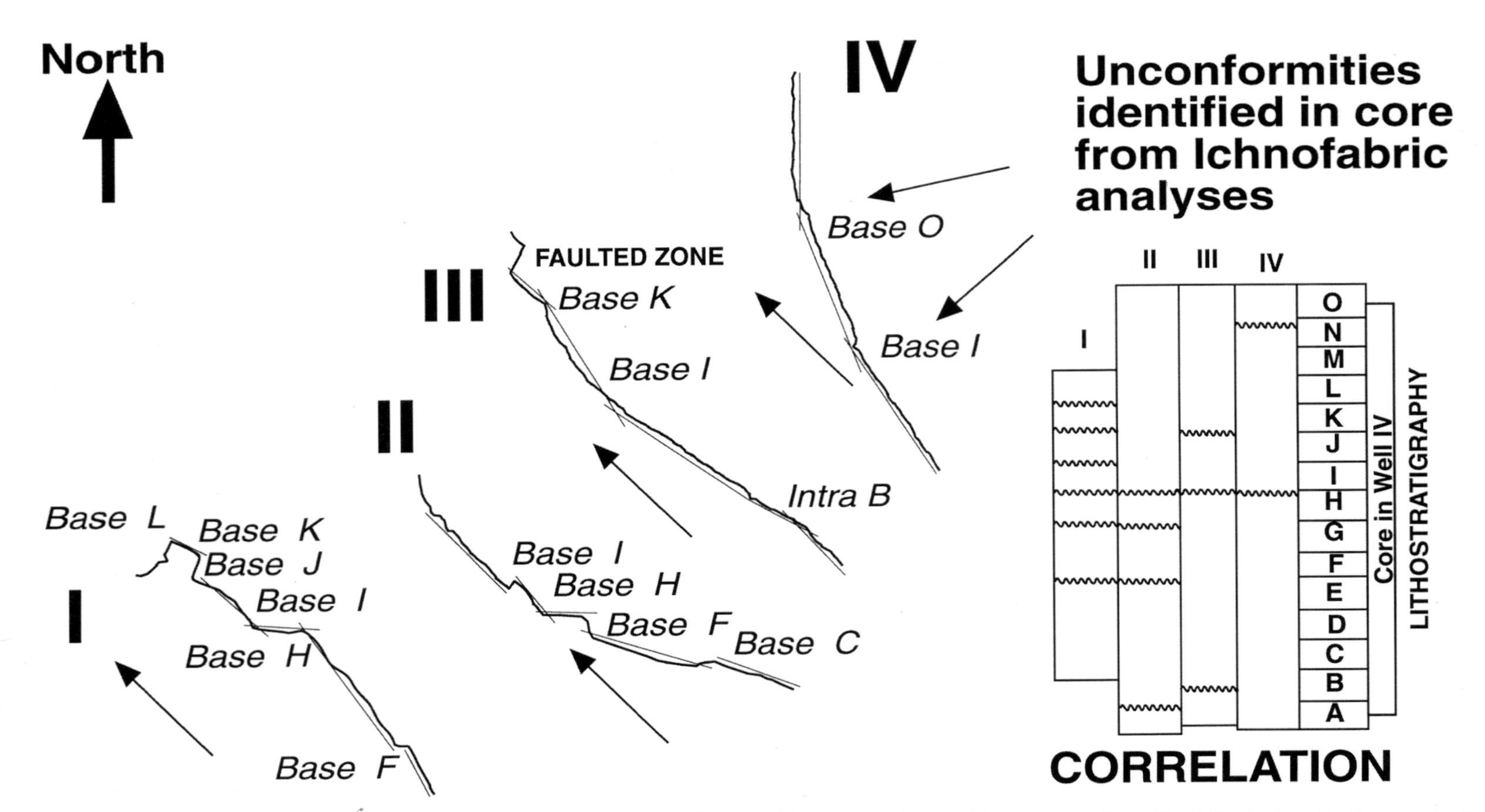

Fig. 18. Cumulative dip azimuth vector plots for shale bedding from four wells. Correlateable unconformities are illustrated, together with a lithostratigraphically determined reservoir correlation based upon matching wireline log trends. Arrows indicate direction of dip trends.

interpreted sedimentary dips). The plots were typically chaotic, revealing few distinct trends. However, in some cases, bioturbated bedding fabrics within shoreface deposits displayed one or two distinctly dominant modes within an effectively 360 azimuthal distribution. These indicated the presence of a dominant sediment transport direction, perhaps in an along shore or an onshore-offshore direction. Where evident, changes in these dominant transport directions were typically, broadly coincident with unconformities interpreted from cumulative shale bedding azimuth vector plots, and hence probably reflect significant changes in the configuration of the palaeocoastline.

Discussion

In successions where intense bioturbation effectively 'reduces' the overall quality of borehole images, and changes in tectonic tilt are manifest only as subtle variations in dip, then image log and dipmeter data may be of limited use for unconformity recognition if interpreted in isolation. Detailed inferences as to the significance of the subtle variations in dip trends observed in some of the examples illustrated would under normal circumstances represent gross over interpretation of data. However, where core data have been available to calibrate image log information, then interpretation confidence is significantly increased, and construction of detailed correlative models can be justified.

A flow chart describing a recommended methodology for the analysis of data from shallow marine settings of the type described is provided in Fig. 19. The method outlined is based upon that applied in this study. The key element in this flow chart is the fact that during initial investigations, analyses of borehole image log data (FMI) were initially carried out in isolation from the core and ichnofabric studies. In this way, the significance of interpretations of image log and dipmeter data could be critically assessed and calibrated with core data, and were not biased by the core interpretation. This enabled confidence limits to be built into the interpretation methodologies, prior to extending them to lower density SHDT data sets. The methodology applied should not be considered 'environment specific' and could easily be adapted for analysis of a variety of sedimentary deposits.

At all times during analysis of borehole image log and dipmeter data, it is essential to be aware of the level of interpretation confidence, and more importantly, how it will effect the reservoir and production models that are generated from the data. This can only be achieved by thorough integration of the different data-sets available, and understanding the limitations of each individual data set, and the inter-dependence of errors associated with interpretations from different data-sets.

Ichnofabrics and lithofacies derived from core photographic data have been of great importance for calibration of lithofacies interpretations made from 'vintage' borehole image logs and dipmeter data. Ichnofabric analyses have been particularly useful for building confidence in the interpretation of unconformities characterized by only subtle changes in dip magnitude and azimuth. Core photographs represent a valuable resource, especially when spliced together to form a continuous 'photographic image log'.

Conclusions

- Dipmeter and borehole image logs clearly contribute valuable information to the development of correlative models for the petroleum exploration and production industry. The analyses described have identified subtle variations in structural dip trends, which in isolation could not be interpreted as of significance. However, with core calibration, interpretation confidence can be significantly increased, allowing construction of detailed correlative models.
- At all stages during evaluation of borehole image and dipmeter log suites, understanding the confidence limits associated with interpretations is crucial to development of reservoir models for production purposes.
- Core photographs provide a valuable resource which, using a workstation, can be utilized for high resolution studies long after the original core may have deteriorated. It is fair to say that the level of interpretation found possible from core photographs was surprising.
- Assumptions that low lateral sampling density will render vintage SHDT data of limited use for detailed sedimentological and structural study, are likely to be flawed. Given a good high density training data set (image log or core), and willingness to integrate and critically evaluate the data, the information contained in SHDT logs can be readily incorporated into high resolution reservoir models.
- Although the study described was restricted to application of the above techniques to

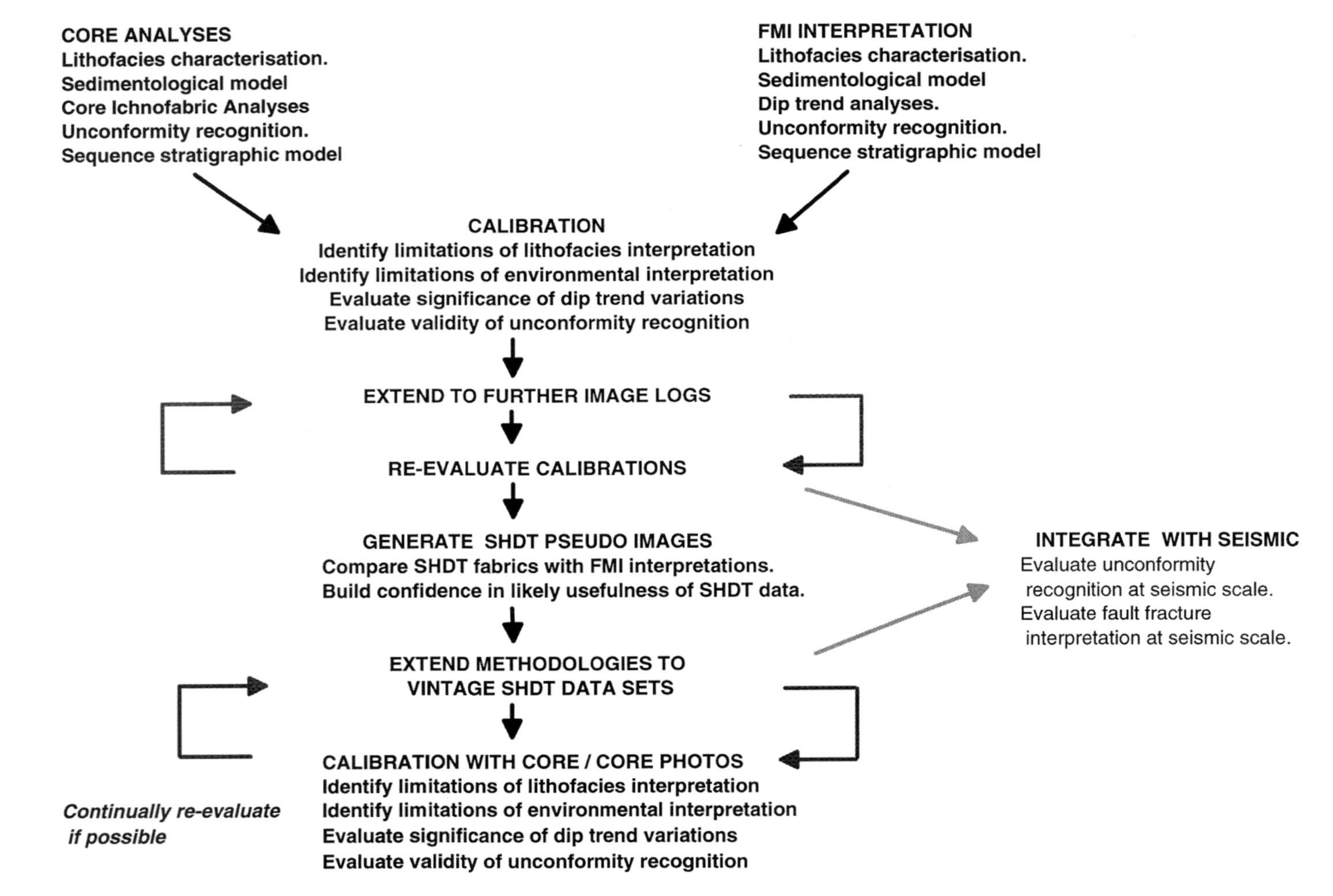

Fig. 19. Flow diagram illustrating methodology for integration of different data sets used in this study.

shallow marine depositional setting, the methodologies used are obviously transferable to other depositional environments.

References

COLLINSON, J. D. 1969. The sedimentology of the Grindslow Shales and the Kinderscout Grit: a deltaic complex in the Namurian of northern England. *Journal of Sedimentary Petrology*, **39**, 194–221.

DALRYMPLE, R. W. 1992. Tidal depositional systems. *In*: WALKER, R. G. & JAMES, N. P. (eds) *Facies Models: Response to Sea Level Change*. Geological Association of Canada, 195–219.

EDWARDS, M. B. 1995. Differential subsidence and preservation potential of shallow-water Tertiary sequences, northern Gulf Coast Basin, USA. *Special Publication, International Association of Sedimentologists*, **22**, 265–281.

EKDALE, A.A, & BROMLEY, R. G. 1983. Trace fossils and ichnofabrics in the Kjolby Gaard Marl, uppermost Cretaceous, Denmark. *Bulletin Geological Society*, Denmark, **31**, 107–119.

FREY, R W. 1975. (ed.) *The study of trace fossils*. Springer, New York.

GALLOWAY, W. E. 1989. Clastic facies models, depositional systems, sequences and correlation: a sedimentologist's view of the dimensional and temporal resolution of lithostratigraphy. *In*: CROSS, T. A. (ed.) *Quantitative Stratigraphy*, Prentice Hall, New York.

HURLEY, N. F. 1994. Recognition of faults, unconformities, and sequence boundaries using cumulative dip plots. *A.A.P.G. Bulletin*, **78**, 1173–1185.

LLOYD, P. M., DAHAN, C. & HUTIN, R. 1986. *Formation imaging with electrical scanning arrays. A new generation of Stratigraphic High Resolution Dipmeter Tool.* SPWLA Tenth European Formation Evaluation Symposium, Aberdeen, April 22–25.

PEMBERTON, S. G. 1992 (ed.). Applications of Ichnology to Petroleum Exploration: a core workshop. *SEPM Core Workshop*, **17**, 429.

PHILLIPS, C., DIFOGGIO, R. & BURLEIGH, K. 1991. Extracting information from digital images of core. *Society of Core Analysts, 5th Technical Conference*, Paper **SCA-9125**, 15.

PRAVE, A. R., DUKE, W. L. & SLATTERY, W. 1996. A depositional model for storm- and tide-influenced prograding siliciclastic shorelines from the Middle Devonian of the Central Appalachian Foreland Basin, USA. *Sedimentology*, **43**, 611–630.

SERRA, O. 1989. *Formation MicroScanner Image Interpretation*, Schlumberger Educational Services, SMP-7028.

SOVICH, J., KLEIN, J. & GAYNOR, N. 1996. A thin bed model for the Kuparuk A sand, Kuparuk River Field, North Slope, Alaska. *37th Annual logging symposium transactions; Society of Professional Well Log Analysts*, Paper **D**, 13.

SWIFT, D. J. P., PHILLIPS, S. & THORNE, J. A. 1991. Sedimentation on continental margins, V: parasequences. *In*: SWIFT, D. J. P., OERTEL, G. F., TILLMAN, R. W. & THORNE, J. A. (eds) *Shelf Sand and Sandstone Bodies-Geometry, Facies and Sequence Stratigraphy*, Special Publication, International Association of Sedimentologists, **14**, 153–187.

VAN WAGONER, J. C., MITCHUM, R. M., CAMPION, K. M. & RAHMANIAN, V. D. 1990. Siliciclastic sequence stratigraphy in well logs, cores and outcrops. *A.A.P.G. Methods in Exploration series*, **7**, 55.

WALKER, R. G. 1992. Facies, facies models and modern stratigraphic concepts. *In*: WALKER, R. G. & JAMES, N. P. (eds) *Facies Models*. Publication of the Geological Association of Canada, 1–14.

WEIMER, R. J., HOWARD, J. D. & LINDSAY, D. R. 1982. Tidal-flats and associated tidal channels. *In*: SCHOLLE, P. A. & SPEARING, D. (eds) *Sandstone Depositional Environments*. American Association of Petroleum Geologists Memoir, **42**, 191–246.

A pre-development turbidite reservoir evaluation using FMS electrical images

MALCOLM RIDER,[1] TIM GOODALL[2] & TIM DODSON[3]

[1] *Rider-French Consulting, P.O. Box 1 Rogart, Sutherland IV28 3XL, UK*
[2] *Production Geoscience Limited, North Deeside Road, Banchory, Kincardineshire, AB31 3YR, Scotland*
[3] *Statoil, Forushagen, P.O. Box 300, N–4001, Stavanger, Norway*

Abstract: FMS (Formation MicroScanner) electrical images in a single well from a familiar, deep marine North Sea Palaeogene section are illustrated. More than 900 m of good quality images were acquired through an interval from the Palaeocene Ekofisk Formation to the Eocene Grid Formation in a well from the centre of the South Viking Graben, Norwegian sector. The information that was obtained from these images is illustrated and discussed in terms of lithology, sedimentary structures, stratigraphy and petrophysics. The information on this deep marine, mainly gravity deposit succession, is considerably enhanced using image interpretation. Lithological information includes an improved net-to-gross ratio and an understanding of lithological heterogeneities. Sedimentary structure information includes the geometrical and lithological characterization of slumped intervals and a detailed description of evolving gravity deposit sequences. In terms of stratigraphy, illustrative images are shown of the Balder, Lista and Sele Formations. Some of the key surfaces of the Palaeogene interval, used for correlation and sequence stratigraphy, are shown in detail, especially the surface at the base of the Balder Formation and the boundary between the Lista and Sele Formations. The effect of hydrocarbons on the images is illustrated and used qualitatively in petrophysical terms to define the hydrocarbon-water contact.

This paper describes the study of electrical images from one well which penetrated a familiar Palaeogene sedimentary sequence in the central part of the South Viking Graben, in the Norwegian sector, North Sea. A hydrocarbon column was discovered. The images from this well are of interest on two counts. The first is that they illustrate the way in which image logs may be used to refine lithological interpretation, to characterize sedimentary sequences and to identify hydrocarbon bearing intervals. And the second count is that the images give new insight into formations frequently encountered in the subsurface and which are familiar on the standard logs and used as tools for reservoir development in correlation, sequence stratigraphy and mapping.

For reasons of confidentiality, some information and the location and name of the well are not given This makes no material difference to the general descriptions and does not affect the conclusions of the paper.

Tool and run information

The images presented here have been processed from the Schlumberger Formation MicroScanner (FMS) tool which has four active pads. Each pad has 16 button electrodes arranged in two rows of eight offset by half a button width or 0.1 inch (Cheung 1999, Boyeldieu & Jeffreys 1988). Electrodes are 0.2 inch in diameter and are slightly courser than the electrodes on the FMI tool, which has 24 buttons per pad (2 rows of 12). The FMS tool was run over the entire interval from the Eocene, Grid Formation to the basal Palaeocene, Ekofisk Formation, a total of over 900 m (Fig. 1). Hole size was 8.5 inch so that coverage of the borehole wall is approximately 40%.

Mud resistivity at the borehole temperature was 0.027 ohm/m, while formation resistivities generally varied between 0.5 and 1.0 ohm/m, but in some intervals were up to 10 ohm/m, a ratio of Rf/Rm = 370. Hole conditions over the run were excellent from the Balder Formation downwards to TD (Ekofisk Formation) but poor to moderate in the Eocene shales. Image quality is excellent where the hole is good, that is from the Balder Formation through to the Ekofisk Formation, but poor over the bad hole intervals of the Eocene shales (Fig. 1). Although tool spin was quite severe over the lower 200 m of the

From: LOVELL, M. A., WILLIAMSON, G. & HARVEY, P. K. (eds) 1999. Borehole Imaging: applications and case histories. Geological Society, London, Special Publications, **159**, 123–137. 1-86239-043-6/99/$15.00.

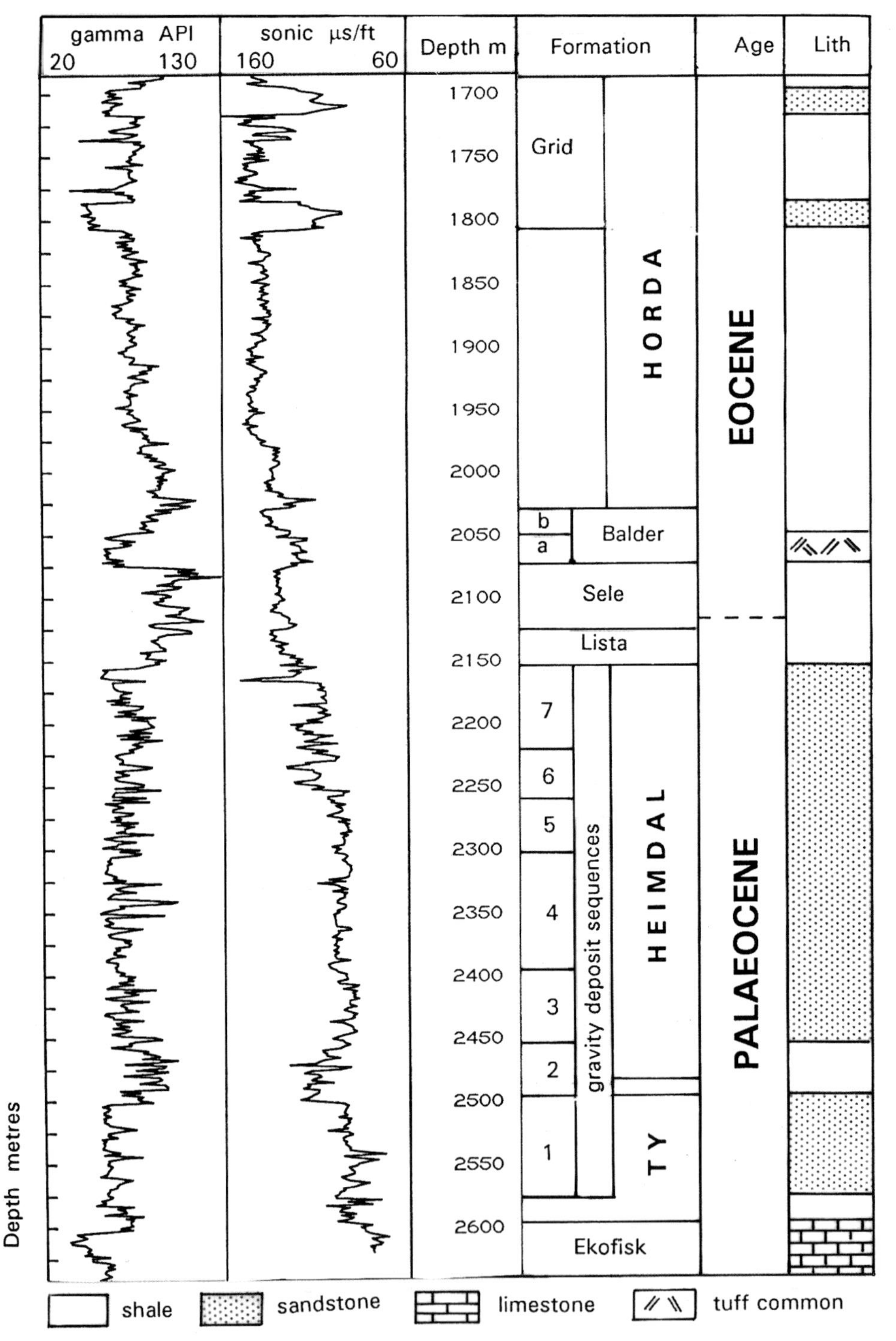

Fig. 1. The stratigraphic interval of the North Sea Palaeogene logged by the Formation MicroScanner tool (FMS of Schlumberger) and illustrated in the paper. The well is in a basinal setting in the Norwegian sector of the South Viking Graben. (gravity deposit sequences and Balder Formation a and b are explained in the text). Main stratigraphy from the operator.

hole, even reaching one rotation every 15 m, near the recommended limit (Carter 1987), it does not appear to have affected image quality (Cheung 1999). Taking the entire run, 76% (700 m) of the image was good to excellent and allowed reliable interpretation, 4% was of moderate quality where interpretation was possible but difficult and 20% was totally unusable.

The image processing was carried out using proprietary PC software, PC ImagePro (from Wireline Reeves). A static normalization was used to produce images and also a dynamic normalization using 2 m, 5 m and 10 m windows. The polarity of the plots in the figures is a dark shade for higher resistivity values. This is the preferred polarity as most intervals have lower resistivities, giving lighter shades, which are more easily interpreted than the dark, opaque shades. For interpretation, on-screen polarities and colours were varied at will.

Lithology

Lithology in general

Electrical images, like standard electrical logs, do not give geophysical responses which are immediately diagnostic of lithology. The electrical response of, for example, a laminated sandstone is the same or can be very similar to the electrical response of a laminated shale, so the electrical image of a laminated sand or a laminated shale are not lithologically diagnostic. This is especially the case with dynamically normalized images (Serra 1989), where colour ranges are limited to short vertical intervals of typically 2 m–10 m (*cf.* Fig. 4). However, as with the standard logs, electrical responses can be used to refine a lithological interpretation (Rider 1996). As an example, sandstone intervals containing salt water tend to give vertically consistent electrical responses in the same formation. Once identified, the salt water sandstones are seen to have a distinct and diagnostic electrical image response (Fig. 2).

Net-to-gross ratio

This aspect, the use of images for lithological interpretation in turbidite successions, is important because the images have a very high vertical (and horizontal) sampling rate (2.5 mm) giving them a much improved definition of bed boundaries and identification of thin beds (Trouiller *et al.* 1989). The log can therefore be used to derive a significantly improved net-to-gross ratio (Sullivan & Schepel 1995). In the present dataset, the net-to-gross is raised by up to 10% in some intervals, a significant amount, principally because shale intervals interpreted from the gamma ray and neutron-density logs include thin sandstones below their detection limit, which contribute to reserves and which are well characterized on the images (Fig. 2). The reverse situation, of course, is also possible, when thin shales are enclosed in thick sandstones and are not seen by the standard logs but are detected on the images. In this case, image interpretation would tend to lower the net-to-gross ratio, but such beds are rare in the present well.

Discontinuous bedding (concretions)

The behaviour of standard log responses can be examined against the images when irregular beds are present. The example shows a possible carbonate concretion set in sand matrix and below it a cemented bed of similar thickness (Fig. 3). The bed (2456.5 m, Fig. 3), is 70 cm thick and is registered on the gamma ray, density-neutron combination, the sonic and the MSFL,LLS,LLD logs. The concretion (2454.7 m, Fig. 3), with a maximum thickness of 50 cm is not registered on the LLD-LLS logs, despite the fact that these measurements are non-directional, that is spread out in all directions from the tool. The reason is that the volume of current affected by the concretion is small, especially as the latter has a high resistivity, and therefore does not contribute a great deal to the final tool measurement. Although the sonic is equally a non-directed response, the quickest (i.e. compressional) wave will have travelled through the thickest part of the concretion exposed on the borehole wall. This is the fastest route. Interestingly, all the directed sensors, that is sensors on pads, being the density, neutron and MSFL, detect the concretion and must have been 'directed' in the right orientation.

Aspects of this example are well known, that irregular bedding such as concretions may or may not be detected by the pad tools. What is often not considered, is the response of the non-directed measurements. As the example shows, it depends on the formation characteristics relative to the geophysical measurement.

Sedimentary structures

Using the electrical image logs it is possible to investigate even quite small sedimentary structures in the subsurface. In the North Sea Tertiary this is important as cores are always limited

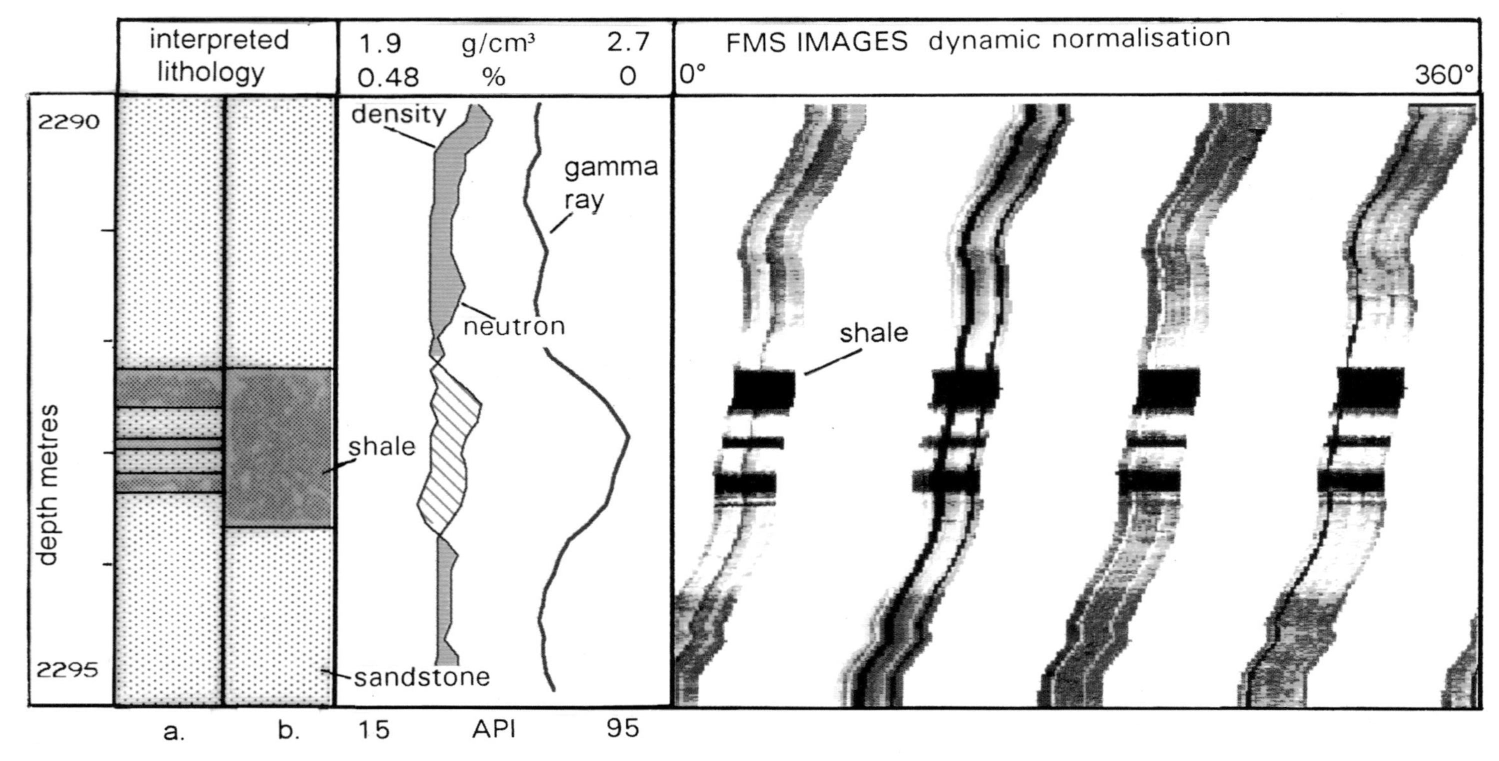

Fig. 2. Example of the bed resolution of the electrical image logs in thin-bedded gravity deposits (salt water fluids). Images provide an improved net-to-gross ratio compared to the standard logs. (a) image log interpretation. (b) standard log interpretation. FMS image using dynamic normalisation, 5 m window, dark colours indicate higher resistivity.

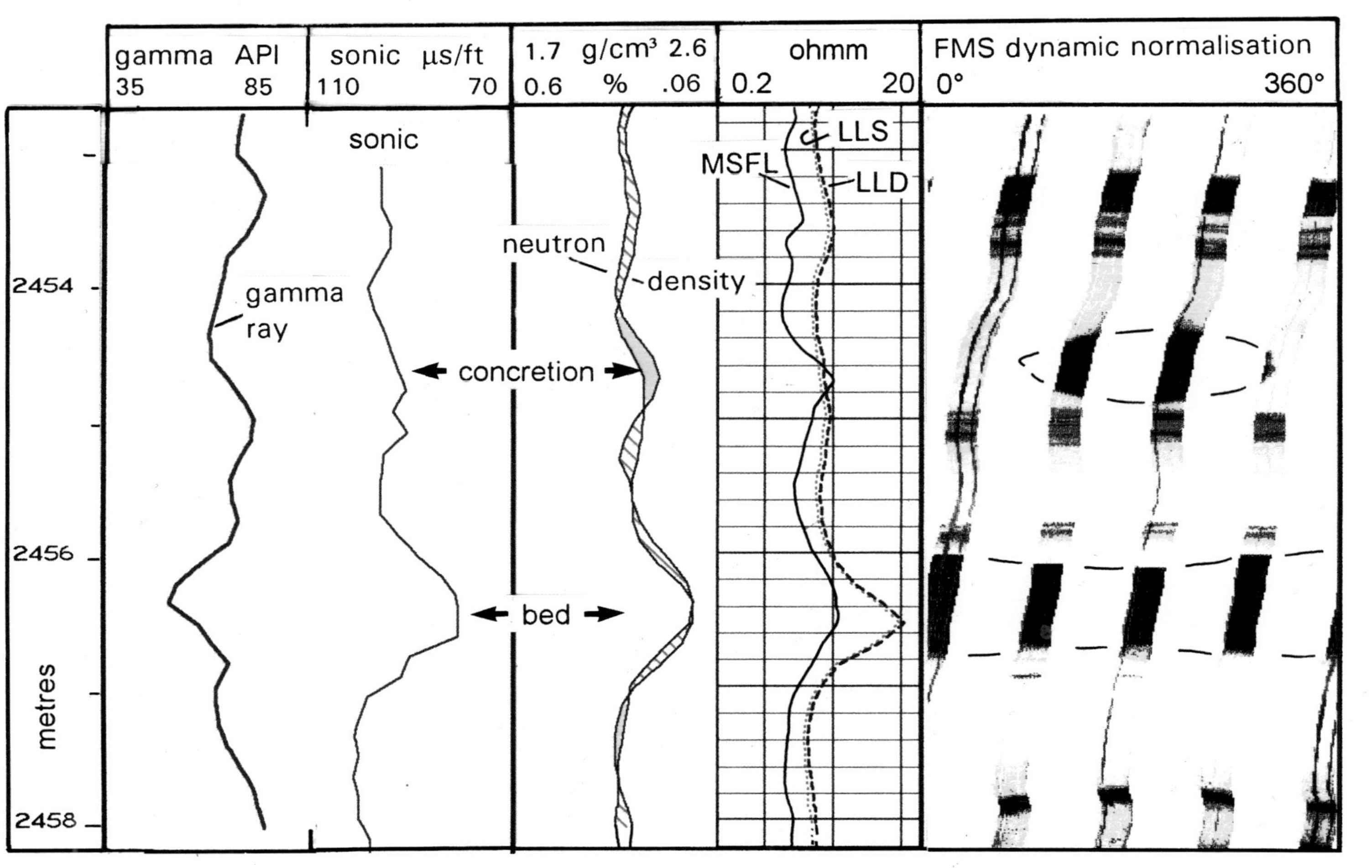

Fig. 3. Comparison of log responses through a continuous bed and a concretion, both cemented sandstones. The density, neutron and MSFL sensors are pad mounted. The gamma ray, sonic and LLD-LLS tools receive signals from all round the borehole. MSFL = microspherically focused log, LLD = deep laterolog, LLS = shallow laterolog – all from Schlumberger. FMS image using dynamic normalisation, 5 m window, dark colours indicate higher resistivity.

and, because the sediments are mainly unconsolidated, core recovery can be quite limited, especially in the cleaner reservoir sands.

Grain size changes

Electrical image logs can only be used to imply grain size variations where there is an associated change in permeability affecting the electrical properties of the formation (cf. Prosser *et al.* 1995). Variations in permeability are associated with differences in the depth of invasion of drilling mud filtrate, which create differences in fluid resistivities close to the borehole. High permeability generally allows greater flushing and filtrate invasion, while low permeability, at least in a hydrocarbon zone, is associated with higher irreducible hydrocarbons (oil). Electrical images are affected by these near-hole, fluid resistivity changes. A second grain size related property which also affects electrical behaviour is clay content, finer grain sizes tending to have a higher clay volume. The two effects work to some extent in parallel.

Changes in near-hole fluid resistivity are best seen in hydrocarbon bearing intervals where the electrical images are responding to the formation just beyond or through the flushed zone. In lower permeability zones where there is a higher volume of unflushed (irreducible) hydrocarbons, image colours show higher resistivities (dark colours in the figures). These lower permeability zones are associated with finer grain sizes. The better permeability, associated with coarser grain sizes, shows as lighter shades, where the hydrocarbons have been flushed (Fig. 4). The clarity of these variations depends on the choice of image processing parameters and changes in dynamic window length are critical (Fig. 4).

In water zones, gradual changes in image colours are more likely to be associated with changes in clay content, themselves linked to variations in grain size (finer grain size more clay) working in parallel. The presence of clay clogs the pore spaces, decreases permeability and increases formation resistivity so giving darker shades. The effects are seen in some of the thinner sandstones of the Heimdal Formation but the changes are often subtle and difficult to detect.

Slumping

Post-depositional slumping is seen to be typical over some intervals of the Heimdal and Ty Formations in this well. All the sediments involved were deposited in deep water, so that slumping and slump orientation are related to local palaeoslope conditions.

Slumps in the Heimdal and Ty Formations occur in intervals 2 m–5 m thick which consist lithologically of thin, irregular, inter-laminations of shale, and silt or sand. The inter-laminated lithofacies is quite distinct on the images and through some intervals is always slumped. In these same intervals, the clean sands occurring below and above the inter-laminated lithofacies, are not slumped.

Slumping is recognized on the images as steep dips in the inter-laminated lithofacies (not as isoclinal folds). That is, a slump is recognised as an interval of inter-lamination with steep 10°–40° dips, occasionally up to 70°, with relatively consistent dip azimuth orientations (Fig. 5). The implication is that the intervals remain internally coherent during slumping. In terms of orientation, the over-steepened slump dips are consistent within individual intervals but change to some extent from one interval to the next. However, taking all the slump related dips from the entire Ty Formation and Heimdal Formation in the well, a dominant west to south-west dip azimuth is evident.

That the internal geometry of the slump intervals is similar vertically through the well, indicates that the same geometry is persistent laterally in individual slumped beds. This observation restricts the possible geometrical interpretations from the dip and azimuth data. For tilting of the inter-lamination to occur only over a particular interval, that is over a limited vertical thickness, and be persistent horizontally, there must be associated internal faulting, as occurs for example in growth fault terrain (Fig. 5). Such an internal slump geometry which can be consistent horizontally despite being limited vertically, is one in which bed rotation occurs on a basal slide plane: which is the interpretation used in this paper (Fig. 5). As there is a tendency for the higher dips to occur at the tops of the slumps, it suggests that in addition, planar, tensional faults, break the slumped beds up into blocks (cf. McClay 1990), which then deform along this basal slide (Fig. 5). The images indicate that the tilted blocks are covered over by undeformed, sandstone gravity deposits. It is possible that the turbidite deposition of these overlying sandstones was actually the cause of the immediately underlying slumping. This would explain why all fine grained, inter-laminated lithofacies in some intervals are slumped, without exception.

This explanation of the internal geometry of the slumped units means that the dip azimuth

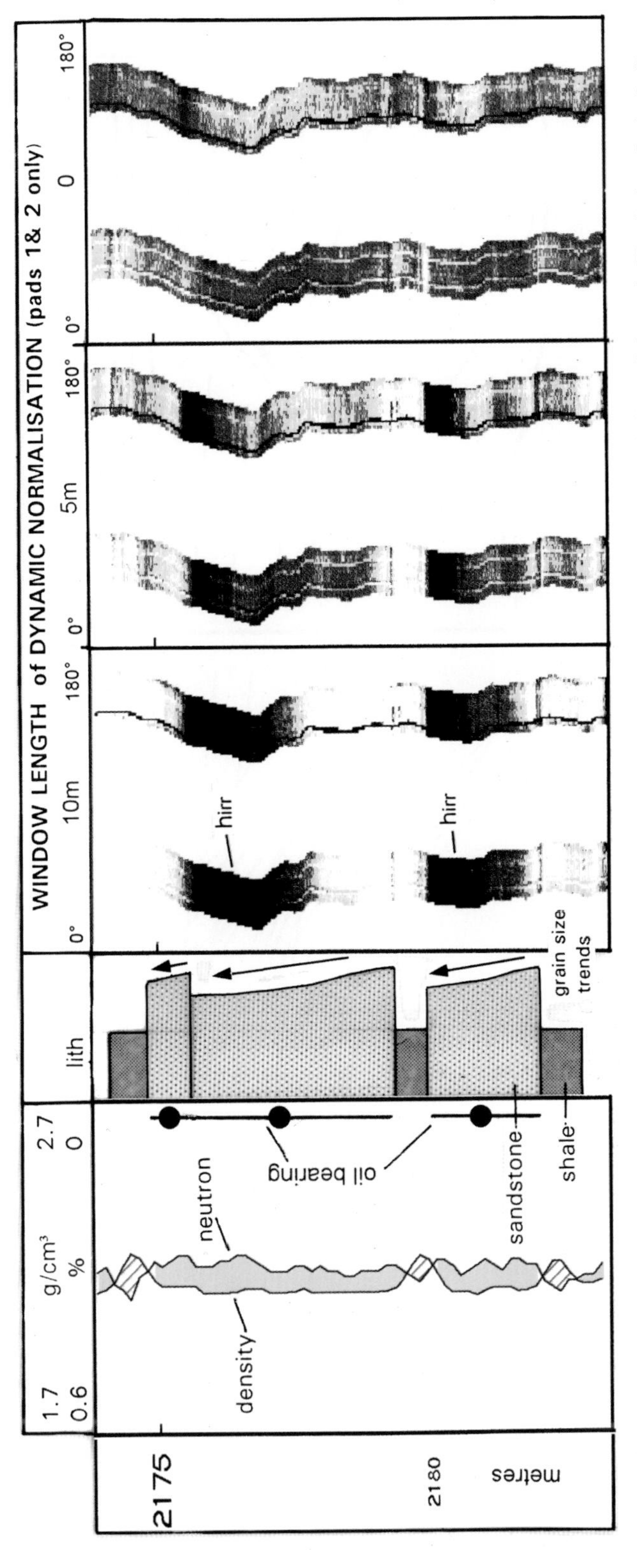

Fig. 4. Sandstone grain size grading reflected in the electrical image log colour changes in the hydrocarbon zone (arrows indicate fining-up). The darker shades indicate a higher volume of irreducible hydrocarbons (Hirr) as a result of lower permeability. Higher permeability sands are flushed. The subtle responses are brought out best using a 5 m dynamic normalization window. Shales are seen as lighter shades, that is lower resistivity (compare to Fig. 2, shale and sand responses in salt water bearing formation).

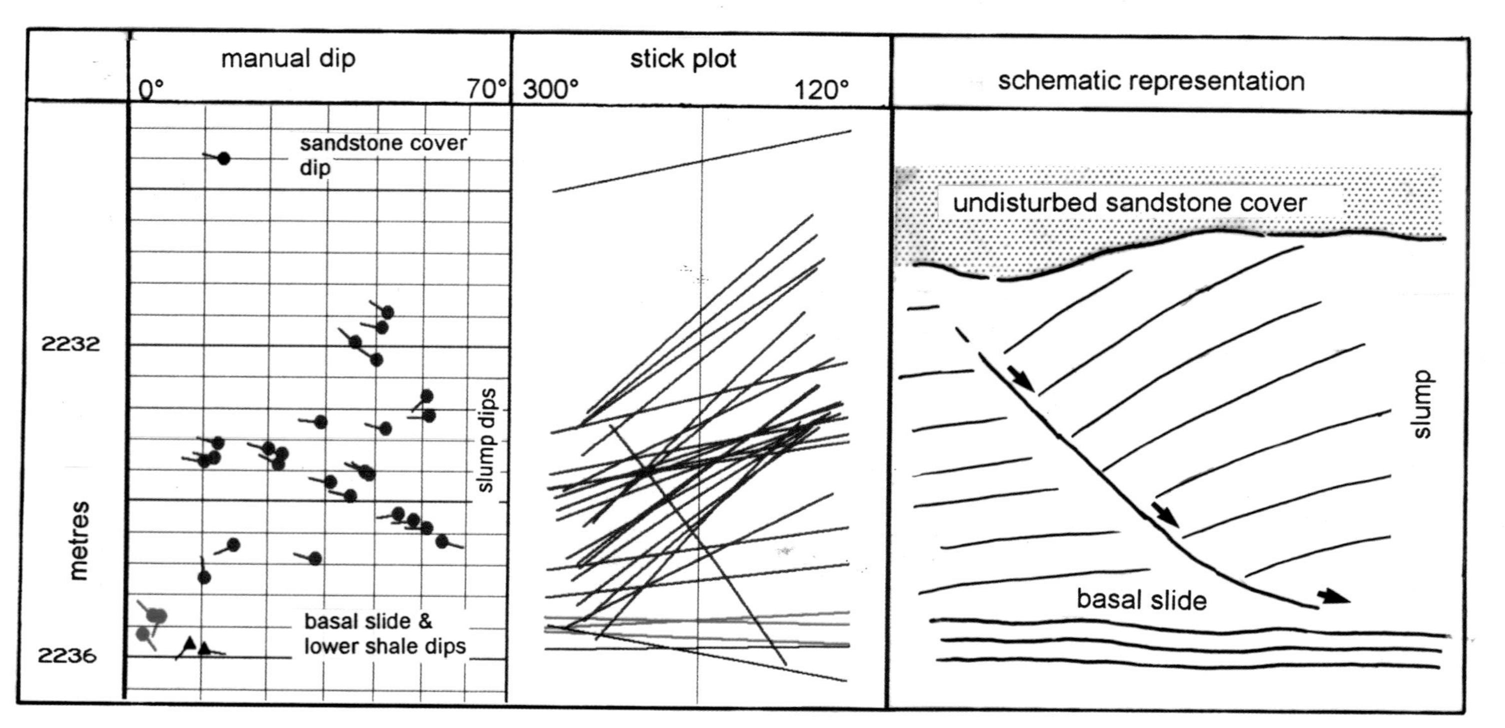

Fig. 5. Dipmeter, stick plot and schematic representation of a slumped interval. The slump lithology is inter-laminated sand and shale. Dips are oversteeped in the opposite direction to that of the presumed (down palaeoslope) slump movement.

of the tilted beds is in the opposite direction to that of the slump movement, that is, up the inferred palaeoslope. For the Heimdal Formation in general, in which dips are over-steepened towards the west and south-west, it means that the palaeoslope direction was down to the east and north-east (Fig. 5). Strictly, such structures can be termed slides rather than slumps (Stow *et al.* 1996), but are called slumps in this text for simplicity.

Gravity deposits

Electrical images present an excellent opportunity for examining, in the subsurface, changes in gravity deposits over long vertical sections. In the present well, a vertical section of nearly 450 m has been examined, grouped as the Heimdal and Ty Formations (Fig. 1).

Individual sandstone beds over the imaged section have been examined for basal contact, internal grain size changes and measurable orientation features, top contact and nature of the beds between the sands. Looking at only the standard wireline logs, it is clear that sandstone responses differ through the examined formations in terms of gamma ray values, interval transit time (Fig. 1) and neutron-density values. From a gross lithology established from the standard logs and using the characteristics of lithofacies, layering and layering geometry, interpreted from the electrical images, seven distinctive sequences are identified, each showing a different style of gravity deposit development (Fig. 1). For example, sequence 1 the lowest (comprising most of the Ty Formation), shows very distinct sandstones generally 15 m thick with 5 m thick shale interbeds. The basal contact of the sandstones is indistinct but the basal 2 m shows normal grading. The main body is massive sand, although there are occasional internal laminae with steep dips to the west. The uppermost 3 m–5 m of each bed consists of irregular, wavy stratified sands with interpreted patchy diagenetic cement. The top of this wavy stratified sand grades up into thinly bedded shales and sands which are always slumped (laminae over-steepened to the west) (Fig. 6).

The details of the six remaining sequences (2–7) are given in graphic form in Fig. 6. From this, the differences between them are evident. For instance, the thickest sands are at the base of the entire gravity deposit section, in sequence 1 described previously. In sequences 1 and 6 all the shale intervals are slumped, in sequences 2, 3, 4 & 5 none of them are. The final, sequence 7 is demonstrably a single, thickening-up, coarsening-up sequence and only the very basal shale interval is slumped.

Internal grading, interpreted from image colour changes, is not common in any of the sandstones from this well but does occur in the bases of sequence 1 sandstones (described previously), in some (30%) of the thinner sandstones of sequence 5 and in some sands in sequence 7 (the single sequence) (Fig. 4). Those working on core from deep water sands in this area (Shanmugam *et al.* 1995) have observed that grading is generally rare in North Sea Tertiary gravity deposits, although the eventual reasons for this are much debated (Hiscott *et al.* 1997).

It is possible to provide some elements of a depositional model from the image characteristics. For example, sequence 1 is interpreted as consisting of high density turbidites with a basal traction carpet (the grading). The topmost irregular, wavy stratification may also be associated with traction (Lowe 1982). The fact that all laminated interbeds in sequence 1 are slumped, suggests, as described, that the overlying sandstone influx caused the slumping in the underlying layers. In contrast, sequence 5 consists of classic, undisturbed, low density turbidites. These are not arranged into upward bed thickening or thinning sequences (Mutti 1977) and so cannot be readily interpreted as part of larger scale depositional features. Sequence 7 however, is a single, 70 m, clearly thickening-up, coarsening-up sequence capped by massive, high-density turbidites and may be interpreted as the deposit of a prograding lobe.

These are necessarily superficial interpretations as it is only when several wells are analysed in the same way as the present one and the information combined, especially with seismic interpretation, that any further conclusions can be warranted (Den Hartog Jagger *et al.* 1993). However, the amount of lithological and sedimentological detail that can be interpreted from the electrical images is demonstrated and contrasts with the usual description of this interval, based on the interpretation of the standard wireline logs alone, as monotonous sand–shale alternations.

Stratigraphy

Characteristic electrical images were registered in the Balder, Lista and Sele Formations. These images include some of the key stratigraphic levels used for correlation in the North Sea Palaeogene and are described and illustrated.

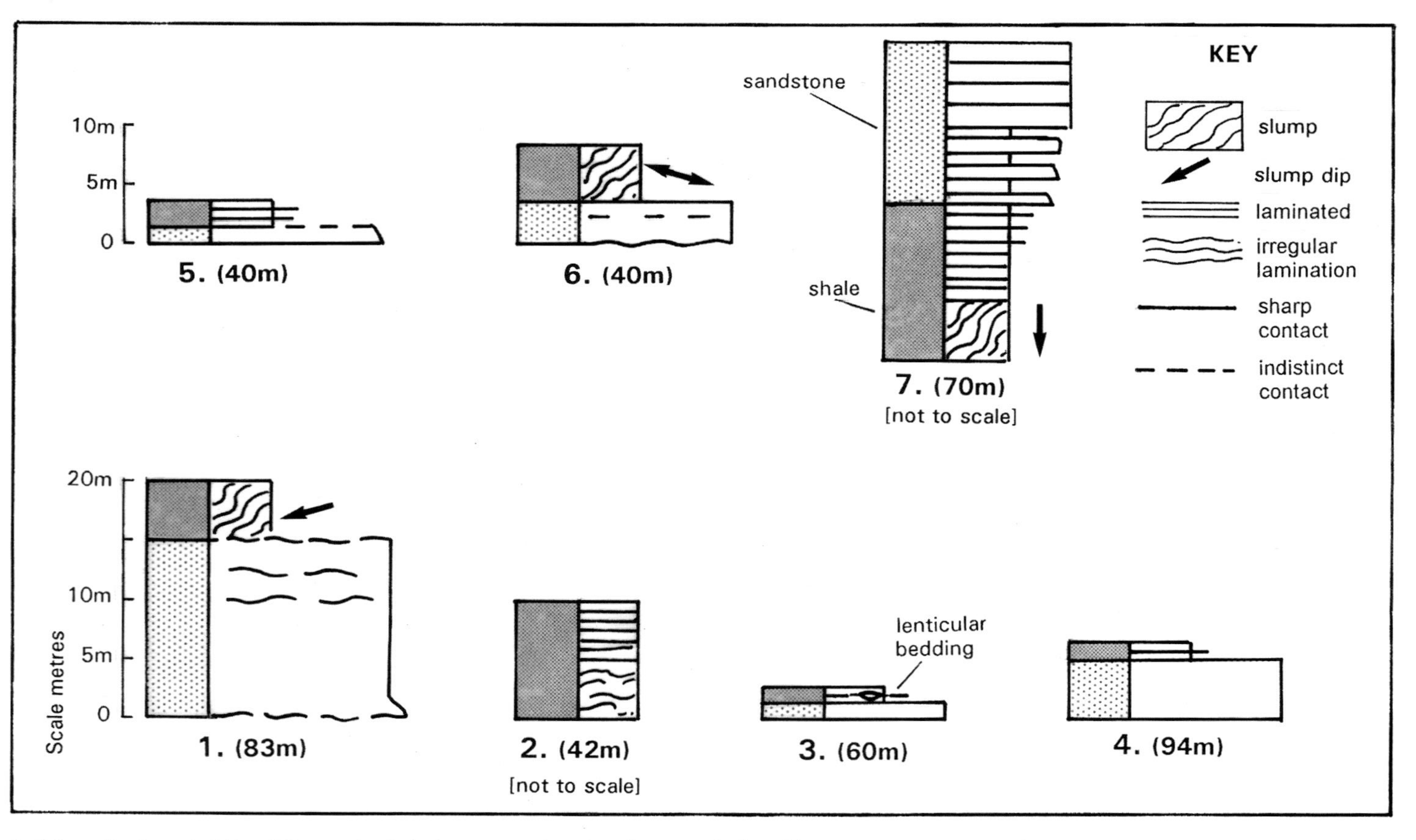

Fig. 6. Schematic characteristics of the gravity deposit sequences based on image analysis and log response. Representation to scale (except where indicated) showing a typical sand interval and following shale or fine grained interval. Numbers refer to sequences shown on Fig. 1, thicknesses in brackets. Orientations are for dips within slumps (cf. Fig. 5).

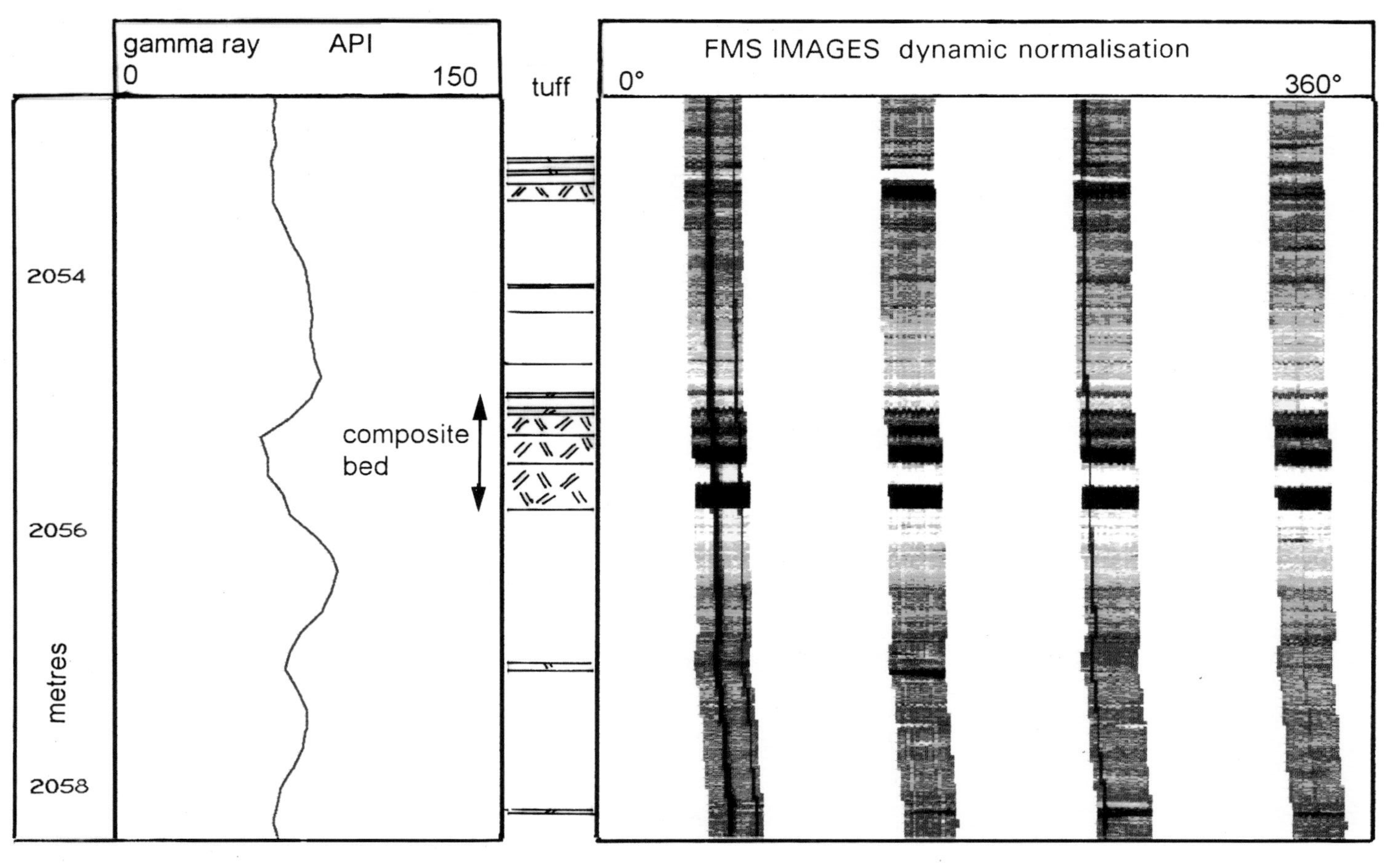

Fig. 7. Image characteristics of the Balder Formation. Tuff layers are calcite cemented and appear as dark colours (high resistivity) The thick bed between approximately 2055 m–2056 m is composite and consists of several, distinct layers. Example from near the top of the tuff bearing interval (Fig. 1, Balder a). FMS image using dynamic normalization, 2 m window, dark colours indicate higher resistivity.

Balder Formation

As expected, the Balder Formation, with abundant tuff layers, shows extremely regular lamination. The tuff layers themselves are easily recognized as they are mainly calcite cemented (Jacqué & Thouvenin 1975, Malm *et al.* 1984) and have a much higher resistivity than the surrounding shale (Fig. 7).

Using the high resistivity beds as indicative, 90 tuff layers have been counted, irregularly distributed through 28.0 m of Balder shales. An image hardcopy of the formation effectively shows the distribution of the tuff with most beds near a 1.0 cm detection limit. The study of a core through part of the Balder Formation in well Norway 30/2-1 (approximately 200 km north of the present well), shows that 30% of the tuff layers are less than 1 cm thick, nearly 50% are between 1 cm and 5 cm and the rest thicker. (Malm *et al.* 1984). Image resolution of the tuffaceous horizons seems to be effective within the 1.0 cm–5.0 cm range. Thus, while the core count is 117 beds (1–28 cm thick) over 16.1 m (i.e. between 7–8 beds per metre), the image count over the equivalent interval only identifies 4–5 layers per metre, suggesting that beds thinner than 1cm are not detected. The image therefore, enables approximately 60% of the layers to be detected compared to the core. At the outcrop in Denmark, over 500 km to the south east of the study well, up to 140 layers are numbered through the main tuff interval (Ølst and Fur Formations) (Schmitz *et al.* 1996) as opposed to 90 from the images, suggesting again that perhaps 60% (90/140) of the tuff layers only are detected on the images. As a general case, when high resistivity (tuff in this well) beds are set in lower resistivity shales, there is a tendency for the images to distort their thickness because of unequal current distributions (Cheung 1999). Beds below 1.0 cm thickness will generally be affected, beds above 6.0 cm show true thickness (Trouiller *et al.* 1989).

There are three intervals seen on the images of much thicker tuff beds. One of these is over 90 cm thick but consists of several tuff layers combined (Fig. 7). It has the same characteristics as the 'major tuff bed' described in well 30/2-1, which is over 1.4 m thick and also consists of several tuff layers (Malm *et al.* 1984).

The top 20 m of the Balder Formation in this study well have no obvious tuff layers (Fig. 1, Balder b). Shale dips change sharply, from dip 4° azimuth 145° in the tuff below (Fig. 1, Balder a) to dip 14°, azimuth 255° in the upper zone (Fig. 1, Balder b). There is layer distortion and a bioturbated interval at the base of zone b at the level where the dip changes. The borehole also behaved differently in the two zones during drilling, showing a slight rugosity on the calipers only through zone b. An upper limit to the main tuff zone is identified in well 30/2-1 (about 200 km north, Malm *et al.* 1984), which shows features similar to the zone a–b boundary seen in the present well. The change in the images at this limit is very clear in the study well, but its significance and the reason for the swing in dip and orientation are not known.

Base Balder Formation – top Sele Formation key surface

The high gamma ray peak below the Balder tuff zone is well known and frequently used in correlation across the North Sea, being equated to the top of the acme of *Cerodinium wardense* (Mudge & Bujak 1996). It is interpreted as a marine condensed sequence, or maximum flooding surface in sequence stratigraphic terms (Milton *et al.* 1990). The interval with highest gamma ray values is 1.0 m thick in this well, the peak having values 25 API above the normal shale background. The images show the high gamma ray interval to be finely laminated and without bioturbation. Thin resistive layers, 3 cm to 5 cm thick, within the interval of the gamma ray high, may be associated with hardground cementation (cf. Baum & Vail 1988). There is a subtle change in the characteristics of the layering immediately above the gamma ray 'spike', they are seen as less continuous on the images, an indication of depositional changes following the interpreted maximum flooding and condensed sedimentation. The first interpreted tuff occurs some 7.0 m above the gamma ray high and is where the base of the Balder Formation is drawn (Fig. 1). Although the Balder Formation itself may be put into a sequence stratigraphic context and defined using key surfaces, the presence and abundance of volcanic tuff would not seem to have a sequence stratigraphic significance, the source of the explosive volcanic material having no relationship to sedimentary basin margin configuration.

Sele Formation

The Sele Formation is generally distinct on the image log responses. When seen in core, the sediments are 'dark grey, carbonaceous, pyritic, fissile mudstones' (Knox & Holloway 1992). In this well the Formation is 50 m thick. The images show the mudstones have fine, planar laminations throughout, with no apparent bioturbation. The layering does not show great image colour contrast (i.e. electrical resistance is similar throughout) and differs from the layering in both the Balder Formation above and the Lista Formation below.

There is no change in either the amount or orientation of dip between the Sele and the Balder Formations.

Lista Formation

The Lista Formation at this locality consists of 27 m of mudstones. From core in other wells, these are described as grey–green, bioturbated, poorly laminated, generally non-calcareous mudstones (O'Connor & Walker 1993). The images show strong but irregular lamination with, over a number of intervals, a speckled image texture interpreted as being caused by small-scale bioturbation.

Lista–Sele Formation boundary and associated key events

One of the most stratigraphically distinct contacts in the North Sea Tertiary is that between the Sele and the Lista Formations, recorded in both lithological and microfaunal changes (Knox & Holloway 1992). A number of other significant biostratigraphic and lithostratigraphic events also occur close to this boundary, all of which are in the vicinity of the Palaeocene–Eocene boundary itself, the precise location of which is still being debated (Knox 1996). Electrical images of some of these important events are illustrated.

In core (O'Connor & Walker 1993), the boundary between the Lista Formation and the Sele Formation is marked by a change upwards from grey–green, bioturbated (oxygenated) shales to dark grey, finely laminated shales (anoxic) and is accompanied by an uphole disappearance of diverse benthic agglutinated foraminifera (Knox 1996). The shale boundary is well marked on the images which show the irregular layering with high electrical contrast of the Lista Formation changing abruptly to the fine, delicate, planar laminae which typify the Sele Formation (Fig. 8). The changes are associated with a marked gamma ray 'spike' (Fig. 8). The manual dip data derived from the image tool also shows an abrupt change, with a low but variable southerly structural dip in the Lista Formation but a consistent, very low structural dip to the north in the Sele Formation above. Similar changes at this boundary, from irregular to regular dips upwards, have already been documented on the dipmeter alone (Mudge & Copestake 1992*a*, *b*).

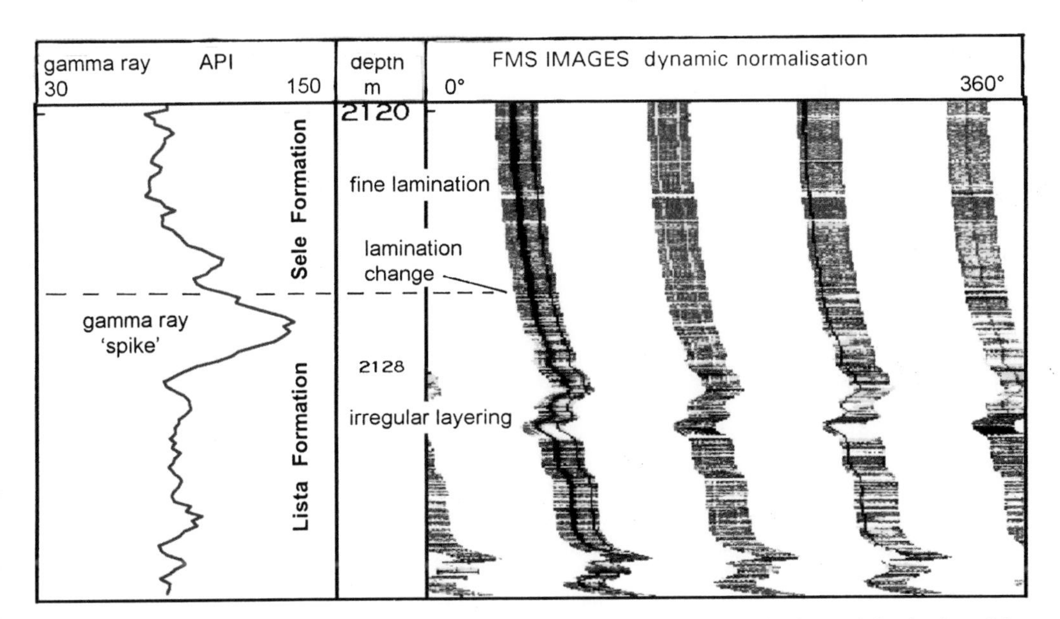

Fig. 8. The Lista–Sele Formation boundary clearly indicated by the FMS images. The fine scale lamination of the Sele Formation is seen to begin only above the gamma ray 'spike' associated with the boundary. FMS image using dynamic normalization, 2 m window, dark colours indicate higher resistivity.

The abrupt change in shale defining the Sele–Lista boundary, as indicated, is associated with a dramatic extinction of benthic foraminifera which, in turn is interpreted to have been caused by a rapid, global warming event and ocean-bottom water temperature change indicated in oxygen isotope ratio values from sites around the world (Corfield & Norris 1996, Mudge & Bujak 1996). In the present well, the images show that the abrupt change in shale lamination and image derived dipmeter character, is above the gamma ray spike and not below. The gamma ray spike could be expected to be associated with anoxia and uranium abundance (Rider 1996) typical of the Sele Formation, but for some reason occurs associated with the oxygenated Lista Formation sediments in this case.

Petrophysics

As is well known, the Schlumberger electrical image tool resistivity responses are not calibrated, being continually modified in response to formation conductivity (Cheung *this volume*), so that quantification of resistivity based water saturations from images is not reliable and no such work has been undertaken in this present well. The images have, however, been used qualitatively to identify hydrocarbon bearing sands and to localize the hydrocarbon–water contact. On many logs, it is clear that the depth of investigation of the FMS (and FMI) is quite deep, possibly deeper than the flushed zone resistivity pad tool the Microspherically Focused Log (MSFL of Schlumberger). In this present well, there is a clear difference between the images in water bearing sandstones and those containing hydrocarbons. The example in Fig. 2 shows salt water bearing sandstones in light shades, with the interbedded shales as dark, resistive shades. The example in Fig. 4 shows the reverse, with the hydrocarbon bearing sandstones in dark, high resistive shades and the interbedded shales (and permeable sands) in light colours. In the present well, the capacity of the images to be affected by hydrocarbons and to 'see' beyond the flushed zone, has been used qualitatively to define the hydrocarbon water contact. Image interpretation allowed far more precision for this, as the lowest hydrocarbons occur in thin beds not detected at all by the standard logs. To date, work on the modelling of electrical image current characteristics is partly unpublished, but the fact that electrical images commonly show the influence of hydrocarbon saturation is a valuable attribute.

Conclusions

In practical terms, interpretation of the FMS electrical images through the Palaeogene of this South Viking Graben, Norwegian well give both original and additional information concerning lithology, sedimentary structures, stratigraphy, key stratigraphic surfaces and petrophysical data.

Of immediate significance for this discovery well, the electrical images have allowed the net to gross ratio to be refined, with an increase of up to 10% in some intervals and the hydrocarbon water contact to be more precisely defined.

Key stratigraphic horizons used for correlation have been examined. The Sele–Lista Formation boundary is seen to be very distinct on the images and possibly more indicative of the formation change than the simple gamma ray 'spike'. The Balder Formation shows very distinctive image characteristics and a clear subdivision into two intervals.

In terms of sedimentary structures and sedimentology, a great deal of information is interpreted from the electrical images but cannot be fully interpreted in isolation. Seven sequences with different types of gravity deposit are described. The differences are expressed mainly in terms of lithological and sedimentary structure detail which cannot be interpreted (without core) from sources other than the electrical images. Of the seven sequences, only three show slumping and in two of these every silty/shale interbed is slumped. The identification, and precise dip and azimuth orientation of these slumps, and hence local palaeoslope information, is entirely based on image interpretation.

The image data on which this paper is based are owned jointly by Statoil, Elf Norge, Norsk Hydro and BP Norway. The permission from the companies for their data to be used in this paper is gratefully acknowledged. However, any of the interpretations and opinions expressed are those of the authors alone and for these they accept the full responsibility. Thanks are also due to the referees.

References

Baum, G. R. & Vail, P. R. 1988. Sequence Startigraphic Concepts Applied to Paleogene Outcrops, Gulf and Atlantic Basins. *In*: Wilgus, C. K., Hastings, B. S., St. C. Kendall, C. G., Posamentier, H., Ross, C. A. & Van Wagoner, J. (eds) *Sea Level Changes – An Integrated Approach.*, SEPM Special Publication, **42**, 309–327.

Boyeldieu, C. & Jeffreys, P. 1988. Formation MicroScanner: New Developments. *Transactions of the Eleventh European Evaluation Symposium. S.P.W.L.A.*, 1–16.

CARTER, C. W. 1987. Effect of tool rotation on the computation of dip. *Transactions of the 28th Annual Logging Symposium., S.P.W.L.A.* paper Q, 1–22

CHEUNG, P. S. 1999. Microresistivity and ultrasonic imagers: tool operations and processing principles with reference to commonly encountered image artefacts. *This volume.*

CORFIELD, R. M. & NORRIS, R. D. 1996. Deep water circulation in the Pacific Ocean. *In*: KNOX, R. W. O'B., CORFIELD, R. M. & DUNAY, R. E. (eds). *Correlation of the Early Palaeogene in Northwest Europe.* Geological Society of London Special Publication, **101**, 443–456.

DEN HARTOG JAGGER, D., GILES, M. R. & GRIFFITHS, G. R. 1993. Evolution of Palaeogene submarine fans in the North Sea in space and time. *In*: PARKER, J. R. (ed.) *Petroleum Geology of Northwest Europe: Proceedings of the 4th Conference.* Geological Society London, 59–72.

EKSTROM, M. P. DAHAN, C. A., CHEN, M. Y., LLOYD, P. M. & ROSSI, D. J. 1987. Formation imaging with microelectrical scanning arrays. *Log Analyst*, **28**(3), 294–306.

HISCOTT, R. N., PICKERING, K. T., BOUMA, A. H. *et al.* 1997. Basin-floor fans in the North Sea: sequence stratigraphic models vs. sedimentary facies: discussion. *Bulletin of the A.A.P.G.*, **81**(4), 662–665.

JACQUÉ, M. & THOUVENIN, J. 1975. Lower Tertiary tuffs and volcanic activity in the Noprth Sea. *In*: WOODLAND, A. W. (ed.) *Petroleum and the Continental Shelf of Northwest Europe: Proceedings of the 1st Conference.* Applied Science Publishers, London, 455–466.

KNOX, R. W. O'B. 1996 Correlation of the early Palaegene in northwest Europe: an overview. *In*: KNOX, R. W. O'B., CORFIELD, R. M & DUNAY, R. E. (eds) *Correlation of the Early Palaeogene in Northwest Europe.* Geological Society of London Special Publication, **101**, 1–14.

—— & HOLLOWAY, S. 1992. Palaeogene of the Central and Northern North Sea *In*: KNOX, R. W. O'B. & CORDEY, W. G. (eds) *Lithostratigraphic nomenclature of the UK North Sea.* British Geological Survey, Nottingham.

LOWE, D. R. 1982. Sediment gravity flows: II. Depositional models with special reference to the deposits of high-density turbidity currents. *Journal of Sedimentary Petrology*, **52**(1), 279–297.

MCCLAY, K. R. 1990. Extensional fault systems in sedimentary basins: a review of analogue model studies. *Marine & Petroleum Geology*, **7**(3), 206–233.

MALM, O. A., CHRISTENSEN, O. B., ØSTBY, K. L. FURNES, R., LØVLIE, R. & RUSELÅTTEN, H. 1984. The lower Tertiary Balder Formation: an organogenic and tuffaceous deposit in the North Sea region. *In*: SPENCER, A. M. (ed) *Petroleum Geology of the North-European Margin.*, N.P.F. Symposium Trans. Graham & Trotman, 149–170.

MILTON, N. J., BERTRAM, G. T. & VANN, I. R. 1990. Early Palaeogene tectonics and sedimentation in the Central North sea. *In*: HARDMAN, R. P. F. & BROOKS, J. (eds) *Tectonic Events Responsible for Britain's Oil and Gas Reserves.* Geological Society of London Special Publication, **55**, 339–351.

MUDGE, D. C. & COPESTAKE, P. 1992a. A revised Lower Palaeogene lithostratigraphy for the Outer Moray Firth, North Sea. *Marine & Petroleum Geology*, **9**, 53–69.

—— & —— 1992*b*. Lower Palaeogene stratigraphy of the northern North Sea. *Marine & Petroleum Geology*, **9**, 287–301.

—— & BUJAK, J. P. 1996. An integrated startigraphy for the Palaeocene and Eocene of the North Sea. *In*: KNOX, R. W. O'B., CORFIELD, R. M. & DUNAY, R. E. (eds). *Correlation of the Early Palaeogene in Northwest Europe.* Geological Society of London Special Publication, **101**, 91–114.

MUTTI, E. 1977. Distinctive thin-bedded turbidite facies and related depositional environments in the Eocene Hecho Group (south central Pyrenees, Spain). *Sedimentology*, **24**, 107–131.

O'CONNOR, S. J. & WALKER, D. 1993. Palaeocene reservoirs of the Everest Trend. *In*: PARKER, J. R. (ed) *Petroleum Geology of Northwest Europe: Proceedings of the 4th Conference.* Geological Society London, 145–160.

PROSSER, D. J., MCKEEVER, M. E., HOGG, A. J. C. & HURST, A. 1995. Permeability heterogeneity within massive Jurassic submarine fan sandstones from Miller Filed Northern North Sea. *In*: HARTLY, A. J. & PROSSER, D. J. (eds) *Characterization of Deep Marine Systems*, Geological Society of London Special Publication, **94**, 201–219.

RIDER, M. H. 1996. *The geological interpretation of well logs. 2nd Ed*, Whittles Publishing, Caithness, **280**.

SCHMITZ, B., HEILMANN-CLAUSEN, C., KING *et al.* 1996. Stable isotope and biotic evolution in the north Sea during the early Eocene: the Albæk Hoved section, Denmark. *In*: KNOX, R. W. O'B., CORFIELD, R. M. & DUNAY, R. E. (eds). *Correlation of the Early Palaeogene in Northwest Europe.* Geological Society of London Special Publication, **101**, 275–308.

SERRA, O. 1989. *Formation MicroScanner image interpretation.* Schlumberger Education Services, Houston, 117.

SHANMUGAM, G., BLOCH, R. B., MITCHELL, S. M. *et al.* 1995. Basin-floor fans in the North Sea: sequence stratigraphic models vs. Sedimentary facies. *A.A.P.G. Bulletin*, **79**(4), 477–512.

STOW, D. A. V., READING, H. G. & COLLINSON, J. D. 1996. Deep seas. *In*: READING, H. G. (ed.) *Sedimentary Environments: Process, Facies and Stratigraphy*, 3rd Ed. Blackwell Science.

SULLIVAN, K. B. & SCHEPEL, K. J. 1995. Borehole image logs: applications in fractured and thinly bedded reservoirs. *Transactions of the 36th Annual Logging Symposium. S.P.W.L.A.* paper **T**, 1–12.

TROUILLER, J-C., DELHOMME, J-P., CARLIN, S. & ANXIONNAZ, H. 1989. *Thin bed reservoir analysis from borehole electrical images.* Society of Petroleum Engineers 19578, 61–72.

Stratigraphic relationships in the upper Rotliegend: interpretations from horizontal and vertical well borehole images, core and logs in the Lancelot area, Southern North Sea, UK

DAVID J. WENT[1] & WILLIAM C. FISHER[2]

[1] *Robertson Blackwatch Limited, 8 Buckingham Street, London*
[2] *Mobil Producing Netherlands Inc., Koningin Julianaplein 30, 2595 AA Den Haag, Netherlands*

Abstract: Interpretation of data from core and borehole images in horizontal and vertical wells in Lancelot area fields has resulted in a clearer understanding of the stratigraphic relationships in the upper section of the Rotliegend reservoir. The complex stratigraphy of the upper Rotliegend stems from the reworking and marine inundation of an erg margin. Extensive erosion, fluidization and reworking of the erg occurred but a remnant aeolian topography was retained. The remnant aeolian dune sequence (Unit 3) shows a mounded geometry with a relief up to 40 m and an estimated wavelength of 2 km. Sets of aeolian dune cross bedding are heterogeneous with low permeability layers associated with foresets and set boundaries. Cross bed set orientation and sizes affect upscaled permeability. Differences in set size may relate to a mixture of dune and draa slipfaces or variously sized dunes migrating over draa without slipfaces. The reworked sequence (Unit 4) is a variably thick, commonly erosively-based interval of slightly to heavily cemented sandstone. It is commonly structureless but locally contains slumped aeolian dune foreset beds, dewatering structures, deformed bedding, and sequences featuring repeated intervals of marine shale (Kupferschiefer). The distribution of the various subfacies is related to a model of remnant aeolian dune relief which influenced the style of sediment reworking. The geometric attributes and heterogeneity distributions associated with Units 3 and 4 have important implications for successful targetting of horizontal wells, particularly those planned for close to the top of the reservoir (ensuring maximum standoff from the field gas water contact), for volumetric estimates and reservoir simulations.

This paper presents descriptions and interpretations of the stratigraphy of the upper Rotliegend in the Lancelot area, using core, borehole images and electric logs as the data sets. The majority of core data was obtained from vertical exploration and appraisal wells, whereas the bulk of borehole image data was obtained from horizontal development wells. The objective of our studies was to better characterize and quantify the geometries and heterogeneities present in the upper Rotliegend and to use the results to obtain more accurate volumetric estimates, build more realistic simulator models and to improve the location and targeting of further development wells.

Geological setting and upper Rotiliegend stratigraphy

This study is based on data from the three Lancelot area producing fields, Lancelot, Guinevere and Excalibur. The fields are located in the Southern North Sea, 60 km off the United Kingdom coast in water depths of approximately 25 m (Fig. 1). The gas accumulations are in Permian Rotliegend Group aeolian and fluvial sandstones developed within faulted anticlines, developed at depths of around 2600 m. The field development wells are predominantly of horizontal design and were drilled through 1992 and 1993.

The subdivision of the Lancelot area Rotliegend reservoir, based on the cored vertical wells, is shown in Fig. 2. For the purposes of this paper three main reservoir units are distinguished, labelled downwards Units 4, 3 and 2. Top Unit 2 is used as the datum in Lancelot and is interpreted as an approximately flat, time equivalent surface. Units 3 and 4 are the target horizon for several of the horizontal development wells and form the subject of this article.

From: LOVELL, M. A., WILLIAMSON, G. & HARVEY, P. K. (eds) 1999. Borehole Imaging: applications and case histories. Geological Society, London, Special Publications, **159**, 139–153. 1-86239-043-6/99/$15.00.

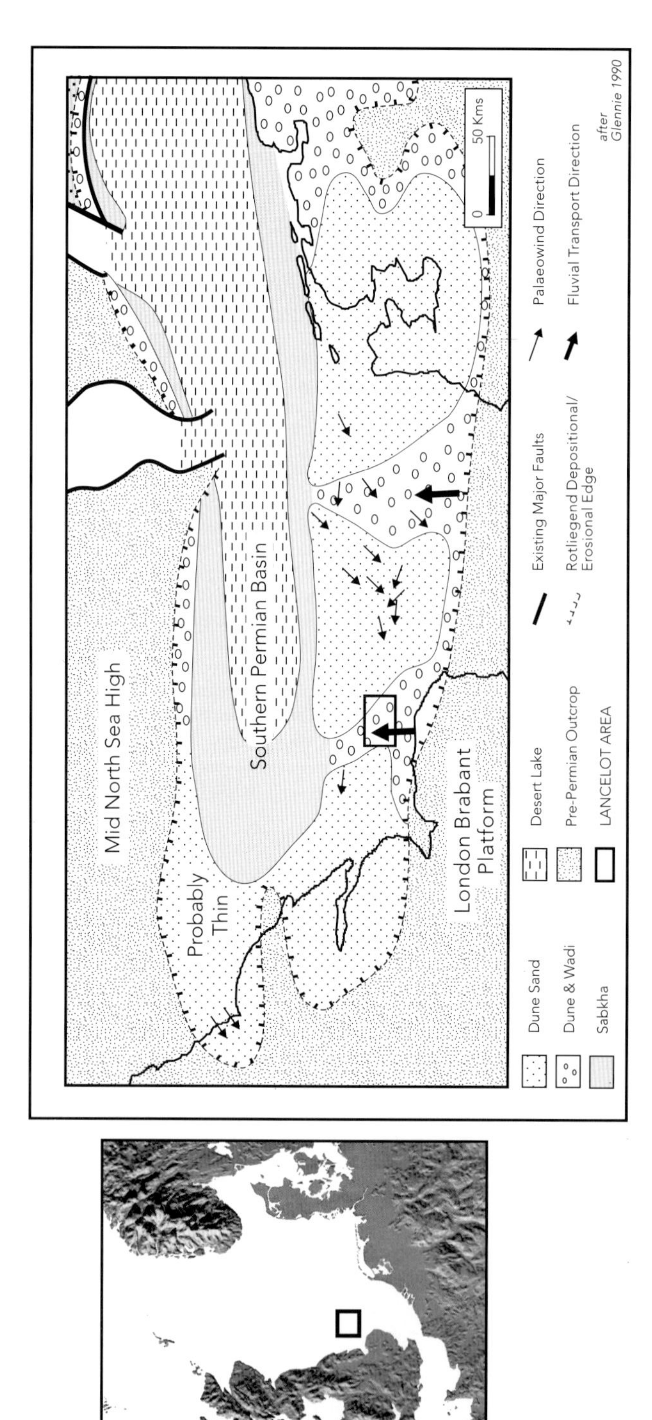

Fig. 1. Location map and early Permian palaeogeography.

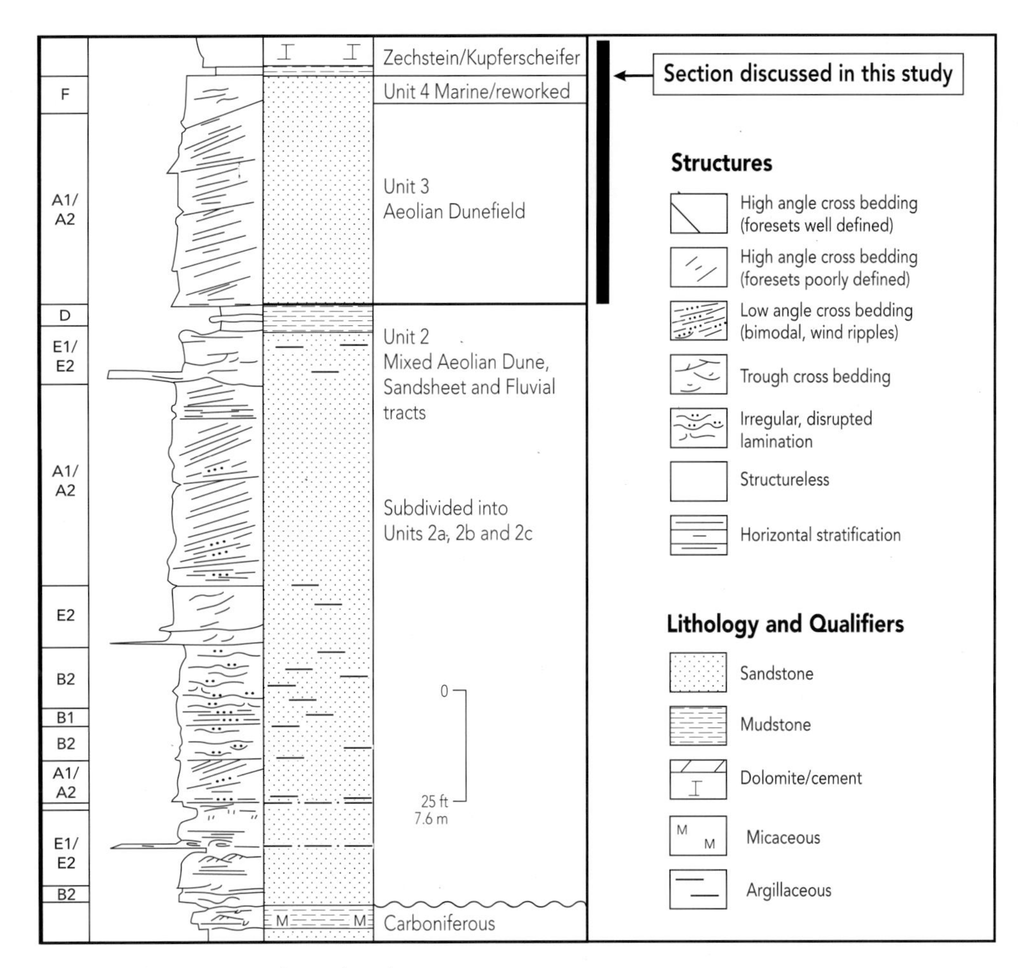

Fig. 2. Lancelot area reservoir stratigraphy.

The most important aspects of the upper Rotliegend stratigraphy in the context of reservoir modelling and field development are as follows:

- Unit 3 (remnant aeolian dune field) is an excellent quality reservoir which shows a mounded geometry with thickness ranging from 0 to 40 m. Prediction of the distribution of this good quality reservoir is important, not least because it is a target for horizontal development drilling.
- Unit 3 shows substantial permeability heterogeneity associated with cross bed foresets and set boundaries. This necessitates an appreciation of the scales and orientation of cross bed sets.
- Prediction of Unit 3 thickness variation is also important because the uppermost reservoir layer Unit 4 (reworked sandstones) only partly infills Unit 3 relief, resulting in an undulating top to the Rotliegend. A lack of appreciation of this effect and its scale of occurrence can lead to a loss of reservoir exposure in well sections that are aimed at near top Rotliegend (to ensure maximum standoff from the field gas water contact) if the well exits the top of the unit.
- Unit 4 sandstones show variation in sedimentary character and facies, directly related to heterogeneity and reservoir quality, sometimes exacerbated by diagenetic processes. This can lead to potential loss of reservoir quality rock in mis-targetted well sections in Unit 4.

The remainder of this paper shows how we obtained and utilized data from core and image logs to better characterize and quantify the scales of reservoir variability described above.

Data acquistion

Exploration and appraisal wells

As much core as possible, in addition to log data, was obtained over the reservoir section (typically 50–90% core coverage with each well) of each of the exploration and appraisal wells in order to maximize knowledge of the reservoir as early as possible. Early studies revealed variable thickness in Unit 3 and facies within Unit 4. These are the uppermost units in the reservoir and are the key pay zones. Hence, it was critical to gain an understanding of their geometry, and lateral, and vertical heterogeneity.

Development wells

Vertical wells yield important information regarding vertical changes in stratigraphy but limited information on lateral variations in reservoir character. Horizontal producers were drilled parallel to the predominant dip direction of foreset beds within the aeolian deposits (i.e. parallel to the palaeowind direction) to minimize the effects of permeability anisotropy. As such they provide considerable lateral information, albeit in a narrow arc and in a limited vertical sense. Early horizontal development wells were extensively logged using electrical borehole image logs. Seven out of nine wells were fully logged using either the Western Atlas sonic-based Circumferential Borehole Imaging Log (CBIL) or Schlumbergers's Formation Micro-Scanner (FMS). In addition 20 m of core was taken in one of the horizontal wells. The aim was to acquire data to better understand lateral changes in stratigraphy and facies in the upper units 3 and 4.

Methods of data interpretation

Cores and logs from vertical wells were used to identify the facies, establish the layering scheme and determine the reservoir parameters in the Lancelot area (Table 1, Fig. 2).

Borehole images from horizontal wells were interpreted and recordings of sedimentary and tectonic structures made in the form of a diagram of the borehole cross section, with associated dip tadpole plots (Fig. 3). The method used to generate the diagrammatic interpretations of the structures present in horizontal boreholes is summarized on Fig. 4. Interpretations of lithologies were made from gamma ray, density-neutron and resistivity logs. Artefact features are particularly common on CBIL logs (FMS to a lesser extent) and need to be distinguished from genuine geological structures. The most common artefacts seen on the Lancelot well CBIL images and their possible causes are listed in Table 2.

Facies interpretations

Unit 3

Two main subfacies may be distinguished on borehole images in Unit 3.

High angle foreset beds (A1). In horizontal wells these are typically multiple, parallel, moderate to high amplitude sine waves, with the nose of the sine wave centered near to the borehole image centre-line. The near perfect centering of many of the sine waves reflects the predominance of foreset dips parallel with the well bore, a feature of pre-planned well design. Less common, off-centre closure, however, is associated with very high amplitude sine curves indicating foreset beds dipping obliquely across the borehole.

These are interpreted to be high angle foreset beds of aeolian dune sandstone origin (Fig. 5a).

Low angle foreset beds (A2). In horizontal wells these are multiple parallel, very high amplitude sine waves, commonly associated with a bright

Table 1. Typical average porosity and permeability values of the Lancelot area Rotliegend facies in reservoir units 3 and 4.

Facies & Unit	Arithmetic average porosity (%)	Geometric average permeability (mD)
Unit 4: Fluvial and marine reworked sandstones	8–13	1–77
Unit 3: Aeolian dune sandstones	7–15	1–54

Data uncorrected for overburden conditions.

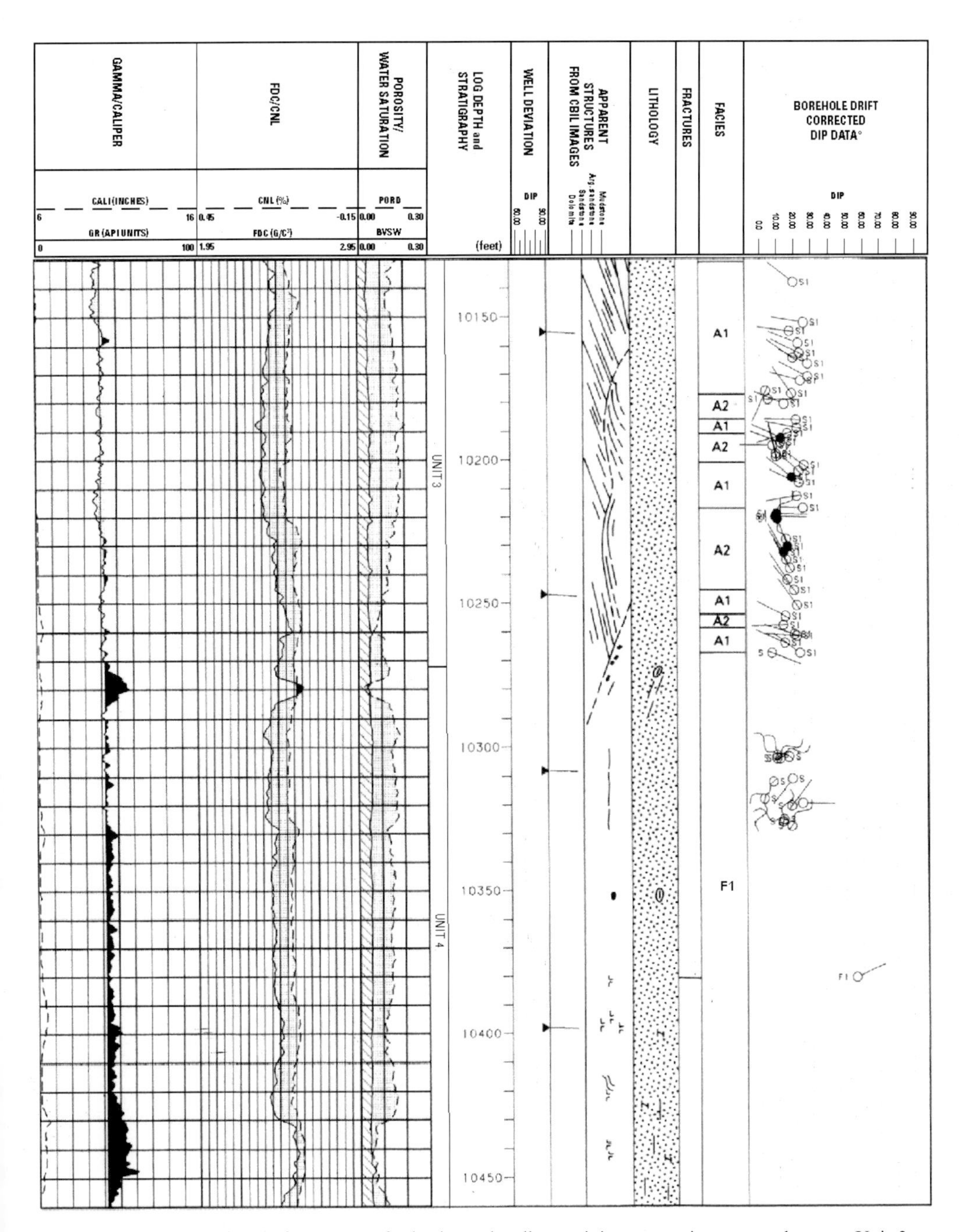

Fig. 3. Interpretation of a 320 ft segment of a horizontal well containing an erosive contact between Unit 3 (10130 10270, cross bedded) and Unit 4 (10270–10450, massive).

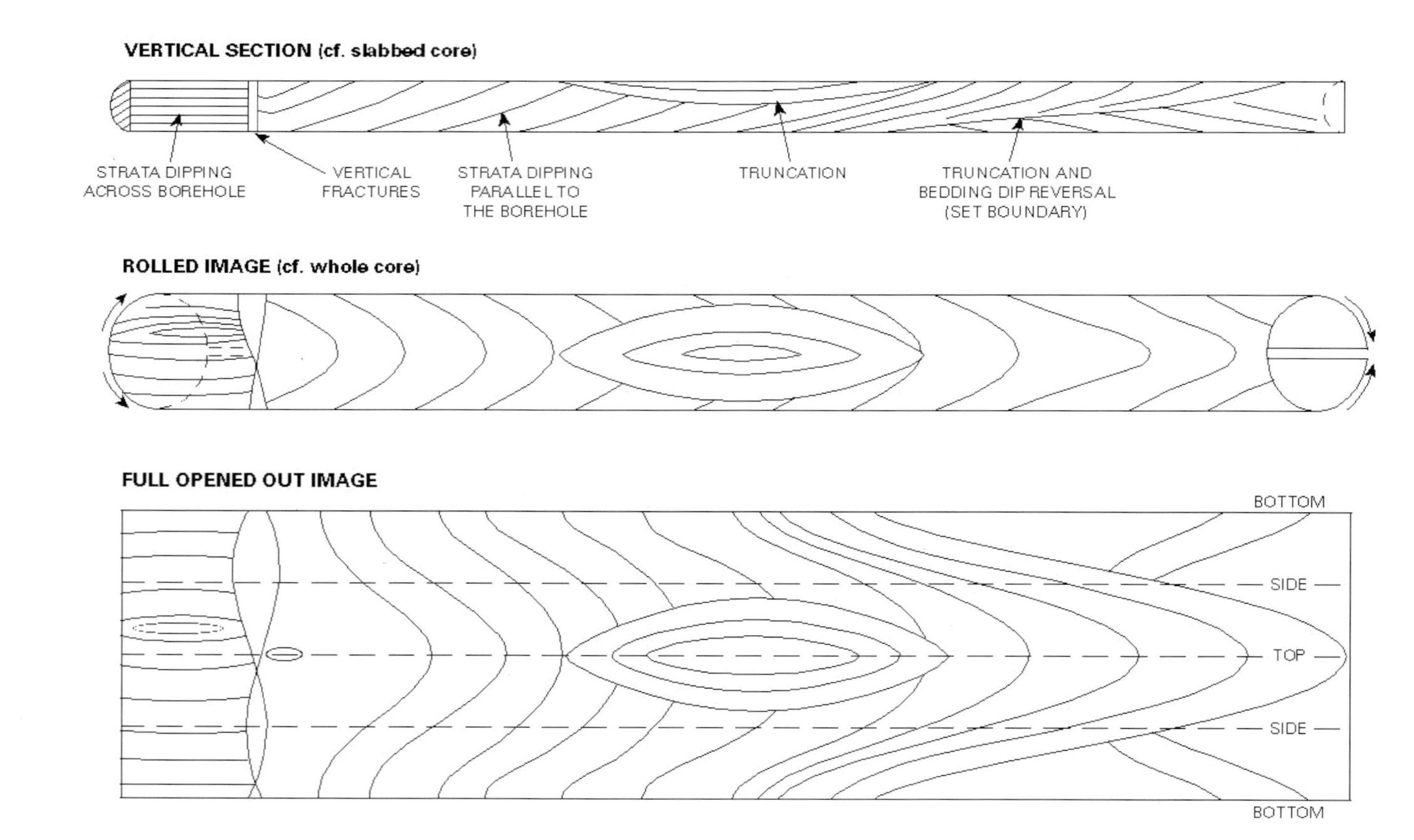

Fig. 4. Method used to identify and illustrate the sedimentary and tectonic structures present in horizontal well CBIL images.

Table 2. CBIL image artefacts present in Lancelot horizontal wells and their possible causes. Logs were pipe conveyed with the image tool typically used on a second logging run.

Image artefact feature	Possible causes	Example
Borehole parallel dark streaks and bands at 090° and 270°	Collapse/breakout of borehole sides, probably along bedding planes	
Discontinuous dark streaks subparallel to borehole	Longitudinal scratches and gouges, probably caused by arm springs and ploughs on pad contact tools in previous log runs	
Diagonal marks and streaks, commonly with (drill bit) 'teeth' marks	Spiral scratches and gouges on borehole wall caused by contact of drill pipe, debris and drill bit with borehole wall, on tool/drill pipe removal.	
Uneven image colouring/brightness contrast: bright top hole, dark base hole and *vice versa*	Tool eccentered and/or borehole elliptical	Fig. 5a (travel time image)
Uneven, irregular, diagonal drifts in image colouring/brightness contrast	Tool eccentered and spiralling and/or borehole deformed due to *in-situ* stresses and drilling induced irregularities	Fig. 5a (travel time image)
Regular 0.20 m borehole-perpendicular 'striping' with stepped sine waves	Tool stick and pull	Fig. 5b (reflectance image)
Abrupt changes in image contrast perpendicular to borehole axis	Visual enhancement of image during log processing	

image response, reflecting a tighter lithology. The high amplitude sine waves show closures close to the centre-line of the image reflecting a low dip to the bedding in this facies.

These are interpreted to be the gently inclined parts of aeolian cross bed sets, representing deposition near the dune base.

Unit 4

Six subfacies or features were recognised on borehole images in Unit 4 (Facies F1).

Erosively based sandstones. These are sandstones, most commonly structureless, showing an unconformable contact with the underlying Unit 3 dune foreset beds. The erosive unconformity is typically marked by the truncation of a set of parallel sine waves by a single sine wave overlain by featureless strata (Figs 5b & 3). In the wells studied this is the most common type of contact between Unit 3 and Unit 4.

Structureless sandstones. These are featureless strata showing no evidence of bedding in the form of sine waves.

Texturally mottled sandstone. These are sandstones showing mottling within the image. The mottling reflects patchy distribution of clay and cement within sandstones. This texture is typically more apparent in the resistivity images (Fig. 6), than sonic images. Comparison with core suggests this facies represents dewatered sandstones and cement mottled massive strata.

Mudstone beds and clasts. These are dark bands and spots respectively. The beds form sine waves, of conductive or low velocity strata found within the sandstones of Unit 4. They are similar in thickness and character to the mudstone that caps the reservoir, named the Kupferschiefer. Comparison with observations made in core suggest that these are occurrences of split Kupferschiefer, that is where the marine mudstone has been interrupted by an episode of sandstone redeposition (Fig. 7). The spots may in some cases reflect shale clasts. Alternatively they may represent plucked holes filled with drill fluid.

Deformed bedded sandstone. These are sandstones showing clearly defined strata in the form of imperfect and irregular sine waves (not

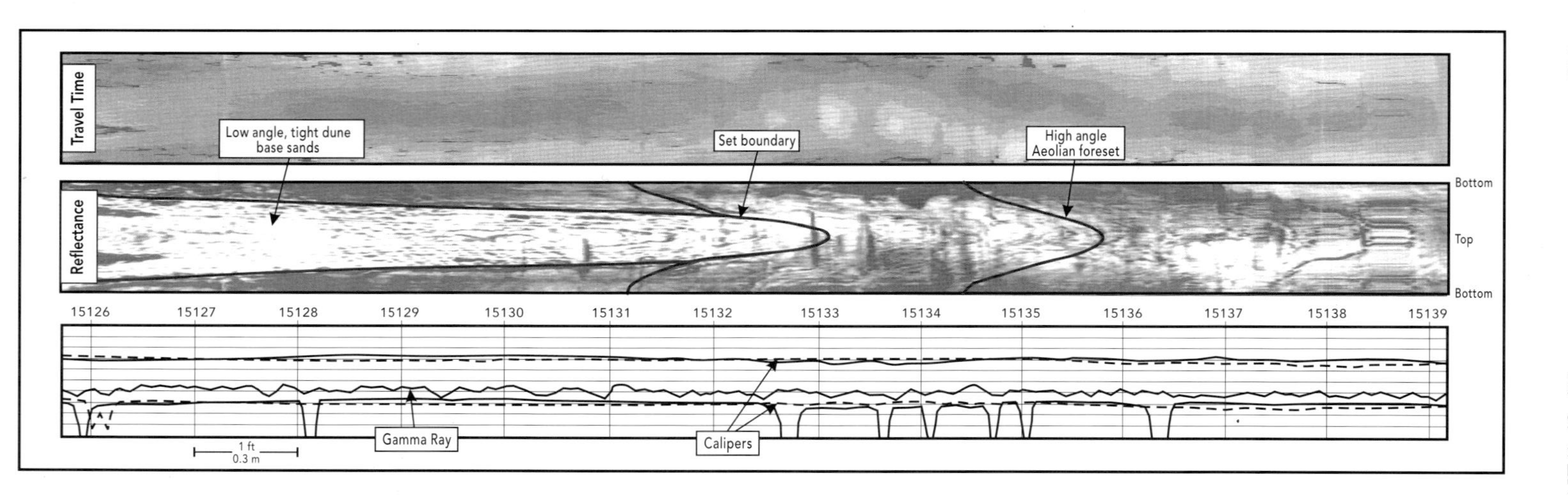

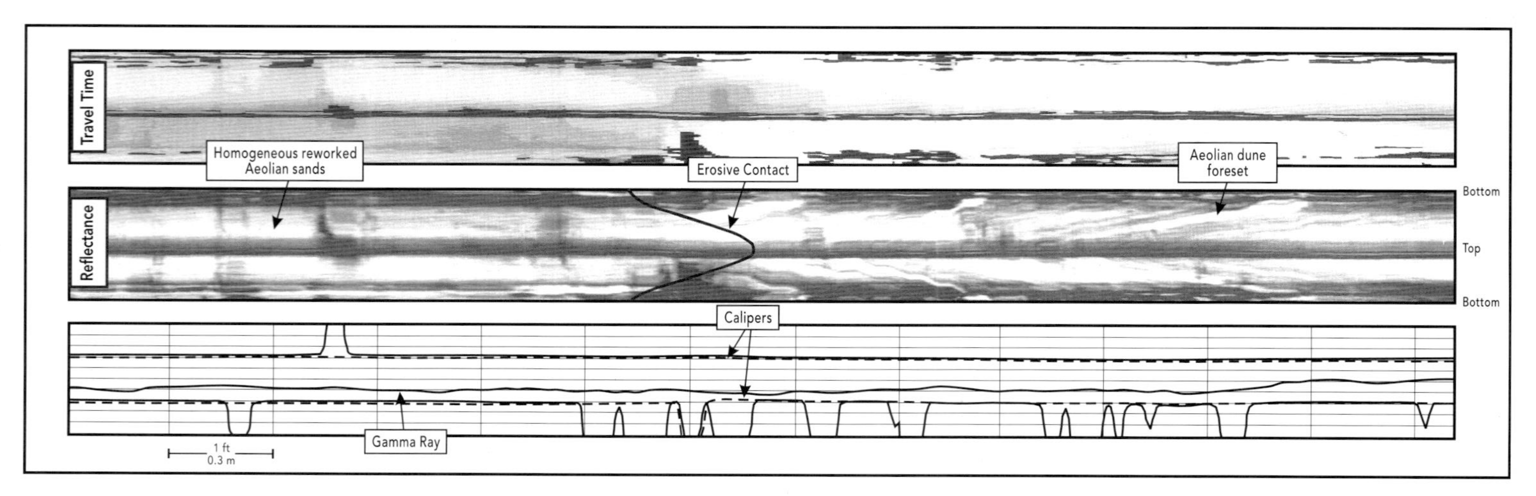

Fig. 5. (a) Borehole image (CBIL) showing foreset dips and set boundary in dune sandstone of Unit 3. (b) Borehole image (CBIL) showing the erosive contact between structureless sandstones of Unit 4 and dune foreset bedding of Unit 3.

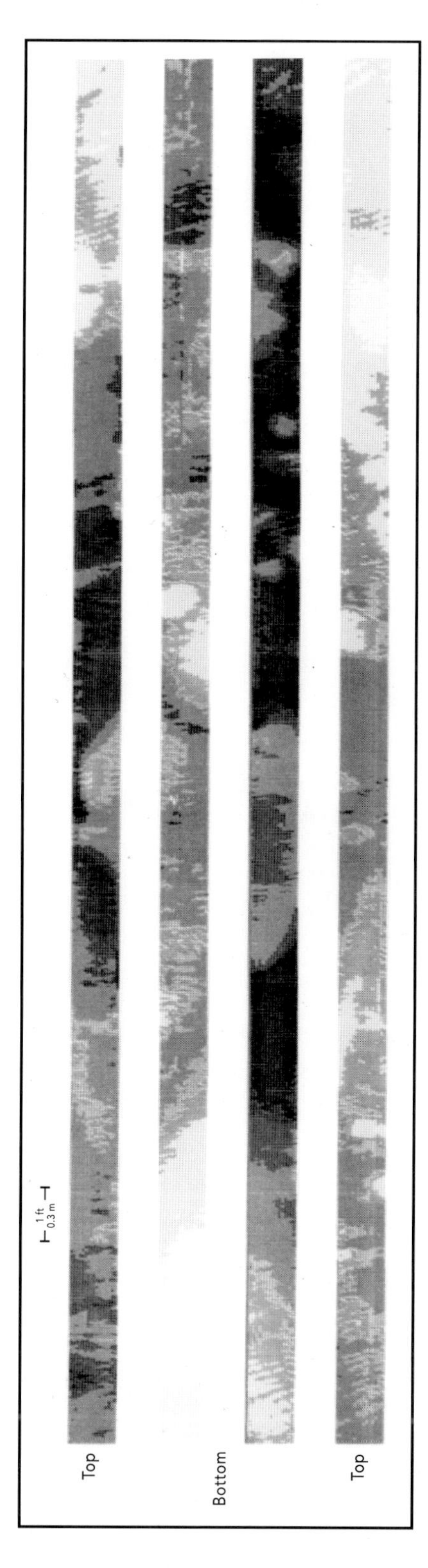

Fig. 6. Borehole image (FMS) showing texturally mottled, fluidized character from Unit 4.

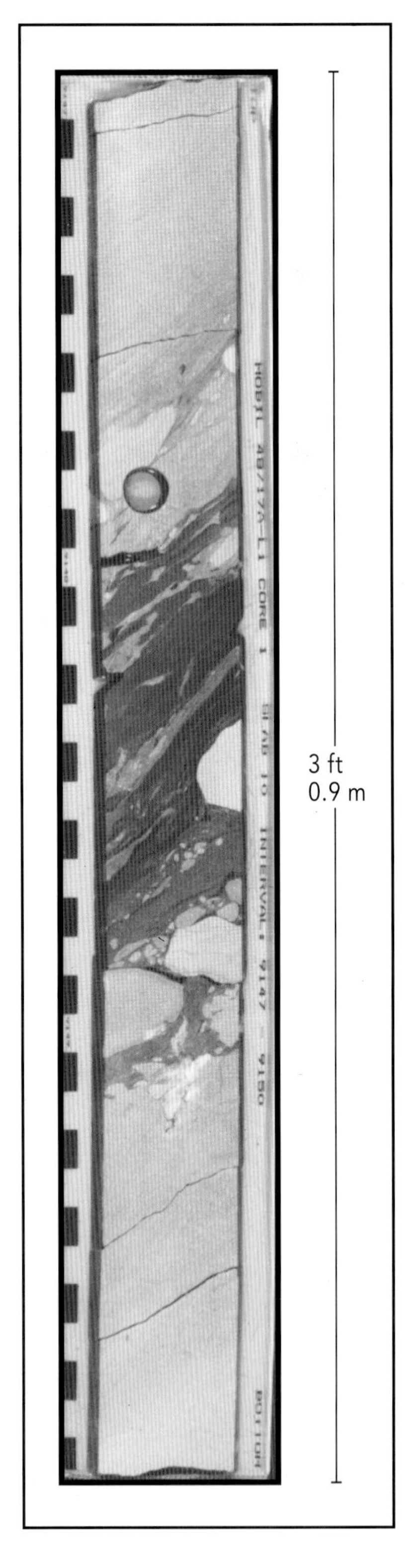

Fig. 7. Core photograph of upper Rotliegend showing reworked slumped Kupferschiefer shale (Kupferschiefer, verified by palynology) overlain by Unit 4 Rotliegend sandstones. Note the shale contains clasts of sandstone, suggesting some early cementation of the dune sandstone shortly after marine inundation.

those resulting from stick and pull of the tool, Table 2). Comparison with observations made in core suggest that where sine waves are closely spaced, well defined and sub-parallel they probably represent slumped aeolian dune foreset beds. Where faint and widely spaced they probably represent dewatered or slumped waterlain deposits.

Faint flat bedded sandstone. These are deposits featuring faint, fairly widely spaced high amplitude sine waves in horizontal wells. These represent flat bedded sandstone.

Characterization of unit geometry and heterogeneity

Unit 3

Geometry. Cross sections were constructed through horizontal wells and projections of nearby vertical wells. All horizontal development wells were drilled approximately parallel to the inferred palaeowind direction (e.g. Fig. 8). A gradual mounding to top Unit 3 was detected based on observations of facies changes within the well bores of all the highly deviated/horizontal wells drilled through Unit 3 and Unit 4. Furthermore, the scale of the longest wavelength mounding appeared consistent in each of the cross sections, suggesting a wavelength the order of twice the average length of the reservoir well sections (i.e. 2×1000 m). Consideration of the thickness of Unit 3 in other well locations confirms a 2 km wavelength to the mound features. The mounds are interpreted as partly eroded large compound sand dunes, termed draa by Wilson (1973).

Facies heterogeneity. Considerable permeability variability is associated with the foresets as well as set boundaries of sets of aeolian cross bedding in the Lancelot area (Went & Fisher 1997). This contrasts with results from aeolian strata from other areas (e.g. Goggin *et al.* 1988). The heterogeneity associated with the sets of cross bedding suggests that the scale and orientation of the cross bed sets will influence large scale permeability (simulator grid block scale) and its directional properties (cf. Sweet *et al.* 1996).

The orientation of cross bed foresets has been recorded in all wells, yielding a mean dip direction to the NW. The dimensions of cross bed sets in Lancelot wells vary considerably (e.g. heights ranging from <0.3 to >20 m). This may reflect deposition of the cross bed sets as either a mixture of draa and dune slipfaces or slipfaces to a wide variety of different sized dunes migrating over a draa body without slipfaces (Fig. 9). We have not been able to distinguish between the two, but have noted the distribution of set sizes and orientation to enable long term comparison of well performance.

Unit 4

Geometry

Observation of thickness variations in Unit 4 through all wells indicate that where Unit 3 is thin, Unit 4 tends to be thick and *vice versa*. However, where Unit 3 is thin or absent, Unit 4 although relatively thick, never approaches the maximum thickness of Unit 3 and Unit 4 combined. In other words, Unit 4 acts to fill some but not all of the relief on Unit 3 and up to 25 metres of relief is still present at the top of the reservoir. This has clear and important implications for the targeting and design of wells aimed at near top reservoir horizons, in areas away from current well locations, and also for reserves estimates. Since the magnitude of this relief may frequently be below seismic resolution this would cause a significant inaccuracy on the top reservoir map. This would be especially so where the top reservoir map is generated by bulk shifting down from a basal Zechstein mapping horizon.

Facies heterogeneity. The distribution of the different subfacies within Unit 4 is difficult to predict with confidence, in part due to the scant occurrences in cores of some of the subfacies (e.g. slumped dune foresets). However, the additional data gained from borehole images in the horizontal wells together with our improved understanding of the geometric relationships of reservoir units 3 and 4 has resulted in a speculative model for the distribution of the different subfacies within Unit 4 (Fig. 10). This model shares some similarities with previously published works (Glennie & Buller 1983, Heward 1991) and features the following key aspects:

- A mounded dune sandstone topography was reduced in elevation by wind or water erosion, as witnessed by the erosive contact between Unit 3 and Unit 4.
- The flanks of the dunes underwent slumping resulting in the local preservation of slumped dune foreset beds. Destabilization of the dune flanks possibly resulted from fluvial erosion and undercutting as the watercourses favoured the interdraa low areas.

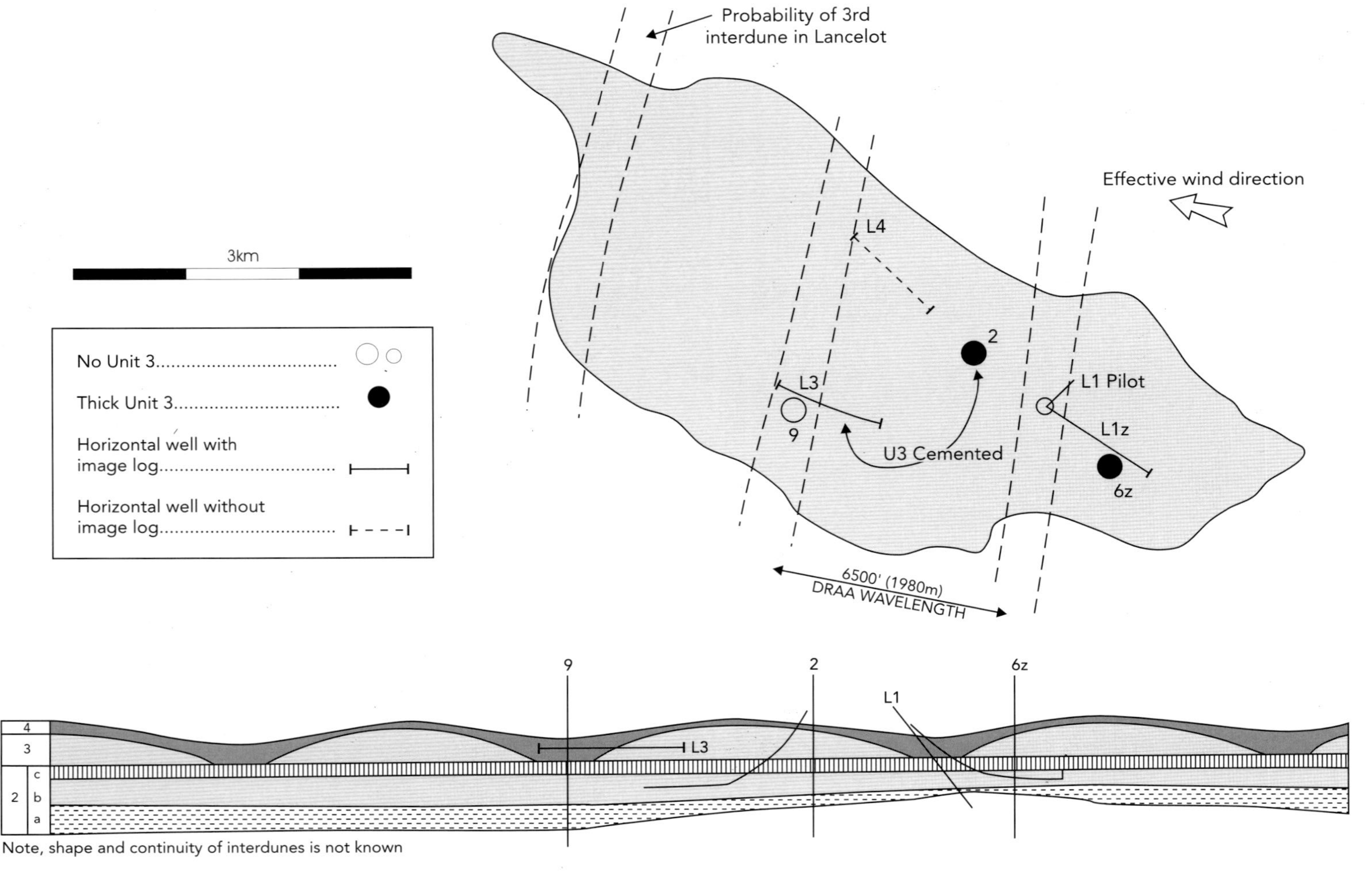

Fig. 8. Cross section illustrating lateral change in facies in a horizontal well section (L3) due to the discontinuous mounded geometry of Unit 3.

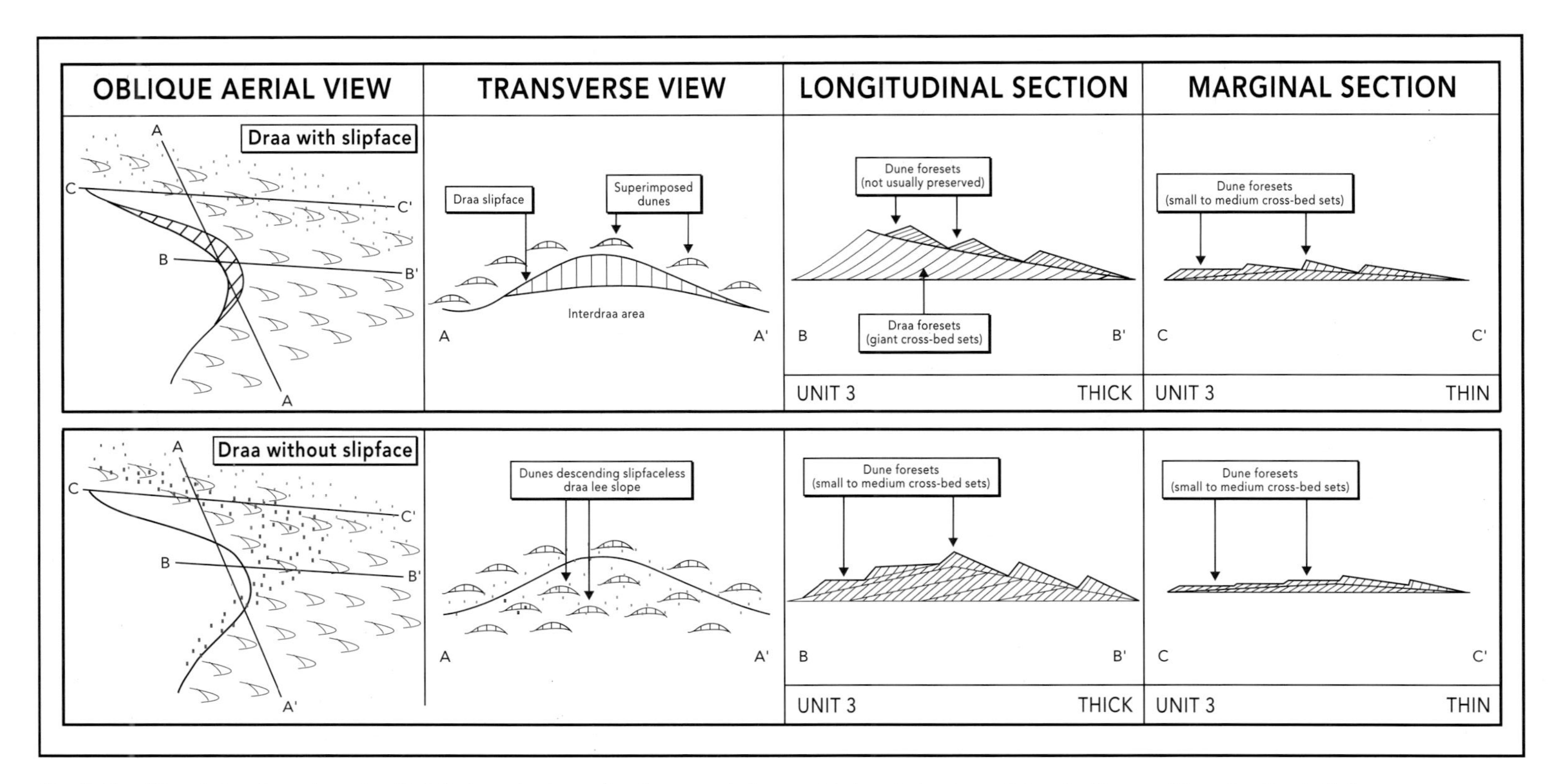

Fig. 9. Possible relationship of varying Unit 3 character to contrasting morphology and internal structure of draa bedforms with and without slipfaces.

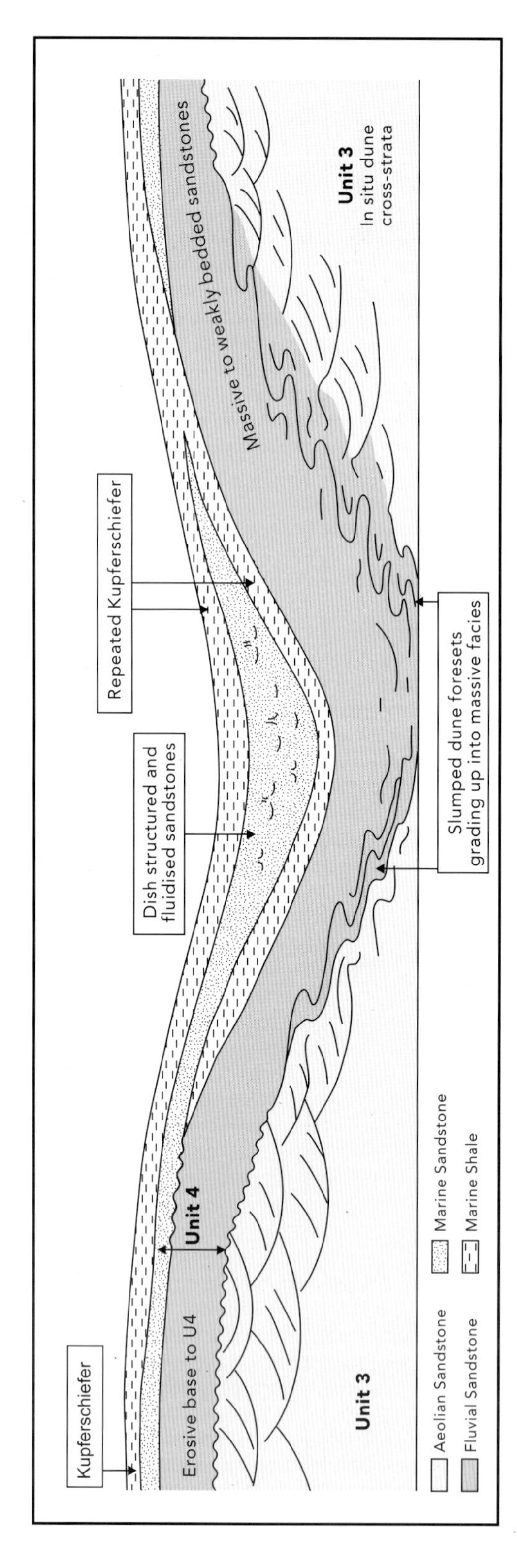

Fig. 10. Cartoon illustrating the possible relationships of different subfacies within Unit 4.

- The interdraa areas were filled with massive to faintly flat-laminated sandstones, probably of ephemeral fluvial flood origin. Rapid sedimentation or subsequent remobilization may have resulted in the formation of water escape and deformation structures.
- Later stages of deposition of Unit 4 are interpreted to have occurred under marine conditions. Initially marine currents may have resulted in some erosion and re-sedimentation of sand around dune crests.
- Deepening of the marine conditions was associated with the deposition of the anoxic Kupfershiefer shale which draped the relic dune relief.
- Instability on the flanks of the dunes resulted in erosion of shale and sand (in part as clasts) from submerged dune crests and re-deposition by mass flows in interdune hollows. This resulted in sands between the split Kupferschiefer sequences.
- Deposition of the Kupferschiefer as a shale capping to the total sequence.

Conclusions

Horizontal wells are a valuable source of geological information. Rigorous integration of geological data derived from image logs in bed parallel horizontal wells with vertical well data results in improved reservoir descriptions. This study has shown that there are considerable depositional and stratigraphic complexities in the upper part of the Rotliegend, which may not be susceptible to seismic character analysis. The careful analysis of an extensive image log and core data set has allowed some quantification of these complexities.

We wish to thank Mobil North Sea and Robertson Research International Limited for permission to publish this paper, and the Geological Society reviewers Andy Morton and Rose Davey for critical reviews that helped improve the article.

References

GLENNIE, K. W. & BULLER, A. T. 1983 The Permian Weissliegend of NW Europe: The partial deformation of aeolian dune sands caused by the Zechstein transgression. *Sedimentary Geology*, **35**, 43–81.

HEWARD, A. P. 1991. Inside Auk – the anatomy of an eolian oil reservoir. *In*: MIALL, A. D. & TYLER, N. (eds) *The three dimensional facies architecture of terrigenous clastic sediments, and its implications for hydrocarbon discovery and recovery*. Society of Economic Palaeontologists and Mineralogists Concepts in Sedimentology and Palaeontology, **3**, Tulsa, 44–56.

GOGGIN, D. J., CHANDLER, M. A., KOCUREK, G. A. & LAKE, L. W. 1988. Patterns of permeability in eolian deposits: Page Sandstone (Jurassic), Northeastern Arizona. *SPE Formation Evaluation*, **3**, 297–306.

SWEET, M. L., BLEWDEN, C. J., CARTER, A. M. & MILLS, C. A. 1996. Modelling heterogeneity in a low permeability gas reservoir using geostatistical techniques, Hyde Field, Southern North Sea. *AAPG Bulletin*, **80**, 1719–1735.

WENT, D. J. & FISHER, W. 1997. Integration of horizontal well geological data into reservoir descriptions: Rotliegende sandstone, Lancelot area, Southern North Sea. *AAPG Bulletin*, **81**, 135–154.

WILSON, I. G. 1973. *Ergs. Sedimentary Geology*, **10**, 77–106.

Application of borehole image logs in constructing 3D static models of productive fracture networks in the Apulian Platform, Southern Apennines

ROBERT TRICE

Enterprise Oil Italiana SpA, via dei Due Macelli, 66-00187 Roma (email: Robert.Trice@rome.entoil.com)

Abstract: Fractured reservoirs of the Apulian Platform carbonates present significant challenges from the perspective of formation evaluation. Evaluation problems manifest themselves at all stages of reservoir development from exploration and production drilling through to reserves calculation and reservoir simulation. A significant aspect of these problems arises from the difficulty in locating and testing productive fractures during drilling operations. Once productive fractures have been identified the problem then becomes how to model them away from the borehole and to assign field scale fracture parameters. A key data set in such evaluations is provided by borehole image logs. This paper discusses how borehole image logs have been used to describe the Apulian Platform carbonates with respect to: (a) identify producing fractures, (b) to understand and model the relationship of fractures to the stress field, (c) to provide input to 3D models of producing fracture systems, and (d) how the 3D models were used to quantify the fracture system. The combination of these elements has led to a practical methodology by which a significantly improved understanding of the Apulian Platform fracture system has been achieved.

The techniques and conclusions of this paper resulted from in-house work designed to interpret a pre-existing data set, to establish whether wireline logs could be confidently used to locate producing fractures in the Apulian Platform and could a testable stochastic model be built of the producing fracture system that would have some practical benefit to reservoir simulation or well placement. Timing of the study was such that conclusions could be used to influence decisions on selecting test intervals on an ongoing exploration well. The results from this work are in large part due, to the application of specialist and research code. Specifically, stress modelling was undertaken by use of the SFIB software (Stress and Failure in Inclined Boreholes), developed by Stanford Rock Physics & Borehole Geophysics Project. 3D fracture modelling was achieved by use of the research software FRACA and TRANSFRAC (written by Institute Francais Petroleum).

The reservoirs of the Apulian Platform are located within the thrust belt of the Italian Southern Apennines, and consist of thick sequences of Mesozoic–Cenozoic platform carbonates. The Southern Apennines are an exposed accretionary wedge that suffered major thrusting in the Miocene . Current structural trends have been influenced greatly by recent extensional processes (Hippolyte *et al.* 1995) associated with regional uplift, probably induced by isostatic rebound. A complex tectonic history complicates the already difficult task of understanding and modelling the producing fracture networks of the Apulian platform carbonates.

Within the Apulian platform it is generally recognized that the reservoir primarily consists of low porosity carbonates and therefore the fracture system plays an essential part in both the storativity and the producability of hydrocarbon. In evaluating the fracture system for input into STOIIP (stock tank oil originally in place) and reserves calculations the formation evaluation problem falls into four basic phases: (a) detection of producing fractures, (b) interpretation of the fracture style and its relationship to any matrix porosity/permeability, (c) extrapolation of the static fracture system away from the borehole, and, (d) modelling of the bulk fracture properties. The following text considers the basic problem of evaluating a fractured reservoir from the perspective of borehole image data and then describes how image data has been interpreted and integrated

From: LOVELL, M. A., WILLIAMSON, G. & HARVEY, P. K. (eds) 1999. Borehole Imaging: applications and case histories. Geological Society, London, Special Publications, **159**, 155–176. 1-86239-043-6/99/$15.00.

within each of these phases of reservoir description to quantify the fracture systems of the Apulian Platform.

Fractured reservoir evaluation, the basic problem

A natural fractured reservoir consists of a large volume of rock and a network of fractures. (Fig. 1a). Within the fracture system will be relatively permeable and relatively impermeable fractures. There will also be fractures unconnected and connected to the part of the fracture population that carries fluid flow. This flowing sub-population of the fracture network is termed the backbone (Suhammi 1994, Odling & Olsen 1997). In addition there will be a range of fluid types within the fractures some of which will contain hydrocarbons in sufficient concentrations to be commercially prospective. In addition there may be faults that can act as local

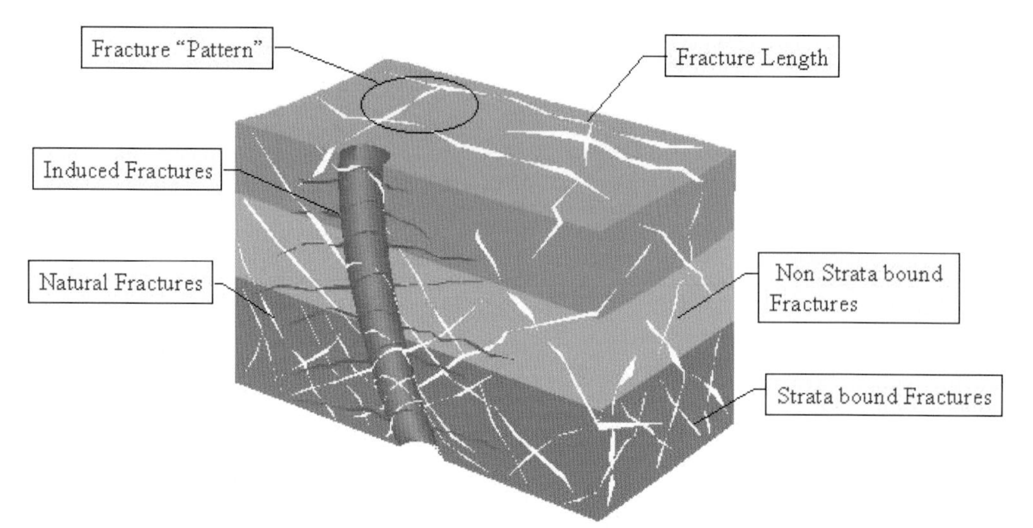

Fig. 1a. Conceptual diagram exhibiting fracture distribution within a formation and in the near well bore environment.

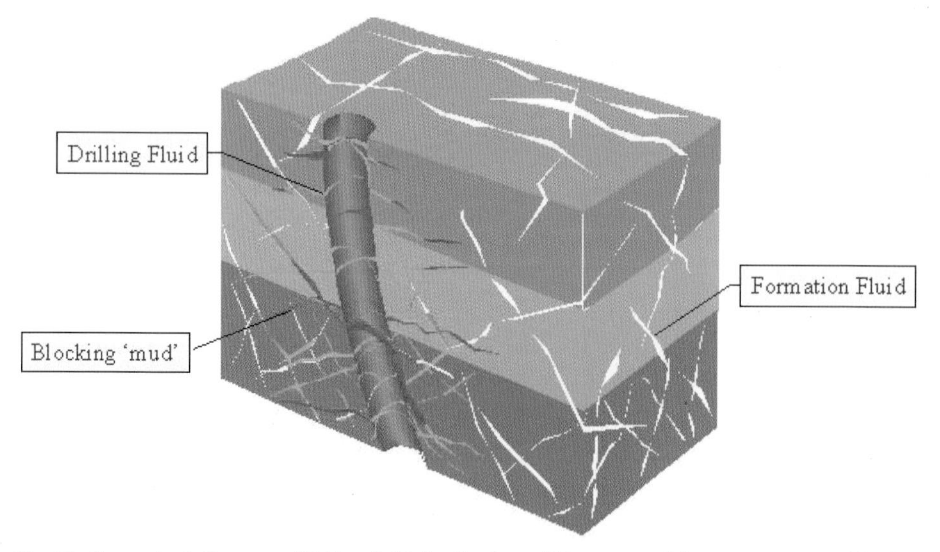

Fig. 1b. Conceptual diagram exhibiting fluid distribution within a natural fracture system and in the near well bore environment.

flow enhancement or flow barriers. The system is therefore complex. The problem of modelling this complexity is made more difficult by the limitations of sampling by boreholes which are effectively one dimensional scan lines which can never adequately sample three dimensional fracture geometry. The application of borehole image data to detecting and modelling productive fractures is further handicapped as, with the exception of down hole video cameras, borehole image logs provide only static information of the fracture subset. This observation is very much a universal 'law' apart from unusual exceptions such as gas bleeding into a well bore recorded by borehole televiewers. The static information recorded by borehole image logs is confined to (a) fracture depth, (b) fracture attitude, (c) apparent aperture, and (d) the fracture morphology, which is the relationship of the fracture and the image fabric. Information concerning (a) fracture length, (b) fracture height, (c) the fracture pattern or template, i.e. how fractures cross and abut, and (d) whether the fractures are strata or non strata bound (Fig. 1a) is not recorded by borehole image logs. The exception to this is within thin and medium beds where strata bound fracture heights can be estimated. Such missed information is crucial to constructing 3D static models and can only be inferred through well correlation or through analogue studies. In addition the drilling process causes a range of stress release and drilling induced fractures in the well bore (Fig. 1a) and in the near well bore environment. Such features typically penetrate the formation a matter of millimetres but have the potential to propagate well into the formation. Further complexity to interpreting natural productive fractures from borehole images is caused by the process of drilling which alters the distribution of formation fluids (Fig. 1b) through a range of processes which include: (a) flushing the original fluid away from the borehole (normal invasion), (b) plugging permeable fractures with mud or filtrate, and (c) altering the effective permeability of the formation. Whilst rarely directly affecting the response of image logs these factors are of considerable significance when interpreting production log and MDT (Schlumberger Modular Dynamic Tester) data, and conventional resistivity logs.

In evaluating the Apulian Platform reservoir productive fractures were defined through the analysis of multiwell data sets that consisted of static and dynamic data. The under sampling problem was overcome by creating statistically valid 3D static fracture models that were well constrained by borehole data. The construction of such models necessitated the integration and interpretation of all available static and dynamic subsurface data. Borehole image data was specifically interpreted in, (a) resolving the reservoir stress field, (b) defining a distribution of producing fractures from individual well bores, (c) creating a series of conceptual fracture models, (d) locating and testing a fracture analogue, and (e) constructing and verifying 3D stochastic realizations of the producing fracture network. Once a satisfactory 3D model had been created it was used directly to ascertain what aspects of the fracture system were contributing most significantly to flow, and to estimate the bulk fracture porosity.

Modelling the stress field

Previously published data (Barton *et al.* 1997) indicate that natural open fractures can have preferential production characteristics if aligned parallel to the maximum horizontal stress in extensional tectonic regions. In addition it has been observed that active fault planes can modify the local stress field (Barton & Zoback 1994). To understand the potential of stress effect to hydrocarbon flow in the Apulian Platform an analysis of the regional and local stress fields was undertaken. From outcrop studies it is clear that the Southern Apennines have a well defined anisotropic stress field, which has a maximum horizontal stress oriented NW–SE associated with NE–SW extension. (Hippolyte *et al.* 1995). This stress field has been confirmed in the subsurface at reservoir levels (Amato *et al.* 1995). To understand the stress field within the reservoir section additional detailed work was carried out using the SFIB software. In summary a combination of density, sonic log and mud weight data was used to establish the components of vertical stress whereas breakout data and tensile wall failure were interpreted with respect to defining the minimum and maximum horizontal stress components. Breakout orientations were derived from the callipers of dipmeter, electrical imaging logs and from borehole televiewer transit time images. Examples of the type of stress features, interpreted from image log data, used to help constrain the stress tensor are seen as Fig. 2. Dark colours are conductive features on electrical images and low transit time/amplitude on acoustic images; by combining these data within the confines of the SFIB programmes it was possible to define a quantified stress tensor. The values of the tensor were then checked by forward modelling of tensile wall failure fractures, identified from electrical imaging logs, in inclined boreholes. This

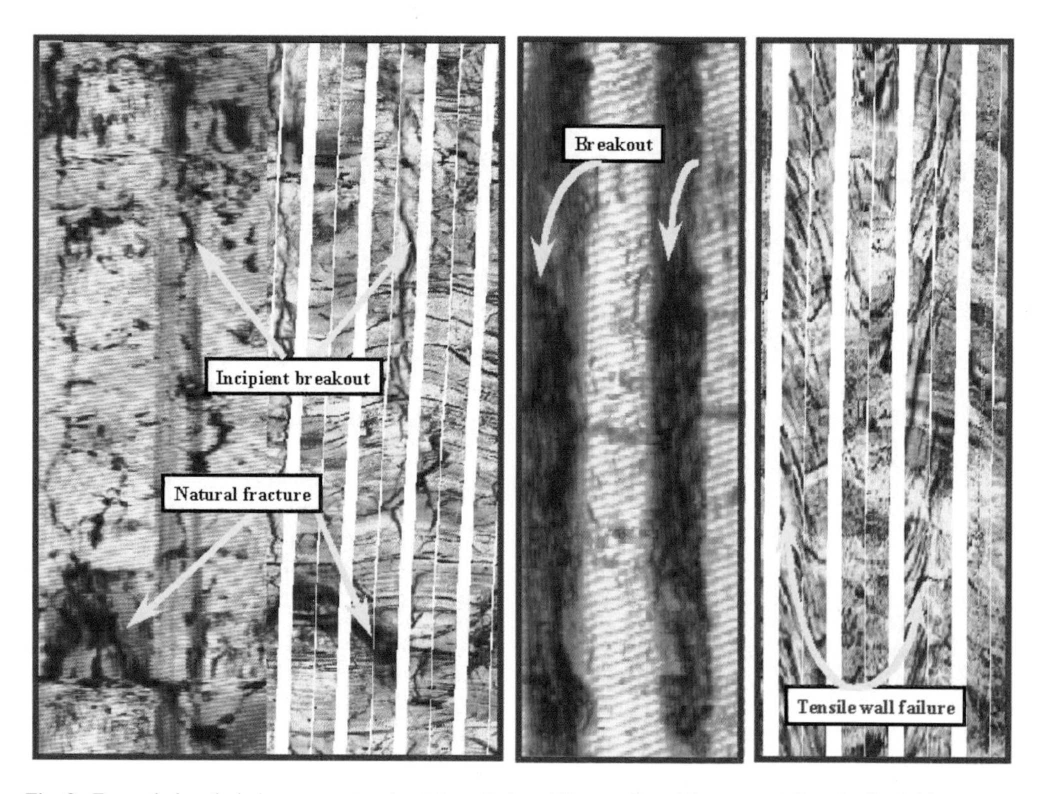

Fig. 2. Example borehole images portraying 'stress-induced features' used in constraining the far field stress tensor within the reservoir. All images are for three metre sections and are dynamically normalized over the presented interval.

forward modelling technique takes the values of the calculated stress tensor and models a known well bore, previously not used in constructing the stress tensor, within a computer generated stress field. Modelled parameters, appropriate to the well-bore, include orientation and inclination, depth, mud weight and rock strength. The result of the analysis is a prediction of where in the well bore and at what angle tensile wall failure fractures should occur (Peska & Zoback 1997). A comparison of the predicted values from the model with actual tensile wall failure seen in the known well was then used as a check on the values of the calculated stress tensor.

From the interpreted stress tensor and from subsurface interpretations of faults from electrical image logs it was clear that many fractures would be a combination of normal and shear processes. Given this observation it was possible to use a Mohr circle to estimate a stress window in which fractures would be aligned preferential to frictional failure. Within the stress window fractures stressed preferentially for frictional failure are effectively 'opened' by the stress field and therefore have a higher potential to flow fluid than those fractures oriented perpendicular to the maximum horizontal stress. The position of a given fracture within the frictional failure envelope is a function of the strike and dip of the fracture. This concept can be appreciated from Fig. 3 where fractures interpreted as having flow potential (based on temperature profiles) and fractures with proven flow from a published geothermal well are critically stressed and have a distribution primarily between the Coloumb lines of frictional failure. A Mohr circle was constructed using the calculated reservoir stress tensor and a population of dips having variable dip magnitudes and azimuth were plotted within its space. Using this technique it was possible to establish the range of fracture attitudes that would be critically stressed within the Apulian Platform reservoir and therefore have a higher productivity potential. This approach is demonstrated by the bottom Mohr circle of Fig. 3. Values have been omitted from the axis, for

Mohr circle diagrams showing relationship between fractu type and stress state. Failure envelopes are drawn for frictional coefficient s (u) of 1.0 and 0.6.

HYDRAULICALLY CONDUCTIVE FRACTURES.

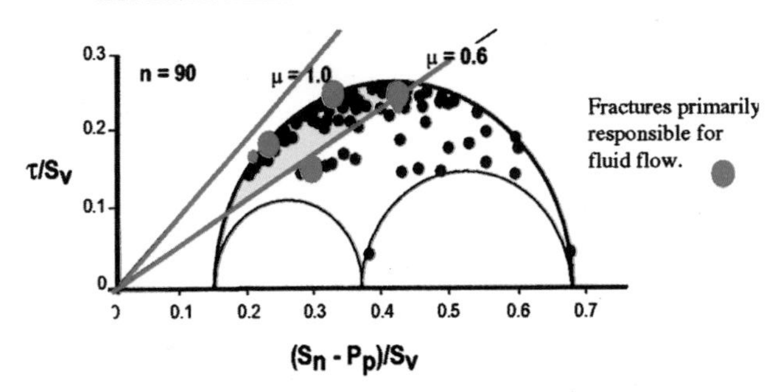

NON HYDRAULICALLY CONDUCTIVE FRACTURES.

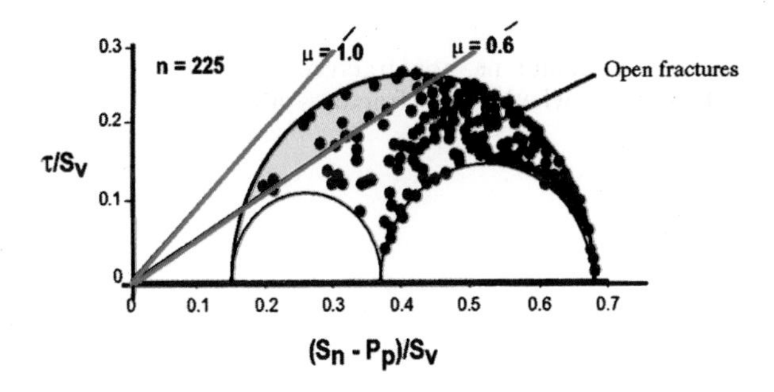

MODELED FRACTURE SETS FOR APULIAN PLATFORM

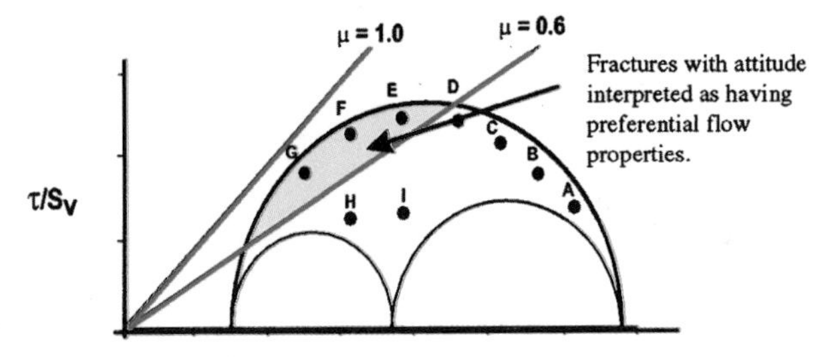

Modified after Barton et. al. 1997

Fig. 3. Modelled fracture sets with respect to the far field stress tensor and potential for fluid flow.

reasons of commercial sensitivity, however the diagram conveys the approach taken. Comparison of fractures from intervals of proven production within the Apulian Platform have confirmed the observation that critically stressed fractures are associated with improved flow.

Having determined a value for the regional or far field stress field it was possible to locate deviations from this 'norm' which were interpreted as resulting from faulting. Mapping the orientation of fractures present within fault blocks 'isolated' from the regional stress field was considered important for both production zone identification and also for fracture modelling away from the well bore. Combining such stress data with bedding and fault dips picked from borehole image logs together with seismic structure maps, a structural and stress 'map' was created for the reservoir. This 'map' was used within the static fracture modelling software as one of the inputs from which sub-seismic faults could be modelled.

Detecting natural fractures from wireline surveys

Virtually all open hole wireline logs have been examined at one time or another in an attempt to locate permeable fractures. In cases of identifying 'large' open fractures in the Apulian Platform the separation between standard deep and shallow resistivity logs proved a reasonably reliable wireline technique. Photoelectric factor responses in barite muds can be particularly effective as can transit time in larger fractures. However for many fractures the sampling density and tool resolution of most wireline logs is not sufficient to effectively locate or characterize permeable fractures. This observation is valid for the Apulian Platform fracture systems and is demonstrated in Fig. 4. From this diagram it is clear that in addition to the FMS the dipmeter micro-conductivity buttons and the lateralog respond to the open fracture whereas neutron, density, caliper and sonic do not. However detecting open fractures is only the first stage and detection is of little value unless information about the fracture's attitude can be gathered. For this reason dipmeter logs have proven of particular use in some formations. With the advent of imaging logs confident fracture detection in the low porosity carbonates of the Southern Apennines has been possible (Arlotto *et al.* 1997). In brief this approach relies on the FMS/FMI as the most important imaging tool as electrical resistivity devises have the capability of detecting all fracture types (induced & natural) as well as responding to subtle changes in rock texture and in analysing bedding frequency and structural dip. Information concerning rock fabric is of particular relevance in predicting fracture distribution with respect to stratigraphic parameters. Borehole televiewer data proved capable of responding to the most obvious fractures but was considered to be poor

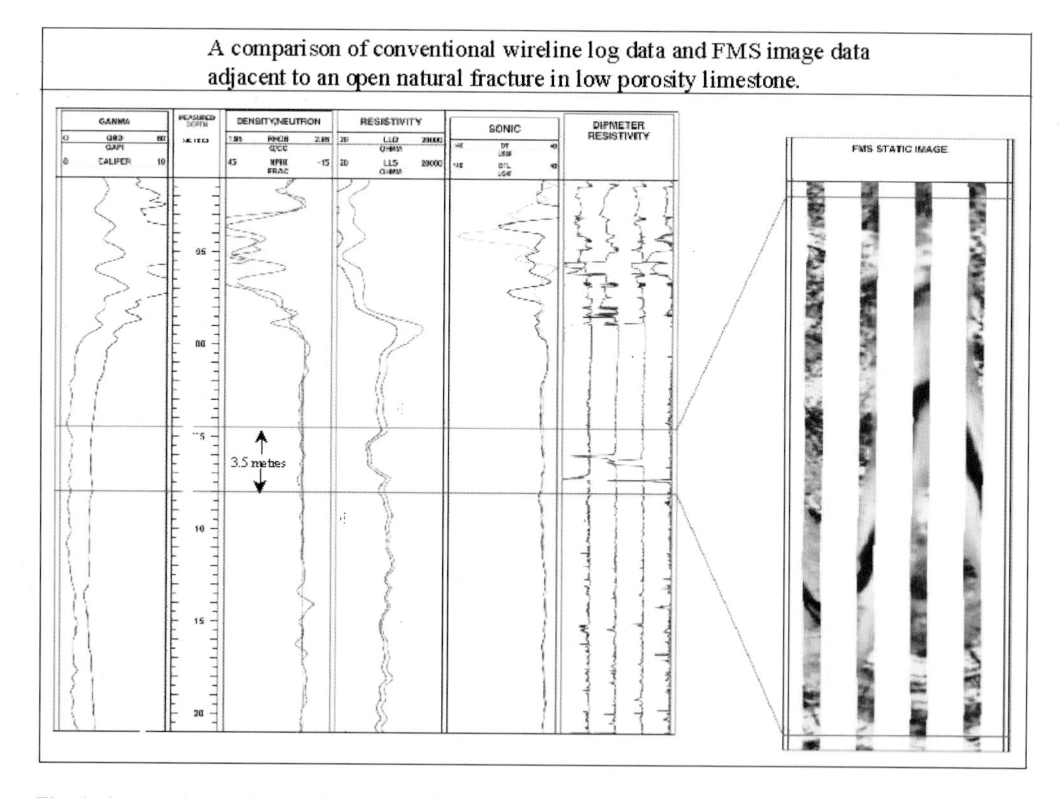

Fig. 4. A natural open fracture is interpreted between five and eight metres. This feature is clearly seen on the electrical image log and as dipmeter microresistivity anomalies. The only conventional wireline data to demonstrate a response to the fracture is the deep and shallow lateralog.

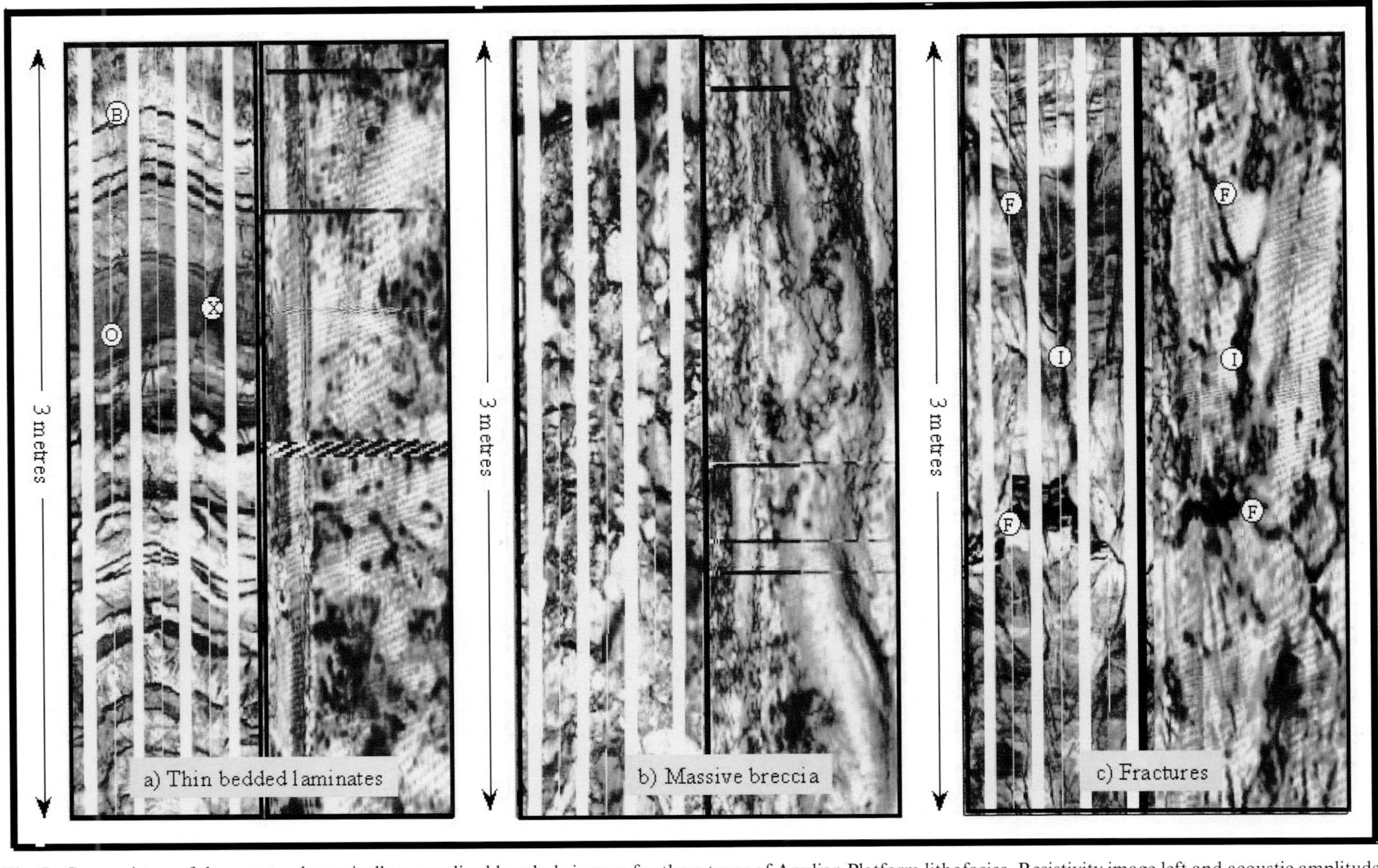

Fig. 5. Comparisons of three metre dynamically normalized borehole images for three types of Apulian Platform lithofacies. Resistivity image left and acoustic amplitude, right. Annotations refer to: **B**, bedding, **O**, bedding offset, **X**, fault plane, **I**, induced fracture, and **F**, natural fracture.

at defining detailed formation character and subtle fractures as clearly seen in Fig. 5. In this figure three example facies are presented from a vertical well section and include: (a) carbonate laminate, (b) carbonate breccia, and (c) fractured interval. The textures, relevant to the geology of the formation, are clearly seen with the electrical imaging log but are obscure on the televiewer data. In Fig. 5(a), planar features have been highlighted these include bedding, (B), small scale fault, (X), and bedding offset (O). In example (b) the electrical image characterizes a brecciated fabric clearly detected on the resistivity image but obscure, in part by stabilizer damage to the borehole, on the amplitude image. Example (c) includes natural fractures, (F), and induced fractures, (I). Both of these fracture types are detected on both the resistivity and acoustic images. Although the 'geological resolution' of borehole televiewer logs, is less than resistivity imaging logs in the environment of the Apulian Platform their application in confirming large open fractures and in establishing appropriate hole conditions for running the MDT for fracture testing is clear (Chellini *et al.* 1997).

Detecting producing fractures from image logs

Despite the acceptance that electrical image logs provide an unprecedented perspective of the Apulian Platform formations initial reliance solely on borehole image log data to locate producing fractures had limitations. The problems originated from the fact that image log based descriptions are limited to the static behaviour of the fracture in the borehole. The image log data can only provide a reliable interpretation of the depth, and attitude of a given fracture, and arguably an apparent fracture aperture. Further integration with core and bedding information can lead to an interpretation of fracture morphology (fracture types) which leads to interpretations of fracture frequency and fracture clustering. The classification of a fracture is further hampered by the fact that a given Apulian Platform fracture is basically a transmissive fabric bound or fabric cutting lamination from an image perspective and may or may not be permeable (Trice 1999). The approach taken in this study included the integration of mudloss, waveform interpretations and azimuthal array resistivity data to help confine the probability of a fracture being permeable and propagated beyond the well bore. Additional information came from the analysis of drilling induced fractures and the relative orientation of interpreted natural fractures to the local stress field. Defining the open fractures that were clearly responsible for transmitting hydrocarbon into the well bore or that had the potential to flow hydrocarbon was achieved by data integration. In practical terms the integration of data is achieved by the assimilation of static and dynamic data on work sheets. Work sheets include the comparison of all available data on a common depth reference and summarize core descriptions, goniometry data, temperature logs, production logs, drill stem tests, mud loss, stress trends, stratigraphy and structural zones, routine petrophysical analyses, fracture aperture and fracture frequencies. Interpretation of such assembled data was achieved by several iterations made by an integrated team composed of reservoir engineers, geologists, image log analysts and fracture specialists. Potentially productive fractures were classified and an apparent producing fracture frequency was calculated for each well. This concept is summarized conceptually as Fig. 6 which demonstrates part of a typical integrated work sheet. Fractures interpreted as being potentially productive included the following characteristics: (a) transmissive fractures described from imaging logs, (b) faults, (c) 'high' angle fractures oriented parallel to the maximum regional horizontal stress, and (d) 'high' angle fractures oriented parallel to local horizontal stress. These characteristics are discussed in brief in the following paragraphs.

Transmissive borehole image fractures. Transmissive fractures within the Apulian Platform are interpreted as being fractures seen on borehole image logs that are either electrically conductive, or with acoustic attributes of low amplitude and slow transit time. Transmissive fractures are often associated with sonic waveforms events which produce the classic 'chevron' pattern (Chelini *et al.* 1997) and conductivity anomalies with respect to the azimuthal array resistivity log (Arlotto 1997). However, chevron patterns are best relied upon as permeable fracture detectors only over intervals of good hole condition. This statement is made due to the observation that, (a) key seats, (b) ledges, and (c) hole rugosity can create waveform responses similar to open fractures. In the initial phases of the study increased confidence in the potential permeability of fractures was given to fractures that could be confidently identified on electrical imaging logs, the sonic waveform and the array resistivity device. However with further analysis and integration, it became apparent that the most useful wireline tool for permeable natural

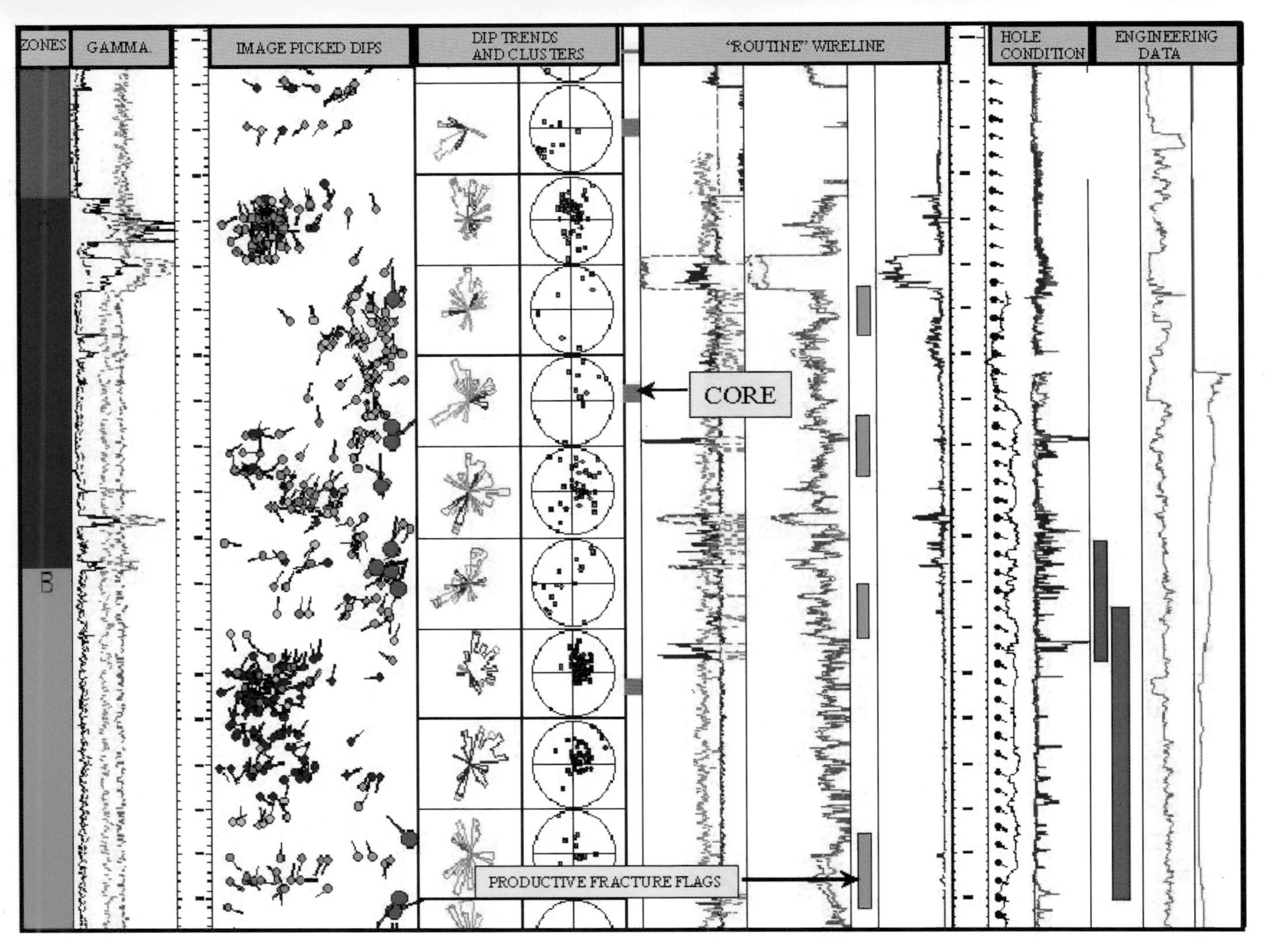

Fig. 6. Example integrated work sheet, applied in establishing intervals of productive and potentially productive fractures.

fracture detection was the resistivity imaging log. With experience the FMI has become the preferred borehole image log from which to identify and classify natural fractures in the reservoirs of the Apulian Platform.

Faults. Faults within this study are classified as natural fractures associated with bedding displacement and or local variations in the structural dip trend. Within the Apulian Platform faults can be identified from a variety of wireline methods including (a) directly from a borehole image log where the fault plane, damage zone and associated bedding offset can be identified or, (b) where bedding offset has been inferred from changes in structural dip and or the association of bedding drag or rollover, and (c) in the instances of large fault zones through specific dip distributions and caliper activity. The characteristics of a fault's core and damaged zone will to some degree control the fault's behaviour with respect to being a conduit or barrier to fluid flow (Forester *et al.* 1997). These fault zone characteristics combined with the fault's orientation relative to the local stress tensor significantly affect the ability to produce hydrocarbon at reservoir conditions. An example of one such productive fault is seen in Fig. 7. This particular fault example has two characteristics that make it an important productive feature. Firstly the fault plane strikes parallel to the far field regional stress and is of a significant dip magnitude that it falls within the critical stress envelope and secondly the damaged zone of the fault is composed of cataclasite 'cut' by a network of open anastomozing fractures.

Relationship to the stress field. The constraint of stress and its effect on fracture producability was determined from the mohr circle analysis and basically indicated that within a specific dip magnitude range fracture strikes with azimuths near parallel to the maximum horizontal stress would be potentially productive with respect to fluid flow (see Fig. 3). In the majority of cases the reference for this was taken to be the direction of maximum horizontal stress described for the field. However in cases where a given borehole crosses into a different fault block stress 'islands' were occasionally encountered. These stress 'islands' were characterized by maximum horizontal stress directions sub oblique or sub parallel to that of the far field stress. In these instances it was the local rather than regional maximum horizontal stress direction that was used as the reference vector for establishing the flow potential of a given fracture.

Classification of productive fractures. Fractures identified as having the necessary characteristics for potential production where then classified as 'producing fractures' and effectively separated from the background fracture population. One of the most problematic aspects of this filtering was the identification of fractures induced by the drilling process. This was only achieved after thorough integration of core photographs and core goniometry studies. This problem is demonstrated in Fig. 8 where a series of high angle electrically conductive 'laminations' and 'partial laminations' can be seen on the electrical imaging log. The image represented as Fig. 8 was recorded over an interval of borehole having 20° inclination. Such events are similar in characteristics to natural fractures and are oriented parallel to the azimuth of maximum horizontal stress. Comparison with core over the same interval demonstrate that only scarce fractures, interpreted as resulting from core damage, are present. This observation is testimony to the induced nature of the electrically conductive features seen on the image log (Fig. 8) which are interpreted as fractures resulting from shear failure. It appears that inclined boreholes within the Apulian Platform have a greater probability of developing such image features. This problem is further discussed in Arlotto *et al.* 1997 & Cesaro *et al.* 1997.

The robustness of the applied technique in identifying producing fractures from static data has been verified by application of detailed MDT (Modular Formation Dynamic Tester, Schlumberger) surveys, run in dual packer mode (see Joseph *et al.* 1992) and by drill stem tests, in exploration wells. The advantage of the MDT over drill stem tests is that individual fractures, rather than bulk intervals, can be measured for fracture producability giving: (a) fracture fluid sample, (b) formation pressure, (c) fracture permeability measurement, and (d) formation fluid gradient. The concept of the MDT testing in dual packer mode is presented as Fig. 9. In this Fig. fractures have been identified from the FMI data during ongoing drilling operations and targeted as potential locations for MDT testing. The MDT is set against the FMI picked fractures after verification of hole quality by borehole televiewer analysis and a small interval of open hole (commonly between 1.5 m–2 m) is isolated by two inflatable packers. A formation pressure and fluid sample from the fractures is then taken by the MDT. This dynamic data proved the interpreted, potentially productive fractures, to be permeable and hydrocarbon bearing adding confidence to the applied technique of productive fracture identification from

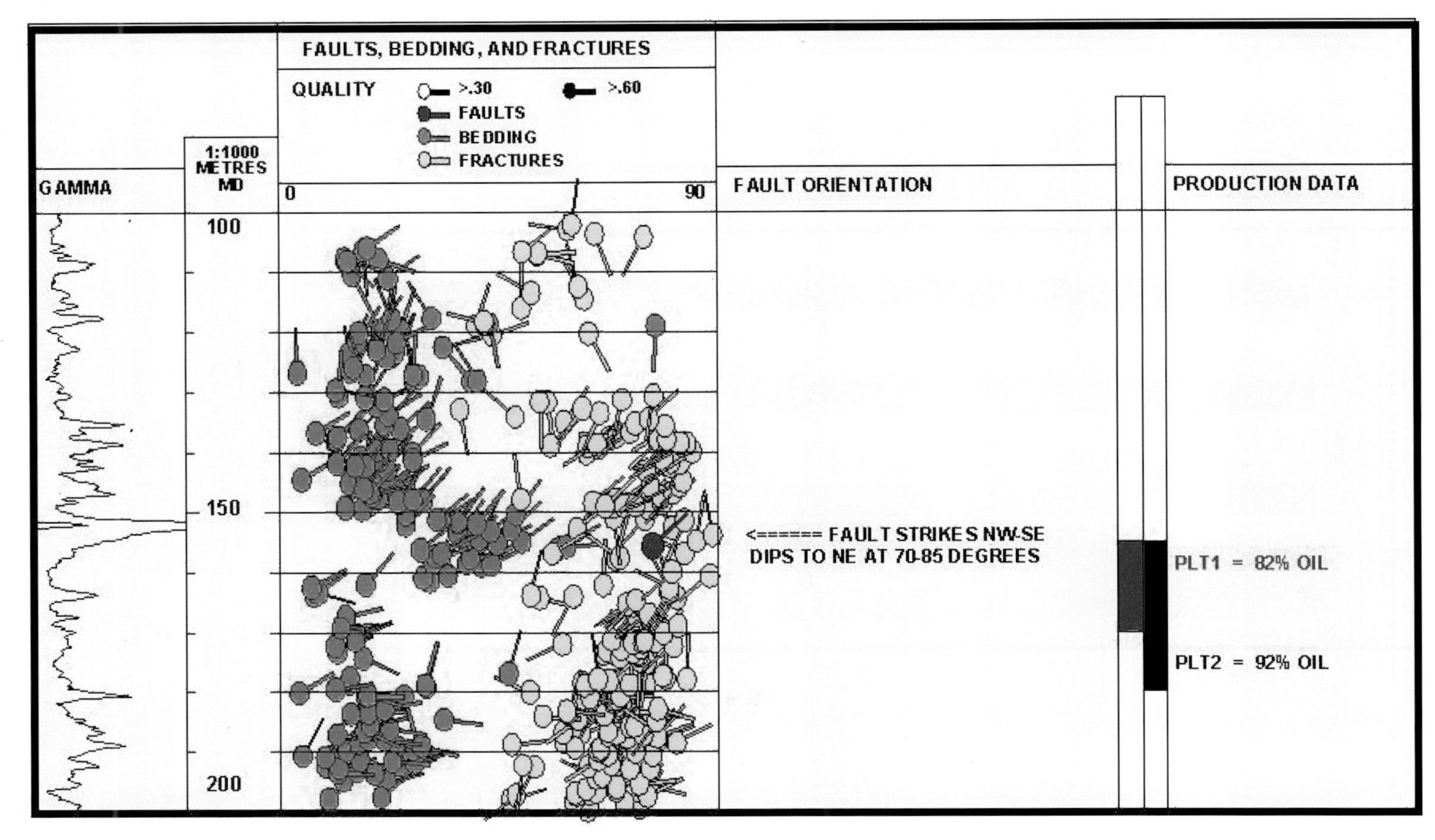

Fig. 7. Example of a productive critically stressed fault, interpreted from resistivity imaging logs (FMS). PLT flags indicate flowing zones.

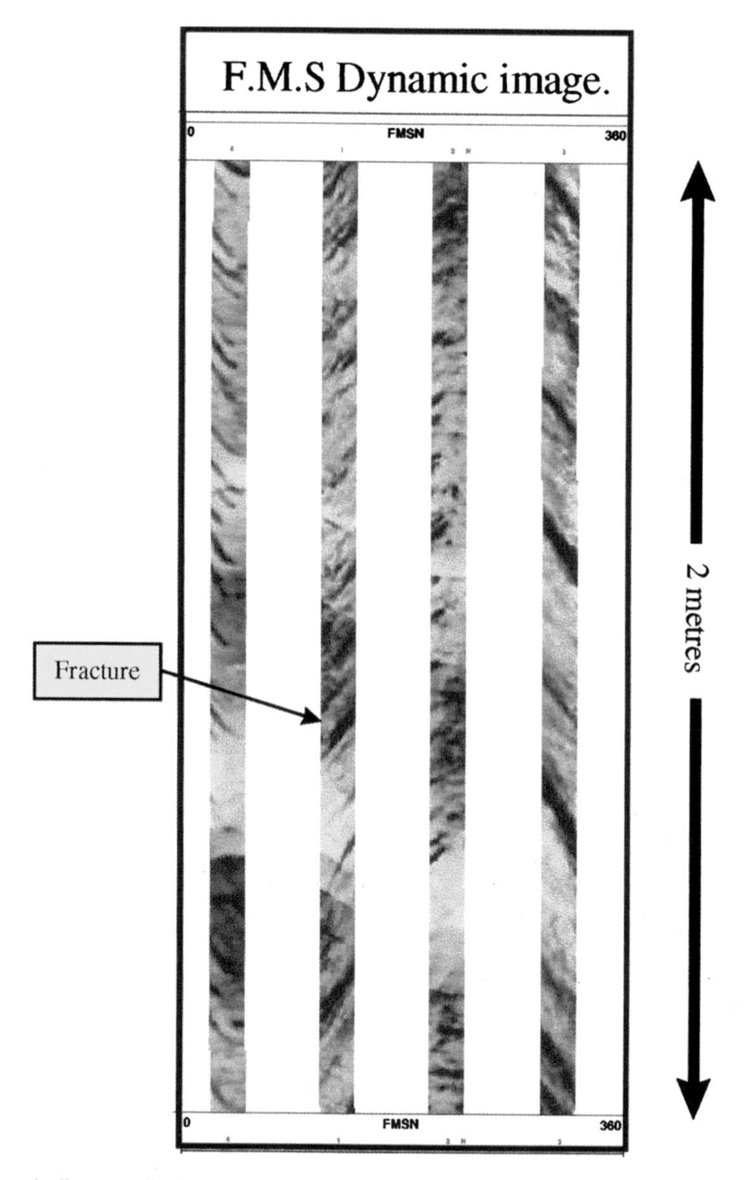

Fig. 8. Dynamically normalized two metre FMS image exhibiting fractures interpreted as resulting from borehole shear failure.

static data. This technique and application to the Apulian carbonate is discussed in detail by Chellini *et al.* (1997).

After the productive fractures and productive interval were identified, correlations between wells were undertaken. Through this application of a multiwell data set it was possible to construct a series of conceptual fracture models. Such models are important as they provide a vehicle for multidisciplinary data assimilation and discussion. Comparison of the various models lead to conclusions with respect to fracture distribution which provided input on key parameters that should be included within the static model.

Fracture analogue – practical characteristics and verification of suitability

Once the fractures interpreted as being productive had been identified a static fracture model was constructed which allowed for these discrete

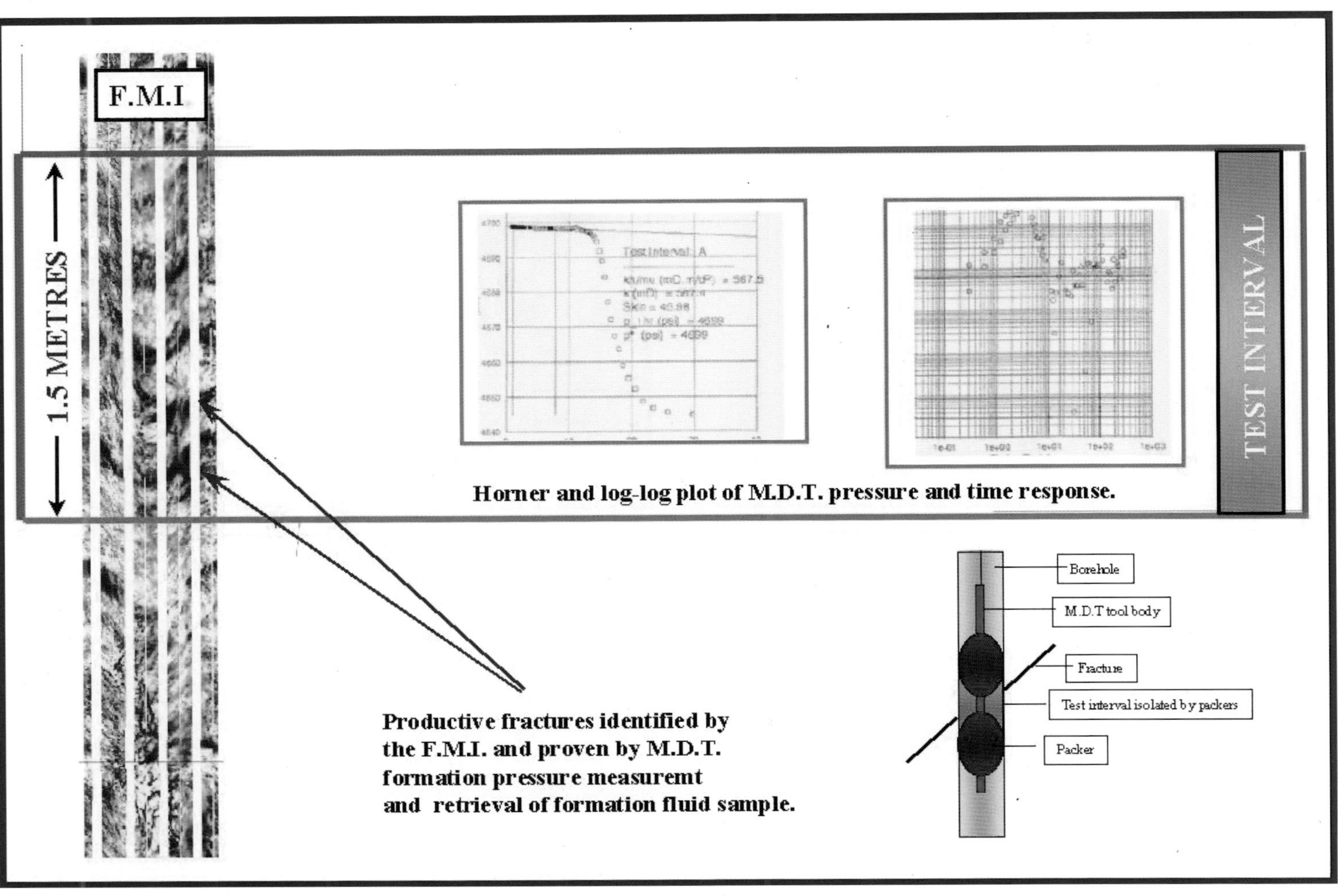

Fig. 9. Combined FMI and MDT data indicating FMI picked fractures that are permeable and hydrocarbon bearing.

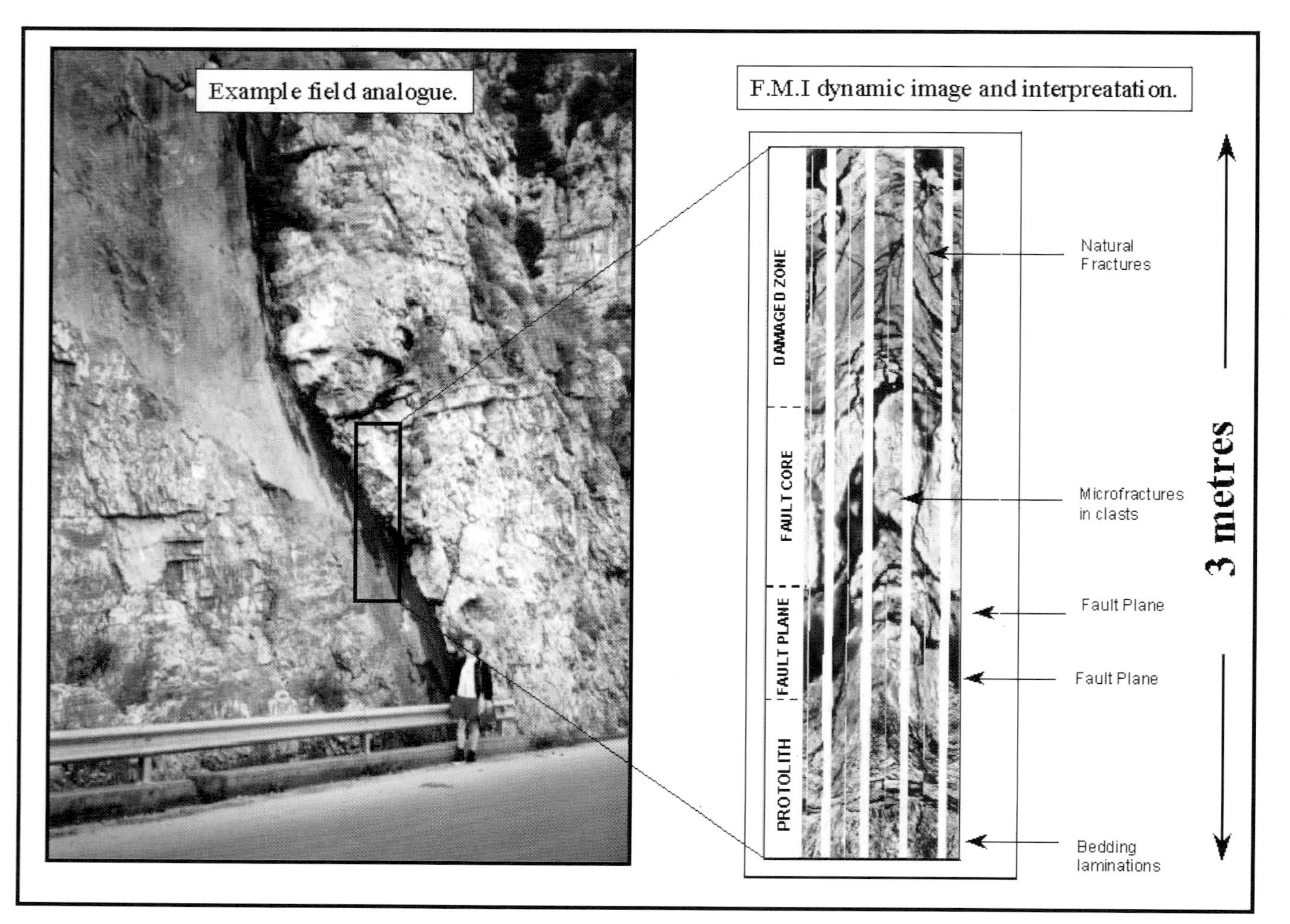

Fig. 10. Ground-truthing the analogue through comparison of faulting styles seen in outcrop and interpreted from borehole image log data.

features to be modelled away from the borehole and for the bulk storage properties of the fracture network to be estimated. A number of parameters not available from well log data are required in the construction of a 3D static model, these include fracture length and the fracture height. In addition the fracture's geometric relationship (i.e. do the fracture sets abut or cross) is also necessary to constrain the stochastic model. Such information is referred to as a fracture template (Genter 1997). The fracture template directly influences the choice of the specific statistical technique applied to generate the fracture network. Establishment of the fracture template was achieved through analysis of a field analogue, chosen after careful scrutiny of a number of sights. Characteristics of the analogue are: (a) it provides outcrop of similar lithology to the reservoir, (b) it provides outcrop of a similar structural setting to the reservoir, and (c) it can be sampled at a range of scales utilizing satellite imagery, aerial photography and field geology to define fracture trace maps from which fracture trace length frequency distributions are obtained.

In order to have confidence in the analogue prior to building the static model verification of the fracture style was undertaken by: (a) comparisons of fracture populations in stereo net space between subsurface and surface fracture clusters, (b) comparison of fault styles between seismic, image logs and outcrop, and (c) comparison of fracture populations close to and within fault zones through a comparison of core and image logs. Examples of field verification can be seen in Fig. 10 and Fig. 11. Figure 10 is an example fault which has similar bedding deformation characteristics and fault zone characteristics to features seen on image logs and from analysis of core. Figure 11 demonstrates the necessity of comparing the smaller scale joints and microfractures and their relationship to lithofacies type through detailed core and image log comparisons. This is particularly relevant to rudist intervals within the Apulian Platform which have the potential to develop significant vuggy porosity in the subsurface. This porosity is often isolated, however vuggy porosity has the potential to be connected to the fracture backbone by microfractures and/or joint systems.

3D static model input parameters

After confirming the analogue's suitability it was necessary to establish input parameters for the 3D fracture model. From outcrop and seismic sources these included: (a) fracture trace length distribution, (b) fracture template, and (c) fracture aspect ratio (height/length). Subsurface data input included: (a) fracture frequency, (b) fracure clusters, (c) fracture attitude, (d) fracture aperture, and (e) fracture porosity.

Fracture length distributions were provided by box counting techniques applied to satellite images of well exposed bedding plane surfaces. Two satellite data sets obtained at different times of the year were used for this, thus allowing for a better control of vegetation effects on trace length analysis. The results of this approach indicated that the distribution of fracture length fitted a power law distribution and was of a similar magnitude to published values (Sahimi 1994). The fracture template was established by comparing the ratio of abutting, crossing and terminating fractures seen in satellite imáge data. Fracture lengths were often of such a magnitude that their corresponding heights were greater than the available vertical exposures. Consequently assumptions of the fracture height distribution had to be made based on the style of fracturing and theoretical joint and fault models. Non strata bound shear fractures were modelled with an aspect ratio of 2:1 whereas strata bound joints were modelled with a log normal height length relationship.

Fracture frequency was established by correcting the mean fracture frequency observed in boreholes for sample bias through the application of a Terzaghi correction to individual well data sets (Terzaghi 1965). A bulk fracture frequency for use in the model was then established by averaging frequencies between four vertical/subvertical wells each of which penetrated in excess of 1km of reservoir section. Fracture clusters where determined from the style of fracturing seen on image logs and core compared with the distribution of the fractures in stereo net space. Fracture attitude was established through stereo net analysis of individual fracture clusters. If the dip data in each population was associated with log normalized eigenvalues attributable to clusters (Woodcock 1977) then the 'average' dip was established through a fit of the first eigenvector. If clustering was not well developed, i.e. the dip population tended to a girdle distribution, a manual great circle fit was applied to determine the population 'average'. Within the static model azimuthal variation was treated by applying a Fisher coefficient (Fisher *et al.* 1987).

Fracture aperture although of little relevance to the geometric distribution of the fracture population is a necessary component in the 3D model as it is used in determining the bulk fracture porosity. Aperture was determined through

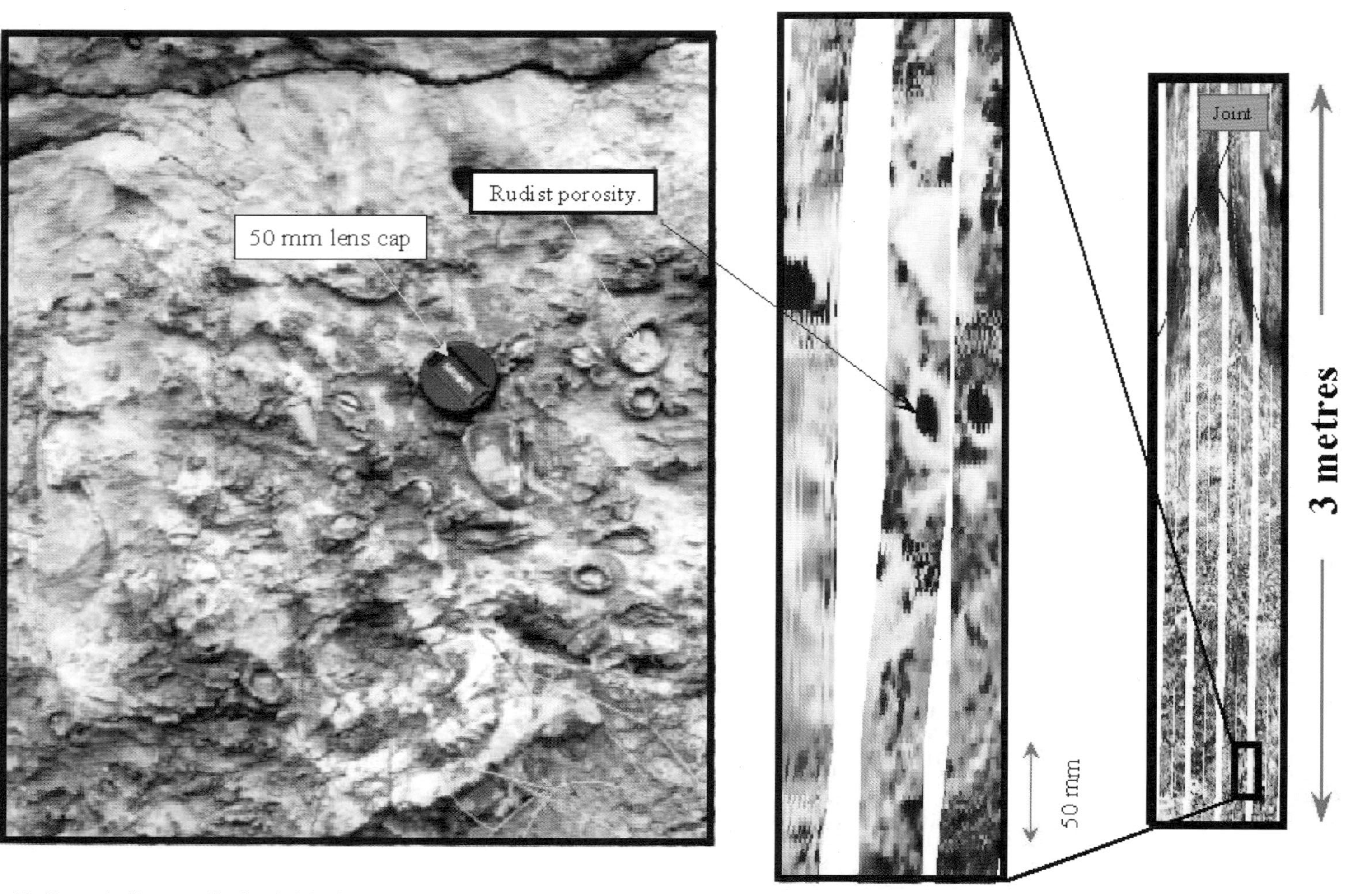

Fig. 11. Dynamically normalized resistivity image compared to outcrop analogue for a 'thick to medium bedded reworked rudist facies'. Dark colours reflect more conductive features, inferred as reflecting local enhanced permeability.

a combination of resistivity image logs (Luthi & Souhaite 1990) in-house techniques and aperture measured from thin sections. Core values were only applied where fractures were clearly bridged by mineral aspirates, the assumption being that aperture was not exaggerated through de-compaction processes. For fault zones aperture is termed width as the permeable and productive interval consists of a complex of open fractures and altered country rock. Widths were established by a combination of conventional wireline logs and resistivity imaging logs. Fracture porosity is often considered to be 100% for a given aperture. This assumption was taken for joints and microfractures, however for seismic and sub seismic faults an average porosity for the fault width was established through detailed analysis of conventional wireline logs. An average value was established and used as a fixed input for all fault widths in the 3D model.

Testing and application of the 3D static model

3D static models were generated using FRACA and TRANSFRAC software. Data input to this model incorporated borehole image results, interpretations from conventional wireline logs, seismic interpretations, and outcrop analogue information. A first and all important check on the statistically generated fracture network model is provided by a provisional model. The provisional model utilizes all the available input parameters apart from fracture frequency. The reason for this is due to the assumption that fracture frequency is a function of: (a) fracture trace length distribution, (b) fracture orientation, and (c) aspect ratio. Should the provisional model result in an unrealistic fracture frequency then a revision of the basic input parameters to the 3D model would have to undertaken. The resultant true vertical spacing however matched the fracture frequencies recorded from individual boreholes, it was therefore reasonable to continue with constructing a definitive 3D model.

A second test on the suitability of the model was provided by the ability of the Transfrac software to numerically simulate apparent flow across the fracture network and to test for fracture connectivity. The Transfrac approach converts all fractures to vertical and horizontal 'pipes' and passes a simulated electrical current across the fracture network. Azimuthal variation in the electrical potential is used to estimate resistance anisotropy across the network. Conductivity variation is equated to the degree of pseudoflow and resulted in a 'picture' of which fractures are connected and which are isolated. Pseudoflow across the 3D static model of the reservoir demonstrated connectivity which indicated such a system in the subsurface had the potential to flow fluid. As the 3D model was based on the distribution of what had been interpreted as producing fractures the connectivity of the modelled fracture network was considered further support for the stochastically generated model.

A third and final test of the static model was to compare the generated pseudoflow tensor to permeability anisotropy determined from inter well tests. Although such an approach is not strictly comparing like with like it was noted that, in orientation and magnitude, permeability anisotropy seen from inter-well tests was of the same order of magnitude as that estimated from the 3D static model. This gave further confidence to using the static model for quantitative applications.

Having vetted the 3D static model through three 'tests' it was decided that the fracture network was suitably robust to estimate a bulk fracture porosity. This aim was achieved by calculating a representative elementary volume (REV) This specific volume, the REV, is defined as the minimum volume for which small changes in size and location result only in small changes in the connectivity of the measured fracture system. It is the size beyond which the measured fracture system can be considered as a continuous medium. This volume was calculated by varying the simulated volume containing the specific fracture network until a specific minimum volume was achieved. The importance of the REV is that it estimates for different fracture scales the volume of rock that can be considered to be representative of a homogenous porosity and permeability system. The REV was calculated for individual fracture populations to determine those clusters most significant for controlling connectivity across the reservoir. An REV was also established for mixed fracture populations. As a range of clusters had been determined from borehole image data and stress analysis it was possible to analyse the importance of each population to connectivity and by inference hydrocarbon flow. From such modelling it became apparent how large scale features such as faults where connected by smaller scale features. This is summarized as Fig. 12 where seismic and sub seismic faults can be seen in comparison to the 'background' smaller joint and fracture network. From Transfrac modelling it became clear that the seismic and

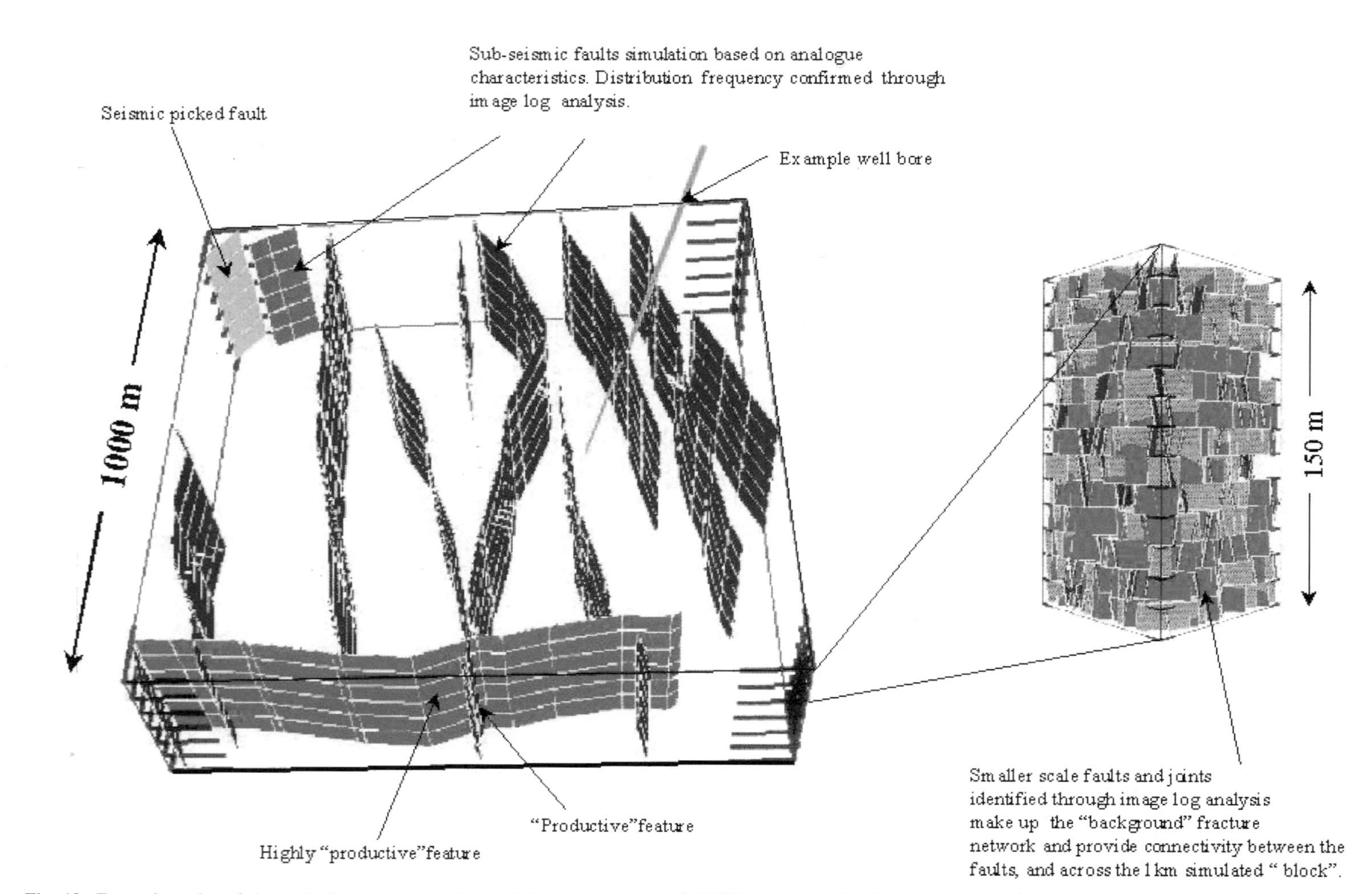

Fig. 12. Example scales of the static fracture network model investigated through REV modelling. Small scale (sub seismic) features provide connectivity to larger scale (seismic) features. This connectivity relationship demonstrates the importance of incorporating all productive fractures, identified from borehole image log data, within the static fracture model.

sub seismic faults were poorly connected and could only 'flow' when linked by the background of smaller scale features. Such conclusions are important in considering well placement and in designing well test strategies.

Once the individual REV's were established estimates of bulk fracture porosity could be calculated, by summing the products of the fracture volumes and their associated porosity. Results using this technique give combined effective fracture porosity ranges comparable to published values. By making assumptions on hydrocarbon saturation within the fracture network, based on production data and capilary curve measurements the REV derived bulk porosities were used in calculating a fracture STOIIP. In addition the REV modelling was used to constrain average grid sizes and grid size orientations for input into field scale dynamic reservoir simulations.

Discussion

The techniques described in this paper represent a point summary of ongoing technical work. Whilst the described approach is workable and has provided sensible results for two separate reservoirs within the Apulian Platform there are deficiencies that come from the fact that: (a) fracture properties we wish to predict and model are not well documented or constrained, (b) it is often the case that the fracture parameters we require cannot be measured directly from a given source, and (c) predicting and modelling fluid flow within field scale fracture systems is in its infancy.

Fracture properties

The majority of up-scaling techniques for fracture modelling have been established based on box counting techniques, of limited exposures, for extensional joints or large scale shear features such as faults. From this published work we are made aware that fracture trace length frequency distributions can be described by a power law (Aarseth *et al.* 1997).This is a key approach taken in modelling the Apulian Platform as described in this paper. The technique is considered a reasonable approach as the areas from which the borehole data were obtained are well constrained and exhibit little structural or stratigraphic variation. The problem occurs when the results from the study area are extrapolated across the field as fracture properties have the potential to vary from the background population in: (a) the vicinity of seismic faults, (b) as a function of relative position on large scale structural features, and (c) as a function of lateral and vertical facies variation. The scale of the Apulian Platform reservoirs in the Southern Apennines are such that future fracture modelling developments will have to overcome some of these issues.

Direct measurement of fracture properties

To understand the properties of a given fracture network in the subsurface it is necessary to apply information recorded from the subsurface and an outcrop analogue. In each instance it is probable that the actual properties we wish to measure are going to be obscured through sampling techniques or through absence of appropriate exposure. A simple example of this problem can be demonstrated by considering large fractures or faults strike orientation. Classical borehole imaging techniques interrogate a given number of related fracture traces using a stereonet and appropriate statistical 'averaging' tools. Results from such an approach give an average dip and a range of azimuth spreads to the fracture cluster. However application of inappropriate azimuth ranges in a 3D static model can cause an unrealistic increase in fracture connectivity. The reason for this is sampling. Borehole image logs sample a given fracture or fault trace at a scale equivalent to that measured by a field geologist. At this scale local subtleties in fracture azimuth can be recorded. This scale dependency on strike azimuth can be readily appreciated by comparing the properties of a fault plane at outcrop with the view afforded of the same fault at the scale of a satellite image. A practical solution to this problem is needed that will describe statistically the relationship between azimuth properties of large scale features recorded at outcrop using aerial images and those measured by the scale of the borehole image.

Predicting and modelling fluid flow

Predicting and modelling fluid flow in fractures at the reservoir scale is not a routine, well established technique. Within this case study four scales of fractures had to be considered from the core scale through to seismic scale and whilst it was possible to effectively model these scale ranges from a static and limited dynamic perspective it was not possible to take the final

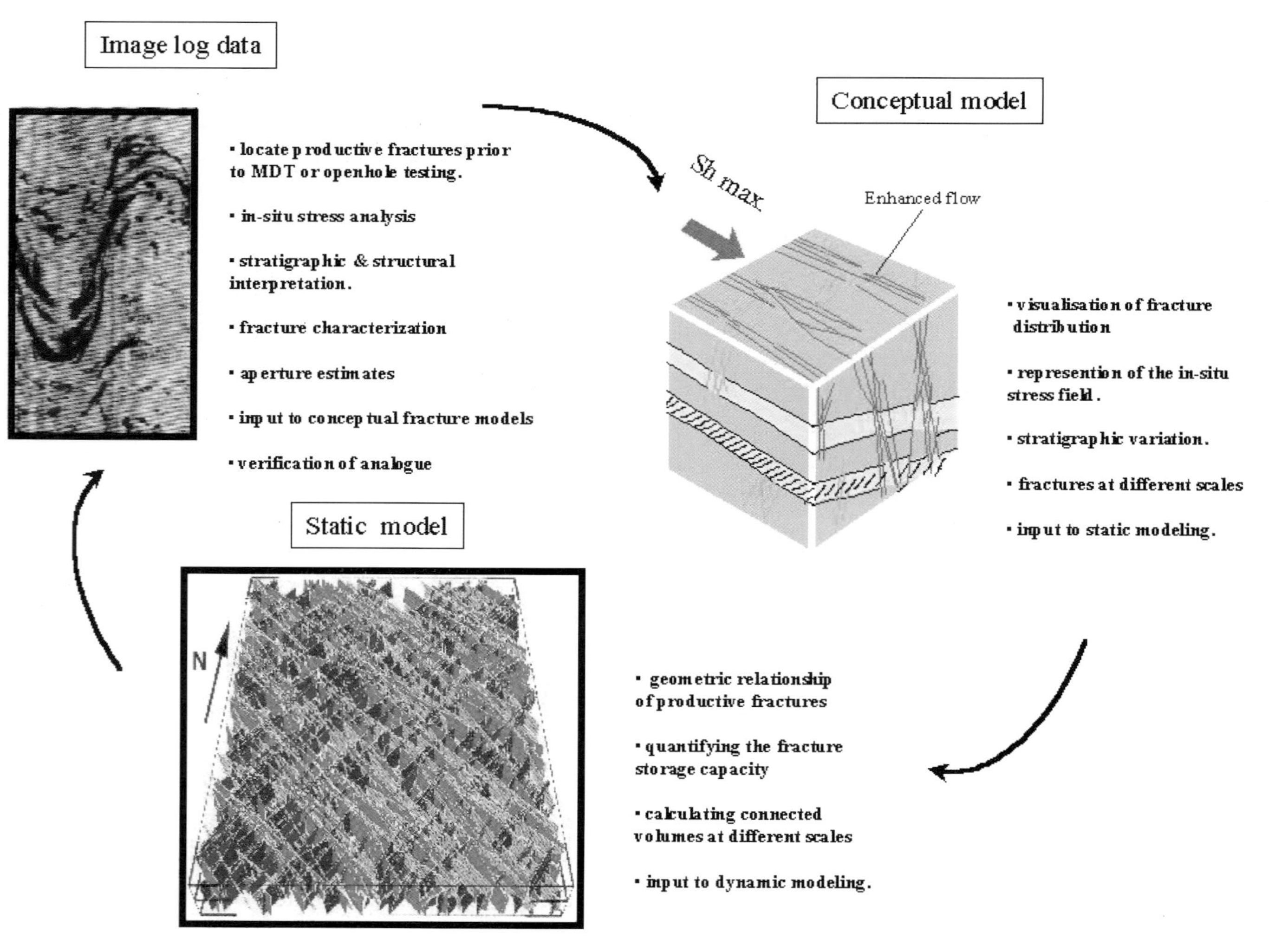

Fig. 13. Summary diagram of the iterative phases and their application in constructing, testing and corroborating 3D static models of productive fracture networks.

model to a full scale true dynamic simulation of the discrete fracture network. Whilst relatively few commercial packages are available to model the dynamic properties of a fracture network attempting to match the dynamic response of a fracture system with that modelled from a discrete fracture network, has specific problems. These problems arise from the experience that well test and production logs respond to the most permeable and productive fractures within the well bore. Constraining a given fracture model based on this data alone, whilst not accounting for any less permeable or less productive fractures that have been detected from other dynamic data such as the MDT may result in a model that only reflects the earliest behaviour of the fracture system. In addition, well test pressure profiles are commonly measured over a period of days and in highly anisotrophic fracture systems, such a time period may be insufficient to characterise the total effective fracture network. Models constrained by well test data alone may therefore underestimate the long term potential of a given system.

A key and increasingly available data source in constraining 3D static models are borehole image logs which when properly integrated with other data can be used to: (a) effectively distinguish producing fractures from the ineffective fracture population, (b) provide clusters and cluster geometry input to the modelling software, (c) provide input on fracture aperture and fault widths estimation, and (d) help constrain interpretations derived from well test data. As more borehole image logs are being interpreted to provide input to fractured reservoir quantification it is hoped that some of the current fracture modelling issues will have the potential to be resolved in the near term.

Conclusion

Borehole image log data has been used to provide input into constraining the bulk properties of a producing fracture network. This has been achieved through the identification of productive fractures in the well bore, the construction of conceptual models and the building of 3D static realizations. This iterative cycle and the associated applications are summarized as Fig. 13. The resulting 3D realizations have been verified as far as possible using data from a variety of sources and are considered to be sufficiently robust as to act as a testable model for quantifying the static characteristics of producing fracture systems present in the Southern Apennines. Testable fracture network models results from an understanding of: (a) the local and regional reservoir stress tensor, (b) the location and characterisation of producing fractures, and (c) the identification, application, and corroboration of a suitable field analogue. Applications of the static realization relate to: (a) predicting fracture geometry and intensity away from available well locations, (b) establishing which fracture groups control fluid flow, (c) estimating fracture permeability anisotropy, (d) calculating a bulk fracture porosity, and (e) providing input to help constrain dynamic models of the reservoir. In the initial stages of fracture detection confidence in the assignment of a given fracture as potentially productive is achieved by the integration of resistivity imaging log picks with sonic waveform and array resistivity logs. However experience has demonstrated that within carbonates of the Apulian Platform electrical imaging logs are the prime data source for locating potentially productive fractures and for providing parameters in the construction of fracture models.

I would like to give special thanks to the management of Agip S.p.A Milan and for Enterprise Oil plc for agreeing to the publication of this paper I would also specifically like to thank Mauro Gonfalini for many fruitful conversations on the subject of wireline applications in fractured reservoirs.

References

AARSETH, E. S., BOURGINE, B., CASTAING C. *et al.* 1997 *Interim guide to fracture interpretation and flow modelling in fractured reservoirs.* European Commision Report, Joule II Contract No. CT93–0334.

AMATO, A., MONTONE, P. & CESARO, P. 1995. State of Stress in Southern Italy from borehole breakout and focal mechanism data. *Geophysical Research Letters*, **22**, **23**, 3119–3122.

ARLOTTO, D., GONFALINI, M., MALETTI, C., TRICE, R. & PONS, U. 1997. *Fracture analysis of carbonates in the Val d'Agri.* Paper 6 in Italy 2000, value added reservoir characterization. Schlumberger, Milan.

BARTON, C. A., HICKMAN, S., ZOBACK, M. D. & MORIN, R. 1997. Fracture permeability and its relationship to in-situ stress in the Dixie Valley, Nevada, *Geothermal Reservoir. Paper D4-1 Stanford Rock Physics & Borehole Geophysics Project*, **63**, June 1997.

—— & ZOBACK, M. D. 1994. Stress perturbations associated with active faults penetrated by boreholes: Possible evidence for near-complete stress drop a new technique for stress magnitude measurement. *Journal of Geophysical Research*, **99**(B5), 9373–9390.

——, —— & MOOS, D. 1995. Fluid Flow along potentially active faults in crystalline rock. *Geology*, **23**(8), 683–686.

CESARO, M., GONFALINI, M., CHEUNG, P. & ETCHECOPAR, A. 1997. Shaping up to stress in the Southern Apennines. Paper 7 in Italy 2000, Value added reservoir characterization. Schlumberger, Milan.

CHELINI, V., SARTORI, G., CIAMMETTI, G., GIORGIONI, M. & PELLICCIA, A. 1997. *Enhancing the image of the Southern Apennines*. Paper 8 in Italy 2000, Value added reservoir characterization. Schlumberger, Milan.

FISHER, N., LEWIS, T. & EMPLETON, B. 1987. *Statistics of Directional Data*. Cambridge University Press, Cambridge, UK.

FORESTER, C. B., CAINE, J. S., SCHULZE, S. & NIELSEN, D. L. 1997. *Fault zone architecture and fluid flow: An example from dixie valley Nevada*. Proc. 22nd Workshop on Geothermal Reservoir Engineering Stanford University, Stanford California, January SGP–TR–155.

GENTER, A., CASTAING, C., BOURGINE B. & CHILÈS, J. P. 1997. An attempt to simulate fracture systems from well data in reservoirs. *International Journal of Rock Mechanics & Mineralogical*, **34**(3–4), Paper No. 044.

HIPPOLYTE, J. C., ANGLIER, J. & BARRIE, E. 1995. Compressional and extensional tectonics in an arc system: example of the Southern Apennines. *Journal of Structural Geology*, **17**(12), 1752–1740.

JOSEPH, J., IRELAND, T., COLLEY, N., RICHARDSON, S. & REIGNIER, P. 1992. The MDT tool: A Wireline Testing Breakthrough. *Oilfield Review*, **4**(2), 58–65.

LUTHI, S. M. & SOUHAITE, P. 1990. Fractures apertures from electrical borehole scans. *Geophysics*, **55**, 821–833.

ODLING, N. E. & OLSEN, C. 1997. Fracture Scaling and fluid flow: Implications for Hydrocarbon Reservoirs. *In*: *The Strategic Importance of Oil & Gas Technology*, **2**, 1183–1201.

PESKA, P. & ZOBACK, D. 1997. *Constraining complete stress tensor using drilling induced fractures in inclined boreholes*. Paper B1-1 Stanford Rock Physics & Borehole Geophysics Project, **63**, June 1997.

SUHAMI, S. 1994. *Applications of Percolation Theory*. Taylor & Francis, UK.

TERZAGHI, R. 1965. Sources of error in joint surveys, *Geotechnique*, **15**, 287–304.

TRICE, R. 1999. A methodology for applying a non unique morphological classification to sine wave events picked from borehole image log data. *This volume*.

WOODCOCK, N. H. 1977. *Geological Society of America Bulletin*, **88**, 1231–1236.

Use of Resistivity At Bit (RAB) images within an Eocene submarine channel complex, Alba Field, UKCS

R. M. McGARVA,[1] C. BELL[2] & J. BEDFORD[3]

[1] *Z&S GeoScience Ltd., Balgownie Drive, Bridge of Don, Aberdeen, AB22 8GU*
[2] *Chevron UK Ltd, Woodhill House, Westburn Road, Aberdeen, AB16 5XL*
[3] *Schlumberger Evaluation and Production Services (UK) Ltd, 1 Kingsway, London, WC2B 6XH*

Abstract: To aid reservoir description a Resistivity At Bit (RAB) tool was run in a series of near-horizontal wells within the Alba Field. The tool characterized lithology types and detailed bed geometry through oriented resistivity images of the borehole. Image quality ranged from good to uninterpretable through zones of image artifact, and image resolution was less than that of conventional resistivity imaging tools. Shale intervals locally display good bedding detail, whereas sandstones display hole spiralling and yield little meaningful information. Shales are conductive, but locally, images show the presence of thin and sharp-based, relatively resistive shale beds that progressively ramp upward into the background shales. These resistivity ramp-defined bedding units may be individual mud-flows, with the resistive beds being possibly coarser grained or cemented. Following correction for structural tilt, bedding surfaces between and within shale successions range from uniform to irregular and possibly indicate some degree of shingling between individual flow units. Tool runs typically encounter long near-uniform sections of sandstones and shales with well defined lithological boundaries. RAB images show that these sandstone-shale transitions are sharp and steep (22°–88°). Such dips imply either steep channel margins, faulted contacts or more likely, the result of borehole bias in horizontal wells drilled into channellized successions. The lateral margins to an individual sandstone body, sampled as entry and exit points along the well path, commonly display discordant strikes, possibly implying lenticular geometries. Sandstone-shale contacts may allow qualitative estimation of the direction of the channel element thickening. Compactional shale drape deformation is often present away from the sandstone masses. Sedimentation is modelled as a two phase system with slope-supplied, mud-dominated gravity-flows temporally disconnected from deposition of shelf-derived turbiditic sandstones. Infrequent depositional events resulted in channels that were infilled with sand from high-density turbidity currents. Fractures and faults are imaged with the RAB tool and indicate a possible topographical control to structural development.

This paper introduces applications of a logging-while-drilling (LWD) resistivity imaging tool in aiding understanding of the oil-bearing Alba Field submarine-channel complex in the United Kingdom North Sea. The successful development of the Alba Field is related to the placement of horizontal wells in the reservoir as high above the oil water contact as possible, and to the minimization of shale penetration. This assists the maximum recovery of reserves. As a result of poor seismic imaging of the top reservoir, and the presence of intra-reservoir shales, the RAB tool has been utilized to help improve geological understanding and thus the development of the field. It was anticipated that this could be achieved through a greater knowledge of the sand and shale bed types, occurrence and geometries, leading to an improved reservoir model.

Alba Field geological background

The Alba oil field (Fig. 1) is located in the Central North Sea (Block 16/26), approximately 225 km NE of Aberdeen. (It occurs within the Mesozoic Witch Ground Graben, south of the Fladen Ground Spur and north of the Renee Ridge (Harding *et al.* 1990, Newton & Flanagan, 1993)). The field was discovered by the well 16/26-5 in 1984 with first oil production in January 1994. The field is approximately 12 km long by 1.5 km in width, trends NW to SE and lies at depth of 6000–6600 ft below sea level.

The reservoir section is of earliest Late Eocene age (Bell *et al.* 1997) comprizing a stacked series of intra-reservoir shales and unconsolidated sandstones, interpreted as being deposited from high-density turbidity currents, within a

From: LOVELL, M. A., WILLIAMSON, G. & HARVEY, P. K. (eds) 1999. Borehole Imaging: applications and case histories. Geological Society, London, Special Publications, **159**, 177–189. 1-86239-043-6/99/$15.00. © The Geological Society of London 1999.

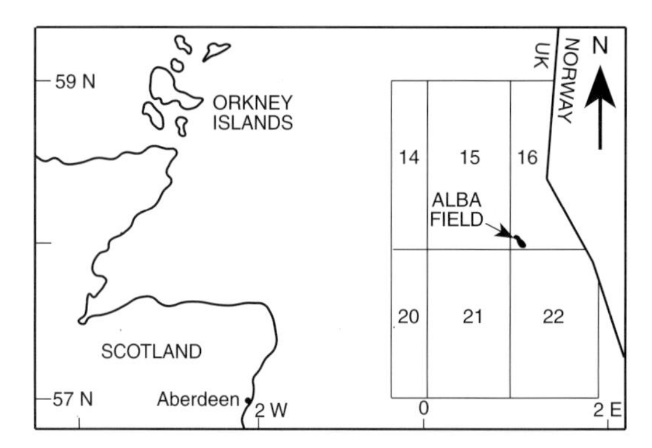

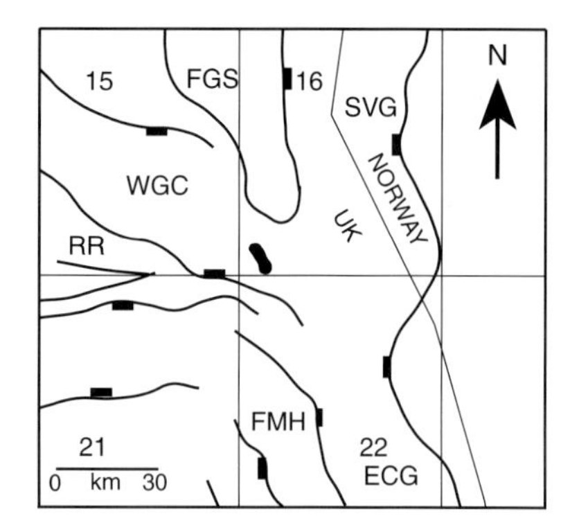

Fig. 1. Alba field location maps and dominant local structural features. FGS, Fladen Ground Spur; WGC, Witch Ground Graben; SVG, South Viking Graben; RR, Renee Ridge; FMH, Forties Montrose High; ECG, East Central Graben.

pre-existing slightly sinuous channel on an intra-slope terrace (Newton & Flanagan 1993). A programme of biostratigraphy, palynofacies, core examination and dipmeter interpretation was completed in parallel with the examination of RAB data, the results of which help constrain interpretation of the RAB images. Seismic and dipmeter data show the presence of steeply dipping bedding surfaces within the Alba Field and cores show mean bedding dips which exceed structural tilt by 5°–20°. Dipmeter data from vertical wells also yields information suggesting consistent zones of relatively steep dips (approximately 10°–25°), with many of the dips striking parallel (NW–SE) to the channel complex.

Core, dipmeter and seismic data show the presence of faulting and fracturing within the Alba Field. Core sandstone samples display a range of features that include normal faults (throws up to 1 ft or more), possible granulation seams, steep bifurcating horse tail splays, shallow dipping shear fractures (typically dipping 20°–25°) and other undifferentiated fracture types. Shales display normal and very rare reverse faults, fractures and injected sandstone dykes and sills. Some sandstone dykes occur as fills to normal fractures. Decompaction of near-vertical, ptygmatically folded injected sandstones from core, imply a median depth of emplacement of *c.* 1 km.

Core examination indicates that a distinguishing feature of the Alba channel fill is that the shales and sandstones are distinct and partitioned by sharp, well defined and locally erosive boundaries. The intra-reservoir shales are slightly silty and sandy, but they appear nearly homogeneous, devoid of organized sedimentary structures that may preferentially contain coarser material, e.g. ripples, graded units and distinct laminations. Sandstones are clean, glauconitic, fine grained, well-sorted and organized into bedded, dish and pillared, laminated, cross-bedded, scoured and visually massive units with rare discontinuous mica-rich laminations. Clastic fines which could be associated with these sandstones, e.g. as the tops of graded units, are absent.

Shales occur throughout the reservoir section, presenting potential small to large-scale barriers to fluid flow. These intra-reservoir shales contain only reworked fossils with a deep marine signature. All are derived from strata older than the channel fill, with the fragmentation, abrasion, segregation and hydrodynamic sorting of microfaunal and floral parts. The intra-reservoir shales were probably emplaced by slope-sourced gravity flows, whereas the sandstones were derived from near-shore to shelf-locations and emplaced by turbidity currents. This results in the characteristic distinct spatial separation between the two main lithology types. The origin and distribution of the shales is poorly understood and individual shale bodies within the reservoir are commonly below seismic resolution.

Resistivity at Bit (RAB) tool

The RAB tool is an electrode resistivity measuring device analogous to wireline laterolog tools. It can make five resistivity measurements and record gamma ray and tool shock data (Bonner *et al.* 1996; Young *et al.* 1996; Prilliman *et al.* 1997). Three azimuthally focused measurements are made by measuring the current leaving 1 inch diameter button electrodes. These three measurements have vertical resolution of approximately 1 inch, but have different depths of investigation of 1, 3 and 5 inches (shallow, medium and deep respectively) and can be used for resistivity profiling. In addition, these buttons are used to provide up to three quantitative resistivity images of the formation. Other resistivity measurements include one through the bit and one through a collar. Within an 8.5 inch borehole each measurement is equivalent to a sector of 0.5 inch. The tool has a vertical resolution of 1–3 inches and provides a 100% coverage of the borehole.

RAB data are acquired in time. Usually one set of five resistivity measurements (three focused electrodes, bit and collar) are made every 5 seconds. For each of the three focused electrodes 56 separate measurements are made over a time span of 2 seconds. These sample times define the rate of penetration (ROP) and revolutions per minute (RPM) required to optimize measurement resolution. Though hole size variation will strictly affect resolution as radial sampling is fixed to 56 bins, this has little practical impact. The 2 second measurement cycle requires a minimum RPM of 30, there is no practical upper limit for standard drilling operations. Though intrinsic resolution is approximately 1 inch the minimum allowable sample interval in depth is 1.2 inches setting an optimum ROP of 72 feet per hour. In practice, optimum ROP is calculated to provide 2 inch resistivity depth sampling which gives an ROP of 120 feet per hour.

Data is partly stored down hole in a 5 MB memory and also transmitted to surface in real time. Those measurements transmitted in real time include resistivity data, quadrant gamma ray and estimated formation dips computed from the medium and deep azimuthal resistivities. From the above, it is apparent that a prior knowledge of anticipated ROP and RPM is required to determine the potential quality of RAB images. The situation is more complex as image data is stored down hole in memory. Given a sample rate of one every 5 seconds, memory would fill after 75 hours. This figure can be doubled to 150 hours with a sample rate set at 10 seconds, though the ROP required for optimum resolution would fall by half from 120 to 60 feet per hour.

The principal acquisition objective in the Alba Field was to provide formation laterolog type resistivity measurements and borehole electrical images for geological interpretation. These images can be used by an interpreter to recognize and orientate geological and artifact features that derive from the borehole. The raw resistivity data and the images can be used for geosteering, formation and fracture evaluation and conventional geological interpretation including stratigraphical, structural and sedimentological studies.

Within the Alba Field the 8.25 inch diameter RAB tool was run in a rotary steerable assembly 28 feet from the bit face mounted behind the motor. The real time geosteering capability was not utilized though data was transmitted in real time via the MWD powerpulse telemetry tool. Real time measurement transmission included bit resistivity, ring, upper and lower quadrant button resistivity, gamma ray and, in two wells formation dips.

Sample rate was set to 10 seconds for all RABs run, and of the three possible resistivity image sets the tool was programmed to save only the deep button arrays together with the other conventional measurements such as ring gamma ray bit data and etc. Though this gave over 100 hours logging time, image quality noticeably deteriorated above 60 feet per hour with quality severely impaired above 90 feet per hour. The five RAB's run through the Alba Field were acquired over a period of ten months during which time significant improvements were made to the service (not all realised for the Alba operation) mainly through improved downhole and surface software development, e.g. potential sample rate increased from 10 seconds to 5 seconds, real time dip computation and tool configuration.

Interpreting RAB images from the Alba Field

Digital versions of RAB images from five near horizontal wells (exact hole orientations are commercially sensitive) were loaded and processed in RECALL (Z&S image interpretation package). Static and dynamic normalized images (20 ft step length) were generated to allow ready visualization of features with interpretation practise following Harker *et al.* 1990, and Lofts *et al.* 1997 and it was found that effective interpretive examination of these images was made over 50 ft intervals. Dips were interactively picked from the images on a workstation in conjunction with open hole logs and classified as shale bedding, sandstone-shale boundaries, internal sandstone bedding and fractures. Image quality in these horizontal wells was affected by a number of artifacts (see Lofts & Bourke 1999) including slide zones (i.e. where the drill string did not turn – no interpretations are possible within these zones), hole spiralling (with poor feature recognition), and low-side mud-invasion profiles within some sandstones. The combination of horizontal well trajectories, tool limitations and facies types limits image quality, feature recognition, bed resolution and dip-picking precision and accuracy. However, even within this difficult logging environment usable data for structural and sedimentological interpretation were derived. The RAB images were often of a high quality, and broadly comparable with images through other horizontal wells, although the density of data is less than would be gathered in a vertical well. RAB images have a lower resolution when compared to conventional resistivity pad-based imaging tools. This results in a wider scatter of manually picked surfaces, implying a lowered confidence in any interpretation.

Borehole bias

Images from horizontal wells are effective at identifying dipping surfaces; however, these features suffer from a borehole bias that locally can be severe. The probability of cutting a geological object is dependent on the intersection angle between the hole and the feature. Boreholes oriented perpendicular to features will have the greatest apparent occurrence, whereas holes drilled sub-parallel to geological features will intersect fewer examples. This results in a sampling bias with an under-determination of features that strike parallel or near parallel to the borehole axis. Horizontal Alba Field wells display an apparently strong sampling bias for sandstone-shale bedding features and fracture orientaions. To reduce sampling bias, corrections were applied to fracture and sandstone-shale boundary dips using the method of LaPointe & Hudson (1985) as outlined in Peacock & Sanderson (1994).

Geological details from RAB images

Lithological observations

RAB images readily discriminate between sandstones and shales, locally with a bed resolution

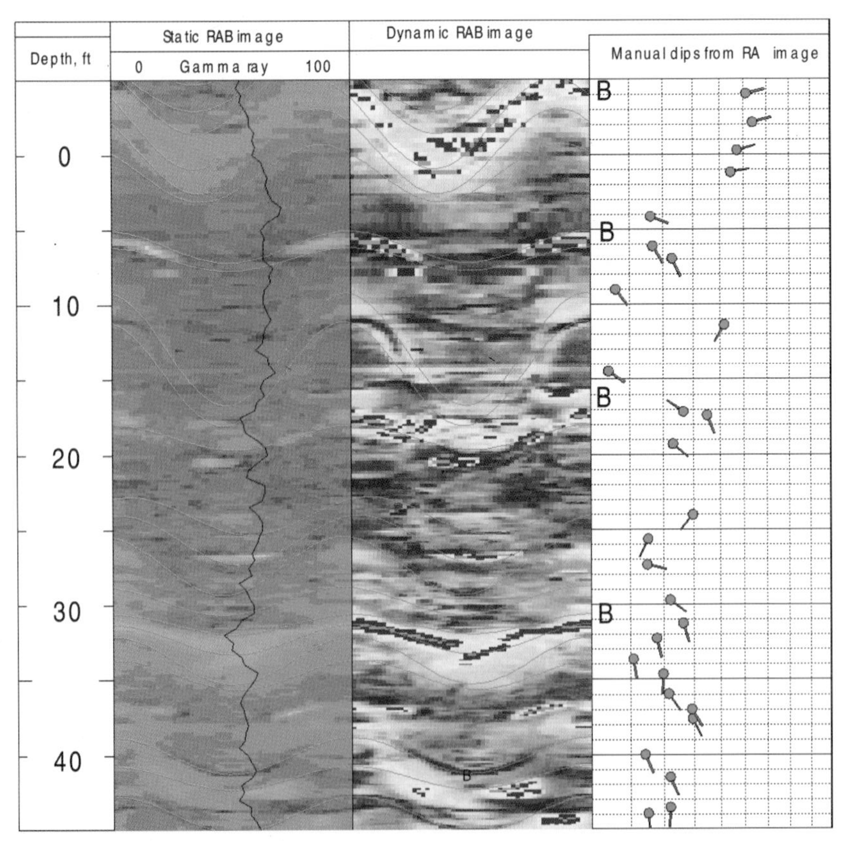

Fig. 2. Typical 50 ft shale section with stacking of relatively resistive beds and possible bases to individual flow units (marked B), stratigraphical up is down hole. Within this shale interval manually picked dips are slightly scattered.

of below 10 cm. Sandstones appear as resistive features (scaled as light coloured, yellow-orange on images), whereas shales appear as conductive (dark) images. Both sandstones and shales typically occur in lengths of 50 to hundreds of feet.

Static normalized images through shale successions are characterized by long intervals of near uniform conductive images (Fig. 2). Bedding and stratal features range from parallel, even, discontinuous to continuous traceable laminae through to slightly wavy, non-parallel and often visually discontinuous forms. Slumps, slides and classic matrix-supported, clast-rich debris flows appear to be absent. Dynamic normalization through these images reveal distinct, sharp-based, thin (10–40 cm), relatively resistive beds that display upward decreasing resistivity ramping into more conductive lithologies, with the basal resistive interval being possibly coarser grained and/or cemented. These shale-dominated, resistivity ramp-defined units, range in stratigraphical thickness from *c.* 0.5 ft to 9 ft with a mean of approximately 2.5 ft, and can appear stacked. Well site biostratigraphical sampling and analysis through these shales commonly results in repeated floods of older reworked faunal (in particular agglutinating formaminiferids) and floral material. These floods occur on a vertical scale of less than ten feet (Braham & Hulme pers. comm. 1997).

From the stereographical analyses of dips through each distinct shale interval, mean local structural tilts were determined (Fig. 3). Bedding dips were then corrected for structural tilt. Following correction, bedding surfaces between and within shale units, range from uniform (planar) to irregular and a number of units have possible eroded bases. Stacked successions display up to 90° swings in bedding dip azimuth between and within units.

The imaged shale intervals are interpreted as stacked, mud-flow units defined by resistivity contrast in which hydrodynamic sorting of fossil material occurred. It is possible that the palaeontological concentrations are associated with individual mud-flow bases. The apparent lack of soft sediment deformation features implies that shale formation was probably not by slumping and/or sliding. The residual dips that are now observed are probably the result of compactional effects across a geomorphological and facies differentiated landscape. Irregularity and variation in bedding surface orientations between flow-units indicate possible shingling of individual flows, suggesting that these were unlikely to have been deposited on a uniformly flat basin floor, and are associated with a slightly irregular local topography.

Static and dynamically normalized images from within the sandstones are dominantly

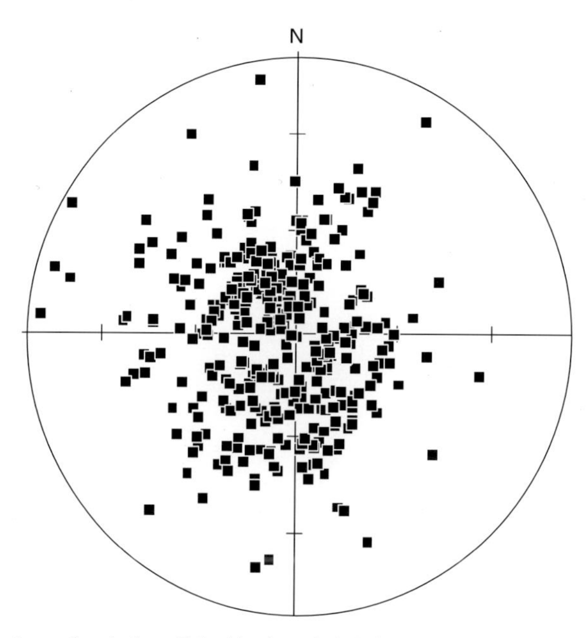

Fig. 3. Poles to planes, lower hemisphere Schmidt plot of shale bedding as imaged by the RAB tool in near horizontal Alba well.

affected by hole spiral artifacts with only localized uniform image character visible. Rare bedding surfaces are usually steeper than the adjacent shale beds and can display variable orientation within a given sandstone body. Large conductive or resistive clasts are not imaged; the tool is possibly not sensitive enough to detect fragments smaller than *c.* 5 cm in diameter. Thin shale bodies are present within a number of long sandstone sections, representing further possible breaks in the sandstone masses. These thin intra-sandstone shale beds are steeply dipping. The difficulty of feature recognition within these Alba sandstones is a reflection of uniform resistivity and the presence of artefacts.

Sandstone-shale transitions

Transitions along the well bore between shales and sandstones are well-defined, sharp and usually steep, with dipping contacts of between 22°–88° (median dip = 67°, Fig. 4). In addition, common shale compactional drape fabrics are often associated with these boundaries and their effects in shale can be observed for up to 30 m away from the margins of some sandstone bodies. The lateral margins to an individual sandstone body, sampled as entry and exit points to one sandstone body along the well path, display discordant strikes as do some conjugate compactional drapes. In the cases of repeated along hole sampling of one sandstone body, bounding tangents often display varying strike orientations.

The sharp resistivity differentiation between sandstone and shale lithology types, with no evidence of mutual gradational contacts, supports the core observations of facies separation. Steep transitions imply either deep erosive channels, faulted contacts, or more likely the effect of borehole bias that occurs when horizontal wells

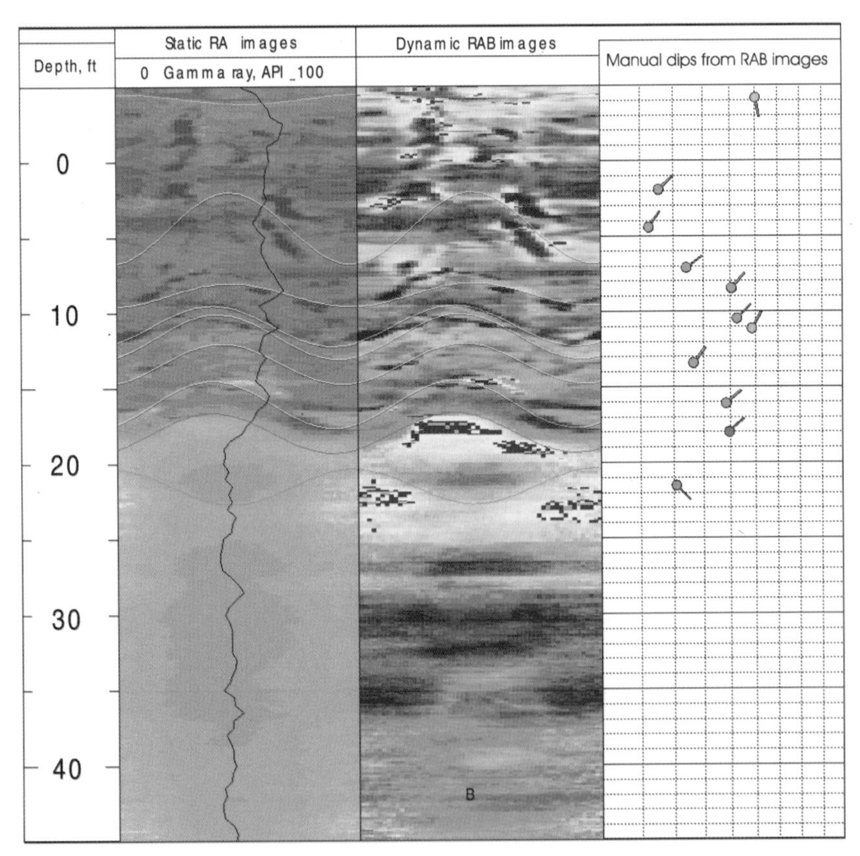

Fig. 4. Static and dynamic normalized RAB images with superimposed gamma ray trace and manually picked dips. Illustrates typical along-hole transition from well-imaged and draped shale section via a steep bounding surface (red tadpole) to poorly imaged sandstone interval.

are drilled through horizontally stratified channelized depositional systems. Horizontal wells preferentially cut the oversteep parts of convex-outward sandstone-shale contacts. This is why Alba's sandstone-shale contacts imaged with the RAB tool in near horizontal wells appear over-steep; it is not by deep erosion or faulting, they are steep by default of sampling bias.

The presence of differential compactional drape (Fig. 4) implies interbedded sandstone-shale lithologies and the detection of an edge to a sandstone body. Discordant entry and exit sandstone-shale strikes indicate lenticular, or irregular geometries in plan-view for the individual sandstone bodies, with a range in sandstone-bounding tangents for any one sandstone body implying the presence of possible sinuous channel elements.

Geometry of reservoir elements

Within turbidite channel complexes, sand-filled channel elements can form discrete pods of sandstone which may or not be connected (cf. Timbrell 1993). The lenticular nature of the sandbodies within the Alba Field appears to be supported by RAB images, with, as noted above, discordant entry and exit strikes observed for any given sandstone mass indicating the presence of non-parallel sandstone margins.

The bulk of the sandstone-shale tangents through all five studied wells strike NE–SW and NW–SE (in the ratio of approximately 2:1), perpendicular and parallel to the Alba channel complex respectively (Fig. 5). These are interpreted as sampling the edges of sandstone masses running parallel and perpendicular to the main Alba channel complex. This dominance of NE–SW striking sandstone-shale transitions, at the expense of those striking NW–SE, is due to a strong preferential sampling bias, and does not indicate an excess of sandstone bodies striking NE–SW across the channel complex axis. Correcting for this bias results in a marked reduction in the ratio of NW–SE to NE–SW striking tangents (Fig. 6), with a revised NE–SW to NW–SE strike ratio of 1:1.6), implying pod like geometries, elongated down the channel complex.

Within individual sandstone bodies, sandstone-shale tangents may tentatively allow qualitative estimation of the direction of the channel element thickening (Fig. 7). In the cases of repeated along hole sampling of one sandstone body this approach can be extended, with bounding tangents implying that the centre of gravity to the channel elements are at least slightly meandering.

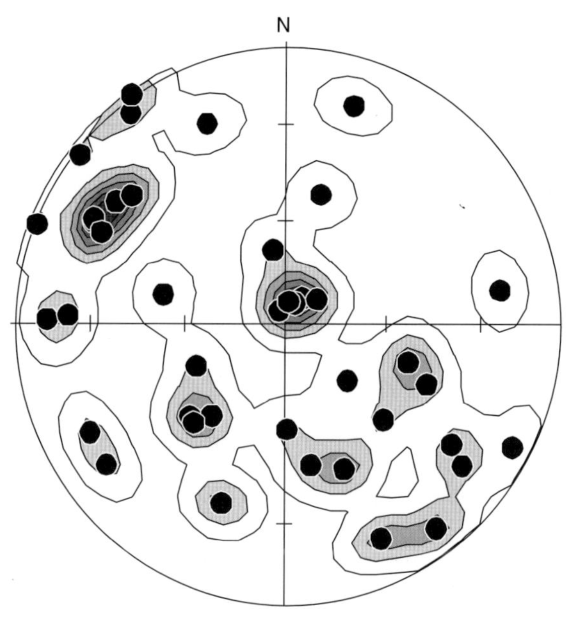

Fig. 5. Raw contoured poles to planes, lower hemisphere Schmidt plot of sandstone-shale bedding contacts as imaged by the RAB tool in near horizontal Alba well.

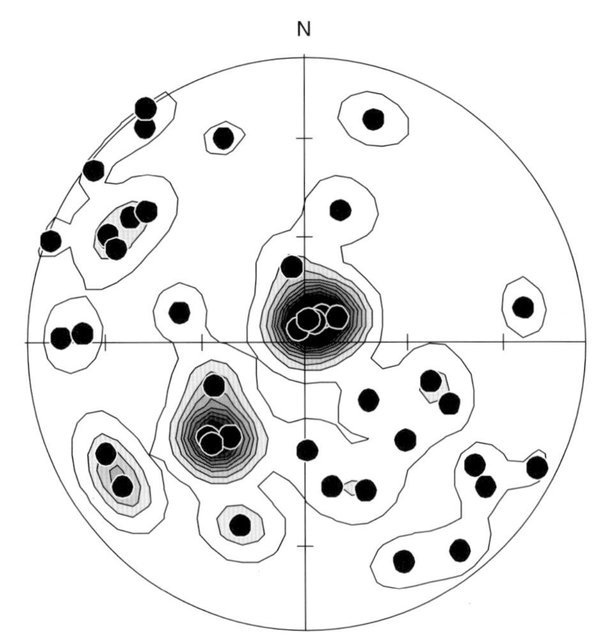

Fig. 6. Contoured poles to planes, lower hemisphere Schmidt plot of sandstone-shale bedding contacts corrected for borehole bias as imaged by the RAB tool in near horizontal Alba well.

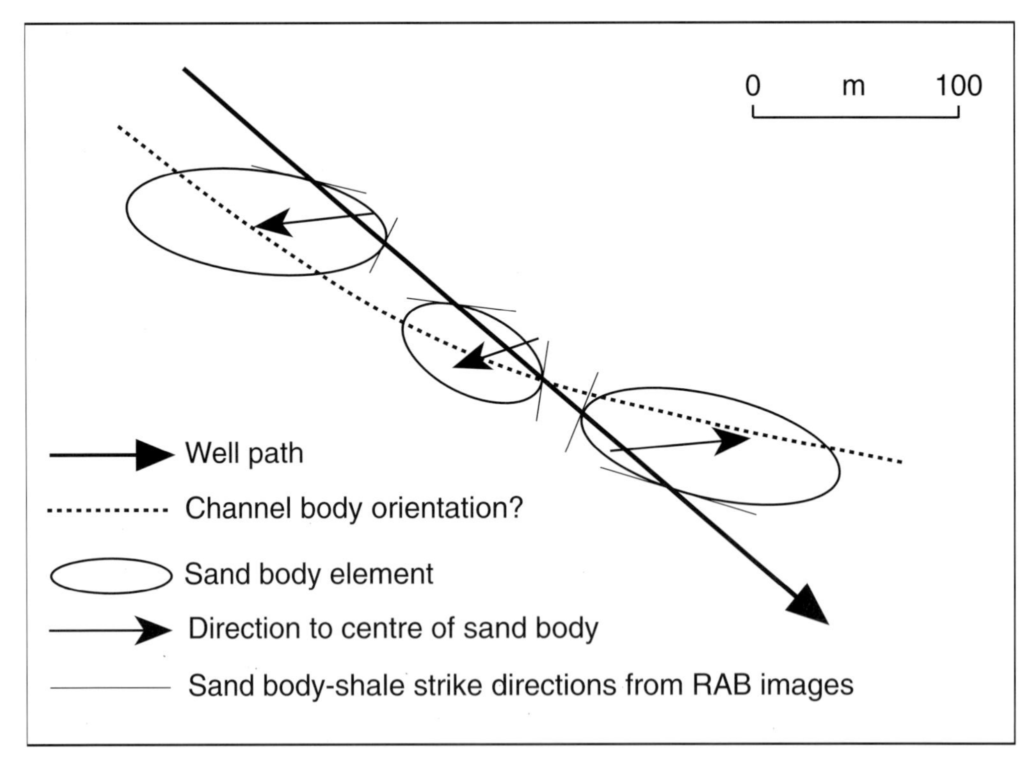

Fig. 7. Interpreted geometrical relationships through one sandstone body imaged three times by the RAB tool. Arrows indicate the direction to the centre of each sampled mass, constrained by sandstone-shale contact strike directions. The relative positions of these centres allow at least one solution for a channel axis to be crudely mapped.

Structural observations

The mean orientation of shale and some sandstone bodies, and their mutual contacts, were derived from the images, with the results tied back to seismic sections. Away from the immediate compactional effect of adjacent sandstone bodies (e.g. Fig. 4) each shale mass appears to have a distinct mean orientation (Fig. 3), with intervals yielding average dip values that range from 8°–24° and are comparable with seismic data. Sections of oversteep bedding are locally common, but are usually confined to short sections of the shale lengths and sandstone bodies rarely yield bedding data that can be used to constrain their structural attitude. Although common within core, sand-stones interpreted as injection features are rarely observed on the RAB images, with only two sandstones identified as possible injection features from five RAB's.

A number of oversteep bedding dip patterns within shales are interpreted as symmetrical and asymmetrical drag features interpreted as faults, although discrete fault surfaces are rarely well imaged, with deformation occurring in hanging- and footwall locations. These interpreted faults dominantly strike N–S and NW–SE. Fault throws have been estimated from the width of the deformed drag zones following Knott *et al.* (1996) and these range from approximately 1 m to 1.7 m (accurate to one order of magnitude).

Fractures are locally common within the shales on RAB image (Fig. 8). They occur as clustered, well-defined (usually traceable around the borehole) and steep (40°–89°), electrically conductive to very conductive features and more rarely as resistive bands within shales (Fig. 9), both show no apparent offset. Fractures were rarely identified with confidence in sandstones, although they are locally common within core.

These interpreted fractures and faults represent a probable mix of Mode 1 hybrid and Mode 2 fracture types concentrated into distinct fracture and fault zones, with under sampling of fault offset due to limited tool resolution. A proportion of the deformation

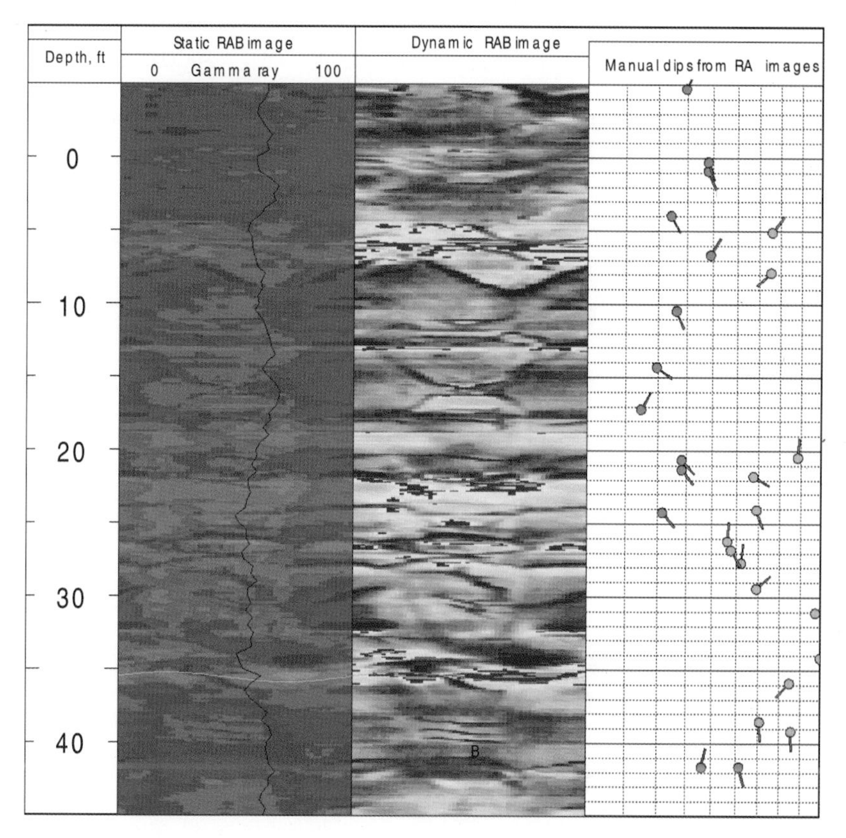

Fig. 8. Shale interval with steep and relatively conductive (dark) planar features discordant to shale bedding (green tadpoles) that are interpreted as fractures (yellow tadpoles).

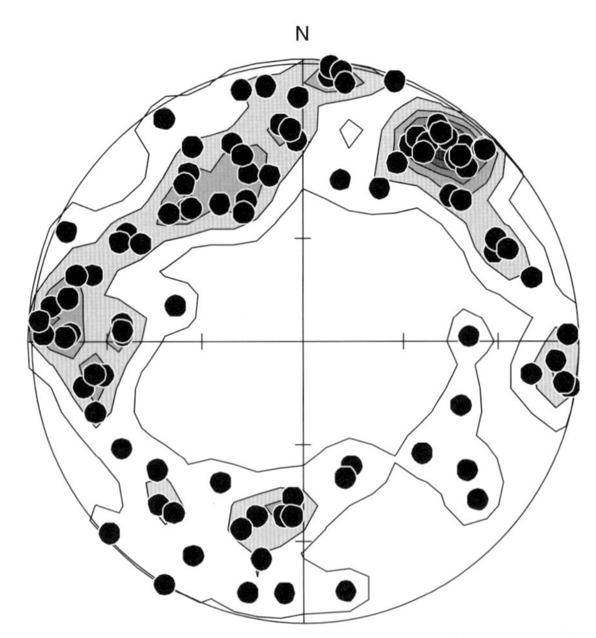

Fig. 9. Raw contoured poles to planes, lower hemisphere Schmidt plot of fractures as imaged by the RAB tool in near horizontal Alba well.

strain has been taken up by plastic deformation within fault drag zones. If these fractures and faults are not shale-layer bound, they may be expected to propagate into and compartmentalize the sandstones.

Horizontal Alba Field wells display an apparently strong sampling bias for fractures, with the development of well defined data shadow zones parallel with borehole axes (Fig. 9). The sampling bias was corrected for and the analysis

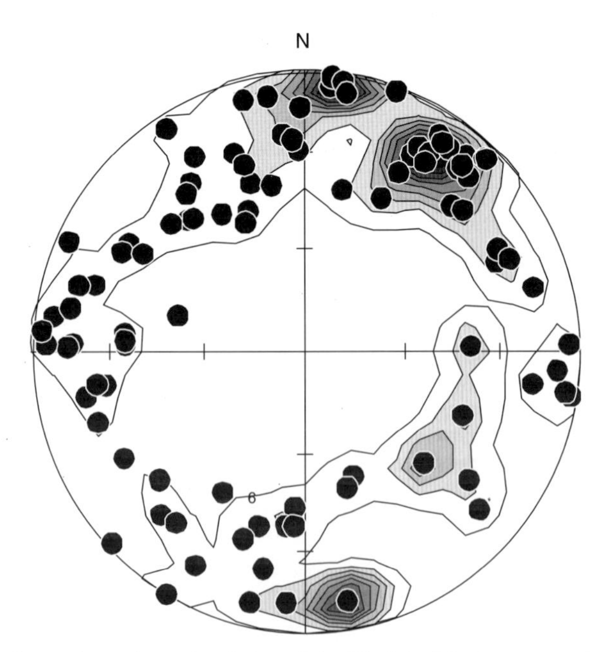

Fig. 10. Contoured poles to planes, lower hemisphere Schmidt plot of fractures corrected for borehole bias as imaged by the RAB tool in near horizontal Alba well.

indicates that under-sampled fracture sets near-parallel to the borehole probably have fracture densities two to eight times greater than that indicated on RAB images (Fig. 10).

Discussion

Sedimentation within Alba is modelled using two depositional end-members, with mud-dominated gravity-flow sedimentation spatially and temporally disconnected from deposition of the sandstones. Data from the RAB images supports this model. Muds were supplied from the slope as reworked deep-marine shales. Sands were derived from the shelf. Mud dominated successions within the Alba field represent stacked gravity flow units deposited rapidly and ponded by the irregular topography within the channel complex. The lack of associated slumps, debrites or shale clasts within the intra-reservoir shales indicates that the flows possibly evolved from slumps to fully disaggregated turbulent flows with accumulation of bioclastic material at their bases. Changes in flow state induced by the channel complex (e.g. passage to a lower slope and/or flow expansion) may have induced flow collapse to a hyperconcentrated state (cf. McCave & Jones 1988; Porebski *et al.* 1991), halting the passage of these flows.

Infrequent, but energetic short-lived flow events resulted in pod-shaped channel elements elongated in the direction of the channel complex and displaying irregular margins. These were infilled with sand from high density turbidity currents shortly after formation, with any associated muds flow-stripped through the channel complex. Although information on the plan–view architecture of fossil turbiditic systems is rare, interpreted data from Alba appears comparable to that described by Bruhn & Moraes (1989) from aerial photographs within the Almada Basin of Brazil. Bruhn & Moraes (1989) detail lenticular, highly irregular to embayed and isolated channelized bodies that are usually elongated parallel to a canyon margin with individual sandstone pods containing sinuous channel paths. The pods are set in shales and the mutual nearest neighbour distance from pod to pod ranges from less than approxmately 30 m–400 m. A down-canyon horizontal well through such a succession could be expected to yield similar image sets as observed within Alba.

The dominant strike orientation of fractures, faults and injected sandstones (from dipmeter data in vertical wells) are near parallel with the axis of the Alba Field channel complex (NW–SE), implying a strong, possibly localized, topographical control on structuration (Fig. 11). The size, geometry and attitude

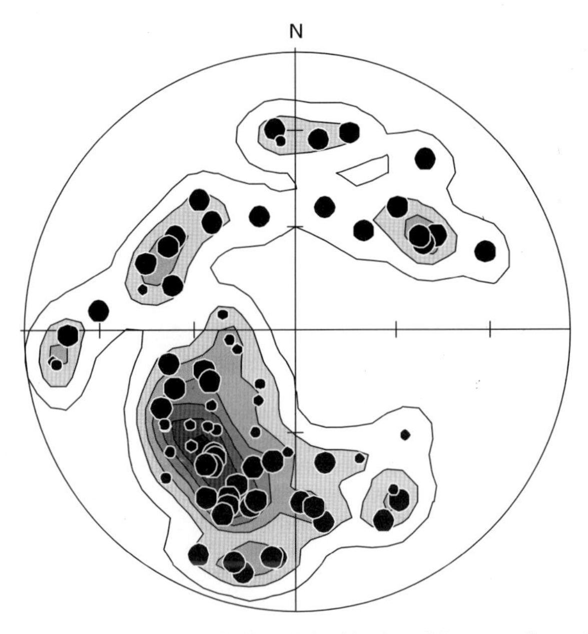

Fig. 11. Contoured poles to planes, lower hemisphere Schmidt plot of fractures (large dots) and injected sandstone's (small dots), manually picked dipmeter data from vertical well with Alba field, features tied to core.

of the sandstone bodies within the Alba Field, and their relationship with adjacent dewatering shales, are strong controls on deformation in and around the Alba Field. The presence of a strong preferred fracture strike within and immediately adjacent to the Alba channel complex, parallel to the field, contrasts with the near random pattern of faulting observed in block 16/26 (Cartwright & Lonegran 1996). This implies that the compacting Alba complex may have experienced a localized stress condition that partially controlled fracture types and their orientations.

Sandstone injection fabrics are common at the top of many Eocene reservoir intervals in the North Sea and have been attributed to mobilization of overpressured sands by liquefaction during (hydro) fracturing (Cartwright 1994). In Alba, it is likely that the presence of rapidly deposited sands and thick mud-flows, will strongly limit dewatering of the succession and help retain interstitial fluids, promoting overpressuring, possibly aiding generation of the injected sands. However, there is a relative lack of injected sandstones through intra-reservoir shales, when compared with reservoir capping muds, indicating that individual pressure cells within isolated sandstone bodies did not form during burial, or that any pressure differential was not sufficient to breach confining intra-reservoir shales, possibly indicating a connected pressure regime within the reservoir.

Conclusions

- RAB images yielded useful data in horizontal wells through a turbidite channel complex, although intrinsic feature resolution is below that of conventional resistivity imaging tools. Lithologies are readily identified and surfaces oriented. The recognition and correction for borehole bias in the interpretation of bedding and fracture surfaces is important.
- A gravity flow origin for intra-reservoir shales is supported from the images.
- Sandstone body-shale boundary orientations can be readily imaged and used to constrain channel-element geometry. Entry and exit points for individual sandstone bodies display discordant strikes, implying a pod-like nature to these masses. Repeated along hole sampling of the same sandstone body allows individual sandstone bodies to be crudely mapped.

The authors would like to thank the Alba Field partner group of Chevron, Conoco, Fina, Petrobras, Statoil, Saga, Unilon/Baytrust and Union Texas for support and permission to publish this paper. In addition, we thank our respective managerial teams for likewise encouragement and permission. The interpretation and opinions are ours alone and do not necessarily reflect the view of our respective management's. Bill Braham and Neil Hulme provided essential descriptive and interpretive insight as to the nature of the intra-reservoir shales. Neil Parkinson and a further anonymous reviewer are thanked for their very useful comments.

References

BELL, C., BRAHAM, W. & HULME, N. 1997. *The recognition and significance of reworked microfossil assemblages and their impact on the development of the Alba Field, UK North Sea.* Conference Abstract. Biostratigraphy in production and development geology, April 9th, Aberdeen University, Scotland.

BONNER, S., FREDETTE, M., LOVELL, J., MONTARON, B., ROSTHAL, R., TABANOU, J., WU, P., CLARK, B., MILS, R. & WILLIAMS, R. 1996. Resistivity while drilling-images from the string. *Oilfield Review*, **Spring 1996**, 4–19.

BRUHN, C. H. L. & MORAES, M. A. S. 1989. Turbiditos da Formação Urucutuca na Bacia de Almada, Bahia: um laboratório de campo para estudo de reservatórios canalizados. *Bol. Geociências da PETROBRÁS*, **3**, 235–267.

CARTWRIGHT, J. A. 1994. Episodic basin-wide hydro fracturing of overpressured Early Cenozoic mudrock sequences in the North Sea Basin. *Marine and Petroleum Geology*, **11**, 587–607.

—— & LONERGAN, L. 1996. Volumetric contraction during the compaction of mudrocks: a mechanism for the development of regional-scale polygonal fault systems. *Basin Research*, **8**, 183–193.

HARDING, A. W., HUMPHREY, T. J., LATHAM, A., LUNSFORD, M. K. & STRIDER, M. H. 1990. Controls on Eocene submarine fan deposition in the Witch Ground Graben. *In*: HARDMAN, R. F. P. & BROOKS, J. (eds) *Tectonic Events responsible for Britain's Oil and Gas Reserves*, Geological Society of London Special Publication, **55**, 353–367.

HARKER, S. D., MCGANN, G. J., BOURKE, L. T. & ADAMS, J. T. 1990. Methodology of formation microscanner interpretation in Claymore and Scapa Fields (North Sea). *In*: HURST, A., LOVELL, M. A. & MORTON, A. C. (eds) *Geological Applications of Wireline Logs*, Geological Society of London Special Publication, **48**, 11–25.

KNOTT, S. D., BEACH, A., BROCKBANK, P. J., LAWSON BROWN, J., MCCALLUM, J. E. & WELBON, A. I. 1996. Spatial and mechanical controls on normal fault populations. *Journal of Structural Geology*, **18**, 359–372.

LAPOINTE, P. R. & HUDSON, J. A. 1985. *Characterization and Interpretation of Rock Mass Joint Patterns*, Special Publication Geological Society of America, **199**.

LOFTS, J. C. & BOURKE, L. B. 1999. The recognition of artefacts from acoustic and resistivity borehole imaging devices. *This volume*.

——, BEWDFORD, J., BOULTON, H. VAN DOORN, J. A., & JEFFREYS, P. 1997. Feature recognition and the interpretation of images from horizontal wellbores. *In*: LOVELL, M. A. & HARVEY, P. K. (eds) *Developments in Petrophysics*, Geological Society Special Publication, **122**, 345–365.

MCCAVE, I. N. & JONES, K. P. N. 1988. Deposition of ungraded muds from high density non-turbulent turbidity currents, *Nature*, **333**, 250–252.

NEWTON, S. & FLANAGAN, K. P. 1993. The Alba Field: evolution of the depositional model. *In:* PARKER, J. (ed.) *Petroleum Geology of North West Europe: Proceedings of the 4th Conference*. Geological Society of London, 161–175.

PEACOCK, D. C. P. & SANDERSON, D. J. 1994. Strain and scaling of faults at Flamborough Head, UK. *Journal of Structural Geology*, **16**, 97–107.

POREBSKI, S. J., MEISCHNER, D. & GÖRLICH, K. 1991. Quaternary mud turbidites from the South Shetland Trench (West Antarctica): recognition an implications for turbidite facies modelling. *Sedimentology*, **38**, 691–715.

PRILLIMAN, J., BEAN, C. L., HASHEM, M., BRATTON, T., FREDETTE, M. A., LOVELL, J. R. 1997. *A comparison of wireline and LWD resistivity images in the Gulf of Mexico*. Society of Professional Well Log Analysts, Transactions of the 38th Annual Logging Symposium, June, Houston, Texas, Paper **DDD**.

TIMBRELL, G. 1993. Sandstone architecture of the Balder Formation depositional system, UK Quadrant 9 and adjacent areas. *In*: PARKER, J. R. (ed.) *Petroleum Geology of Northwest Europe: Proceedings of the 4th Conference*, Geological Society of London, 107–121.

YOUNG, R. A., LOVELL, J. R., ROSTHAL, R. A., BUFFINGTO, L. & ARCENAUX, C. 1996. *LWD borehole images/dips aid offshore California evaluation*. April World Oil.

Insights from simultaneous acoustic and resistivity imaging

T. HANSEN[1] & D. N. PARKINSON[2]

[1] *Ordrupues 160, 292 Charlotteuluud, Denmark (e-mail: tomas.hanson@virgin.net)*
[2] *39 Sydney Road, Richmond, Surrey, TW9 1UB, UK*
(e-mail: neil.parkinson@dial.pipex.com)

Abstract: We describe a method of thin-bed petrophysical analysis that exploits the availability of both acoustic and resistivity images of the same rock. In developing our method we have gained a number of insights into the contrasting response of these images to rock properties, which may be of general interest to the imaging community. The case-study we present is from a thin-bedded, deep-water siliciclastic depositional environment. The resistivity image was used to increase the resolution of a conventional resistivity device. Comparison of bedding dips between the two images suggests a resistivity depth of investigation of around three inches. However, the best fit between image and conventional resistivity was obtained using a somewhat deep-reading resistivity measurement. The acoustic image was used to increase the resolution of our porosity measurement. The best fit to acoustic image amplitude was with the neutron porosity log, rather than with density, slowness, or with any combination of two porosity logs. Given high-resolution resistivity and porosity measurements from the images, and the precise information on bed boundaries that images provide, we proceeded with petrophysical analysis at a much higher resolution than obtainable from the conventional logging suite.

Interpreters of satellite imagery are familiar with the ideas that false-colour images are not conventional photographs; that different image types respond to different physical properties; and also that different image types can be combined to extract quantitative information. With the advent of simultaneous acoustic and resistivity imaging we have the opportunity to acquire two complimentary images of the same rock. This enables us to begin to explore the rock properties being imaged in a more quantitative way. In this paper we report some early experiences with such dual images. The core of our paper is a case study of petrophysical analysis from an interval of thin-bedded siliciclastic rocks interpreted to have been deposited in a deep-marine setting. In the course of this analysis we present some insights into the physical rock properties to which the two images appear to be responding.

Data acquisition

The images used in this study (Fig. 1) were acquired using Western Atlas' STAR™ logging device. At the bottom of the tool string is an acoustic imaging section based upon Western's CBIL™ logging tool (e.g. Parkinson *et al.* 1999). Compared to the CBIL, transducer rotation rates have been increased to 12 rps, but this enhancement is generally used to increase logging speed rather than resolution. Resolution remains the same as with the CBIL, with a vertical sample distance of approximately 0.3 inches and a horizontal sample distance of approximately 0.14 inches in a 12.25 inch borehole.

The resistivity image is acquired by a pad-type device which is typically run some 7.5 m above the acoustic device in the same tool string. There are six pads, with two staggered rows of twelve buttons on each pad (Fig. 2). Radial button spacing is 0.1 inches and vertical sample rate is approximately 400 samples/metre (122 samples/foot). Radial coverage of the borehole surface is approximately 42% in a 12.25 inch borehole.

Both devices share a common orientation device that comprises triaxial accelerometer and magnetometer arrangements, from which borehole deviation, borehole drift azimuth, and the orientation of the tool in the borehole can be derived. Accelerometer data also allow the images to be corrected for uneven travel of the tool up the borehole. An interesting feature of simultaneous imaging is that, whilst tool sticking obviously occurs at the same *time* on both images, the acoustic and resistivity sections will be logging different depth intervals at that time, so it is probable that any given interval of rock will be correctly imaged by at

From: LOVELL, M. A., WILLIAMSON, G. & HARVEY, P. K. (eds) 1999. Borehole Imaging: applications and case histories. Geological Society, London, Special Publications, **159**, 191–201. 1-86239-043-6/99/$15.00.

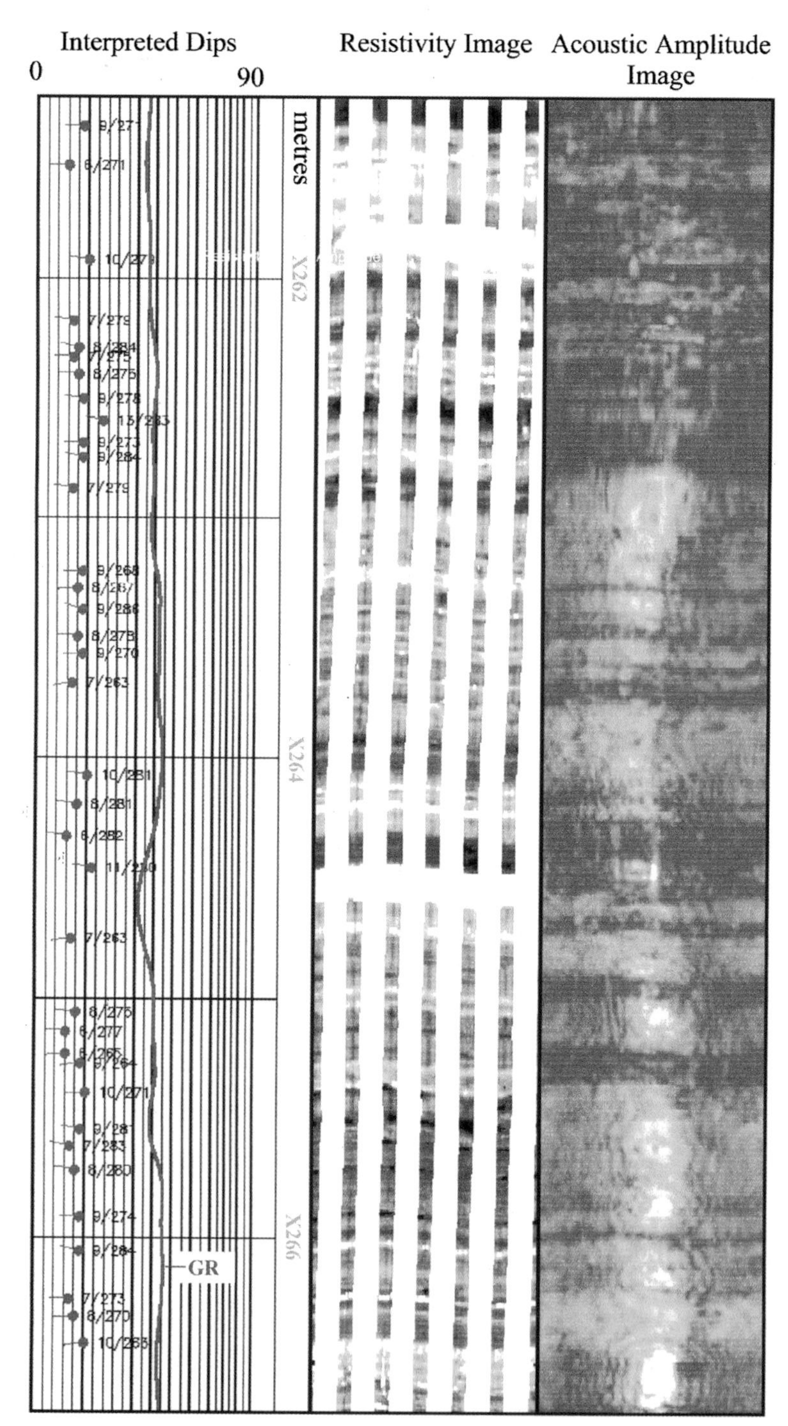

Fig. 1. Typical STAR data from the studied data set. Low circumferential coverage from the resistivity tool is due to 12.25 inch diameter hole size. Brighter colour on the resistivity image corresponds to higher formation resistivity. Brighter colour on the acoustic image corresponds to higher acoustic impedance contrast between the mud and the formation. Stratigraphic dip has been picked on both images.

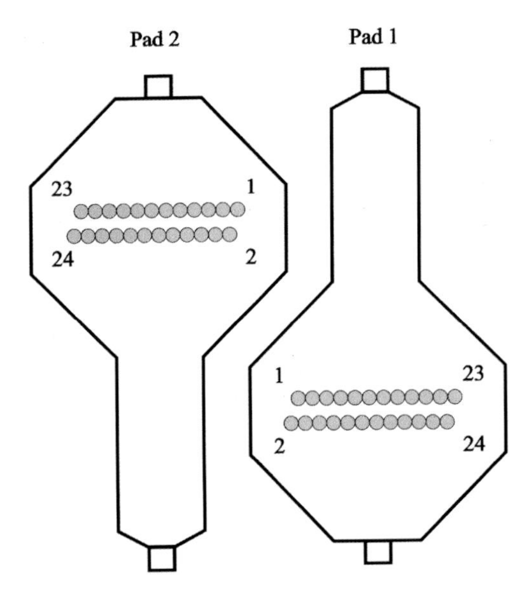

Fig. 2. Pad and button geometry of two of the six pads on the STAR resistivity section. Pads 2, 4 and 6 face up, pads 1, 3 and 5 face down.

least one of the two image types, giving potential for sophisticated accelerometer correction by combining both images.

In addition to the STAR data, a conventional suite of open-hole logs was available for our case study, comprising Compensated Density (ZDEN), Compensated Neutron (CNCF), Acoustic (DT), deep- and shallow-reading ('Dual') Laterolog (RD, RS), Microlaterolog (RMLL) and Natural Gamma-Ray (GR) data.

Image data preparation

If we seek to calibrate an image, then a robust relationship between the image log and a conventional petrophysical log is required. We sought to explore possible relationships by cross-plotting resistivity and acoustic image data with various conventional log measurements. Before attempting this it was important to correct for differences in the character of image and conventional data sets. Data acquired from different logging tools are basically incompatible (Runge & Powell 1967). The key data incompatibilities are:

- Image data are two-dimensional (i.e. they vary both around the circumference of the borehole and in depth)
- Image data have a higher resolution and sample rate
- Image data are affected by variable gain during acquisition
- Each logging tool surveys a different rock volume

We first transformed the image data such that they were represented by a one-dimensional measurement. The simplest method of achieving this is to average all the values at a given depth. We chose this method for cross-plotting with conventional logs since it most closely mimics conventional logging tools, which either investigate the whole circumference of the borehole or, in the case of pad-type tools, investigate a large and arbitrarily orientated rock volume. Note that we chose a more complex method that includes structural dip removal, when deriving the final high-resolution resistivity and acoustic image curves for thin-bed analysis (see below).

Having achieved a one-dimensional representation of the image data, we then filtered it to achieve common vertical resolution with the conventional logs. Ideally, we prefer deconvolution of the coarser measurement to the fine scale and detail of the resistivity image log, but this is obviously not feasible. In practice we average and smooth the image data to the resolution of the variable with the coarsest vertical resolution (Doveton 1994). During acquisition, most conventional logs are brought to similar vertical resolution by the application of a common noise reduction filter. We applied this same filter to the one-dimensional curves derived from circumferential averaging of the resistivity and acoustic imaging data. In addition, we resampled the acoustic and resistivity images from their field sample rates to 100 samples/m in order to reduce the total data volume.

We did not correct the data for gain variations before the data exploration stage, but gain variations are removed in the calibration stage. Incompatibility in the volumes of investigation can not be addressed and must be ignored if any combination of logging tools are to be analysed simultaneously.

Following these data conditioning steps, the image-derived logs have similar apparent vertical resolution as the conventional logs (Fig. 3). We also observe some correlation between the shape of the resistivity image-derived log (RHEXF) and the dual laterolog (RD, RS), and between the acoustic image-derived log (UIPF) and the compensated neutron log (CNCF). We will further explore these relationships below.

Data exploration

We used scatter plots (Figs 4 & 5) to explore the relationships between the conventional log

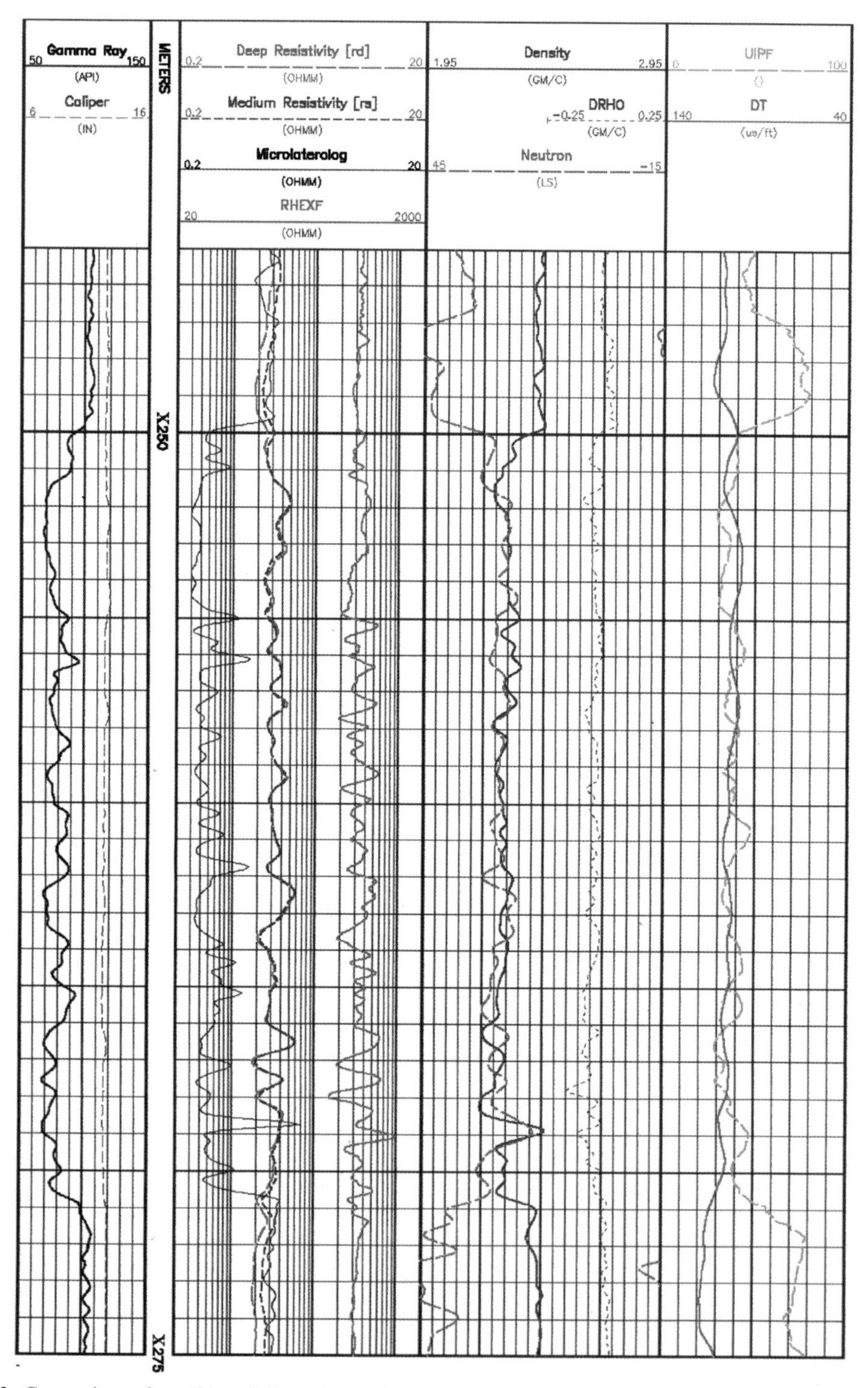

Fig. 3. Comparison of conditioned circumferentially-averaged image log data (UIPF, RHEXF) with conventional logs. Note correlation between UIPF and CNCF, and between RHEXF and RD, RS.

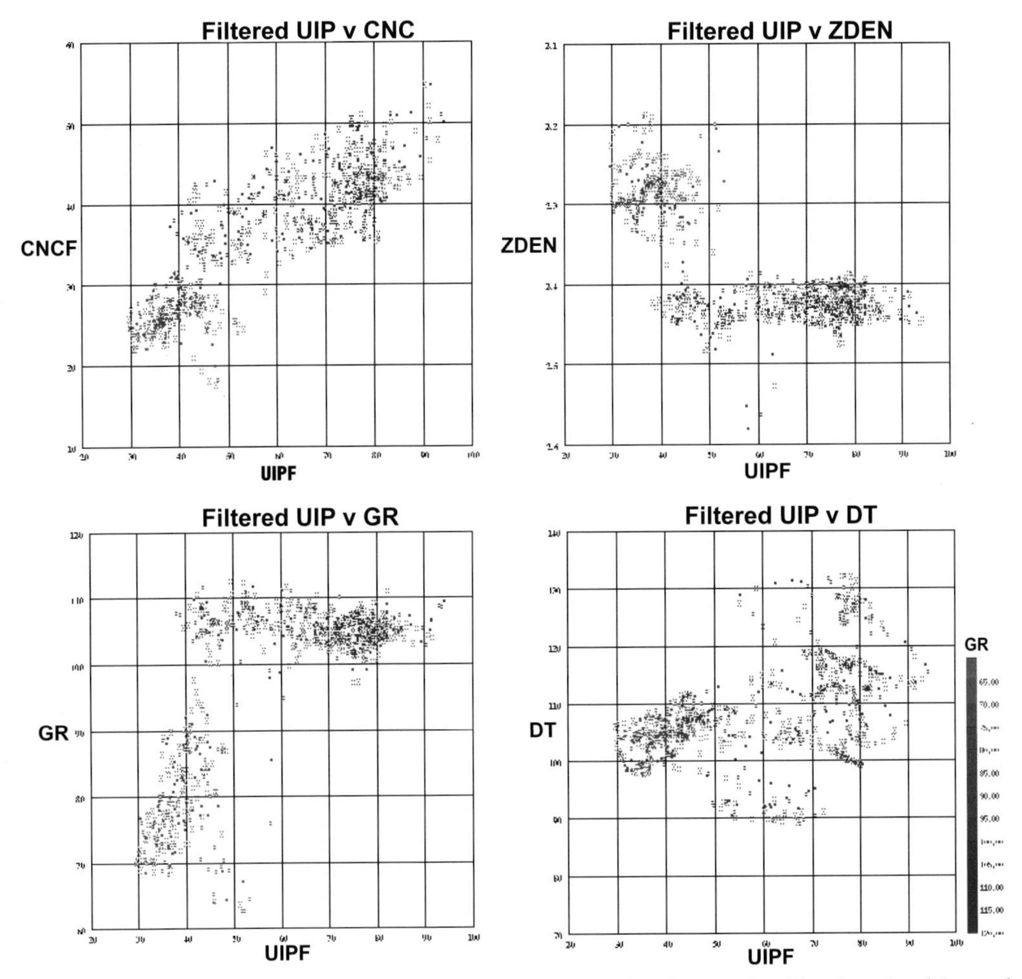

Fig. 4. The relationship between transformed, filtered acoustic (UIPF) and conventional logs is explored. Interval plotted is shown in Fig. 3. Note the best correlation is between UIPF and compensated neutron (CNCF).

data and the filtered one-dimensional curves extracted from the image data.

It is possible to write the expected amplitude of an acoustic image using the following equation (J. Priest, pers. comm. 1997):

$$R_T = R_f - R_r - (D_b - D_t)\alpha - R_t$$

Where R_T is the total return reflection amplitude, R_f is the reflectivity of the interface between the drilling fluid and the rock, R_r is the reflectivity loss due to rugosity and consequent dispersion at that interface, D_b and D_t are the borehole and tool diameters, α is the acoustic attenuation of the mud and R_t describes the various amplitude losses inside the logging tool.

D_t and R_t are constant. D_b is approximately constant if the borehole hole is in-gauge. α varies with temperature and pressure but is essentially constant over a normal logged interval.

R_f is a function of the acoustic impedance contrast between the borehole fluid and the formation. For example, where rocks have low slowness values and high densities there will be a relatively strong reflection from the borehole wall, producing a 'bright' image. We therefore expected to find a strong correlation between UIPF and ZDEN or DT. In fact (Fig. 4), a much stronger correlation was observed between UIPF and CNCF. This suggests to us that, at least in this case, R_f is not the prime control on reflectance amplitude. Rock texture and porosity are obviously related and we suspect that

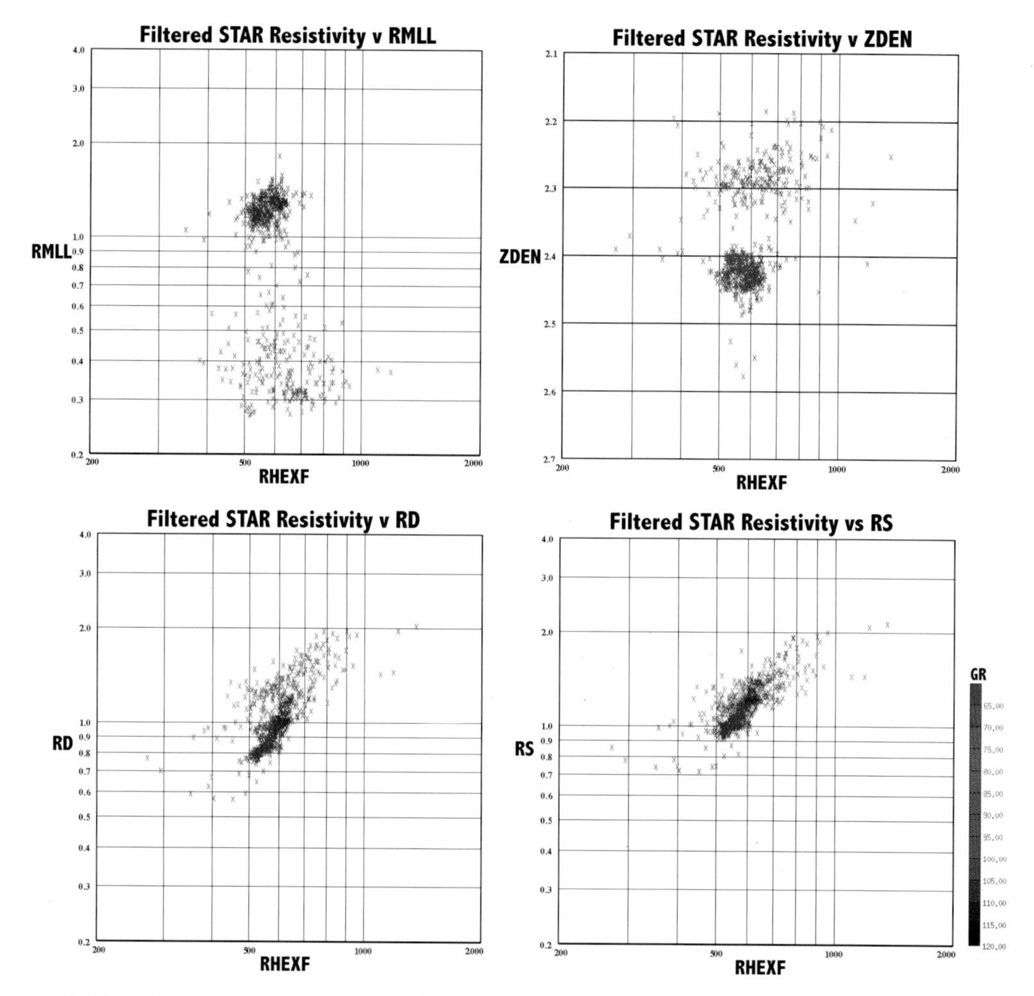

Fig. 5. The relationship between transformed, filtered galvanic STAR image (RHEXF), filtered resistivity and conventional logs. Interval plotted is shown in Fig. 3. Note the best correlation is between RHEXF and RS, the dual laterolog shallow resistivity. RHEXF has poor correlation to the microlaterolog (RMLL).

textural attributes are expressed directly in the image through acoustic dispersion, the R_r term in the above equation. Crudely, high-porosity rocks are probably rougher and more dispersive, so lesser signal is returned directly to the transducer. This appears to be fertile ground for future experimental work.

Unlike an acoustic image, a resistivity image is acquired using a method directly analogous to conventional logging tools: it is fundamentally an array of galvanic resistivity devices. One would expect the two types of measurement to be directly comparable. Current flows from the electrodes through the rock to a distant return electrode. The voltage drop between each electrode and return is a function of the nature and conductivity of the material in the pore space opposite the electrode. The precise current path varies due to the heterogeneity of the rock. However, conventional resistivity logging tools are designed to have somewhat specific and controlled depths of investigation. The nominal depths of investigation of the resistivity logs in this study are 60 inches, 20 inches, and 2 inches for the Deep (RD) and Shallow (RS) Laterologs and Microlaterolog (RMLL) respectively.

Dip- and image-log analysts must assume some depth of investigation in order to calculate the dip of geological features from their tools (Fig. 6). The availability of acoustic and

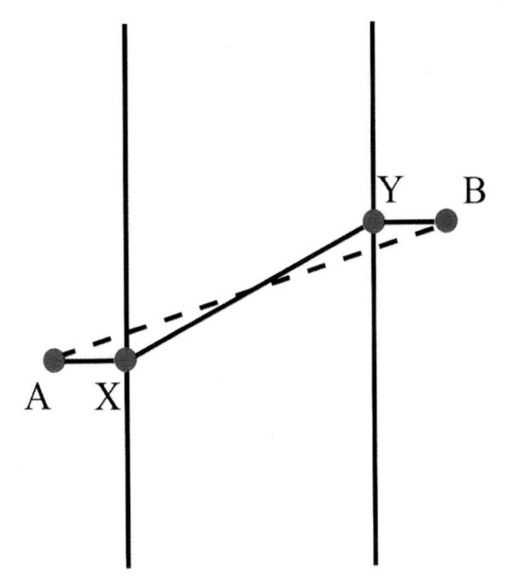

Fig. 6. The impact of assumed depth of investigation on inferred dip. Dip XY is steeper than AB.

resistivity images over the same interval allows this assumption to be directly investigated. We chose an interval of uniform dip, with bedding clearly evident on both the acoustic and resistivity image. Figure 1 shows part of this interval. The acoustic image was 'picked' on an interpretation workstation and a mean bedding dip of six degrees was calculated assuming a zero depth of investigation for the acoustic image. The resistivity image was then picked and the dip computed assuming depths of investigation of zero, one, three and six inches (Fig. 7). The six degree dip derived from the acoustic image was best replicated assuming a three inch depth of investigation for the resistivity image, which is towards the high-end of what would be expected from the geometrical properties of the electrode arrangement.

Returning to our exploration of the relationship between the image data and the conventional logs, we might expect, from the above discussion, that the resistivity image data would correlate most closely with the data from the

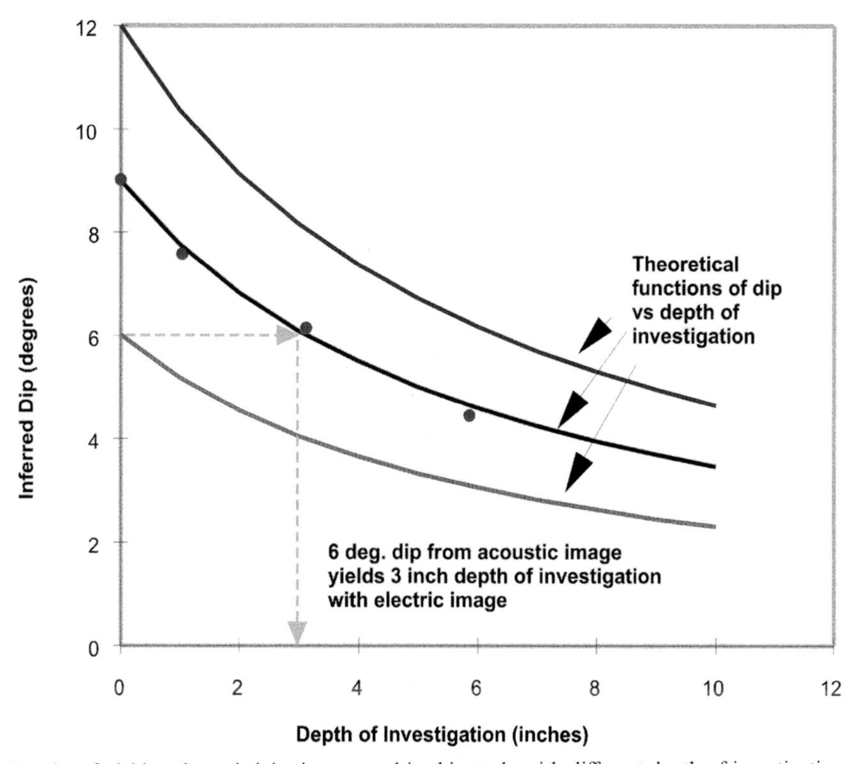

Fig. 7. Results of picking the resistivity image used in this study with different depth of investigation assumptions. The slight scatter reflects statistical variation in the picking process.

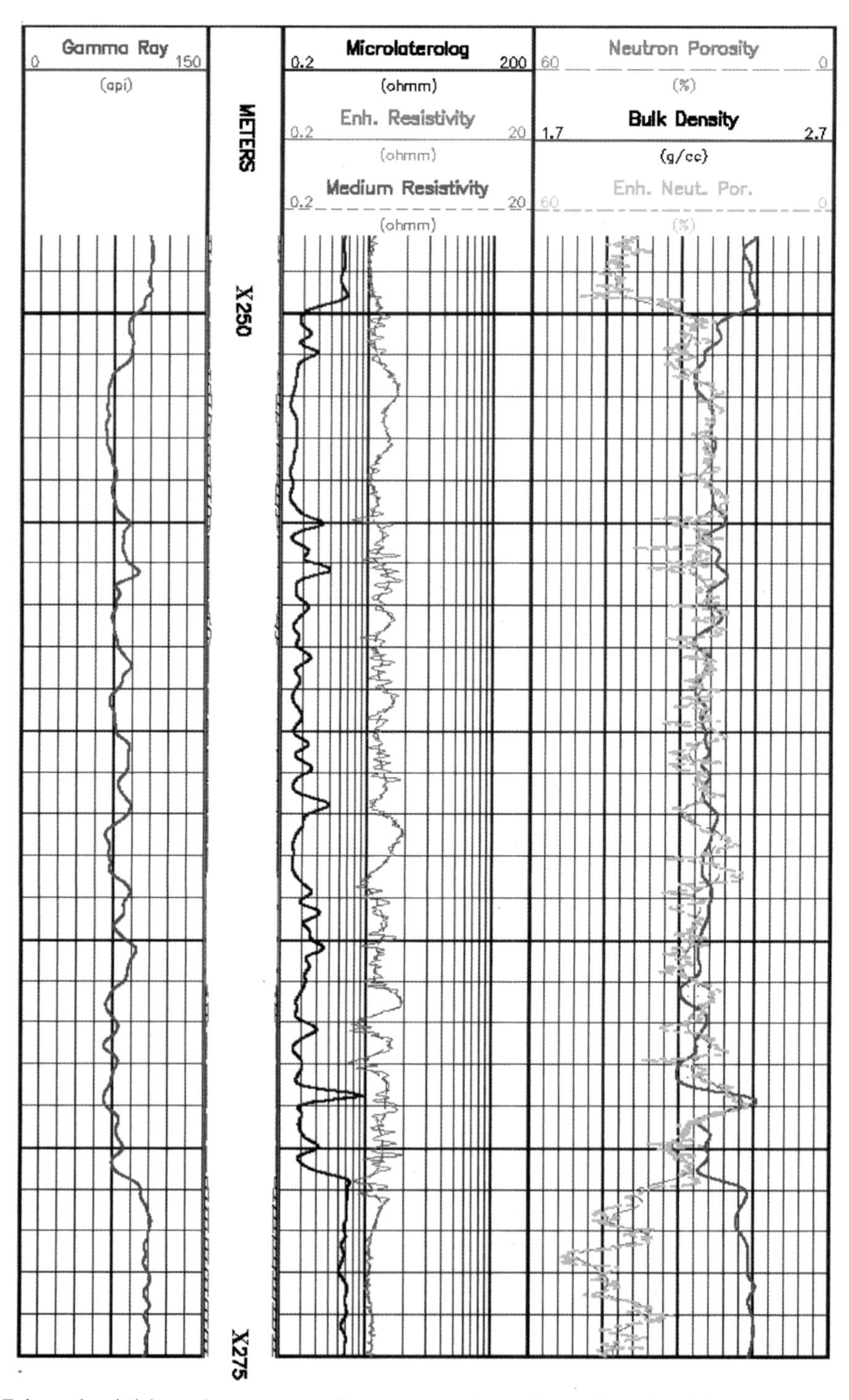

Fig. 8. Enhanced resistivity and neutron porosity are presented superimposed on the original logs. Note noise free signal in sand stones and in overlaying and underlying shales. Interbedded shales are clearly laminated. High resolution logs identify sands within these shales.

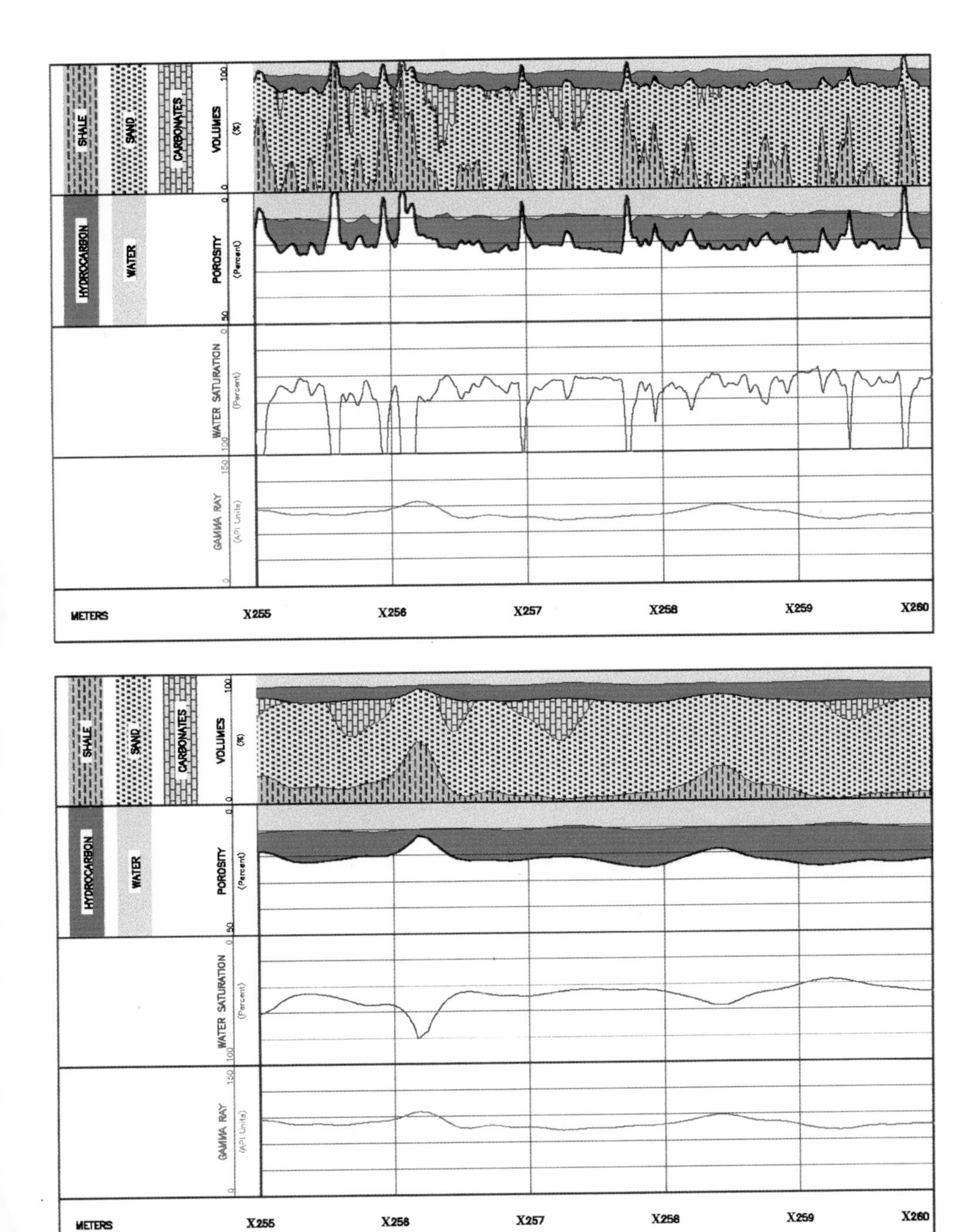

Fig. 9. This template illustrates the effect of using high resolution logs when performing a basic petrophysical analysis. The results using the conventional logs is shown to the left, and the same analysis using high resolution logs is presented to the right. Each horizontal grid line corresponds to 1 metre. Note that features such as permeability barriers and sandstones in the order of a few centimetres are delineated. The petrophysical properties of each layer are predicted.

shallowest-reading resistivity tool, the RMLL. This is not the case (Fig. 5): both the RS and RD curves exhibit a better correlation, the best correlation being with RS. Once again, this opens up an interesting line of investigation. Our suspicion is that it is necessary to distinguish the absolute value of resistivity, which is measured using the total current path, from the inflection points in the data, which are the record of bed boundaries. Further modeling work is required.

Calibration and resolution enhancement

As a result of the data exploration phase, we determined that there are robust relationships between the RHEXF and UIPF curves derived from the images and the RS and CNCF curves derived from the conventional logs. The next stage in our analysis was to use these relationships to calibrate the unfiltered image data, thus providing images that can be used quantitatively as measures of the resistivity and porosity of the rock.

Two aspects of our approach are of note. Firstly, the high-resolution image-derived curves to be calibrated (RHEX and UIP) are not simply the RHEXF and UIPF curves prior to filtering. As discussed above, when using these curves at high resolution it is important that the effects of bedding dip are removed otherwise circumferential average properties mix values from different beds. In the case of the acoustic image, mean structural dip was removed at the pixel level before circumferential averaging. In the case of the resistivity data, the mean value of the four central buttons on each pad was extracted and then data from each pad were carefully depth-matched before averaging.

Secondly, the calibration was not performed using a global model based on simple regression of UIPF on CNCF and RHEXF on RS. Rather, we computed the ratio between the filtered image-derived log and the selected conventional log at depth increment, for example:

$$X = UIPF/CNCF$$

The resolution-enhanced neutron porosity was then derived as:

$$CNCFE = UIP/X$$

Resolution-enhanced resistivity (RSE) was computed in an analogous manner.

This method results in a dynamic calibration based on the established relationship. By dynamic we imply that the relationship is normalised at the sample scale of the conventional log. Both automatic changes in signal gain (which are typically applied over several meters) and changes in hole size and shape are on a coarse scale compared to changes in the signal due to the formation. The dynamic calibration, which is performed at 7 cm intervals, thus neutralises any signal amplitude changes that may be attributed to changes in image signal gain or changes in hole geometry.

The enhanced logs show satisfactory overall agreement with the original conventional logs, but a significant improvement in vertical resolution (Fig. 8).

Petrophysical analysis

Having derived our high-resolution logs, a number of conventional petrophysical analysis routes were available to complete our analysis. We chose a deterministic method, in which the density log was the primary total porosity device. The enhanced neutron porosity was used to compute shale volume and subsequent shale correction of the total porosity to produce effective porosity. Finally we used the enhanced resistivity to derive water saturation from the Archie equation. Distinct beds – both permeability barriers and thin sandstones – down to 5 cm can be recognized on the high-resolution interpretation (Fig. 9). This is likely to lead to much greater accuracy, both in reservoir modeling and reserves estimation.

Conclusions

Acoustic and resistivity imaging devices respond to different rock properties. These responses can be used quantitatively and in combination to produce high resolution petrophysical analyses. Further investigation of the rock and fluid attributes to which images respond is required, but our preliminary results, based upon this case study, suggest that acoustic images are highly sensitive to porosity, perhaps via the acoustic-dispersive effect of rock fabric. Resistivity images, when used quantitatively, appear to investigate somewhat deeper into the rock than might be assumed from the way in which they record formation dip.

This paper is published with the permission of The BP and Statoil Alliance who provided the data. We would like to thank Nick Hill of Western Atlas for his assistance with the data processing.

References

DOVETON, J. H. 1994. Geologic Log Analysis Using Computer Methods. *AAPG Computer applications in Geology*, **2**.

PARKINSON, D. N., DIXON, R. J. & JOLLEY, E. J. 1999. Contributions of acoustic imaging to the development of the Bruce field, Northern North Sea. *This volume*.

RUNGE, R. J. & POWELL, N. J. 1967. The effect of sampling on sonic log span adjustment. *Transactions of the SPWLA 8th Annual Logging Symposium*. Paper D.

A sedimentological application of ultrasonic borehole images in complex lithologies: the Lower Kimmeridge Clay Formation, Magnus Field, UKCS

IAN GOODALL,[1] JEREMY LOFTS,[2]* MATTHEW MULCAHY,[2] MICHAEL ASHTON[1] & SAM JOHNSON[3]†

[1] *Badley Ashton & Associates Ltd, Winceby House, Winceby, Horncastle, Lincolnshire, LN9 6PB, UK (e-mail: igoodall@badley-ashton.co.UK)*

[2] *Schlumberger Wireline & Testing, Kirkton Avenue, Pitmedden Industrial Estate, Aberdeen, AB21 0BE*

[3] *BP Amoco Exploration Operating Co. Ltd, Farburn Industrial Estate, Dyce, Aberdeen, AB21 7PB, UK*

* *Currently Baker Atlas GeoScience, Kettock Lodge, Campus 2 Aberdeen Science & Technology Park, Balgownie Drive, Bridge of Don, Aberdeen, AB22 8GU, UK*

† *Currently BP Amoco Petroleum Development (Norway) Ltd, PO Box 197, Forusbeen 35, 4033 Forus, Norway*

Abstract: Ultrasonic borehole image tools are generally considered to be of limited value as an interpretative sedimentological tool because thick packages of a single lithology have similar acoustic properties resulting in resolution of very little diagnostic detail. In addition, build-up of mudcake in porous sandstones typically masks geological information by impeding the acoustic signal. In contrast, in heterolithic sediments, differences in acoustic impedance between intercalated lithologies enable clear distinction of sandstones and mudrocks from ultrasonic amplitude images. Whilst resolution of internal detail is still limited, valuable diagnostic data may be obtained for the sandstones by careful examination of the geometries and orientations of their lower and upper contacts with enclosing mudrocks. This is the case in the Lower Kimmeridge Clay Formation reservoir in the Magnus Field, UKCS, using the Schlumberger Ultrasonic Borehole Imager tool (UBI). Here, the bulk of the reserves are held in relatively thin, depositionally flat-lying, high-density turbidite sandbodies which are donors to variably inclined and complexly bifurcating sandstone injections. Recognition of sandstone injections and an understanding of their orientations is crucial to future development of this reservoir because they may be the principal means of communication between sandbodies and also have pay potential. Whilst this is not possible using conventional openhole logs it is easily achieved using UBI amplitude images.

The Schlumberger Ultrasonic Borehole Imager (UBI) tool was introduced in the early 1990's as a replacement to the little used Borehole TeleViewer (Zemanek *et al.* 1969) and has both openhole and cased hole applications (Hayman *et al.* 1994). In common with other acoustic tools (e.g. Western Atlas CBIL, Faraguna *et al.* 1989) the UBI tool is essentially a focused, rotating transducer which emits a high-frequency (250 or 500 kHz) ultrasonic pulse into the borehole and measures both the peak amplitude and transit time of the signal reflected back from the borehole wall. Most geological information is provided by the amplitude data, whilst the transit time measurement provides data on borehole radius and condition. The latter is therefore particularly useful for borehole stress analysis and cased hole applications such as evaluation of internal casing corrosion and detection of holes (Hayman *et al.* 1994).

The amplitude of the returned signal is dependent on a number of factors including acoustic impedance contrast between the formation and borehole fluid, angle of incidence of the formation which controls the degree to which the signal is reflected, and attenuative effects of mudcake on the borehole wall. Therefore, the amplitude signal is critically sensitive to the hole

From: LOVELL, M. A., WILLIAMSON, G. & HARVEY, P. K. (eds) 1999. Borehole Imaging: applications and case histories. Geological Society, London, Special Publications, **159**, 203–225. 1-86239-043-6/99/$15.00.

condition and, in consequence, geological data are commonly masked by drilling artefacts such as bit-whirl, breakouts and washout of mechanically weak sections. In addition, in deviated wells the development of prominent 'keyseat' furrows overprint geological information on the low side of the borehole. The ability of the amplitude signal to resolve geological data relies on the well-bore remaining in-gauge and the tool being centralized within the hole, and is dependent both on acoustic heterogeneity of the formation and on the subtle changes in borehole radius or hole rugosity. The latter typically respond to variations in mechanical strength at lithology changes, or along bedding surfaces or are associated with tectonic features (Hayman *et al.* 1994). Acoustic impedance is closely related to sediment composition, texture and porosity, and is therefore partially lithology dependent. For example, compacted low-porosity mudrocks and, in particular, tightly-cemented sandstones have a higher acoustic impedance than porous sandstones. In addition, within highly permeable sandstones build-up of drilling mudcake is common, and further attenuates the amplitude of the returned ultrasonic pulse.

Within compositionally homogeneous sediments it is typical for diagnostic sedimentary features such as bedding surfaces and ichnofabrics to be either poorly resolved or absent due to the lack of acoustic impedance contrast. Therefore the UBI tool is generally considered to be of limited value as an interpretative sedimentological tool. In consequence, micro-resistivity image tools (e.g. FMI (Schlumberger) & STAR (Western Atlas)) are typically included in logging runs in preference to acoustic tools because they resolve a significant amount of sedimentological detail (Harker *et al.* 1990, Bourke 1992). However, micro-resistivity tools require a conductive fluid in order to function and hence cannot be used successfully in oil-based muds.

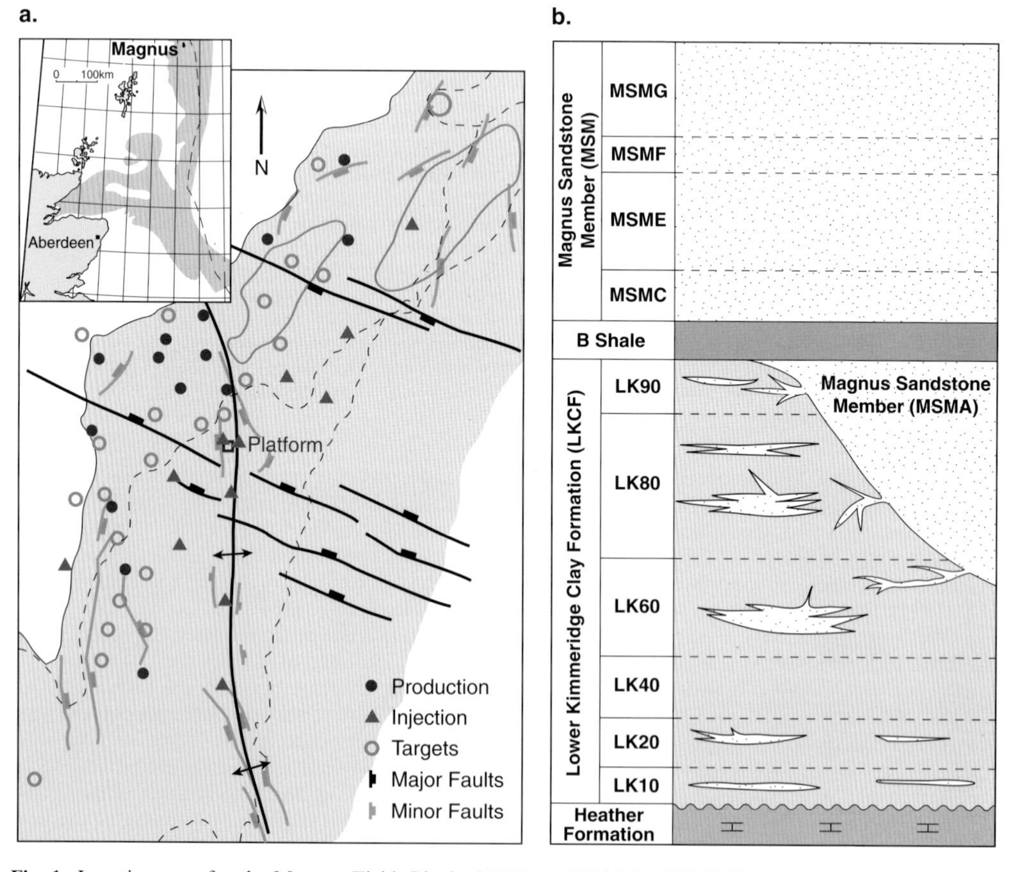

Fig. 1. Location map for the Magnus Field, Blocks 211/7a and 211/12a, UKCS illustrating well sites and the principal structural elements (a), and a summary reservoir zonation for the Lower Kimmeridge Clay Formation (LKCF) and overlying Magnus Sandstone Member (MSM; b).

In the northern North Sea, the development of reservoirs with low mechanical rock strengths and/or containing water sensitive clays commonly necessitates the drilling of wells using oil-based muds. Therefore, in these reservoirs borehole image data have to be acquired by acoustic tools. It is the intention of this paper to demonstrate that in heterolithic reservoirs where there is good acoustic impedance contrast between the intercalated lithologies, the UBI tool can provide valuable sedimentological data. The example chosen for this purpose is the Lower Kimmeridge Clay Formation in the Magnus Field.

Magnus Field background

The Magnus Oil Field is situated close to the north-western boundary of the east Shetland Basin, 160km northeast of the Shetland Isles in Blocks 211/12a and 211/7a, UKCS (Fig. 1a). Discovered in 1974, it came into production in 1983 with recoverable reserves of light crude (39 API gravity) of 724 mm STB stored in overpressured Upper Jurassic deep marine sandstones at a depth of approximately 3000 m SS (Shepherd 1991, Leonard *et al.* 1992). The Upper Jurassic section is dominated by calcareous mudstones of the Heather Formation (Oxfordian) overlain by organic-rich mudstones of the Kimmeridge Clay Formation (Oxfordian to Kimmeridgian). In the vicinity of Magnus Field the Kimmeridge Clay Formation contains a lower, low net : gross succession, the Lower Kimmeridge Clay Formation (LKCF), which is the focus of this paper, and the overlying Magnus Sandstone Member (MSM), a high net : gross reservoir interval which currently sustains most of the oil production from the Magnus Field (Leonard *et al.* 1992).

The Magnus Field structure lies on the footwall of the Magnus High Boundary Fault, a major early Cretaceous normal fault with a throw of approximately 4000 m, to the northwest, into the North Shetland Trough (Evans & Parkinson 1983). The Magnus structure is an east–southeast dipping fault block bounded by normal faults. The crest of the field is eroded and unconformably overlain by Cretaceous mudrocks. The trapping mechanism is a combination of truncation by the unconformity and stratigraphic pinch-out, and reservoir seal is provided by mudrocks of the Kimmeridge Clay Formation and younger Cretaceous section (Shepherd 1991).

Initial development focused on the thick (maximum approx. 186 m), high net : gross (approx. 80%) MSM reservoir which contains approximately 80% of the fields STOIIP (Leonard *et al.* 1992). In contrast, the geologically complex, low net:gross (generally <40%, average <25%) LKCF reservoir section was, until recently, considered to be too complex to risk major capital expenditure (Leonard *et al.* 1992). However, because the field is now nearing the end of plateau production, the LKCF is now been actively developed through a phased water injection scheme which is calculated to raise recovery factors from 10% to 25%, giving a yield of 60 mm STB (Leonard *et al.* 1992).

Lower Kimmeridge Clay Formation geological model

The low net : gross LKCF reservoir is 110–130m thick containing thin (typically 1 to 4 m), interbedded sandstones of variable reservoir quality. The LKCF overlies a thick section of Heather Formation calcareous mudrocks and is separated from the overlying MSM by the B Shale which is a prominent field-wide 'hot shale' marker horizon during which sand supply to the basin was significantly reduced. Locally, part of the LKCF section is eroded out by the MSM Zone A, a sandstone-prone interval which underlies the main, sandstone-rich MSM reservoir section (Fig. 1b). Lithostratigraphic zonation of the LKCF is extremely difficult because sandstones are thin, typically of limited lateral extent and have little correlative, diagnostic openhole log character. A biostratigraphic approach to reservoir zonation is used in which the LKCF section is divided into six high-resolution biozones labelled LK10 to LK90 in ascending stratigraphical order (Fig. 1b).

Depositional model

The LKCF is a sand-poor deep marine clastic system deposited over approximately 5 to 6 Ma on a low relief slope dissected by several discrete, but actively evolving sub-basins (Fig. 2) which follow structural lineaments (Leonard *et al.* 1992). The principal source of the sediment is considered to be from the west.

The mudrocks are largely background, mm-to-cm-scale laminated hemipelagic and dilute low-density turbidite deposits. Intercalated sandstones comprise a variety of sediment gravity flow deposits, dominated by sharp-based, structureless and dewatered high-density turbidites, which form the principal reservoir lithology. These sandbodies form either single event beds

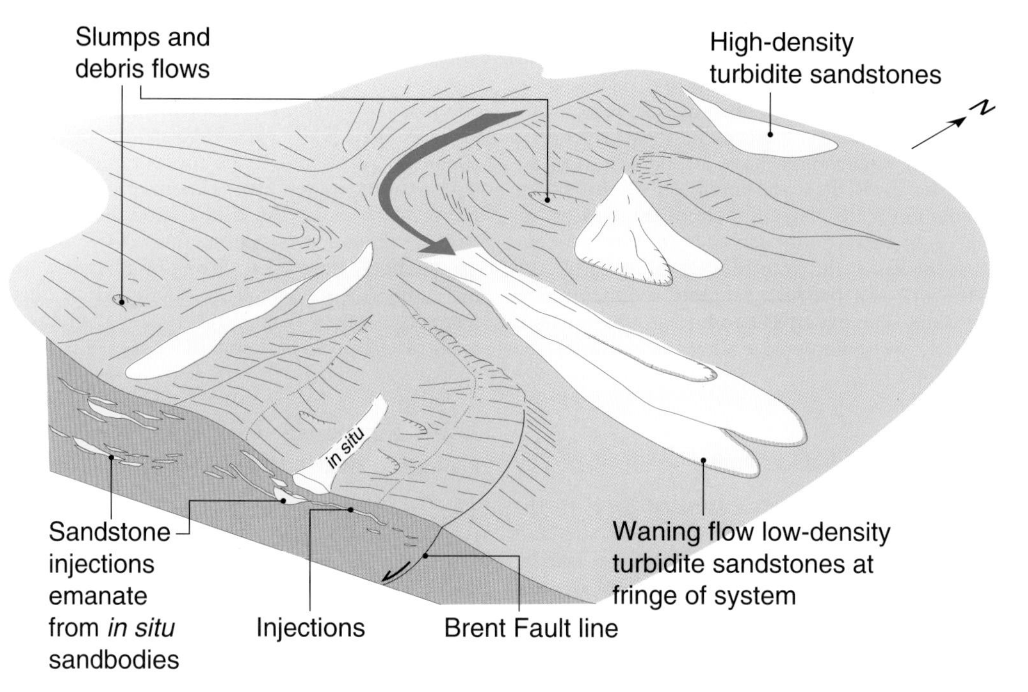

Fig. 2. Schematic depositional model for the low net : gross Lower Kimmeridge Clay Formation reservoir section in Magnus Field illustrating the distribution and geometries of sandbodies.

1 to 2 m thick or amalgamated sandbodies up to 8 m thick. Thin (typically $\ll$1 m), structured low-density turbidites are uncommon and are considered to represent the distal and lateral fringes to the main areas of sand deposition.

The Magnus Field basin was active throughout the Kimmeridgian and controlled sand dispersal pathways by initiating repeated switching of sand deposition between sub-basins. In addition, active tectonism resulted in inherent and continued slope instability, the consequence of which was remobilization of sediment principally in the form of slide blocks, cohesive slumps and unattached debris flows.

Lower Kimmeridge Clay Formation reservoir architecture

The LKCF reservoir architecture is closely related to location in the Magnus Field. In crestal areas sand supply was relatively continuous throughout deposition of the LKCF, and hence high-density turbidite sandstones are typically layered on a fine scale and vertical sand-on-sand amalgamation is relatively uncommon. In contrast, in the sub-basins sand was received more episodically and therefore individual sandbodies are more commonly discrete entities isolated within thick intervals of enclosing mudrocks. There is significant variation in sandstone stacking patterns and geometries across the sub-basins:

- sand supply was concentrated in axial areas which contain a high proportion of high-density turbidite sandstones with a higher potential for sand-on-sand amalgamation. In the 'best case' relatively thick (<8 m) erosively-amalgamated sandbodies result, whereas
- in marginal areas, sections are characterized by limited development of high-density turbidite sandstone and hence sandbodies are typically thinner (1–2m), and more intimately associated with slumps and debris flows.

Post-depositional processes

Core studies have demonstrated that a significant proportion of the sandbodies are post-depositional sandstone injections. These are most common in sub-basins typified by thin isolated sandbodies. They develop during rapid

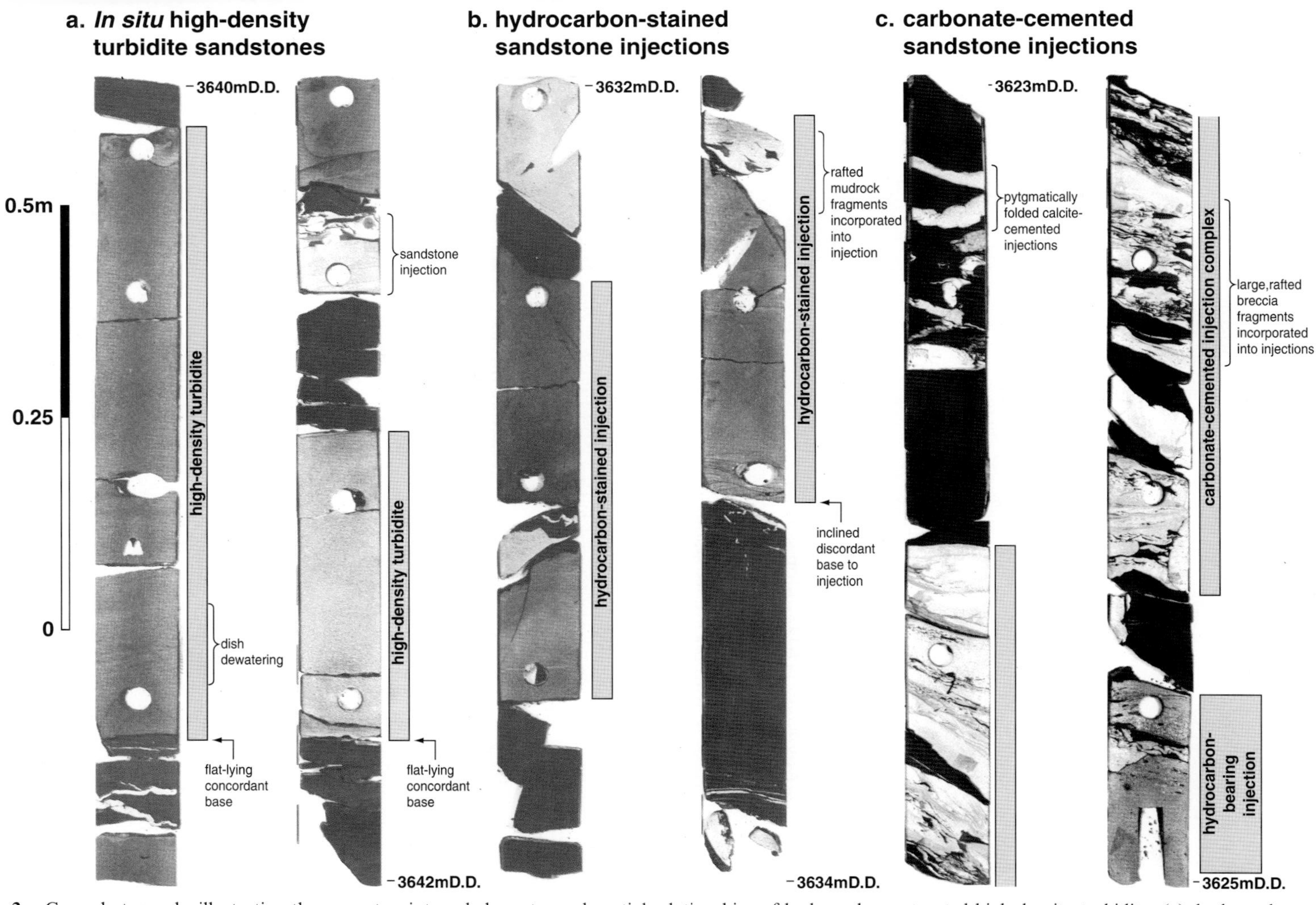

Fig. 3. Core photographs illustrating the geometry, internal character and spatial relationships of hydrocarbon-saturated high-density turbidites (a), hydrocarbon-saturated sandstone injections (b) and carbonate-cemented sandstone injections (c) in the Lower Kimmeridge Clay Formation, Magnus Field.

burial as a result of overpressuring of depositionally isolated and poorly-connected, partially consolidated sandstones. They inject sand up-section into mudrocks, exploiting lines of weakness such as bedding planes and tectonic fractures (cf. Hiscott 1979).

Donor sandbodies are largely thin, unconfined high-density turbidites sourced from within the LKCF section. However, more extensive sandstone injections are potentially sourced from erosively downcutting MSM Zone A canyon-fill sections (Fig. 1b).

Characteristics of the key sandbody types

Key sandbody types making-up the LKCF reservoir are *in situ* high-density turbidite and injection sandstones. The core and hydrocarbon-staining (as a proxy for reservoir quality) characteristics of these sandbodies are illustrated by Fig. 3.

High-density turbidites (HDTs) form sandbodies typically 1 to 2m thick, locally stacked into amalgamated sandbodies up to 8m thick. They have sharp, flat-lying bases and gradational tops that are concordant with enclosing mudrocks (Fig. 3a). The sandstones are fine to medium-grained and structureless, but dewatering features including dish and pipe structures and consolidation laminae do occur. They mostly contain little or no carbonate cement and have porosities of 15–25% and permeabilities between 50 and 2000 mD.

Sandstone injections have a highly variable character in terms of thickness, geometry, cementation and reservoir potential (Fig. 3b & c). Two key types of sandstone injection are developed:

(1) thick (<2.5m), high-quality, low-angle sandstone injection sills which are characterized by sharp and discordant bases and tops which clearly inject into enclosing mudrocks and commonly contain rafted, autobrecciated mudrock fragments around their margins (Fig. 3b). These injection sills typically have porosities of 15–25% and permeabilities of 10–50 mD.
(2) thin (typically <1 to 50 cm maximum), tight carbonate-cemented injections which have both low-angle, sill-like and steeply-inclined dyke-like geometries (Fig. 3c). These are typically developed in swarms and steeply-dipping injection dykes may be pytgmatically folded (Fig. 3c). Depending on the extent of carbonate cementation, porosities range up to 15% although permeabilities are less than 1 mD. These sandstone injections are non-reservoir lithologies and are likely to thicken and become less tightly cemented close to the donor sandstones from which they are sourced.

The geometries of *in situ* high-density turbidites and the thick sandstone injections are very different and hence they need to be treated differently in the LKCF reservoir model, both in terms of their dimensions and petrophysical parameters. However, as illustrated by Fig. 4:

- their openhole log responses are virtually identical, both having a blocky to spiky gamma-ray motif and wide, ‘clean sandstone’ neutron-density separations. For example, compare the openhole log responses of the high-density turbidite sandbody at *c.* 3629 m with that of the sandstone injection at *c.* 3635 m.
- sandstone injections commonly exploit weak planes at the boundaries between high-density turbidites and enclosing mudrocks forming complex, composite sandbodies, for example at *c.* 3640 m.

Lower Kimmeridge Clay Formation reservoir development issues

Description of the LKCF reservoir relies heavily on openhole log and biostratigraphical data. Core coverage is sparse and the current seismic data do not resolve the internal LKCF architecture (Leonard *et al.* 1992). For these reasons and, in particular, because it is not possible to identify accurately and consistently, high-density turbidite sandbodies using openhole log data alone, the geometry and interconnectivity of sandbodies remains poorly understood.

The available pressure data, obtained from Repeat Formation Tester (RFT) measurements, clearly shows that reservoir pressure has declined steadily during production from approximately 6650 psi at field start-up in 1983 to approaching 3200 psi in late 1996 prior to the first injection (Fig. 5). The fact that the sandstones all tend to be approaching 3200 psi, independent of their location in the Magnus Field, provides clear evidence that there is a mechanism of pressure communication, not only between adjacent sandbodies, but laterally across the field (Fig. 5; Leonard *et al.* 1992). These data are surprising as it is extremely difficult to correlate between *in situ* high-density turbidite sandbodies, even over short distances and with accurate, high-resolution biostratigraphical control (Fig. 6).

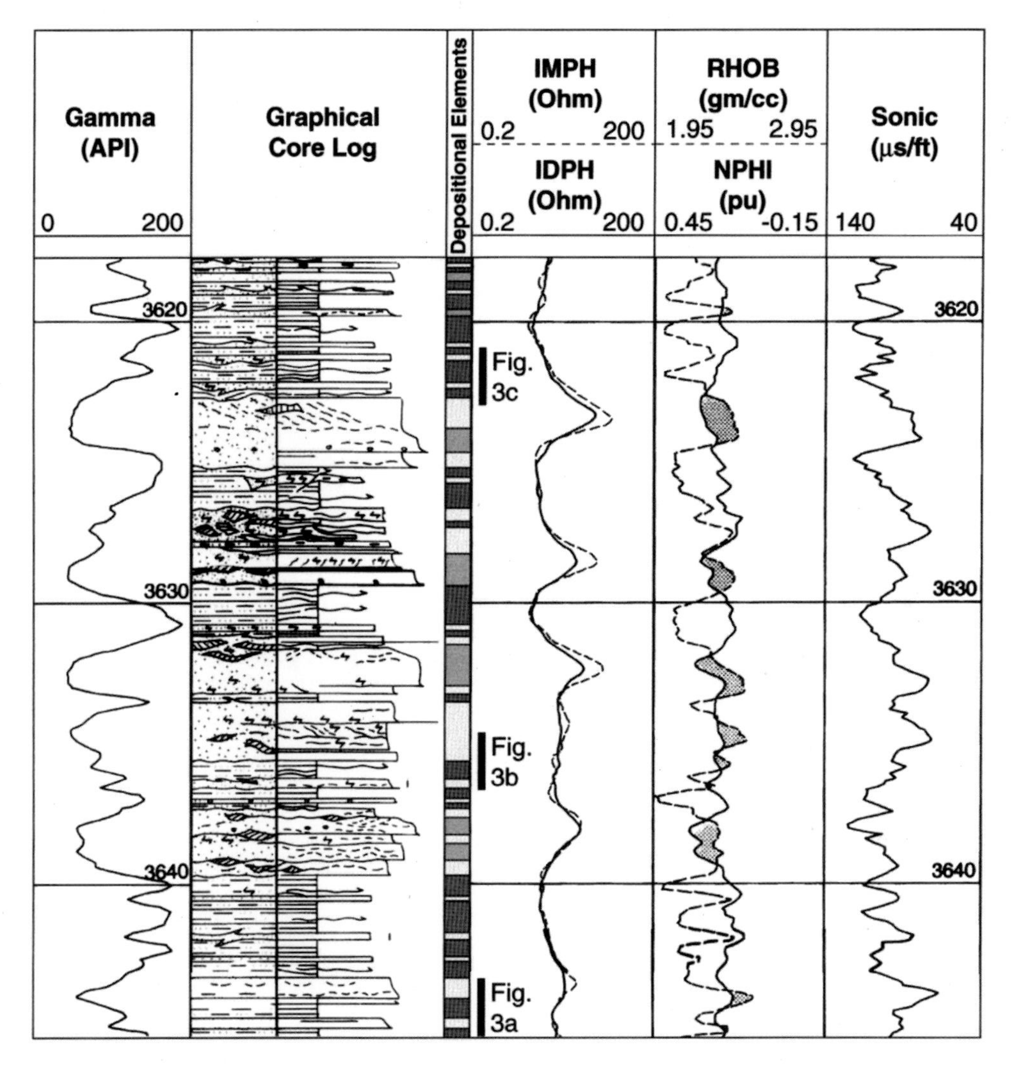

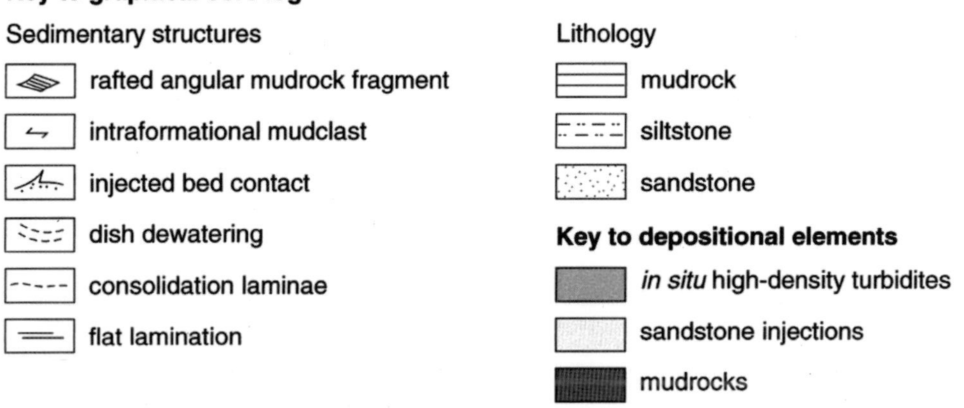

Fig. 4. Openhole log responses and depth-matched core illustrated by a graphical core sedimentology log for a typical interval of the Lower Kimmeridge Clay Formation reservoir in Magnus Field illustrating the character of key depositional elements.

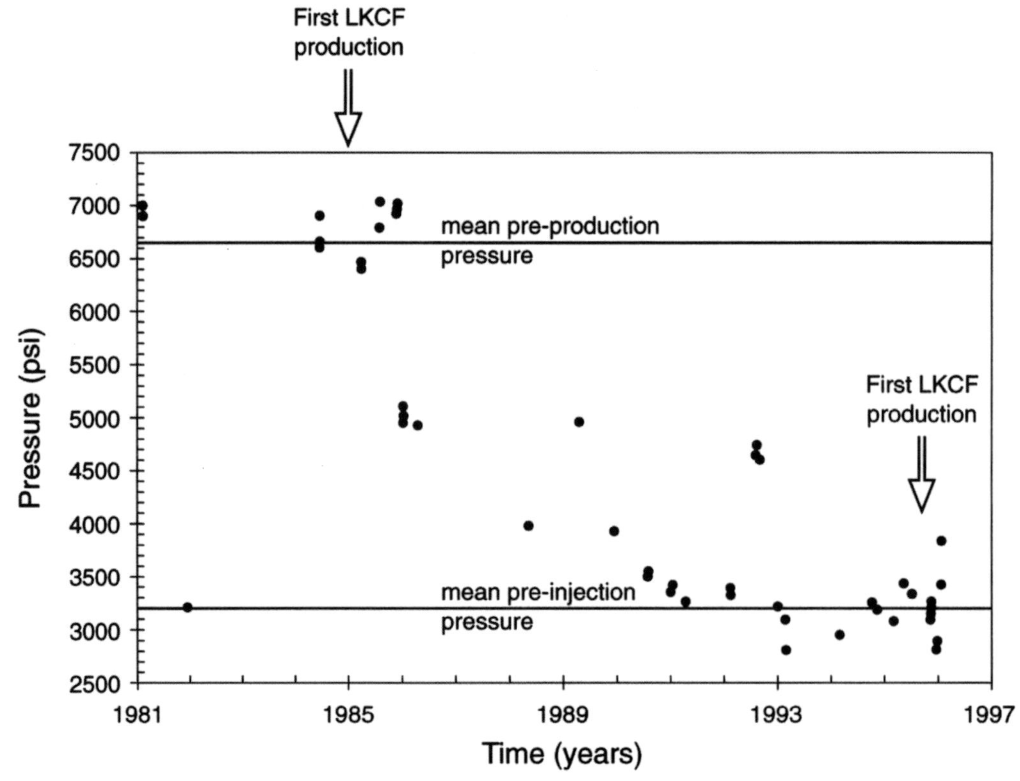

Fig. 5. Repeat Formation Tester (RFT) pressure data *vs* time for all wells in the Magnus Field completed in the Lower Kimmeridge Clay Formation reservoir section, illustrating the gradual decline in reservoir pressure from *c.* 6700psi prior to first production in 1986 to approaching *c.* 3000psi prior to the first injection in 1996.

Whilst this can be achieved locally by erosive amalgamation of sandbodies and through fracture networks within fault damage zones, the current models consider that sandstone injections provide fieldwide pressure communication.

Further evidence for communication is the high proportion of uncemented, cored, sandstone injections that are hydrocarbon-stained and have high porosities and permeabilities not dissimilar to the donor high-density turbidites (Fig. 7). The hydrocarbon stain provides clear evidence for communication with donor high-density turbidite sandbodies and raises the possibility that these sandstones have pay characteristics (Fig. 7). If this is the case it is important to identify them before perforating the well.

In order to plan successfully the siting of future injection/producer development wells and to design completion strategies to maximize sweep efficiency it is vital to have an accurate understanding of reservoir architecture in terms of the distribution and geometries of both *in situ* high-density turbidites and the sandstone injections.

Additional sedimentological data provided by UBI images

High-resolution UBI images have been included in the logging programme in most recent LKCF development wells in order to test whether or not development uncertainties can be minimized. Work has focused on evaluating their suitability to facilitate accurate discrimination of *in situ* high-density turbidite *vs* injected sandstones, to identify sandstone injection directions in order to gain an understanding of fluid migration pathways and to resolve sub-seismic scale faults and fractures.

The UBI amplitude and transit time image data acquired in the LKCF reservoir of the Magnus Field are typically of high-quality and

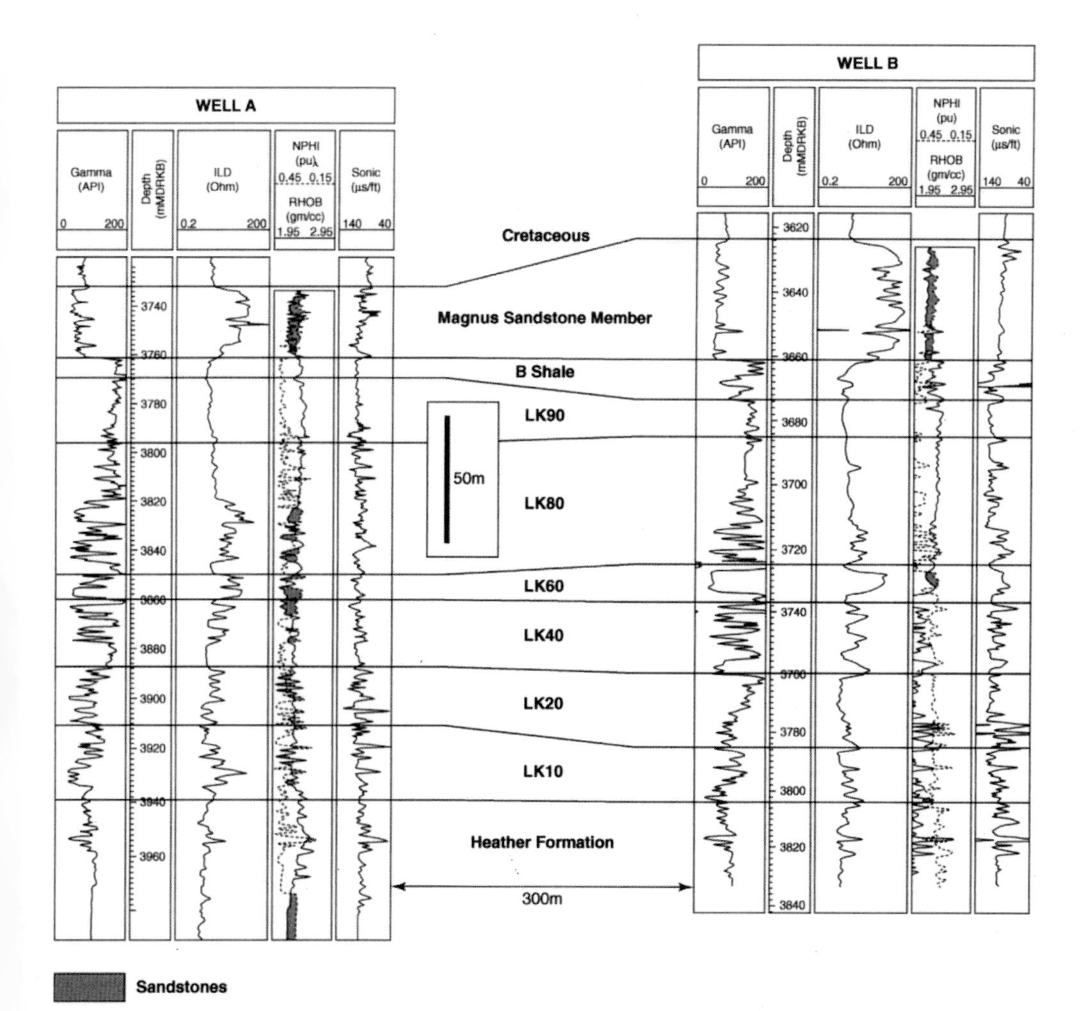

Fig. 6. Correlation panel of the Lower Kimmeridge Clay Formation and overlying Magnus Sandstone Member sections for two, closely-spaced, producer wells located close to the crest of the Magnus Field. High-quality sandbodies, indicated by a wide sandstone neutron-density separation (shaded), are thin, have an apparently random vertical distribution and cannot be correlated across even short distances implying that they have a restricted aerial extent.

resolve significant sedimentological and structural information, even in highly deviated wells (<60) in which prominent keyseats and minor tool sticks commonly overprint geological features (Figs 8 to 11). Detection of geological features as small as 0.2–0.4 inch is possible because the holes are typically in gauge minimizing the impact of drilling/acquisition artefacts and because the heterolithic LKCF section has considerable acoustic heterogeneity. Sedimentological information resolved by the images includes the distribution, thickness and geometries of thin sandstones revealed by the orientation of lithology contacts and the presence/absence and orientation of bedding surfaces within sandbodies. In addition, textural data may be resolved which aids identification of mudclast conglomerates and slide/slump blocks, and pytgmatic folding in sandstone injections.

Methodology

The methodology used to interpret the processed UBI image dataset is broadly similar to that outlined for micro-resistivity images by Bourke (1992) and maximizes the value of the UBI data by integration with any available core and openhole log data. Of particular importance in any detailed evaluation of high-resolution borehole images is calibration of UBI image

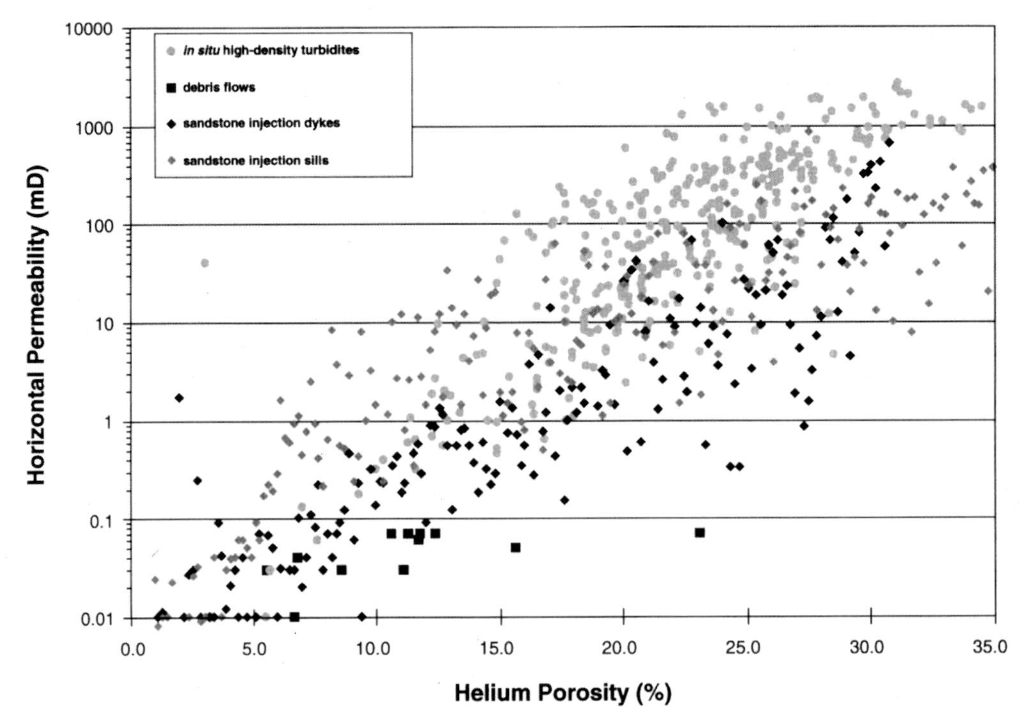

Fig. 7. Cross-plot of helium porosity *vs* horizontal permeability for conventional core analysis data acquired for the entire Lower Kimmeridge Clay Formation core dataset in Magnus Field illustrating a strong facies control on sandstone quality.

response to any available core control in order that a full appreciation of the origin, geometry and significance of features resolved by the processed images can be made. Unfortunately, to date, it has not been possible to calibrate the response of the UBI images in the LKCF section directly to core control because core is typically not included in the LKCF development logging programme. However, it has been possible to make extensive calibrations to core control in the overlying MSM section in which the facies types developed are similar.

All interpretations presented have been undertaken on workstations using appropriate proprietary borehole image visualization and interpretation software. Whilst use of such software is crucial to the calculation and display of bedding surface and structural feature orientations it also is an extremely valuable interpretative tool because its functionality typically allows the user to adjust many display parameters including the vertical scale, image reference and image normalization parameters in order to maximize the detail resolved in user-defined intervals, for example to check whether or not a package of apparently homogeneous, high-amplitude, cemented sandstone is bedded. In addition, a fuller appreciation of internal package and/or lithology contact geometries may be obtained by interactively removing structural dip from bedding surface orientation data and viewing these data on wrapped 3-D image displays.

The interpretative approach used to evaluate the LKCF reservoir is rigorous and involves the following key steps:

(1) orientation vectors are calculated for all bedding surfaces and tectonic features resolved by the processed static- and dynamic-normalized UBI amplitude images by hand-picking of features.
(2) the LKCF section is rationalized into high-quality sandstones, mud-prone sandstones and mudrocks by interpretation of available conventional openhole logs (principally gamma-ray, RHOB, NPHI and DT), drawing on caliper curves to provide quality control of log responses. At this stage no attempt is made to discriminate between *in situ* high/low-density turbidites and injected sandstones.

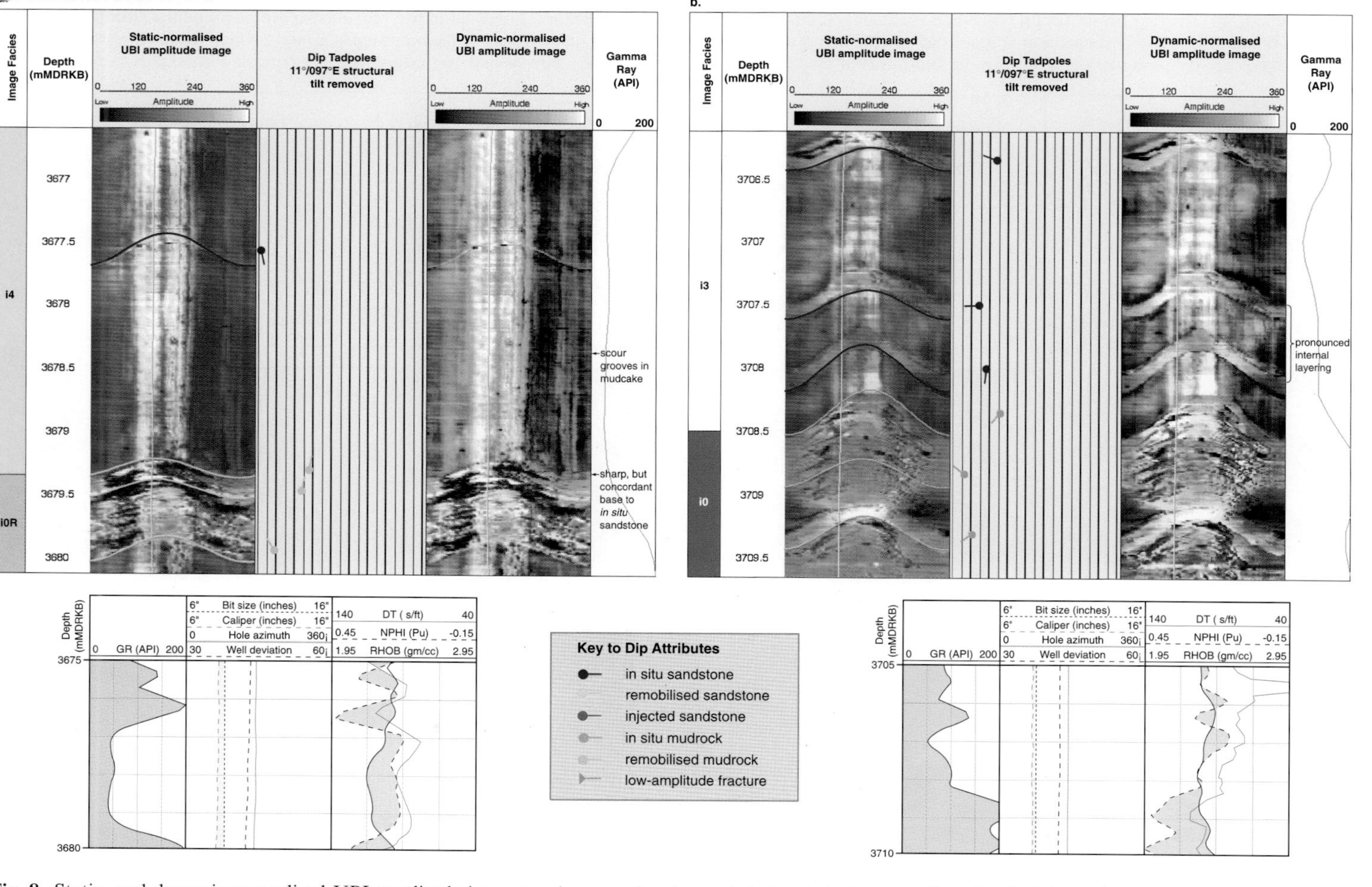

Fig. 8. Static- and dynamic-normalized UBI amplitude images and conventional openhole log data for examples of turbiditic sandstones which form the principal reservoir lithologies in the Lower Kimmeridge Clay Formation. These sandstones have sharp, but concordant bases and are discriminated on the basis of internal organization which shows little character in high-density turbidites (image facies code i4; a) and is structured in low-density turbidites (image facies code i3; b).

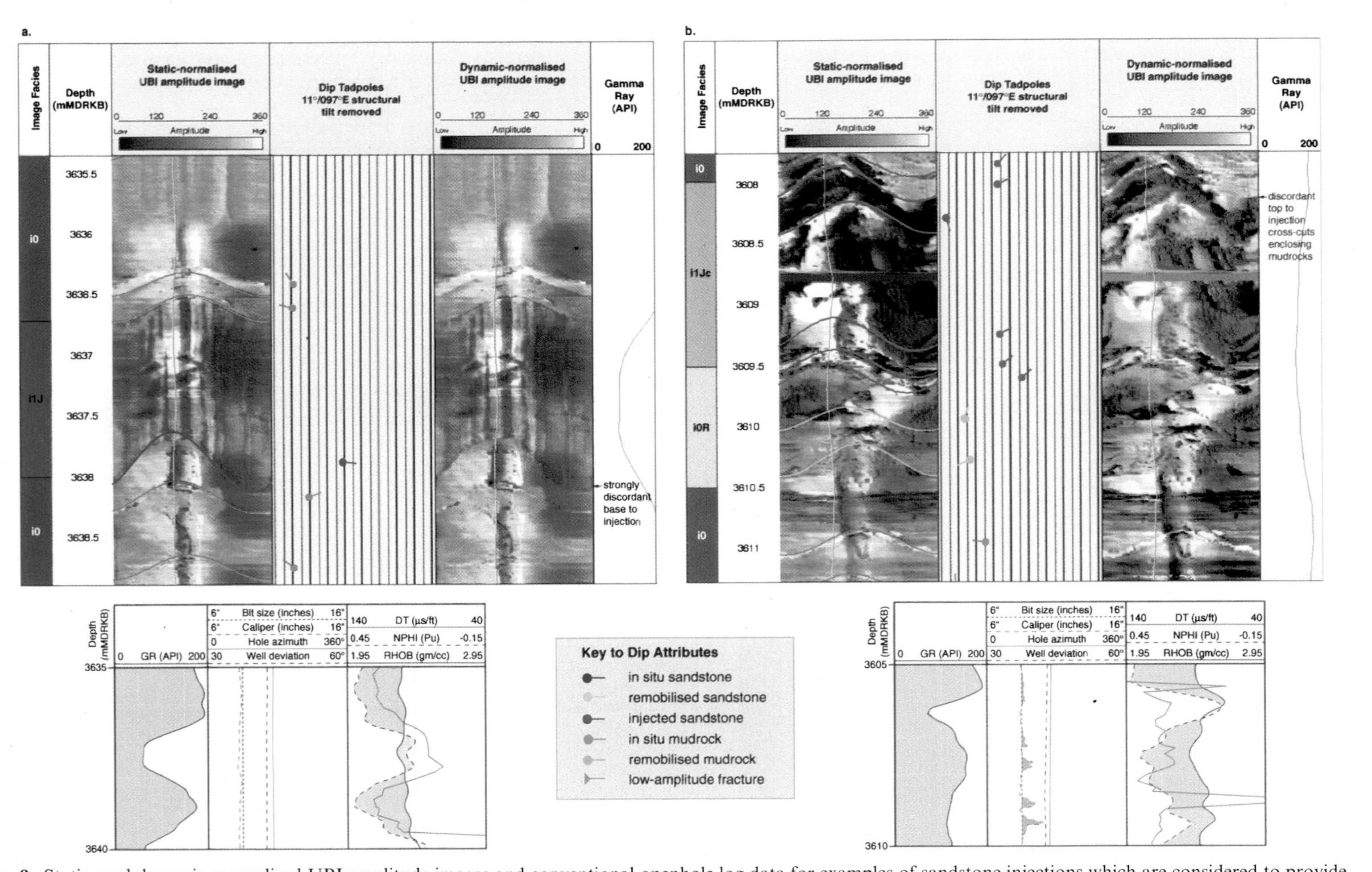

Fig. 9. Static- and dynamic-normalized UBI amplitude images and conventional openhole log data for examples of sandstone injections which are considered to provide communication between depositionally isolated turbidite sandbodies. Sandstone injections are characterized by strongly discordant lower and upper surfaces which are commonly steeply-dipping and are categorized into high-quality (image facies code i1J; a) and tight, carbonate-cemented injections on the basis of openhole log response (image facies code i1Jc; b).

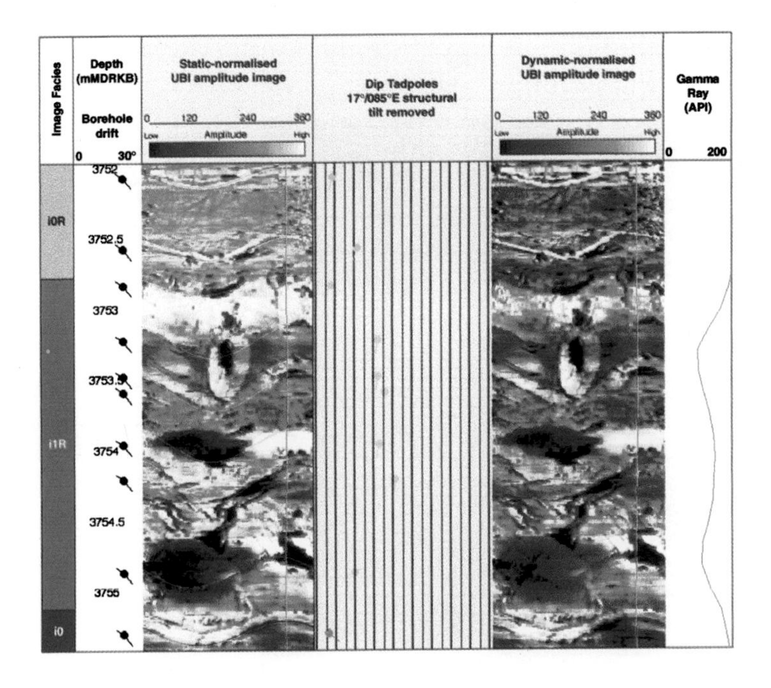

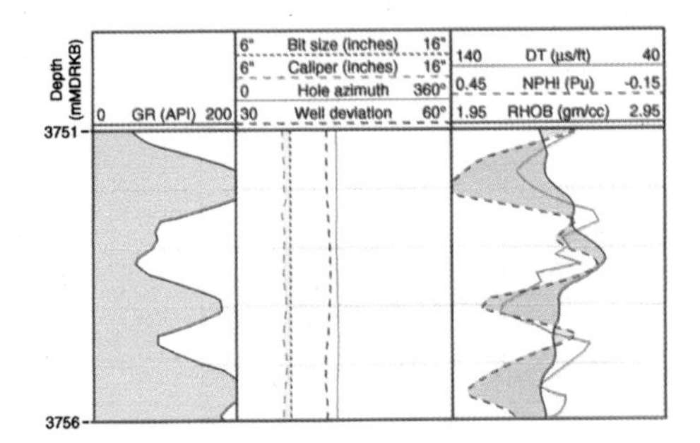

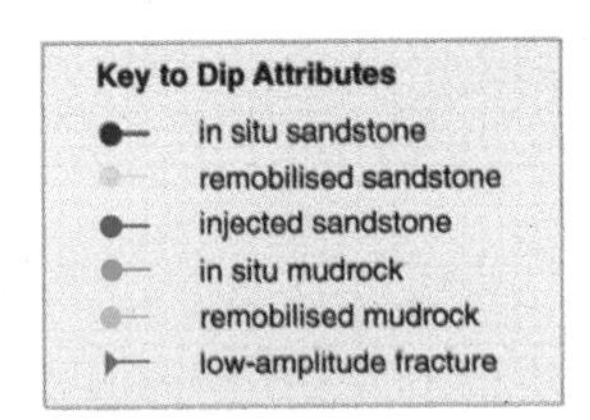

Fig. 10. Static- and dynamic-normalized UBI amplitude images and conventional openhole log data for an example of a poor-quality, mud-prone, remobilized sandstone (image facies code i1R). It is encased by mudrocks and has a highly-disorganized UBI amplitude texture with steeply-dipping internal surfaces.

(3) a provisional interpretation of structural dip is made using bedding surface dip data in mudrock-prone intervals which were defined on the basis of their openhole log response. Structural dip is removed in order to aid discrimination of sandstone injections from *in situ* high-density turbidites and to aid systematic identification of *in situ vs* remobilized mudrocks.

(4) detailed determinations of facies types are then made on the basis of features resolved by the static- and dynamic-normalized UBI

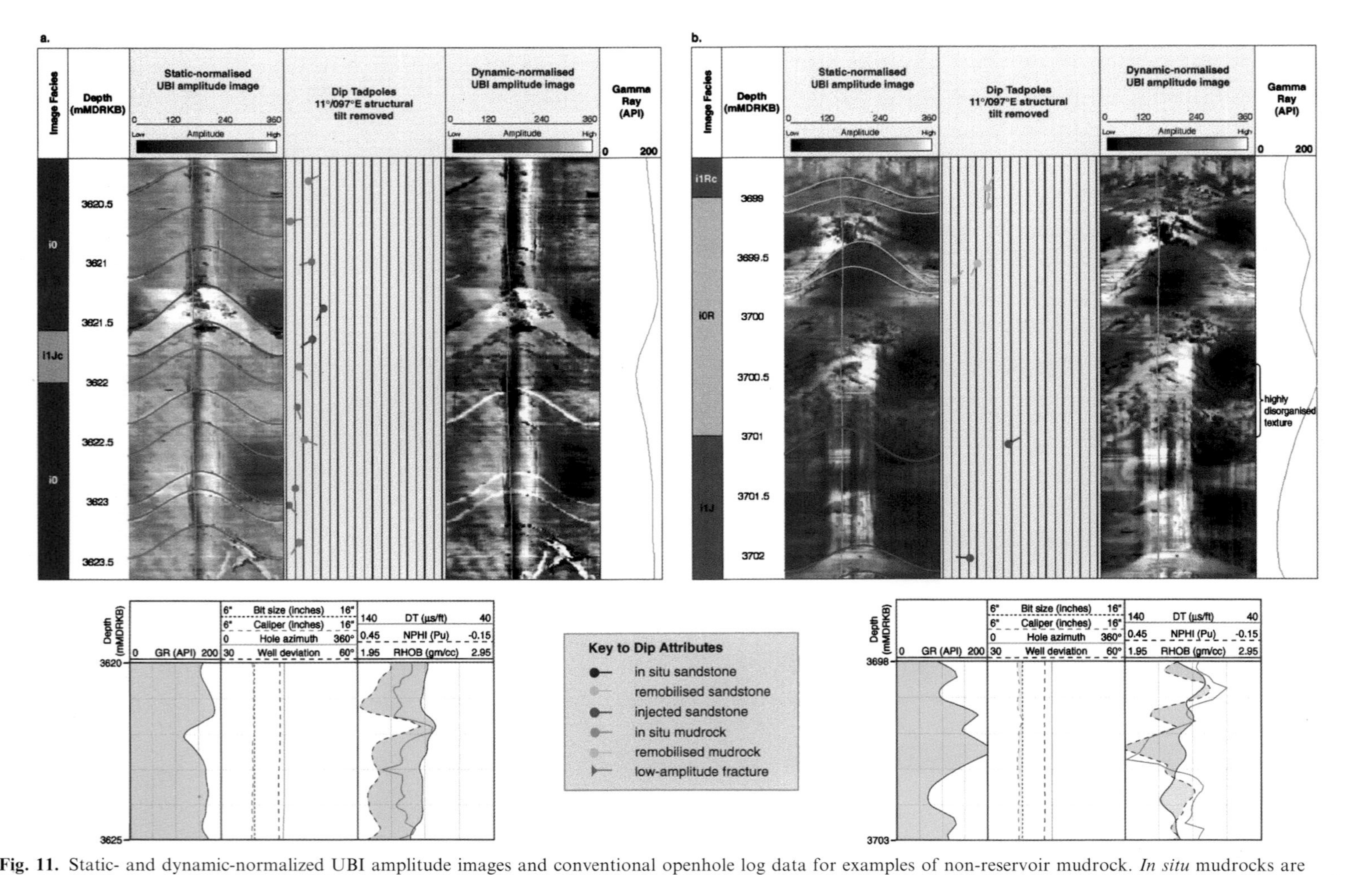

Fig. 11. Static- and dynamic-normalized UBI amplitude images and conventional openhole log data for examples of non-reservoir mudrock. *In situ* mudrocks are characterized by well-defined, sub-parallel and near-horizontal bedding surfaces (image facies code i0; (a), whilst remobilized mudrocks typically have a chaotic internal amplitude character and bedding surfaces are steeply dipping with highly variable azimuths (image facies code i0R; (b).

amplitude images, paying particular attention to the geometry and orientation of the lower/upper surfaces of packages and any internal bedding resolved.

(5) the bedding surface dip data are recoded using a suitable scheme in order to discriminate *in situ vs* remobilized mudrocks, *in situ* high- and low-density turbidites, injected sandstones and remobilized sandstones. This approach is vital in order to facilitate accurate calculation of structural dip from *in situ* mudrocks only, which is then used to interpret sandstone injection orientations and to predict the location of donor sandstones.

Image facies

Each of the key depositional elements making-up the LKCF reservoir in the Magnus Field, namely high- and low-density turbidites, debris flows, *in situ* mudrocks, and mud-prone slumps and rotational slide blocks are easily and consistently identified using high-resolution UBI amplitude images together with conventional openhole logs. The depositional elements are described in terms of six image facies that are illustrated by Figs 8 to 11. They have the following characteristics:

(1) *Massive/homogeneous high-density turbidite sandstones* (image facies i4) typically form packages 1–2 m thick, but where amalgamated are up to approximately 8 m thick. They are characterized by blocky gamma-ray responses, wide, 'clean sandstone' neutron/density separations and are commonly developed in undergauge sections as a consequence of mudcake build-up which is resolved by the UBI amplitude images as faint, hole-parallel lines made by tool stabilizers scouring grooves into the mudcake (Fig. 8a). A key diagnostic feature of these sandstones is that they have sharp, and concordant bases which are either flat-lying or slightly inclined due to the compaction (typically <10°) of the underlying mudrocks and, with the exception of scarce, m–scale bedding surfaces (interpreted as flow tops), they have little/no internal character implying that the sandstones are structureless. Dewatering fabrics observed in core are not resolved. These high-porosity sandstones have a low acoustic impedance and therefore have a dull, low-amplitude, image character (Fig. 8a).

(2) *Structured low-density turbidite sandstones* (image facies i3) are uncommon and occur as thin packages (typically <1 m) intercalated with mudrocks. They comprise variably high-quality to muddy sandstones and are characterized by wide, 'clean sandstone' to near coincident neutron/density separations (Fig. 8b). The key diagnostic features of these sandstones are sharp, but concordant bases which are depositionally flat-lying, but may locally be inclined in response to compaction or loading of the underlying mudrocks and the presence of closely-spaced, flat-lying or low-angle bedding surfaces indicating that these sandstones are structured (Fig. 8b). Again low-amplitude static-normalized images result from loss of the acoustic signal in these high-porosity sandstones (Fig. 8b).

(3) *Sandstone injections* (image facies i1J) are typically 0.5–1 m thick and are largely developed within thick mudrock sections. However, they also exploit mechanically weak contacts between turbiditic sandstones and enclosing mudrocks and have been recorded cutting through sandbodies. They are recognized by blocky to spiky, low gamma-ray responses and a variable neutron-density separation reflecting the presence or absence of carbonate cements. Porous sandstone injections have openhole log responses identical to donor high-density turbidite sandstones, whereas tight carbonate-cemented sandstone injections have low interval transit times (typically <90 μs/ft). The key features resolved by the UBI amplitude images are strongly discordant bases and tops which are typically, but not necessarily, steeply dipping, and highly irregular and commonly bulbous bed shapes (Figs 9a and b). Many steeper injection dykes are pytgmatically folded. Thus far, no clear evidence for the presence of incorporated, angular mudclasts which are an unmistakable feature of cored sandstone injections, has been observed in the UBI images. Evaluation of sandstone injection orientations enables discrimination of steeply-dipping injection dykes from low-angle to near horizontal injection sills, both of which are common in the LKCF reservoir.

(4) *Remobilized sandstones* (image facies i1R) comprise slumps/slurries and debris flows which are both attached to high-density turbidites (image facies i4) as dirtying-up tops or cleaning-up bases, identified by high gamma responses, or are unattached, forming thin packages within mudrock sections. They are typically thin, but can

range up to 1–2 m in thickness and are characterized by a serrate, mud-prone gamma-ray response and variable, but near coincident neutron-density separations resolving lithological variation (Fig. 10). Both static- and dynamic-normalized amplitude images resolve a disorganised, 'broken-up' fabric and highly irregular bedding surfaces (Fig. 10). The orientation of bedding surfaces ranges from steeply inclined in rotational slide blocks (Fig. 10) to near horizontal in attached debris flows. Carbonate cements are developed, but are not common and are resolved by bright, yellow-white, stratabound or patchy, high-amplitude static-normalized amplitude images (Fig. 10).

(5) In situ *mudrocks* (image facies i0) comprise background hemipelagic muds and are typically developed in intervals of slightly overgauge hole. They form packages up to 5 m thick and are characterized by a high gamma-ray response (>150 API) and wide, mudrock neutron-density separations (Fig. 11a). These low-porosity, organic-rich mudrocks have a high acoustic impedance and therefore are characterized by bright, yellow-orange static-normalized amplitude images (Fig. 11a). Bedding surfaces are typically well resolved, are closely-spaced, dip at a low-angle and, crucially, clustering of both dip magnitudes and azimuths implies that they are depositionally flat-lying (Fig. 11a). It is these mudrocks that provide the only reliable indication of structural dip.

(6) *Remobilized mudrocks* (image facies i0R) form packages up to 5.5 m thick and are likely to represent mud-prone slurries and slumps. They also have a high gamma-ray response and mudrock neutron-density separation, but both these are highly serrate and higher density responses than the *in situ* mudrocks imply more porous, silt-prone lithologies which result in duller, yellow-orange, reduced static-normalized acoustic amplitudes on the UBI images (Fig. 11b). Bedding surfaces are typically steeply-dipping and, crucially, are inclined relative to structural dip and both dip azimuths and magnitudes are highly variable and dispersed. In addition, bedding surfaces are typically less common and poorly-defined in comparison to *in situ* mudrocks and the UBI amplitude images commonly have a disorganized, broken-up appearance. This facies is transitional to the remobilized sandstones (image facies i1R).

To date, the image facies scheme outlined here has been successfully applied to six LKCF producer and injector wells in the Magnus Field without modification indicating that it is robust and describes the key depositional elements.

Application of UBI images to the reservoir model

In order to choose well sites which have a high likelihood of contacting reservoir sandstone and to design appropriate completion strategies which maximize sweep efficiency, and hence improve recovery factors in the LKCF reservoir, an accurate understanding of the vertical and spatial distribution of donor high-density turbidite *vs* injected sandstones and their likely communication pathways is vital. In this context the processed UBI amplitude images are invaluable because they facilitate accurate discrimination of the key depositional elements making up the LKCF reservoir. In addition, the bedding surface orientation data obtained from these images may be used to calculate structural dip and sandstone injection orientations and from this to make estimates of the distribution of donor sandstones adjacent to, but not contacted by, the well bore and of their likely sandbody communication pathways.

Calculation of net sandstone and design of completion strategy

Historically, calculations of net pay and design of completion strategy in the LKCF reservoir have largely been based upon an evaluation of conventional openhole log responses integrated with dynamic data and sparse core control. However, as demonstrated in this paper, although the geometries and potential performance of high-density turbidites and sandstone injections are very different, their log character is commonly identical. For this reason net pay is likely to be overestimated, and therefore actual test results may bear little resemblance to predictions.

As demonstrated previously, by careful examination of the textures resolved by the processed UBI amplitude data and by evaluation of the orientations of bedding surfaces hand-picked from these images, it is possible to make confident interpretations of depositional element type to a higher degree of accuracy than is possible from conventional openhole logs alone. These image facies data provide information on the vertical distribution, thickness and spatial relationships of the key sandbody types, from which

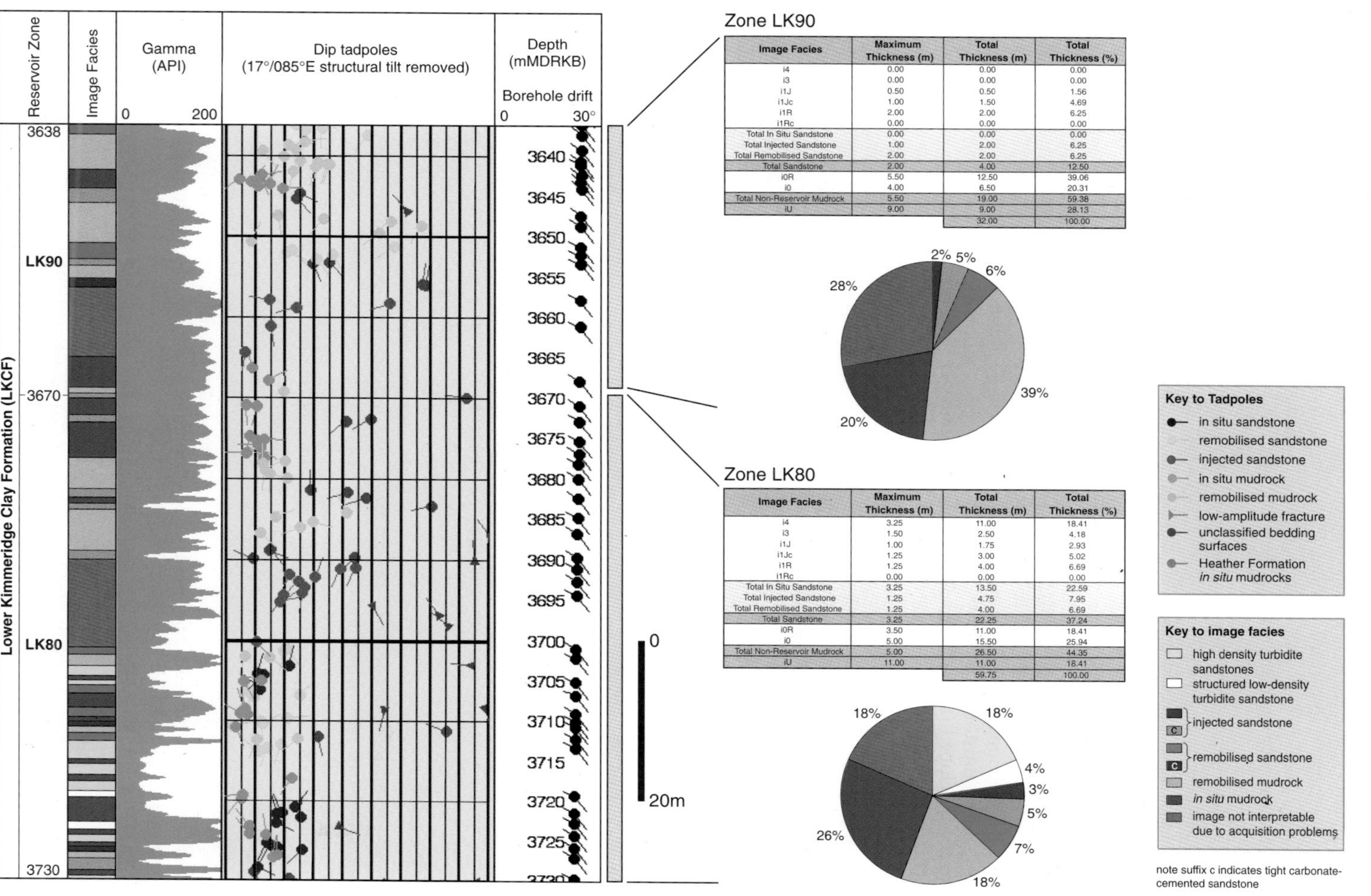

Zone LK90

Image Facies	Maximum Thickness (m)	Total Thickness (m)	Total Thickness (%)
i4	0.00	0.00	0.00
i3	0.00	0.00	0.00
i1J	0.50	0.50	1.56
i1Jc	1.00	1.50	4.69
i1R	2.00	2.00	6.25
i1Rc	0.00	0.00	0.00
Total In Situ Sandstone	0.00	0.00	0.00
Total Injected Sandstone	1.00	2.00	6.25
Total Remobilised Sandstone	2.00	2.00	6.25
Total Sandstone	2.00	4.00	12.50
i0R	5.50	12.50	39.06
i0	4.00	6.50	20.31
Total Non-Reservoir Mudrock	5.50	19.00	59.38
iU	9.00	9.00	28.13
		32.00	100.00

Zone LK80

Image Facies	Maximum Thickness (m)	Total Thickness (m)	Total Thickness (%)
i4	3.25	11.00	18.41
i3	1.50	2.50	4.18
i1J	1.00	1.75	2.93
i1Jc	1.25	3.00	5.02
i1R	1.25	4.00	6.69
i1Rc	0.00	0.00	0.00
Total In Situ Sandstone	3.25	13.50	22.59
Total Injected Sandstone	1.25	4.75	7.95
Total Remobilised Sandstone	1.25	4.00	6.69
Total Sandstone	3.25	22.25	37.24
i0R	3.50	11.00	18.41
i0	5.00	15.50	25.94
Total Non-Reservoir Mudrock	5.00	26.50	44.35
iU	11.00	11.00	18.41
		59.75	100.00

Fig. 12. Interpreted facies breakdown for reservoir zones LK80 and LK90 in an injector well together with statistics and summary pie charts illustrating the maximum and total thickness of the key depositional elements in these reservoir zones. These interpretations are incorporated into the Lower Kimmeridge Clay Formation reservoir model in order to improve the understanding of well performance and to aid design of completion strategy.

statistics of net pay can be calculated and then incorporated into the LKCF reservoir model (Fig. 12). The example illustrated by Fig. 12 is for Zones LK80 and LK90 in an LKCF injector well, and serves to demonstrate that:

- there is a clear difference in net pay between the two zones,
- *in situ* high-density turbidites are only developed in Zone LK80 (11m total thickness), but are associated with high-quality sandstone injections (2.93 m total thickness) which also have pay potential, and
- in Zone LK90, whilst most sandstones comprise tight, non-reservoir carbonate-cemented sandstone injections (total thickness 4.69 m), high-quality sandstones with pay characteristics are developed (total thickness 1.56 m), but comprise exclusively sandstone injections. These clearly need to be sourced from donor turbiditic sandstones which are considered to be located close to the well bore.

This detailed understanding of sandbody types can clearly be incorporated into the LKCF reservoir model and used to aid design of completion strategy and planning of future well placement, and is potentially very different from interpretations based on openhole logs alone.

Calculation of structural dip

It is important to make accurate calculations of structural dip in any reservoir because this not only helps with seismic correlation and identification of sub-seismic scale tectonic setting, it also:

- helps to establish the orientation of reservoir sections and cap rock which aids choice of sidetrack trajectory and future well planning,
- aids definition of the geometry of faults and associated damage zones, and
- allows accurate determinations of palaeotransport directions.

Assessment of structural dip magnitudes and azimuth from interpreted UBI amplitude images are considered reliable and accurate because they are based upon hand-picked bedding surface dip data that have been subject to rigorous quality control. The data can be filtered in order to ensure that calculations are made only using depositionally flat-lying sediments. In addition, a clear benefit of structural dip values calculated from UBI, or indeed any high-resolution borehole image dataset, is that they are independent of seismic data and therefore can be used to complement these interpretations.

In the LKCF reservoir of the Magnus Field, the only depositional element that provides a reliable indicator of structural dip are the *in situ* background hemipelagic mudrocks. However, as discussed previously there is clear seismic evidence for active syndepositional faulting in the Magnus Field during deposition of the LKCF section (Leonard *et al.* 1992). The consequence of faulting was the evolution of depositional subbasins and contemporaneous remobilization of sediment close to active faults producing mudprone slurries and slide blocks. These are the remobilized mudrocks which are recognized on the UBI amplitude images by steeply-inclined dip magnitudes and highly variable dip azimuth directions. These data are clearly not representative of structural dip and must be filtered from the dip dataset. In this study this was achieved by classifying the LKCF bedding surface dip data into five groups, namely *in situ* sandstone, remobilized sandstone, injected sandstone, *in situ* mudrock and remobilized mudrock (Fig. 13). This classification not only facilitates calculation of structural dip, but also forms the basis for evaluation of sandstone injection directions.

The example illustrated by Fig. 13 is for the same LKCF injector well shown by Fig. 12. In this example structural dip is calculated for both the LKCF reservoir and the underlying, partly drilled Heather Formation section. Evaluation of the LKCF structural dip in the example shown is largely made in the mudrock-prone sections making-up Zones LK10 to LK40, because it is here that compactional effects of loading by intercalated sandbodies are minimized. However, even the filtered dip data contain a number of steeply-dipping bedding surfaces that are compacted around thin, intercalated sandbodies in the LKCF section and are displaced by early carbonate-cemented concretions in the Heather Formation interval (Fig. 13a). Examination of the *in situ* mudrock dip data, plotted as true dips (i.e. corrected for hole deviation, but not for structural dip) on both dip tadpoles and on a northern hemisphere Schmidt stereonet clearly demonstrate that the LKCF reservoir in this well dips to the east (Fig. 13a and, b). The angle of structural dip is calculated at 17/085E (Fig. 13c), and whilst that for the underlying Heather Formation interval is similar, close examination of the modal position of the bedding surface dip data for depositionally flat-lying calcareous mudrocks making-up this section shows a slight southward shift in azimuth directions (Fig. 13b and c).

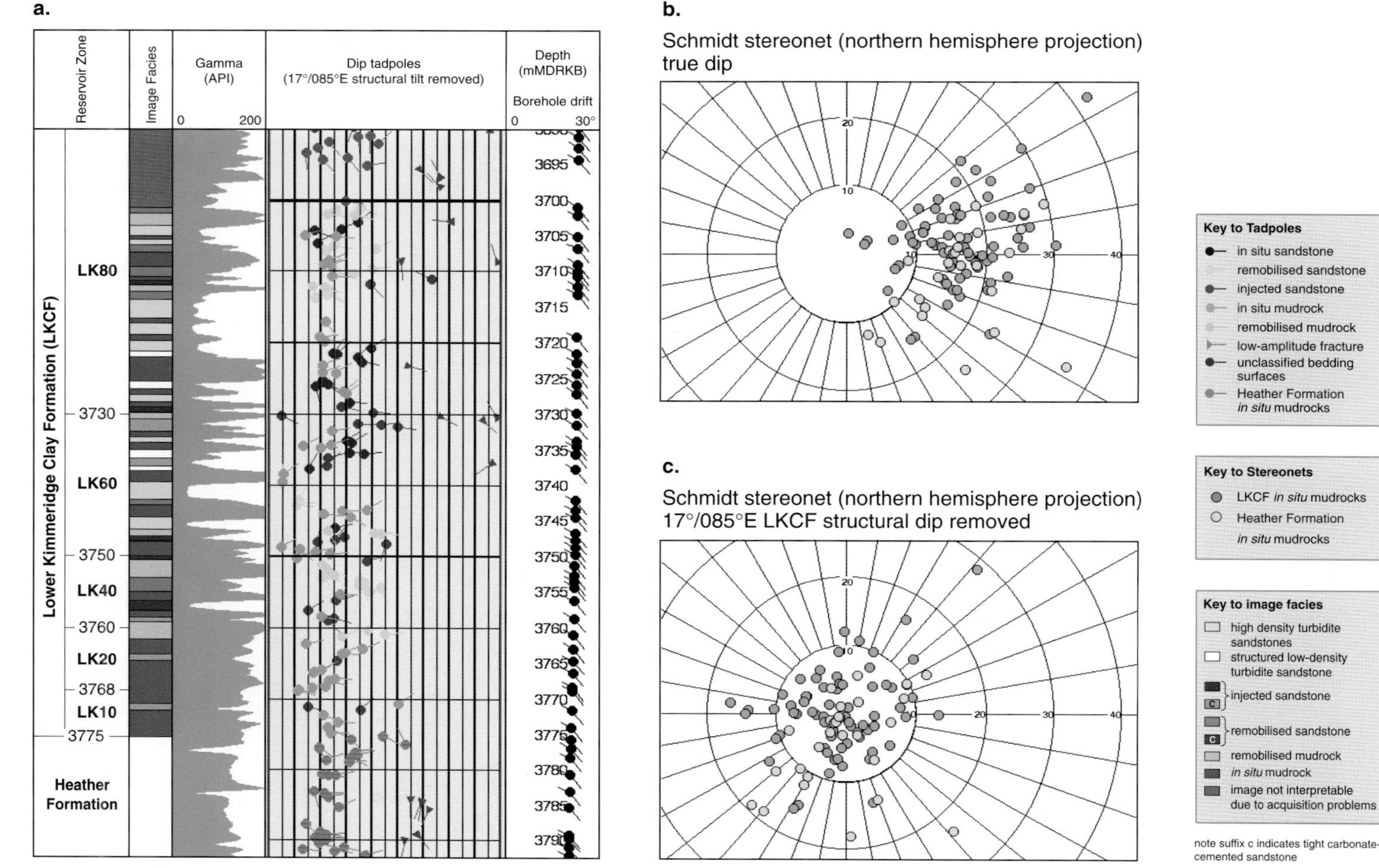

Fig. 13. Interpreted facies breakdown and hand-picked bedding surface dip tadpoles illustrating the distribution and stacking patterns of key depositional elements in both the Lower Kimmeridge Clay Formation and underlying Heather Formation (a), together with Schmidt stereonets for the *in situ* mudrock data used to calculate structural dip plotted as true dips (b; corrected for hole deviation only) and corrected for a structural tilt of 17°/085°E (c).

Evaluation of sandstone injection orientations

Having accurately identified and classified the sandstone injections in terms of their vertical distribution, thickness and reservoir potential (porous *vs* tightly cemented by carbonates) it is possible to examine the orientations of their lower and upper surfaces more closely in an attempt to better understand their geometries and likely injection directions in the immediate vicinity of the well bore. These data may then be used to make assessments of the potential distribution of donor high-density turbidite sandbodies close to, but not contacted by, the well bore and of potential flow directions of either injected water or produced oil which will help provide a more accurate indication of sweep efficiency and potentially gives a valuable insight into any directionality of fluid migration.

In order to calculate accurately sandstone injection directions it is important to have an understanding of their timing relative to structural rotation. In the LKCF reservoir there is clear evidence both in core and on features resolved by the UBI amplitude images for ptygmatic folding of steeply-inclined injection dykes which implies that injection of sandstones was a relatively early process, predating significant burial (Figs 3c and 14a). Therefore, in order to fully understand sandstone injection orientations and to utilize the directional data they provide it is vital to remove structural dip.

A further complexity which needs to be taken into account when evaluating the orientation of steeply-inclined injection dykes is the impact that their highly-irregular, bulbous geometries and the effects of any ptygmatic folding have on the orientation data. For these reasons the lower and upper surfaces of a small proportion of steeply-inclined injections dip in very different directions (Fig. 14b) which are misleading unless their origin is clearly understood. However, when evaluating the orientations of several injections, for example for the entire LKCF reservoir section in a well or for a specific reservoir zone it is apparent from the orientations of the bulk of the data that there are definite preferred injection directions (Fig. 15).

An example of sandstone injection geometries and orientations is shown from an injector well which is drilled through the damage zone of an east–west striking normal fault in the western part of the Magnus Field (Fig. 15). This dataset is typical for the LKCF reservoir and illustrates that there are two distinctive styles of sandstone injection:

(1) low-angle injection sills (typically dipping <20°) are developed throughout the reservoir section and have orientations which are considered to relate to the distribution of donor sandstones (Fig. 15a). Studies of outcrop and core have shown that most sandstone injections are injected up section, away from their donor sandstones in an attempt to reduce overpressure (Hiscott 1979). Therefore, statistically it is likely that the donor sandstones are located in the direction towards which the injections are dipping. For example, in Zones LK10 to LK60 they dip both to the west and east implying that donor sandstones are located both to the west and east of this well location (Fig. 15b). In contrast, in Zones LK80 to LK90 they only dip to the south-east implying that donor sandstones are only developed in that direction (Fig. 15c). Whilst a high proportion of these injections are thin (<1 m) and tight due to pervasive carbonate cementation, relatively thick (<2.5 m), porous injections are also developed. These have pay characteristics and hence potentially contribute to well performance, whereas

(2) steeply-inclined injection dykes (typically dipping 30–80°) only occur in tectonized zones either containing a high fracture population density or actually cut through by sub-seismic scale faults, for example in Zones LK80 to LK90 in the injector well illustrated (Fig. 15b). In the example shown these sandstone injections show sub-parallel alignment to fracture strikes which implies that their orientation is influenced by the local stress regime (Fig. 15b).

With an accurate knowledge of sandstone injection orientations it is possible to gain a more complete understanding of well performance and to identify potential fluid flow pathways.

Conclusions

It has been the intention of this paper to show that high-resolution UBI images are of value as an interpretative sedimentology tool in complex heterolithic reservoirs despite the fact that they typically resolve little internal detail in sandstones with little internal amplitude contrast. This has been demonstrated using the Lower Kimmeridge Clay Formation reservoir of the Magnus Field in which:

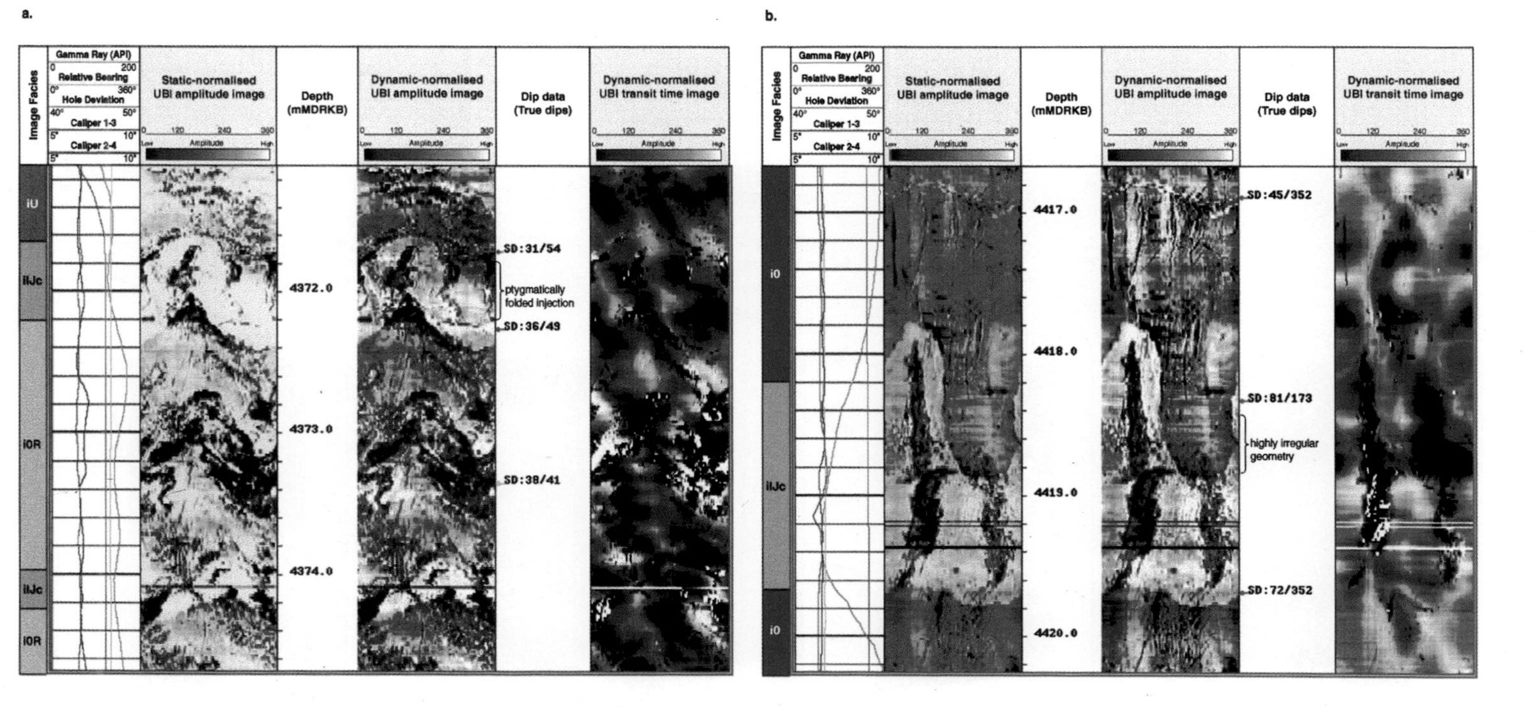

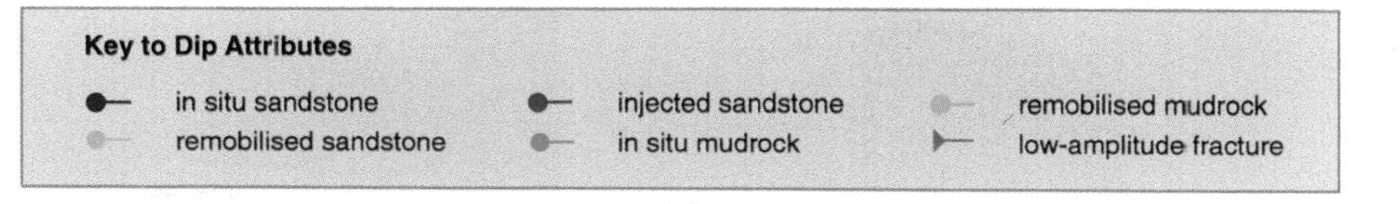

Fig. 14. Static- and dynamic-normalized UBI amplitude and interval transit time images illustrating a pytgmatically folded sandstone injection (a) and a highly irregular sandstone injection dyke in which the upper and lower surfaces of the injection apparently dip in opposite directions (b).

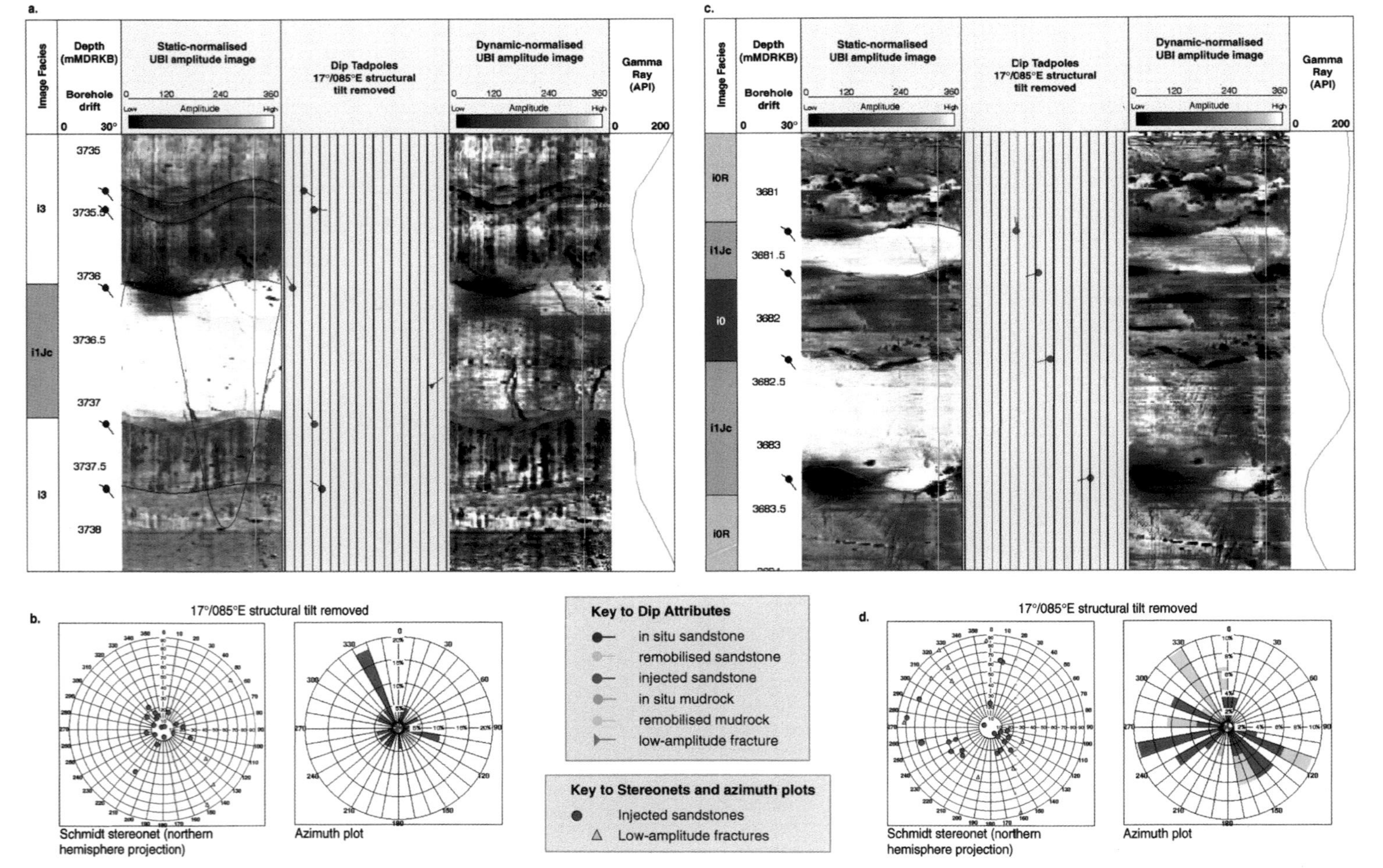

Fig. 15. Static- and dynamic-normalized UBI amplitude images illustrating examples of tight, carbonate-cemented sandstone injections (image facies code i1Jc), together with both Schmidt stereonets and azimuth plots illustrating sandstone injection and tectonic fracture orientations in a structurally simple reservoir interval containing near-horizontal sandstone injection sills (a, b) and in a highly fractured interval cut by a normal fault containing both low-angle, sill-like and steeply-inclined, dyke-like

(1) mudrocks and thin, intercalated sandstones are clearly resolved by UBI amplitude images due to marked differences in acoustic impedance and whilst little internal detail is observed, careful evaluation of the geometries and orientations of the lower and upper contacts of the sandstones has enabled accurate and consistent discrimination of high-density turbidites from sandstone injections. This is important because high-density turbidites form the principal pay lithology in the LKCF reservoir, whilst sandstone injections may form the mechanism by which depositionally isolated sandbodies are in pressure communication. Hence it is crucial to identify sandstone injections prior to completing development wells, in particular if they have pay characteristics.
(2) bedding surfaces hand-picked from features resolved by the UBI amplitude images have been used to make determinations of structural dip in the context of the LKCF depositional model and independent of seismic data. These interpretations are made by filtering of the data to include only depositionally flat-lying, *in situ* hemipelagic mudrocks that are easily identified from the UBI amplitude images and are therefore reliable.
(3) the vertical distribution, stacking patterns and orientations of sandstone injections identified from the UBI amplitude images may contribute to an understanding of the likely flow pathways of both produced hydrocarbon and injected water, and to a better understanding of well performance in terms of the distribution of donor high-density turbidite sandbodies close to, but not contacted by, the well bore.

These interpretations form a valuable part of the LKCF reservoir model. They are not possible from conventional openhole logs alone, and they therefore form a valuable tool for evaluating the reasons behind discrepancies between predicted well performance and actual production data in this reservoir.

BP Amoco Exploration and the Magnus Field partners Brasoil UK Ltd, Goal Petroleum plc, Sun Oil Britain Ltd and Repsol Exploration (UK) Ltd are acknowledged for giving permission to publish this work. Badley Ashton & Associates Ltd and, in particular David Kemp and Heather Nickson who drafted the figures are thanked for their assistance in preparation of this paper.

References

BOURKE, L. T. 1992. Sedimentological borehole image analysis in clastic rocks: a systematic approach to interpretation. *In*: HURST, A., GRIFFITHS, C. M. & WORTHINGTON, P. F. (eds) *Geological applications of wireline logs II*. Geological Society Special Publication, **65**, 31–42.

EVANS, A. C. & PARKINSON, D. N. 1983. A half-graben and tilted fault block structure in the Northern North Sea. *In*: BALLY, A. W. (ed.) *Seismic expression of structural styles*. Studies in Geology Series, **15/2**, American Association of Petroleum Geologists, Tulsa.

FARAGUNA, J. K., CHACE, D. M. & SCHMIDT, M. G. 1989. *An improved borehole TeleViewer system: image acquisition, analysis and integration*. Society of Petroleum Well Log Analysts 13th Annual Log Symposium, June 11–14 Denver.

HARKER, S. D., MCGANN, G. J., BOURKE, L. T. & ADAMS, J. T. 1990. Methodology of Formation MicroScanner image interpretation in Claymore and Scapa Fields (North Sea). *In*: HURST, A., LOVELL, M. A. & MORETON, A. C. (eds) *Geological Applications of Wireline Logs*. Geological Society Special Publication, **48**, 11–25.

HAYMAN, A. J., PARENT, P., CHEUNG, P. & VERGES, P. 1994. Improved borehole imaging by Ultrasonics. *Bulletin of the Society of Petroleum Engineers*, **28440**, 977–992.

HISCOTT, R. N. 1979. Clastic sills and dykes associated with deep-water sandstones, Tourelle Formation, Ordovician, Quebec. *Journal of sedimentary Geology*, **49**, 1–10.

LEONARD, A. J., DUNCAN, A. E., JOHNSON, D. A. & MURRAY, R. B. 1992. Development planning in a complex reservoir: Magnus Field UKCS, Lower Kimmeridge Clay Formation (LKCF). *Bulletin of the Society of Petroleum Engineers*, **25059**, 427–437.

SHEPHERD, M. 1991. The Magnus Field, Blocks 211/7a, **12a**, UK North Sea. *In*: ABBOTTS, I. l. (ed.) *United Kingdom Oil and Gas Fields, 25 years commemorative volume*. Geological Society Memoir, **14**, 153–157.

ZEMANEK, J., CALDWELL, R. L., GLENN, E. E., HOLCOMB, S. V., NORTON, L. J. & STRAUSS, A. J. D. 1969. The Borehole TeleViewer – a new concept for fracture location and other types of borehole inspection. *Journal of Petroleum Technology*, **25**, 762–774.

FMS Images from carbonates of the Bahama Bank Slope, ODP Leg 166: Lithological identification and cyclo-stratigraphy

TREVOR WILLIAMS[1] & CARLOS PIRMEZ[2,3]

[1] *Department of Geology, University of Leicester, Leicester, LE1 7RH, UK (e-mail: tw7@le.ac.uk)*

[2] *Borehole Research Group, Lamont Doherty Earth Observatory, Palisades, NY, USA*

[3] *Now at: Exxon Production Research Co., PO Box 2189, Houston, TX 77252, USA*

Abstract: Ocean Drilling Program (ODP) Leg 166 cored a transect of holes through the prograding carbonate sequences that form the western slope of the Great Bahama Bank, with the aim of detailing the relationship between the sequences and changes in sea-level over the last 25 Ma. A total of 1200 m of FMS resistivity images from Site 1003 (lower slope) and Site 1005 (mid-slope) were divided into three image facies types, with the aid of calibration against the recovered core. Type 1 was conductive (poorly cemented) sediment dominated by pelagic components, Type 2 was resistive (well cemented) sediment dominated by platform (neritic) components, and Type 3 was highly resistive (very well cemented) sediment, usually calci-turbidites but occasionally hardgrounds. Much of the section is composed of metre-scale alternations between Type 1 and Type 2 sediment. We have used the cycle thicknesses in the Middle Miocene to obtain a sedimentation rate curve and to refine the biostratigraphy. The cyclicity is modulated by the precessional astronomical cycle. The FMS images were used to evaluate the lithostratigraphic position and significance of prominent isolated uranium peaks. The peaks tend to occur just below the tops of calci-turbidite-rich units, sometimes coincident with sequence boundaries and maximum flooding surfaces.

Leg 166 of the Ocean Drilling Program sailed from February to April 1996, and drilled a total of seven sites, including a transect of four deep sites, on the western slope of the Great Bahama Bank (Figs 1, 2) (Eberli *et al.* 1997, Anselmetti *et al.* in press). The four deep sites were logged, providing nearly 2 km of FMS images and over 3 km of conventional logs (Table 1).

The sea-level objectives of ODP Leg 166 were

(1) to document the sedimentary record of Neogene and Quaternary sea level changes,
(2) to determine the ages of the major unconformities,
(3) to compare the sedimentary record with the oxygen isotope record of glacio-eustacy (from Site 1006 in the Florida Strait).

The sedimentary facies variations represent the sedimentary response of the carbonate environment to sea-level changes, and together with diagenesis produce petrophysical differences in the sedimentary succession. Thus the petrophysical logs can be interpreted in terms of facies and sea-level. The average core recovery for the leg was 55.3% – sufficient to document the broad facies succession, but the continuous log data are required to pin-point boundaries, make an analysis of the evolution of the cyclicity, and to identify turbidites in low recovery intervals.

In this paper we present a classification of the sediments based on their FMS image facies, calibrated to core, and then present two applications of the FMS using the classification. In the first, a sedimentation rate curve is derived by counting Type 1 – Type 2 alternations and measuring their thickness (cyclo-stratigraphy), and in the second, the potential significance of uranium log peaks is assessed by their location in the FMS-derived facies succession.

The relationship between FMS image and lithology in this setting

The FMS images from Sites 1003 and 1005 were classified into three image facies types (Table 2). The sedimentary processes by which the FMS images are related to lithology and sea-level are illustrated in Fig. 3, and examples of each type are illustrated in Fig. 4 (902–916 mbsf at Site 1003). Details of the sedimentological analysis of the cores are descibed in Betzler *et al.* (in press) and Eberli *et al.* (1997).

From: LOVELL, M. A., WILLIAMSON, G. & HARVEY, P. K. (eds) 1999. Borehole Imaging: applications and case histories. Geological Society, London, Special Publications, **159**, 227–238. 1-86239-043-6/99/$15.00.

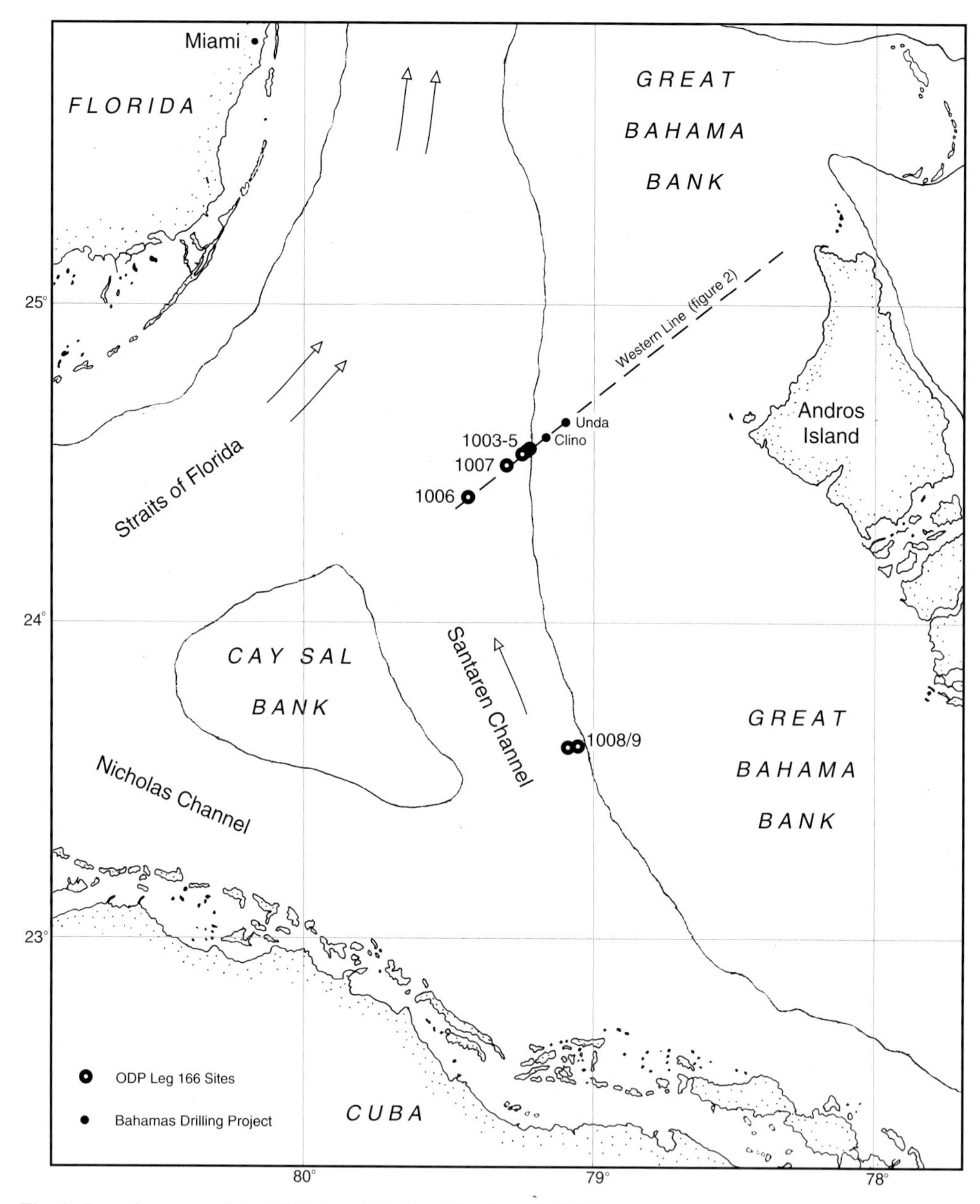

Fig. 1. Location map of the ODP Leg 166 sites (Eberli *et al.* 1997)

Type 1

The Type 1 image facies is conductive (dark on the images), and often has a mottled appearance. It corresponds to dark grey packstones and wackestones, and contains a relatively high proportion of pelagic components. Type 1 sediment contains up to 20% aragonite – it has not been dissolved, remobilized and re-precipitated as cement, hence Type 1 has a high porosity, and burrows which were flattened during sediment compaction. The bioturbation causes the FMS image to be mottled.

Type 2

The Type 2 image facies is resistive (light on the images), and, like Type 1 often has a mottled appearance. It corresponds to well-cemented

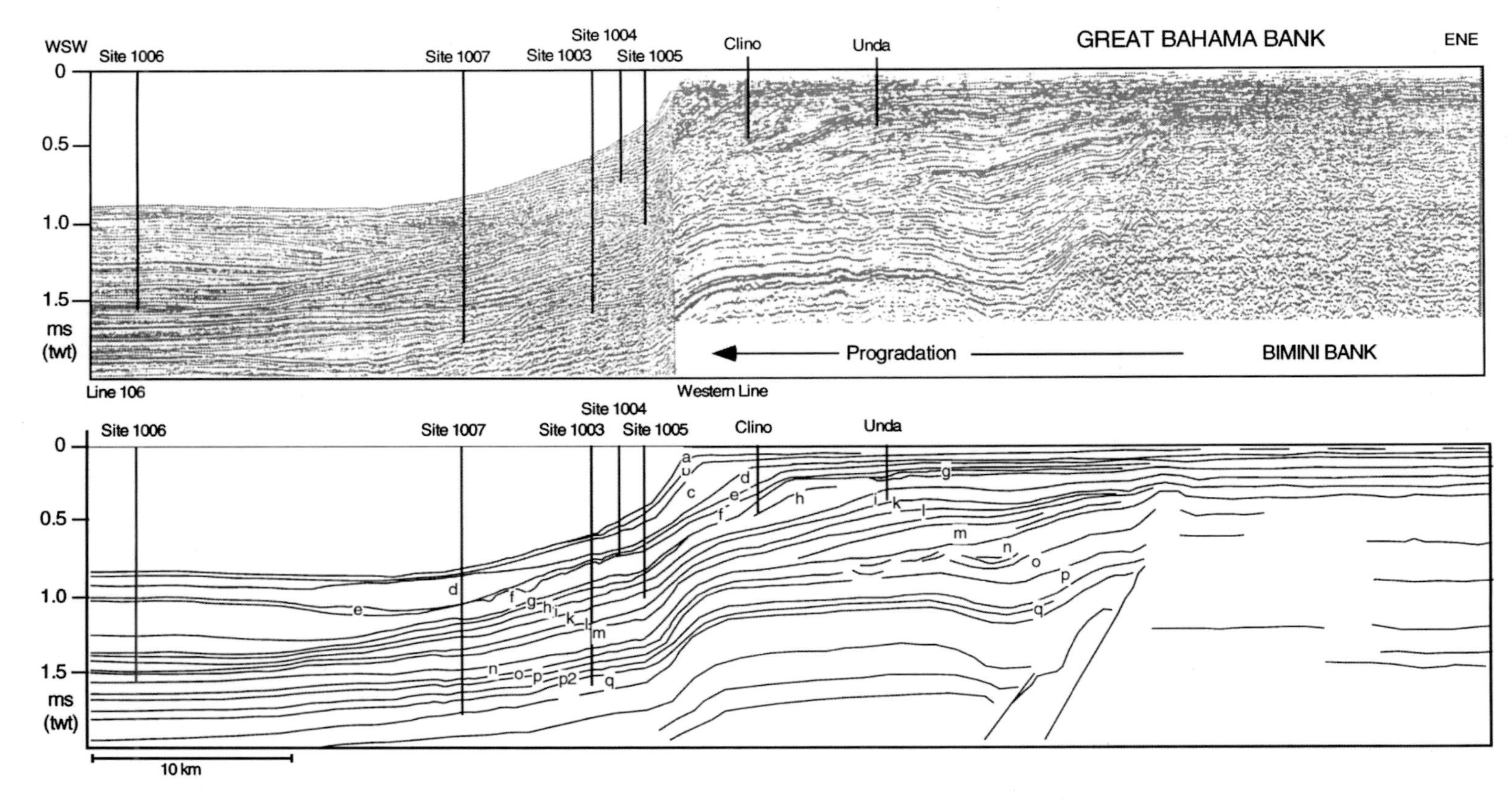

Fig. 2. Seismic section along the Western Geophysical line and its high resolution extension (Fig. 1), with locations of the ODP Leg 166 sites and the Bahamas Drilling Project holes Unda and Clino (drilled in 1990). Below, seismic sequence boundaries and sub-surface geometries. (Eberli *et al.* 1997).

Table 1. *Logging operations during ODP Leg 166*

ODP Hole	1006A	1007C	1003D	1005A, C
Water depth	688 m	648 m	483 m	351 m
Logged interval:				
FMS	103–716	no FMS	107–1050	383–613
conventional	103–718	112–1158	107–1051	89–691

Table 2. *Image, lithological and log character of the three main image facies types.*

	Type 1	Type 2	Type 3
FMS character	conductive, heterogenous (bioturbated)	resistive, heterogeneous (bioturbated)	highly resistive, homogenous or laminated, thin (<1m) beds.
Lithology	dark grey bio-wackestones up to 20% aragonite 5–7% clay flattened burrows	light grey bio-wackestones well cemented 96–99% carbonate burrows retain original shape	calci-turbidites (maybe with current re-working), or occasionally hardgrounds. very well cemented
Log response*			
resistivity	<1 Ω m	1–5 Ωm	>5 Ωm
sonic velocity	2–3 km/s	3–4 km/s	4–6 km/s
density	1.9–2.2 g/cm^3	2.2–2.4 g/cm^3	2.4–2.8 g/cm^3
uranium	4–10 ppm	1–4 ppm	0.5–3 ppm
porosity	40–50 %	20–40 %	25–5 %

* The ranges for the log responses are for the lithified sediment and are a general guide. Most show an overall downhole compaction trend.

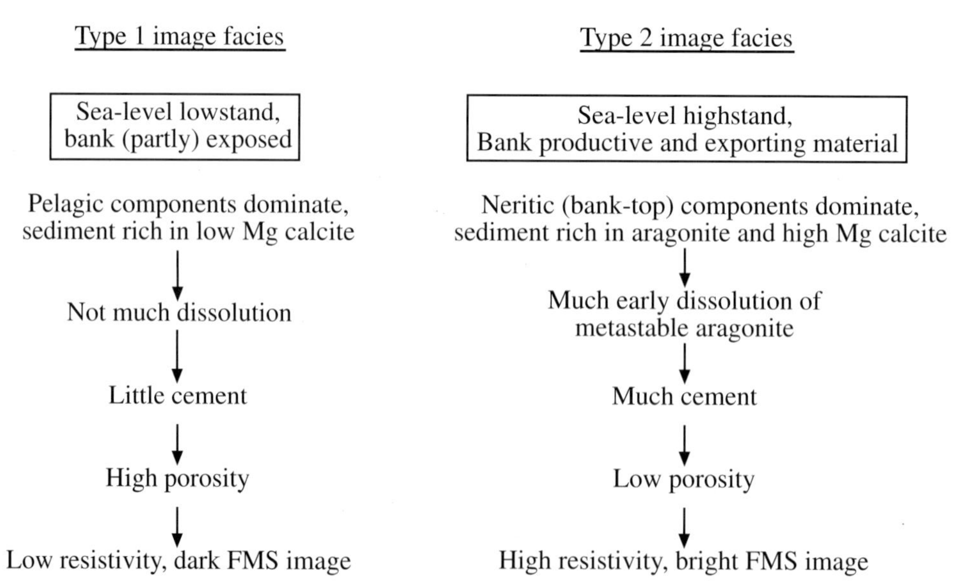

Fig. 3. Schematic diagram showing how the FMS image facies types 1 and 2 are related to lithology and sea-level. Calci-turbidites (Type 3) follow the Type 2 scenario except that they are more cemented and can be deposited during sea-level lowstands as well as highstands.

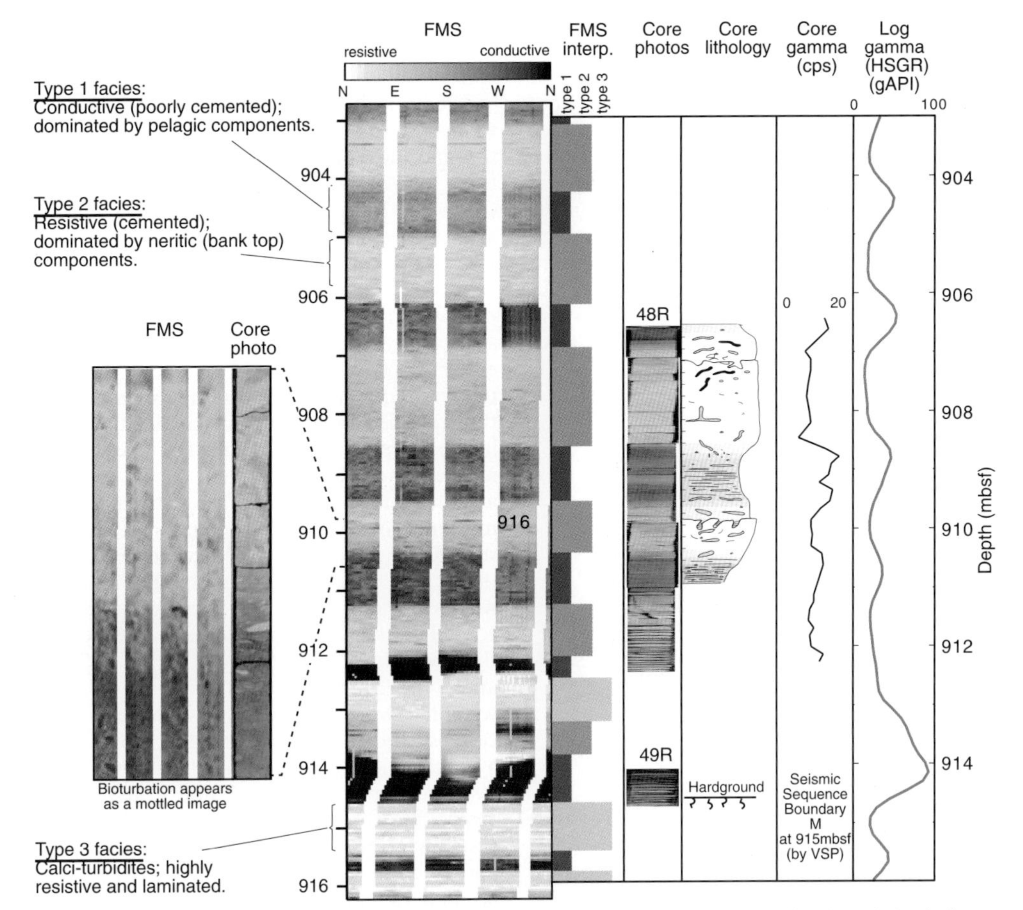

Fig. 4. FMS image from Site 1003, 902–916 mbsf, showing examples of cyclic alternation (902–912 mbsf), calci-turbidites (914.6–916.2 mbsf), a sequence boundary (≈914 mbsf), and bioturbation (detail in the expanded image to the left). The FMS has been divided into Type 1, 2 and 3 sediment in the Interpretation column (see text). Core photographs (compressed), the core lithological log, and the core natural gamma for core 48R are displayed alongside the FMS image. The FMS coverage has been exaggerated (true borehole diameter ≈12 inch).

light grey packstones and wackestones, and is dominated by neritic components from the bank top. Cementation occurred soon enough after deposition to lend the sediment rigidity and thus for the burrows to retain their original shape.

Type 3

The Type 3 image facies is highly resistive (white or very light on the images), either homogenous or laminated, and corresponds to calci-turbidites in the core. Hardgrounds would have a similar FMS response (though they are much rarer), and so are included within Type 3.

Calci-turbidites. The calci-turbidites are usually less than 1 m thick, and generally have well defined tops and bases. Their high resistivity is due to the very high degree of cementation. Less well cemented calci-turbidite layers are distinguished from Type 2 sediment by the lack of mottling (bioturbation) in the Type 3 image.

Hardgrounds. Hardgrounds (and firmgrounds) are recognized in the core as cemented paleosurfaces which have been burrowed down into. They represent periods of non-deposition. In the FMS image, they should appear as a highly resistive layer, however this characteristic alone will not distinguish a hardground from a calci-turbidite, hence in our FMS classification scheme they are both Type 3 (indeed the top surface of a calci-turbidite could also form a hardground). Only two hardgrounds were found in core in the intervals covered by FMS at Sites 1003 and

1005. This is partly due to their preferential non-recovery: the hardness of the hardground relative to the overlying sediment causes a slow-down of the drilling rate, and often associated poor core recovery and a widening of the hole (and hence poor FMS pad contact, for example at 920 mbsf in Fig. 4). Close inspection of the FMS images at depths where hardgrounds were observed in the core showed that hardgrounds and calciturbidites are very difficult to distinguish from each other in the FMS, probably because the FMS image character is partly diagenetic in origin, thus obscuring the detail of the original depositional features.

Early Pliocene sediment

Early Pliocene sediment (120–250 mbsf) has featureless log responses: uniformly low velocity and low resistivity. The hole is too wide in this interval for good FMS pad contact with the borehole walls, but for completeness, we note that this sediment differs from the types described above, in that it is a rapidly deposited, fine-grained, neritic-rich sediment that is uncemented.

Cyclicity in the Middle Miocene

Cyclic alternation was observed in many intervals of the Leg 166 logs and cores. At Site 1003, this cyclicity was particularly well developed in the Middle Miocene seismic sequence m between 738 and 912 mbsf (Fig. 5). The sediment in this interval alternates between Type 1 and Type 2 beds, with occasional interuptions by calci-turbidites. We have defined each cycle as comprizing from the base of a Type 1 bed through to the top of the overlying Type 2 bed.

If each cycle represents a constant amount of time, the relative age of the sediment and its sedimentation rate can be determined simply by counting the cycles and measuring their thickness, in the same way that the age of a tree can be found by counting its yearly tree rings. It is well known that the astronomical cyclicities (orbital eccentricity, axial obliquity (tilt), and precession of the equinoxes) control the glacial-interglacial alternations of Plio-Pleistocene climate, and that precessional periodicity is present in Miocene Mediterranean sediments (Hilgen *et al.* 1995, 1997). Given the regularity with depth of the cycles found at the Leg 166 sites, it seems reasonable that each cycle was deposited in a constant duration, at one of the astronomical cyclicities. The astronomical periodicities are known to be reasonably stable at least back to the Miocene (Laskar 1993), but which of those cyclicities is controlling the lithology at Site 1003 is not known *a priori*.

The statically normalized FMS image was used to analyse the bedding. Firstly, the depth to the top and base of the resistive beds were picked, using the 'Borview' module of Schlumberger's Geoframe log interpretation software (dip and azimuth were also determined for each boundary). Between 738 (seismic sequence boundary (SSB) L) and 915 mbsf (SSB M), 122 resistive beds were identified.

Secondly, the resistive beds were classified as either Type 2 or Type 3 sediment, based on the resistivity amplitude, and the internal structure revealed by dynamic normalization of the image. Type 2 were assigned if the bed was mottled (bioturbated sediment), and Type 3 were assigned if the bed was highly resistive and lacked mottling, or if the bed contained laminated surfaces; mostly, Type 3 beds were thinner than the Type 2 beds (Fig. 5). Though the distinction was clear is most cases, there were some beds that could have been classified either way. Sometimes there was insufficient contrast between Type 1 and Type 2 sediment, and sometimes the beds were thin; where this occurred we have chosen to collapse those alternations into 'sub-cycles' of a main cycle. Of the original 122 resistive beds, 19 were Type 3 (calci-turbidites), and 18 were Type 2 but sub-cycles, leaving 85 full cycles; this assignment of the beds is 'version-1'. Sub-cycles and calci-turbidites were not used when calculating ages and sedimentation rates.

Astronomical periodicities were then assigned to the cycles. If each cycle was 41 ka long (the obliquity periodicity), the resulting sedimentation rate would be incompatible with the biostratigraphic dates, being far too slow. Thus the precession of the equinoxes is likely to be the controlling factor, with its characteristic periodicities of 19 and 23 ka. Mean sedimentation rates calculated from these version-1 cycles were 9.1 cm/ka for 23 ka cycles, and 11.0 cm/ka for 19 ka cycles, which are both lower than the sedimentation rate derived from the biostratigraphy (13 cm/ka) (Fig. 6).

A possible cause for the mis-match could be incorrect classification of either Type 3 sediment (calci-turbidite) as Type 2 or a sub-cycle as a full cycle. So in a second attempt (version-2) five of the 85 version-1 cycles were re-classified as calci-turbidites, and two as sub-cycles. This version-2 yields what can be regarded as an upper limit to the cyclo-stratigraphic sedimentation rates (9.8 cm/ka for 23 ka cycles, and 11.8 cm/ka for 19 ka cycles). The version-2 rate is still slower than the rate from biostratigraphy (Fig. 6).

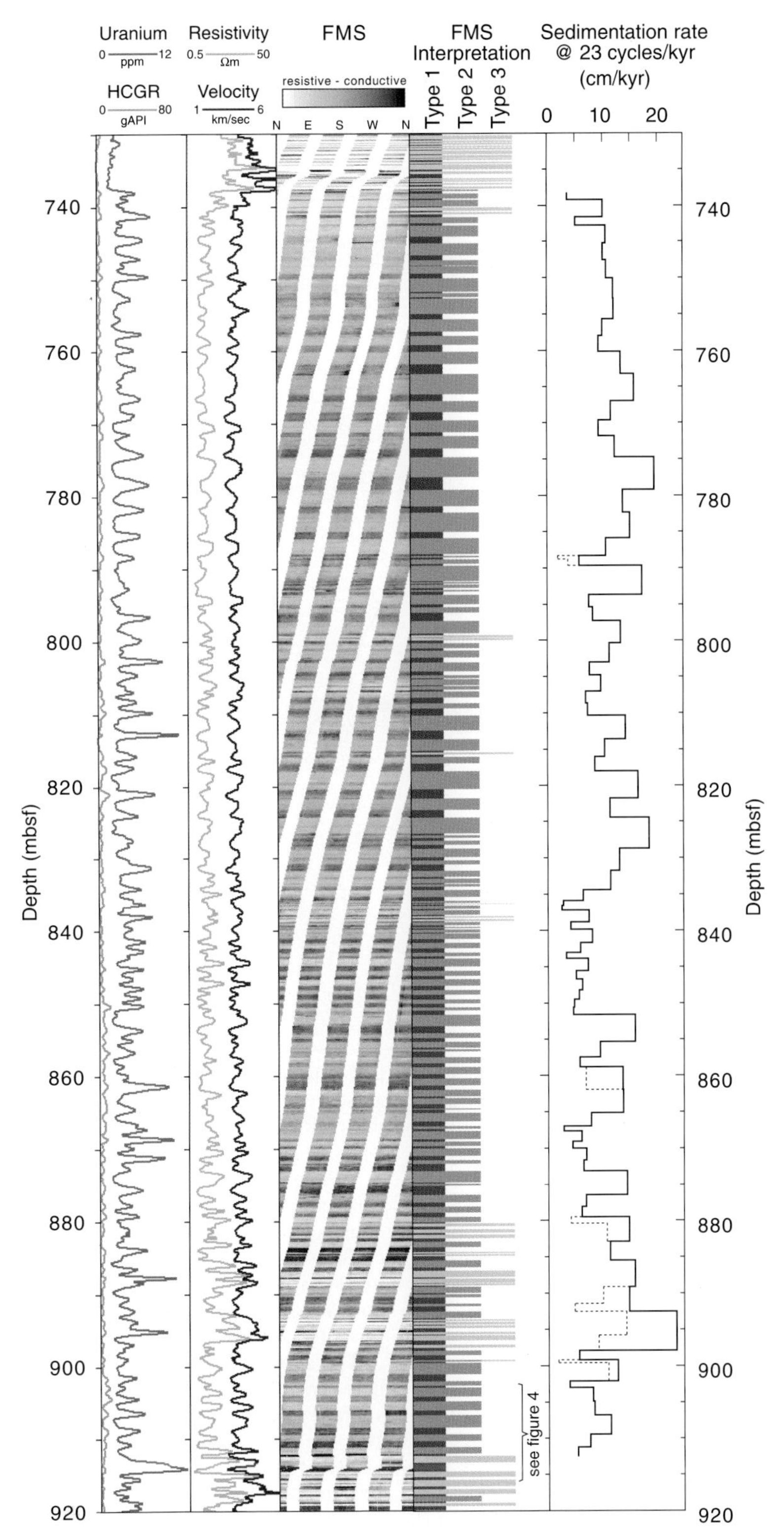

Fig. 5. FMS image of Type 1–Type 2 alternations at Site 1003. Sedimentation rate is derived from the cycle thicknesses in the FMS interpretation (base of Type 1 to base of the next Type 1), and assuming a periodicity of 23 ka/cycle. The solid sedimentation rate curve corresponds to version 2 of the cycles, the version 1 curve is dotted (see text). The FMS coverage has been exaggerated (true borehole diameter ≈12 inches).

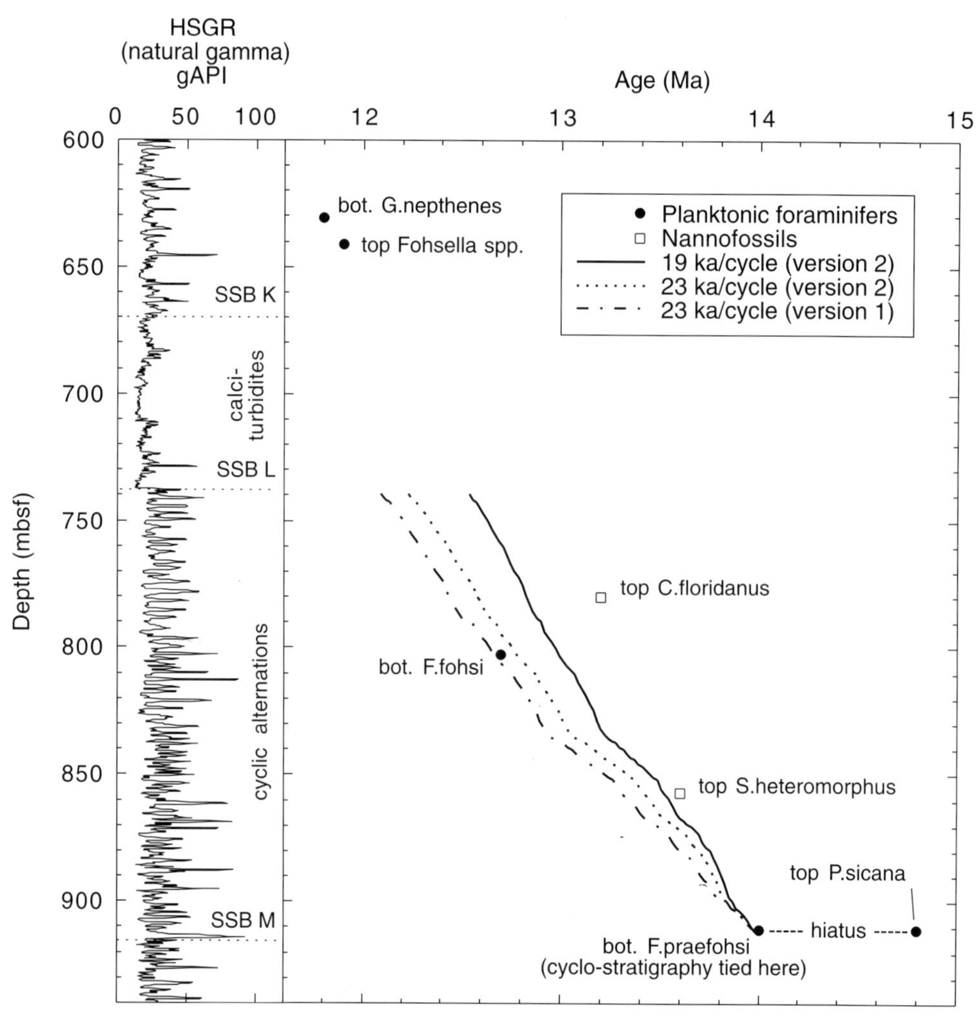

Fig. 6. Age *vs* depth for the Middle Miocene of Site 1003. The sedimentation rates derived from the cyclic alternations are fixed to the end of the hiatus (sequence boundary M). The version 2 curves (19 and 23 ka periodicities) represent the faster end-member of the possible cyclo-stratigraphic sedimentation rates.

We have chosen to fix the floating cyclo-stratigraphy to the end of the hiatus at 912 mbsf, at the first occurrence of the foraminifer *Fohsella praefohsi*, at 14.0 Ma (this is the oldest possible age for the end of the hiatus) (Fig. 6). The 23 ka/cycle cyclo-stratigraphies (version-1 and version-2) are both close to the next foraminifer datum, the first occurrence of *Fohsella fohsi*. However, they are much older than the nannofossil datums (top *Cyclicargolithus floridanus* and top *Sphenolithus heteromorphus*). In the biostratigraphic synthesis (Eberli *et al.* 1997), it is noted that the *C.floridanus* marker (13.2 Ma) occurs at the top of a slumped interval (not apparent in the FMS image), yet it was prefered to the *F.fohsi* foraminifer age (12.7 Ma) when constructing the biostratigraphic age model. The cyclo-stratigraphy indicates that the *F. fohsi* marker is the better, and that both *C. floridanus* and *S. heteromorphus* are unreliable markers at this site.

The slower than expected sedimentation rate between 738 and 912 mbsf requires a higher than average sedimentation rate between 738 mbsf and the next biostratigraphic ages at 630 and 642 mbsf. This is plausible, as this interval is very rich in calci-turbidites, which are likely to have been quickly deposited.

An alternative possibility for the slower than expected cyclo-stratigraphic sedimentation rates is that the Bergren *et al.* (1995) timescale, on which the biostratigraphy is based, assigns too little time to the middle Miocene: if the middle Miocene were longer, this would enable more cycles to be fitted in. However, this possiblity is beyond the scope of this paper, and the few biostratigraphic ages in this interval and the uncertain accuracy of the cyclo-stratigraphy are not sufficient to make that argument.

A complementary study of the cyclicity at Site 1003 also concludes that the periodicity is precessional (Bernet & Eberli, in press), as do Pirmez & Brewer (1998) for Site 1006.

Uranium peaks and their lithostratigraphic position

Several prominent isolated peaks in the uranium logs from Holes 1003D and 1005C occur coincidently with lithostratigraphic boundaries. The lithostratigraphic position of the uranium peaks, and the possible processes by which uranium becomes concentrated are discussed below to try to explain the occurrence of the peaks and to determine their significance. It has been previously suggested that uranium peaks can mark unconformities, intervals of condensed sedimentation rates, and hardgrounds (Rider 1996).

In the Leg 166 sediments, the natural gamma signal is almost entirely due to uranium, with potassium and thorium (present in clay minerals) contributing only a minor proportion to the total natural gamma log. The CGR log (total natural gamma with the uranium contribution subtracted) is shown in Fig. 5.

The uranium peaks are nearly always located in Type 1 sediment (Figs 4, 5 & 7), and usually occur interbedded between the top calci-turbidites in calci-turbidite-rich units. Quite often a relatively thick layer of Type 1 (pelagic) sediment overlies the calci-turbidite units (Fig. 7 Table 3). The majority of the calci-turbidites are a result of shedding during sea-level highstands, but lowstand calci-turbidites are also significant (Bernet *et al.* 1997) and these cannot be distinguished using the FMS images alone.

Below 740 mbsf at Site 1003, a slightly different regime seems to have been in operation, as the uranium log shows repeated moderately high peaks in the cyclic carbonates. However, a few more prominent peaks that stand out from the background and have the same relationship to the calci-turbidites are still identifiable.

In terms of sequence stratigraphy, uranium peaks can be associated with both sequence boundaries (SSB) (e.g. SSB H, Fig. 7a), and with maximum flooding surfaces (e.g. at 646 mbsf at Site 1003). However, not all the ODP Leg 166 sequence boundaries and flooding surfaces have a corresponding uranium peak.

The few hardgrounds and firmgrounds that were recovered in core at Sites 1003 and 1005 are sometimes accompanied by a uranium peak (e.g. at 914 mbsf at Site 1003 and 640 mbsf at Site 1005), but this is not always the case (e.g. the firmground at 505 mbsf in 1003 has no corresponding uranium peak (Fig. 7b)). The unconformity at 365 mbsf at Site 1003 is marked by a uranium peak, while the corresponding unconformity at Site 1005 is not.

Uranium can concentrate in a variety of ways: in confined, reducing sediments, by reduction of U^{6+} (soluble as the uranyl ion, UO^{2+}) to insoluble U^{4+}; adsorbed onto particles of clay and organic matter; by incorporation into carbonate-fluorapatite, possibly at hardgrounds (Serra 1984). Also, uranium is not included in the lattice of re-precipitated calcite, and is therefore expelled from the more cemented beds.

Colley & Thomson (1985) investigated uranium peaks in recent carbonate-rich distal turbidites on the Madiera abyssal plain. They found that oxidation of the organic carbon in the turbidites mobilizes the uranium (originally adsorbed onto the organic matter) as uranyl carbonate, which is reduced and concentrated into a peak below the oxidation front in the pelagic sediment underlying the turbidite. More than 60% of the uranium originally in the turbidites was re-located in this way. Thus these peaks represent relict oxidation fronts. This appears to be analogous to the occurrence of uranium peaks in Type 1 sediment observed in the Leg 166 logs, except that the gamma peaks occur just below the top of calci-turbidite units, rather than under each individual calci-turbidite. It appears that a change in sedimentary conditions, possibly with a reduced sedimentation rate, is necessary for uranium peak formation. By this process, the uranium peaks might coincide with sequence boundaries (just below the top of the highstand calci-turbidites) or maximum flooding surfaces (just below the top of the lowstand calci-turbidites).

Thus the probable general mechanism for generating the main uranium peaks is re-location of uranium from Type 2 and 3 sediment into the Type 1 layers upon oxidation of organic matter and carbonate diagenesis. Those uranium peaks not associated with the top of calci-turbidite units tend to coincide with other lithostratigraphic changes, or stage boundaries e.g. the mid-late Miocene boundary at 581 mbsf at Site 1005 (Fig. 7c), where the uranium concentration could be in the form of phosphates.

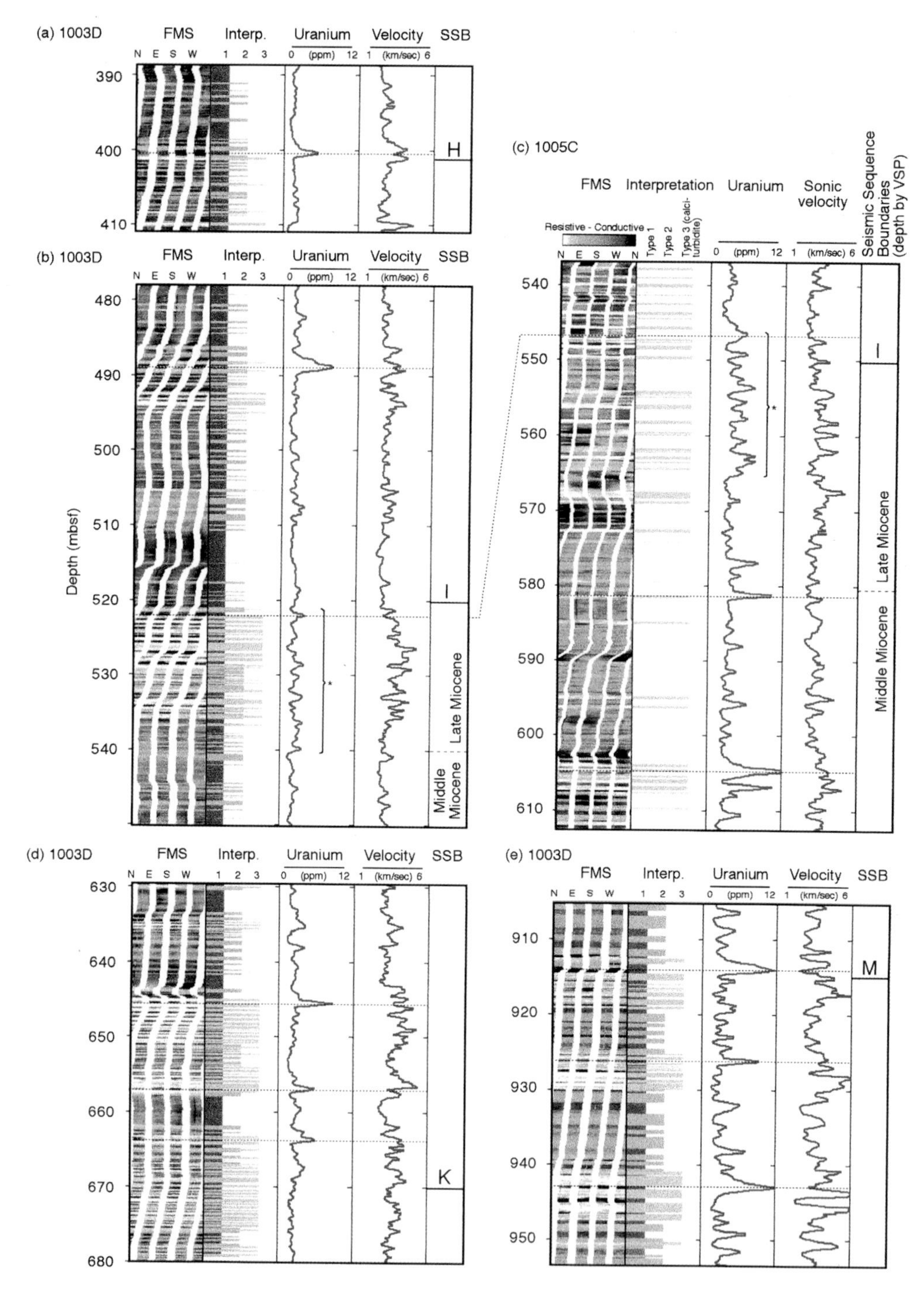

Fig. 7. The position of uranium peaks relative to FMS image facies types and sequence boundaries. In (a) the prominent uranium peak clearly corresponds to SSB H. In (b) (Site 1003) and (c) (Site 1005), there is no single uranium peak corresponding to SSB I but there is a similar uranium pattern (marked *) beneath SSB I; other prominent uranium peaks are marked, one of which is coincident with the Mid–Late Miocene boundary at Site 1005. In (d), SSB K could correspond to the uranium peak and velocity contrast. In (e) there is a contrasting uranium regime to the other examples, but certain peaks still stand out against the background, one of which corresponds to SSB M and a hardground in core. The FMS coverage has been exaggerated (true borehole diameter ≈12 inch).

Table 3. *Position of uranium peaks relative to Type 1 and Type 3 FMS image facies*

Depth to uranium peak (mbsf)	Type 3 sediment below?	Type 1 sediment above?	Lithological setting	Figure no.
Site 1003				
280	no	no	Mudstone to wackestone transition. MFS (maximum flooding surface) ?	
365/370	one	no	Miocene–Pliocene transition. Hiatus ?	
400	yes	yes	Sequence boundary H (402 mbsf).	7a
488	yes	yes	MFS ? Unit boundary.	7b
525–540 (* in Fig. 7)	yes	some	Sequence boundary I (522 mbsf), firmground at 523, Halimeda layer at 532. Interval composed of calci-turbidites.	7b
619	some	yes		
646	yes	yes	MFS, unit boundary.	7d
664	yes	yes	Sequence boundary K (670 mbsf)	7d
729	yes	yes	MFS ?	
915	yes	some	Sequence boundary M (915 mbsf), 2 Ma hiatus, hardground.	4, 7e
943	yes	some	Layer of Halimeda and peloids	7e
974	yes	some	Sub-unit boundary	
984	yes	some		
Site 1005				
491	yes	yes	Sequence boundary H (485 mbsf).	
510	yes	yes		
546–565 (* in Fig. 7)	yes	some	Sequence boundary I (550 mbsf). Interval composed of calci-turbidites.	7c
581	single	yes	Mid-late Miocene boundary	7c
604	yes	yes		7c

One possible cause of the greater frequency of uranium peaks below 740 mbsf at Site 1003 is a higher concentration of organic matter for the uranium to adsorb onto. In Type 1 sediment below 740 m, C_{org} reaches as high as 4%, thought to be mostly enrichment due to hydrocarbon migration (Eberli *et al.* 1997). At these elevated C_{org} contents, complexation with organic matter may also contribute to the uranium peaks.

Conclusions

- FMS images, calibrated to core lithology, can be used to identify pelagic-rich sediment, neritic-rich sediment, and calci-turbidites at the Leg 166 sites.
- Cyclic alternations between pelagic-rich and neritic-rich sediment in the middle Miocene at Site 1003 are controlled by the climatic precessional astronomical period (≈23 ka). Cycle counting yields a cyclo-stratigraphy which suggests that the sedimentation rate derived from biostratigraphy for the interval 740–910 mbsf is too fast.
- Uranium peaks tend to occur just below the top of calci-turbibidite-rich units, and probably represent relict oxidation fronts. They can correspond to sequence boundaries, maximum flooding surfaces, or other lithological boundaries.

We wish to thank Alison Mabillard and Bruce Sellwood for their reviews, Karin Bernet and Christian Betzler for reading early versions of the paper, and Flavio Anslemetti for figs 1 & 2. We also thank Schlumberger engineer Steve Kittredge, and the ODP and SEDCO crews aboard the JOIDES Resolution for their help in obtaining the logs.

References

ANSELMETTI, F. S., EBERLI, G. P. & ZAN-DONG DING (in press). From the Great Bahama Bank Into the Straits of Florida: A Margin Architecture Controlled by Sea Level Fluctuations and Ocean Currents. *Geological Society of America Bulletin*

BERGREN, W. A., KENT, D. V., SWISHER, C. C. III & AUBRY, M. -P. 1995. A revised Cenozoic geochronology and chronostratigraphy. *In*: BERGREN, W. A., KENT, D. V., AUBRY, M. -P. & HARDENBOL, J. (eds.) *Geochronology, Time Scales, and Global Stratigraphic Correlation.* Special Publication Society of Economic Paleotologists and Mineralologists, **54**, 129–212.

BERNET, K., EBERLI, G. P., BETZLER, C. & GILLI, A. 1997. Highstand Versus Lowstand Shedding of Carbonates – new Data for an old Controversy from the Western margin of Great Bahama Bank. *EOS: Transactions, American Geophysical Union,* **78**(46), F359.

—— & —— (in press). Spectral analysis on resistivity data along the western Great Bahama Bank (Leg 166, Site 1003). *In*: EBERLI, G. P., SWART, P. K., MALONE, M. J. *et al. Proceedings of the Ocean Drilling Program, Scientific Results*, **166**: College Station, TX.

BETZLER C., ANSELMETTI, F., BERNET, K., EBERLI, G., FRANK, T. & REIJMER, J. (in press). Sedimentary variations in space and time along the leeward flank of the Great Bahama Bank (ODP Leg 166). *Sedimentology*.

COLLEY, S. & THOMSON, J. 1985. Recurrent uranium relocations in distal turbidites emplaced in pelagic conditions. *Geochimica et cosmchimica acta*, **49**, 2399–2348.

EBERLI, G. P., SWART, P. K., MALONE, M. J., *et al.* 1997. *Proceedings of the Ocean Drilling Program, Initial Reports*. **166**, College Station, TX.

HILGEN, F. J., KRIJGSMAN, W., LANGEREIS, C. G., LOURENS, L. J, SANTARELLI, A. & ZACHARIASSE, W. J. 1995. Extending the astronomical (polarity) time scale into the Miocene. *Earth and Planetary Science Letters*, **136**, 495.

——, ——, —— & —— 1997. Breakthrough made in dating the geological record. *EOS*, **78**(28), July.

LASKAR, J., LOUTEL, F. & ROBUTEL, P 1993. Stabilization of the Earth's obliquity by the Moon. *Nature*, **361**, 615–617.

PIRMEZ, C. & BREWER, T. S 1998. Borehole electrical images: recent advances in ODP. *JOIDES Journal*, **24**, 14–17.

RIDER, M. 1996. *The Geological Interpretation of Well Logs*. Caithness, Whittles Publishing.

SERRA, O. 1984. *Fundamentals of Well-Log Interpretation (Vol. 1): The Acquisition of Logging Data. Development of Petroleum Science*, **15A**(1) Elsevier, Amsterdam.

Geology and reservoir description of 1Y1 reservoir, Oso Field, Nigeria using FMS and dipmeter

LAIRD B. THOMPSON & J. W. SNEDDEN

Mobil Exploration and Producing Technical Center, 13777 Midway Road, Dallas, Texas 75244-4390 (e-mail: lbthomps@dal.mobil.com)

Abstract: Mobil Producing Nigeria's Oso Field produces 120 000 barrels of oil per day from late Tertiary sands in the Niger Delta offshore. The reservoir consists of anastomozing fluvial deltaic sands which are difficult to correlate from well to well using traditional log correlation techniques. Image log data from Formation MicroScanner (FMS) logs and dipmeter data tied directly to cores permit detailed analysis of channel thickness, orientation, and distribution. This geographic understanding of the sand bodies enables geoscientists to construct a detailed 'plumbing' diagram of this complex reservoir, thus allowing for better exploitation.

Studies describing the sandstone reservoirs in Nigeria are relatively scarce. Key publications by Weber (1971) and Doust & Omatsoli (1990) describe the sandstones producing in the central Niger delta region. With the exception of a brief discussion in Odior (1992), little has been published on the eastern portion of the Niger delta offshore. One of the major constraints upon investigation of these sandstones has been a lack of full-diameter core and modern image logs. However, over 4000 ft of core in eight wells was taken in the Oso Field, and FMS or dipmeter logs were run in 14 wells for the purpose of reservoir characterization (Fig. 1). Analysis of the cores, combined with dipmeter and borehole image data, palaeontology, and well-log correlations from Oso Field, has led to development of a highly constrained depositional and sequence stratigraphic model for the major Oso Field pay zone (the 1Y1 reservoir). This model provides a foundation for reservoir evaluation, zonation, volumetric estimation, and description of fluid flow/gas injection potential in this new producing field.

General description of the Oso Field

Depositional setting

The Oso Field consists of two reservoir horizons, the upper 2Y2 and lower 1Y1 zones. These sands occur in the Miocene Biafra Member of the Agbada Formation (Fig. 2). The 2Y2 zone consists of shallow marine sands which are continuous across the field and thin from 175 ft in the north to about 50 ft in the south (Fig. 3). The 2Y2 is of minor importance to the Oso Field volumetrically and will not be described in detail.

In contrast, the 1Y1 reservoir interval (Fig. 4), with typical thicknesses of 500–800 ft (TVT), formed in a deltaic system intermediate in terms of wave, tide, and fluvial processes. Characteristic depositional facies include distributary channel-fill, tidal creek, tidal flat, lagoon, mouth bar, delta front, and prodelta. These facies were initially described from conventional cores and were integrated with both standard wireline data and more detailed wireline log data sets available from dipmeter and Formation MicroScanner logs. The facies are summarized on Table 1 and briefly described below.

The highest reservoir quality is found within distributary channel sands (facies 1), with pebbly-massive lithologies exhibiting 10-plus darcies of permeability; cross-bedded variants have somewhat lower values. Channel-fills are typically 30–90 ft thick successions with sharp, erosional bases and sharp to gradational tops.

Other depositional facies include tidal creek, mouth bar, delta front, and tidal flats sands. These reservoirs exhibit a strong tidal signature, with rhythmic alternations of shale and sand reflecting daily variations, and obvious bundling of sets related to spring/neap variations. Porosities are slightly higher, but permeabilities are lower in these units than in facies 1.

Non-reservoir lagoonal/interchannel shales often exhibit soft-sediment deformation, implying rapid deposition and topographic relief related to steep channel-margin slopes. While these shales often register horizontal permeabilities

From: LOVELL, M. A., WILLIAMSON, G. & HARVEY, P. K. (eds) 1999. Borehole Imaging: applications and case histories. Geological Society, London, Special Publications, **159**, 239–257. 1-86239-043-6/99/$15.00.

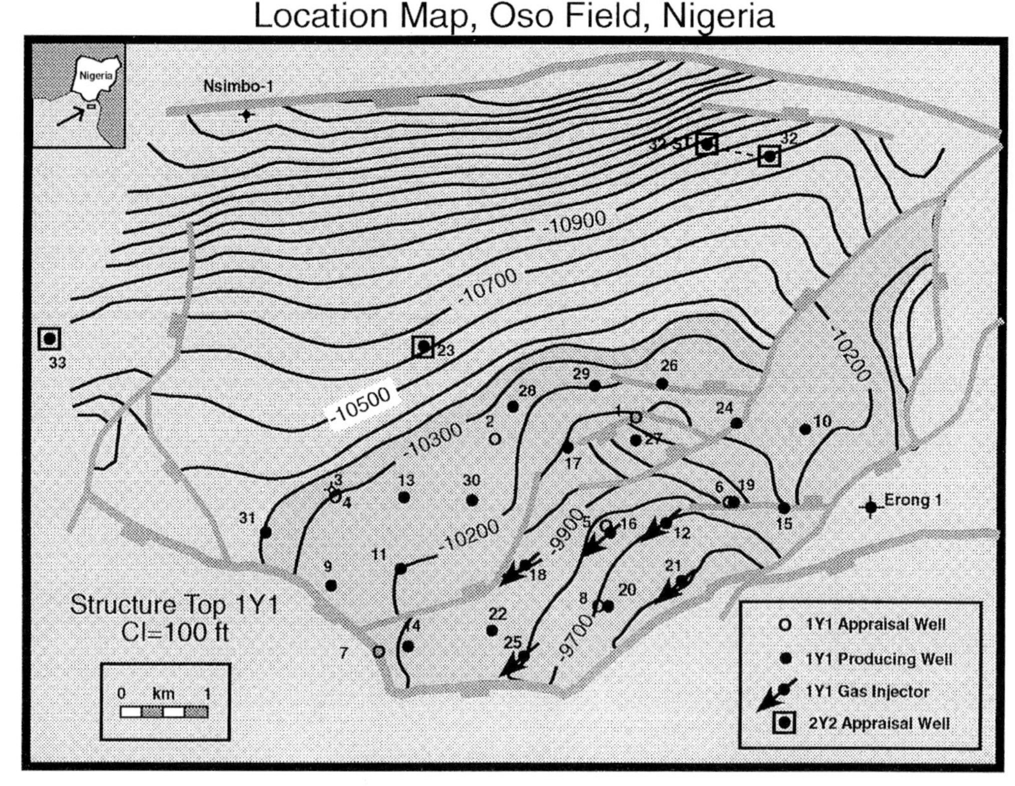

Fig. 1. Structural contour map on top of the 1Y1 reservoir zone and well locations, Oso Field, Nigeria. Contour interval is 100 ft.

in the range of 10–100 md, special core analyses indicate that these shales are effectively impermeable in a vertical sense. Log correlation indicates that these thick (5–50 ft) shales are generally laterally discontinuous, due to incision and cutouts by distributary channel-fills.

This ancient deltaic system probably had some resemblance to the modern Niger delta, which is also a mixed tide-and wave-dominated system. However, the dominance by distributary channel-fills and coarse grain size of the channel bedload suggests that the ancient 1Y1 delta was more fluvially-dominated than its modern counterpart. This may be due to a higher sediment supply in the Late Miocene than at present. In fact, many of the eastern Niger delta distributaries today are inactive or less active than earlier in the Holocene and Pleistocene (see NEDECO 1961; Oomkens 1974).

In addition to the depositional complexity noted above, several intra-1Y1 faults have been penetrated, and associated fractures and deformation bands described. These features do not appear to be initially affecting producibility of the field but may become more important as the field reaches later stages of production.

Dipmeter and borehole image data sets

As the Oso Field was drilled, the stratigraphic complexity was recognized after difficulty in correlating logs from the first few wells. Once a commitment was made to understand this reservoir in detail, several wells cut conventional core, and logging with electrical imaging tools (notably, the Formation MicroScanner (FMS – of Schlumberger)) was done to extrapolate environments away from cored intervals. As the reservoir model was developed, the later stage drilling did not need the extra detail of these dense data sets. More economical drilling was done with oil-based muds, and the Oil Based Dipmeter Tool (OBDT) of Schlumberger, was used to get directional data in order to define sand body geometries.

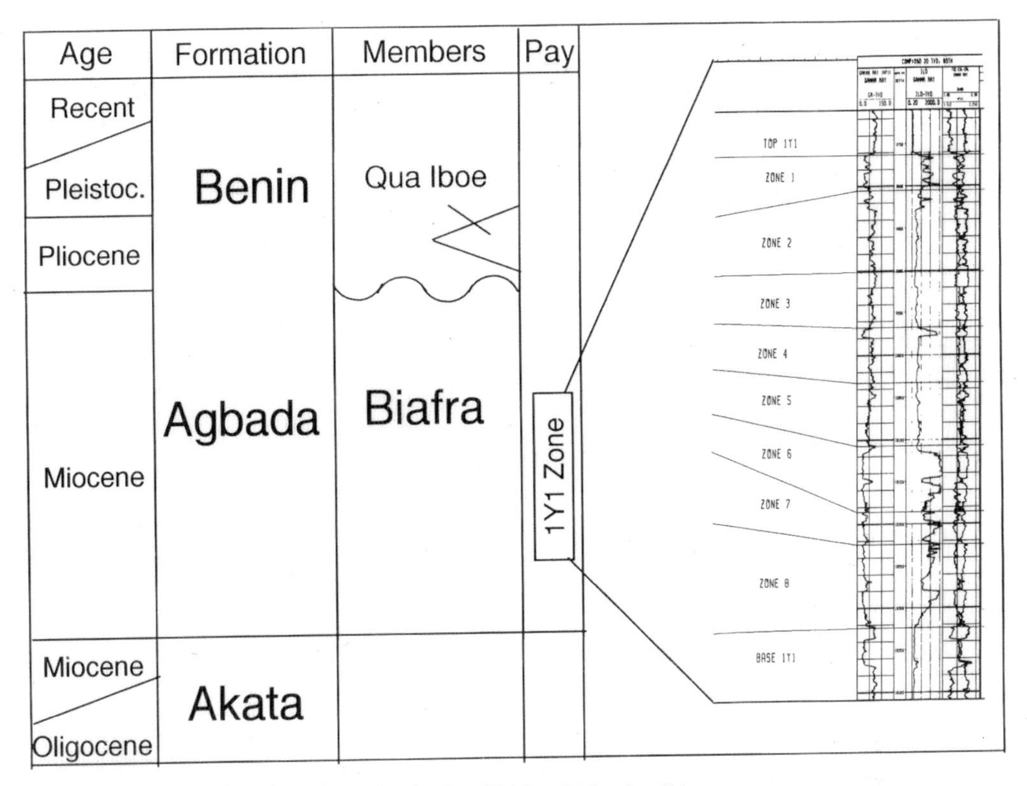

Fig. 2. Summary stratigraphic column for the Oso Field and Nigeria offshore.

The 'standard' dipmeter tools used by the logging industry gather the same data as the electrical imaging tools; they just have fewer sensor buttons and so gather less data. Typical data collected for the Oso Field by these tools are shown on Fig. 5. The first four columns show a data set collected by the OBDT tool; only four curves, each orthogonal to its neighbors, are gathered. The fifth column shows the raw FMS data for one pad – 16 button curves are displayed. The sixth column shows a false-color image of the FMS data. The right-hand column shows several dips which were hand picked from the image data and presented as a 'tadpole' plot. The 'body' of the dip 'tadpole' indicates the dip magnitude from 0–90°, and the 'tail' points to a compass direction indicating the azimuthal orientation of the dipping plane.

Any OBDT dips picked from this interval would be generated by mathematical algorithms which look for planar surfaces intersecting the four button curves. Such algorithms work well in sediment where there are consistent resistive patterns such as the lower half of the display on Fig. 5. When the sediments consist of more complex patterns of electrical conductivity, the computed dip results become less reliable. In the upper portion of Fig. 5 there are some intervals of apparent soft sediment deformation (for about 2 ft around 12 716 ft). Interpretations can be made from the image which are impossible from the OBDT curves.

Oriented dense data sets from dipmeters and imaging tools can be used to investigate a broad range of geologic features – structural as well as sedimentological. When used in combination with other well data such as cores and conventional logs, these data can be invaluable in defining complex depositional systems such as the Oso Field.

Dipmeter/FMS Interpretion

Structural Information

Regional structure dip, faults, and fractures are examples of the types of structural data which can be derived from dipmeter and image data sets. Figure 6 shows the lower part of an

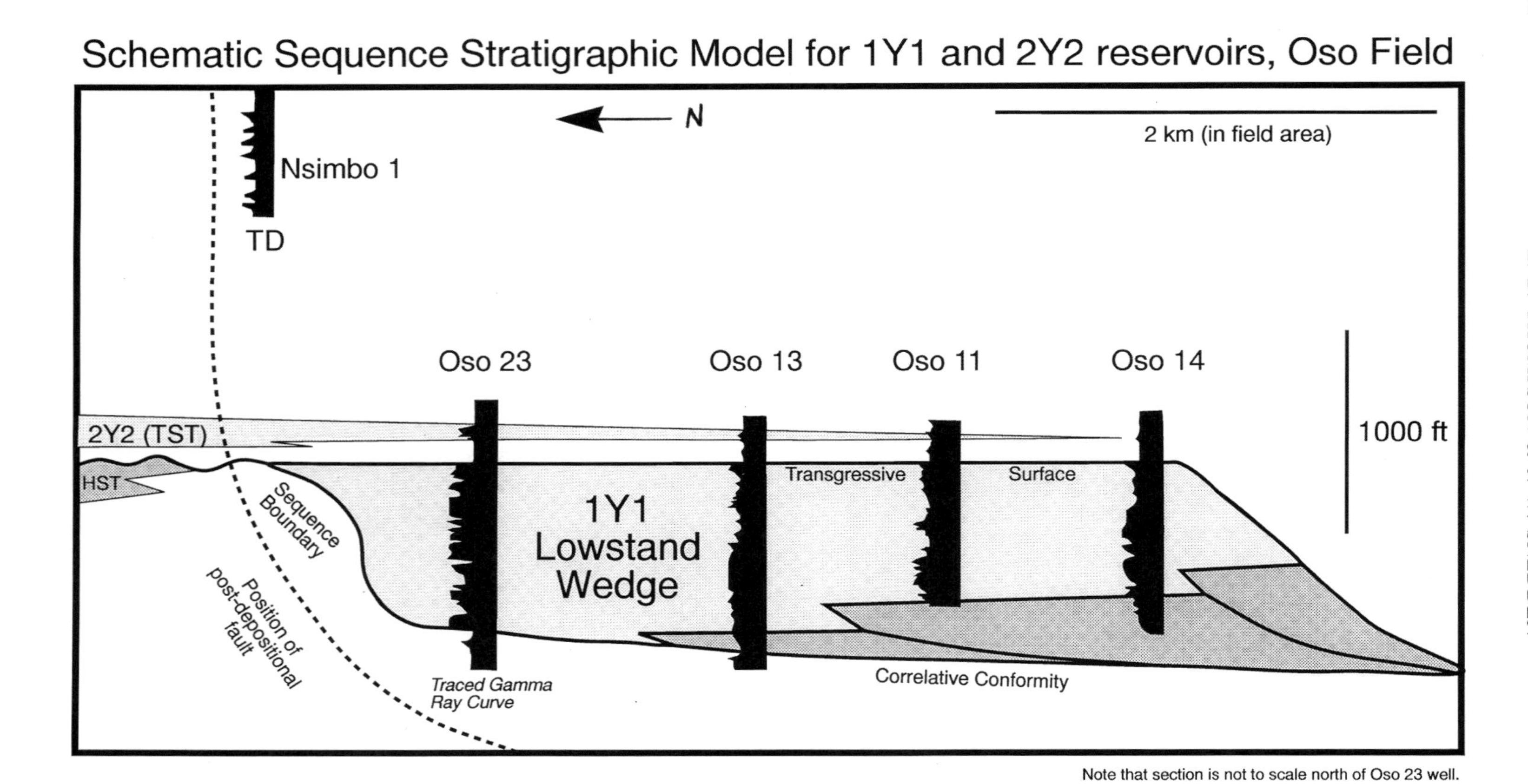

Fig. 3. General cross section for the 1Y1 and 2Y2 reservoirs.

1Y1 Reservoir Depositional Model, Oso Field

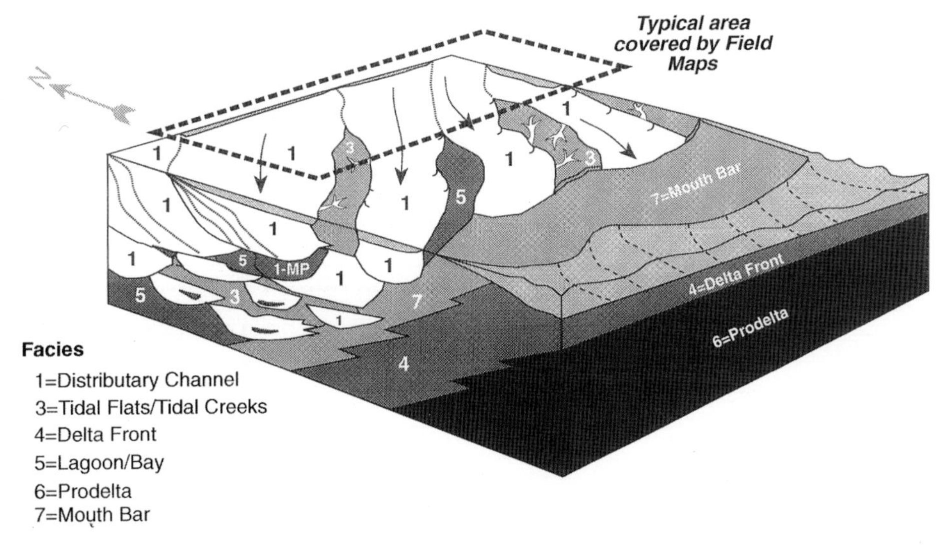

Fig. 4. 1Y1 reservoir depositional model.

Table 1. *Oso 1Y1 facies and characteristics*

Facies	Zone	Characteristics	Palaeoenvironment
1	1Y1	Course-medium-grained sandstone, locally conglomeratic. Cross-bedded variant known previously as Facies 2. High permeability and moderate porosity.	Distributary channel-fill
3	1Y1	Fine-to very-fine grained sandstone, often interbedded with shale in rhythmic fashion. Shales usually 20% of gross interval. Heterolithic, occasional high permeability.	Tidal creek channel (TC) And sandy tidal flat (TF)
4	1Y1 & 2Y2	Very fine-grained sandstone, interbedded with Shale (especially at base). Flat and low angle. Bedding dominant physical structure. Moderate porosity and permeability.	Delta front/shoreface
5	1Y1	20–50% shale. Highly laminated in rhythmic fashion. Few burrows, abundant slumping and soft-sediment deformation. Low net to gross. Non-reservoir.	Lagoon/bay
6	1Y1	Shale, with thin siltstones. Upward coarsening pattern. Non-reservoir.	Prodelta
7	1Y1	Medium-to fine-grained sandstone, well-sorted, compound unidirectional cross-bedding, climbing ripples. High porosity and moderate permeability.	Mouth bar

amalgamated channel sequence (from 10 855–10 875 ft) underlain by thin-bedded silts and shales (10 875 to bottom of image). These thin-bedded, originally horizontal sediments are usually best for using indetermining regional structural dip. In the illustrated case, however, these dips should not be used as indicators of structural dip, since they occur directly under a channel sand. The deposition of the sand was rapid and may have differentially compacted the underlying silts and mud, thus giving an incorrect orientation for structural dip. The preferred location for picking an unbiased regional dip is in thin-bedded material which is deposited on top of the main deltaic section and is thus unaffected by the channel sands.

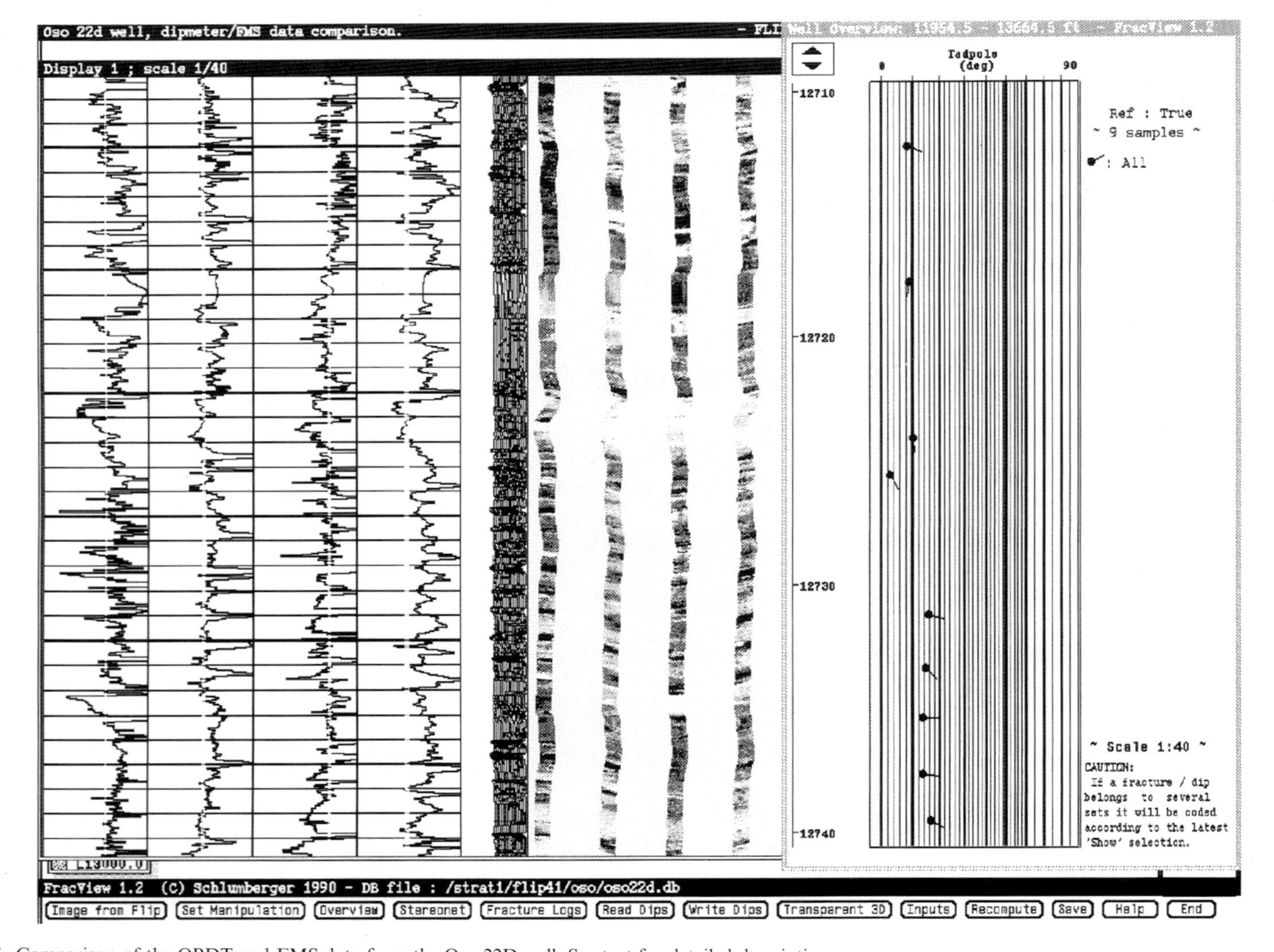

Fig. 5. Comparison of the OBDT and FMS data from the Oso 22D well. See text for detailed description.

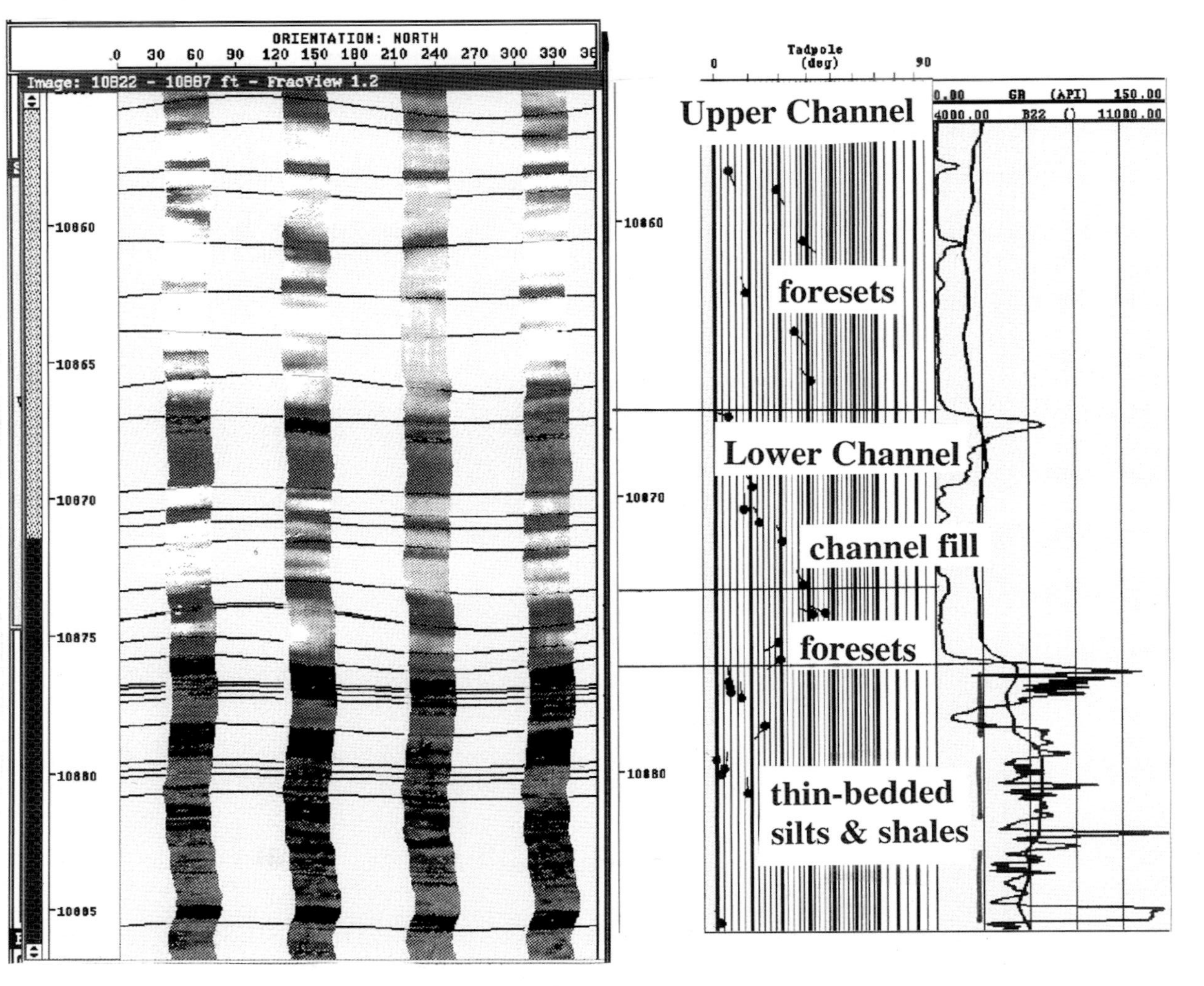

Fig 6. FMS image of amalgamated channel sands and underlying thin-bedded silts and shales. See text for details.

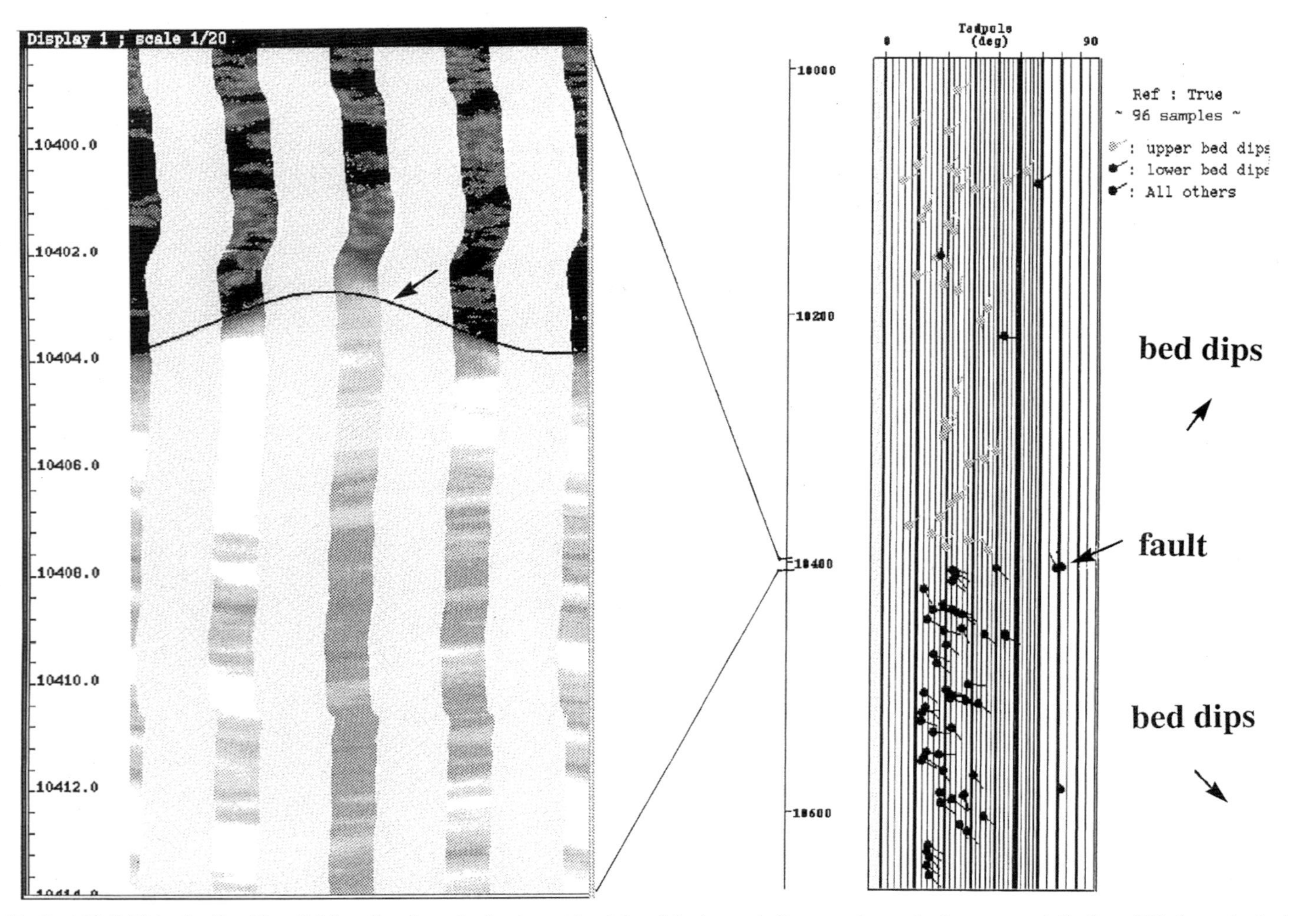

Fig. 7. Fault at 10 404 ft in the Oso 15 well. Note that the tadpole plot to the right of the image indicates a change in the structural dip from NE above the fault to SE below. The arrows indicate the fault on the image and its' orientation as shown by the dip 'tadpoles'.

Table 2. *Structural dip for 3 Oso wells*

Well number	Structural dip (above 2Y2 zone)	Structural dip (below 2Y2 zone)
24C	3 degrees at azimuth 15	9 degrees at azimuth 350
28E	2 degrees at azimuth 310	2 degrees at azimuth 310
31D	1 degree at azimuth 300	8 degrees at azimuth 30

Faults are often seen by the FMS (Fig. 7). The faulted surface is highlighted by a continuous line on the image, and its orientation is represented on the tadpole plot by the red features at about 10 404 ft which indicate an orientation of 70° dip to the NNW. The sediments below the fault are generally dipping to the southeast, whereas the sediments above the fault are dipping to the northeast. While image data allow for direct observation of a fault, such a feature often affects sediments for several tens to hundreds of feet around and thus is frequently found by a dipmeter.

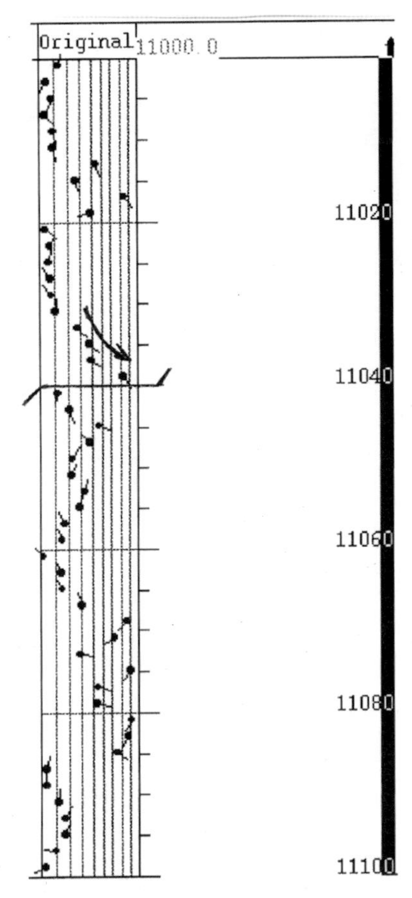

Fig. 8. Fault at about 11 040 ft in the Oso 24C well. Note the increasing dip magnitude above the fault highlighted by the large arrow. Low angle bedding above is flat, but below the fault is dipping northward (below 11 085 ft).

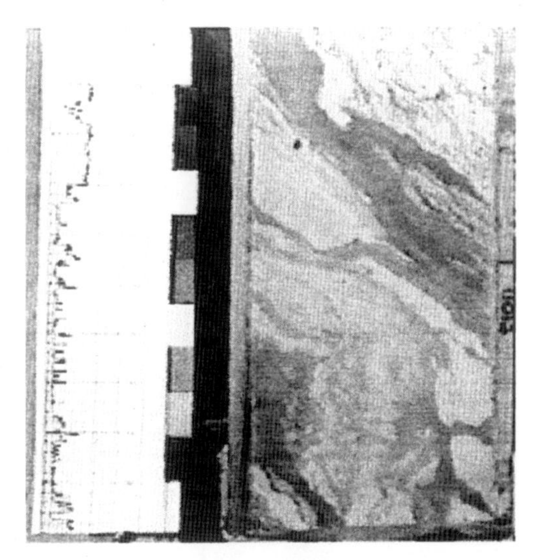

Fig. 9. Fault indicated by OBDT on Fig. 8 seen in core. Note that adjacent sediments have been disturbed by fault movement.

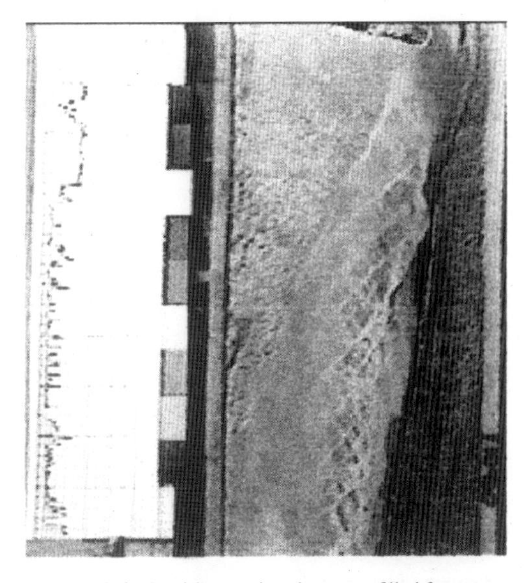

Fig. 10. A fault with associated cement-filled fractures, Oso 24C well. Dark sand on the right side of the core is oil soaked. The sand in the upper left portion of the core contains no oil.

OBDT data from three wells in the Oso Field show faults (Table 2). In the 24C well, two faults can be documented with the OBDT. The lower fault is also present in the conventional core (Figs 8 & 9). Note that the dipmeter fault indicators are necessarily more indirect than the image indicators; one needs to observe a change in structural dip or a steepening of bed dips adjacent to the fault (note arrow on Fig. 8) to make such an interpretation with dipmeter data.

Figure 10 shows a fault and associated cement-filled fractures in the core from the Oso 24C well. The fault is a thin fracture zone in which the individual sand grains are crushed as a result of the shear stress. This type of fracture can occur in a course-grained sand and results in a local permeability barrier. Note that the sand is oil-soaked to the right of the deformation band and shows no oil to the left. This type of feature should show up on an FMS electrical image, since the microresistivity of the sand should be quite different across the deformation band. The OBDT data consist of too few microresistivity curves to provide an interpretable pattern. As a result, the OBD would not show these fractures.

Sedimentological information

Depositional environments

Depositional facies are often seen easily with the FMS images but must be carefully inferred from the dipmeter and are best corroborated with conventional core. An interval in the 18 well (Fig. 6) shows an FMS image containing higher values of microresistivity above 10 875 ft and lower values below that point. The gamma ray curve seen in the rightmost column shows a blocky sand signature consistent with a channel deposit. Conventional core shows that the sandy

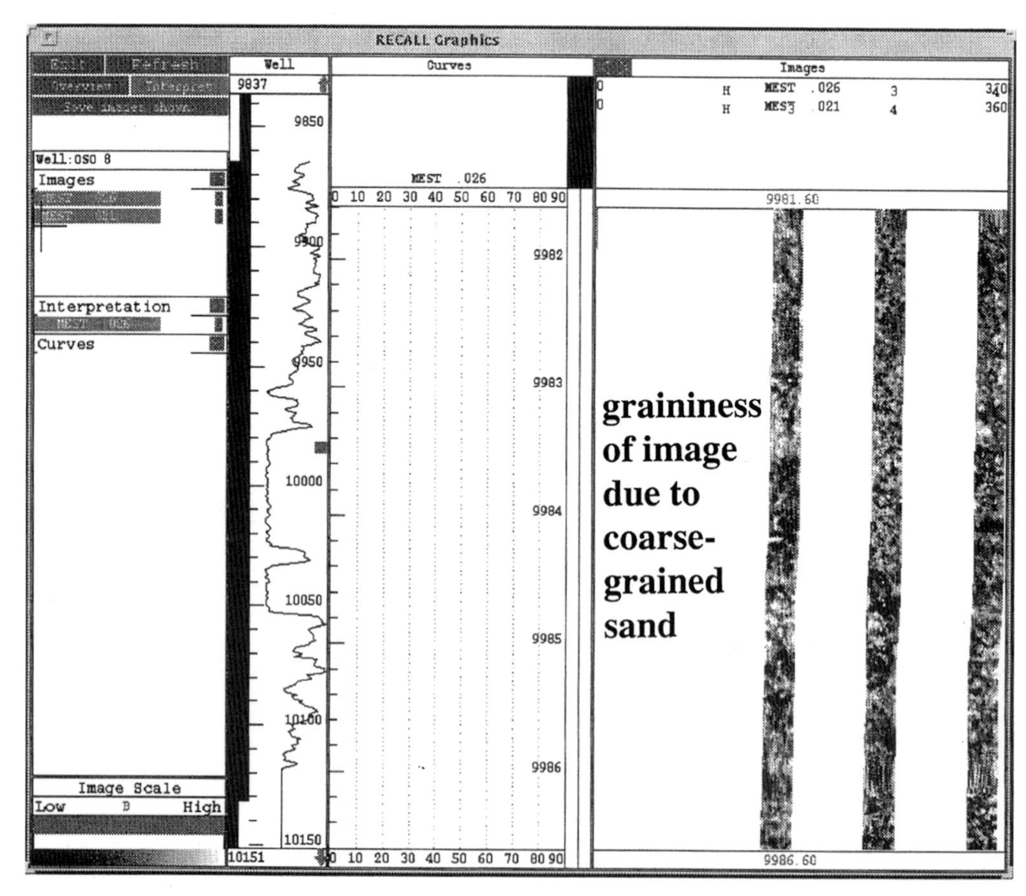

Fig. 11. Coarse grained, pebbly sand in upper channel of the 1Y1 zone, Oso 8 well. The coarse 'graininess' of the FMS image reflects the sediment texture.

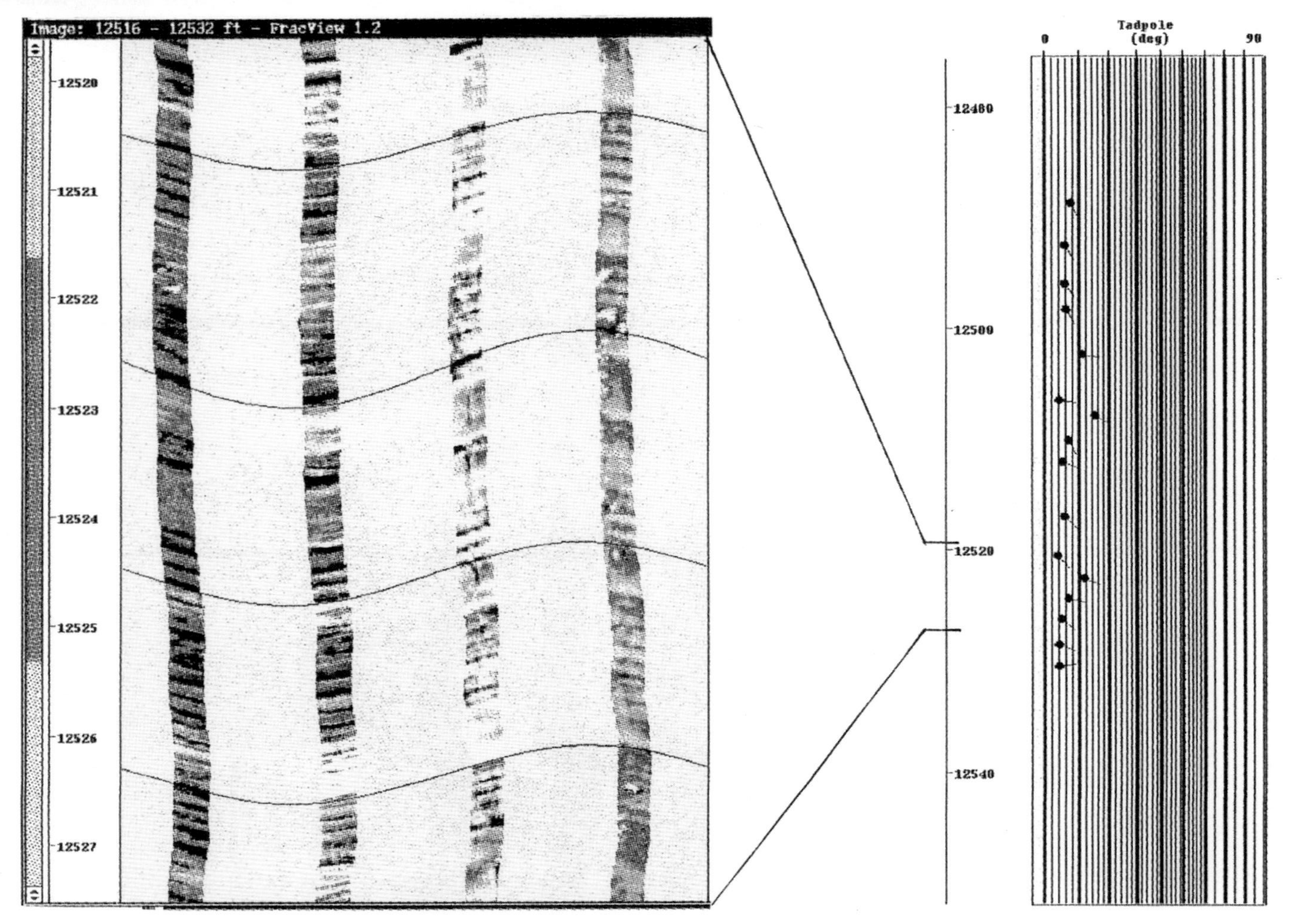

Fig. 12. Tidal bundles in a tidal creek deposit, Oso 22D well.

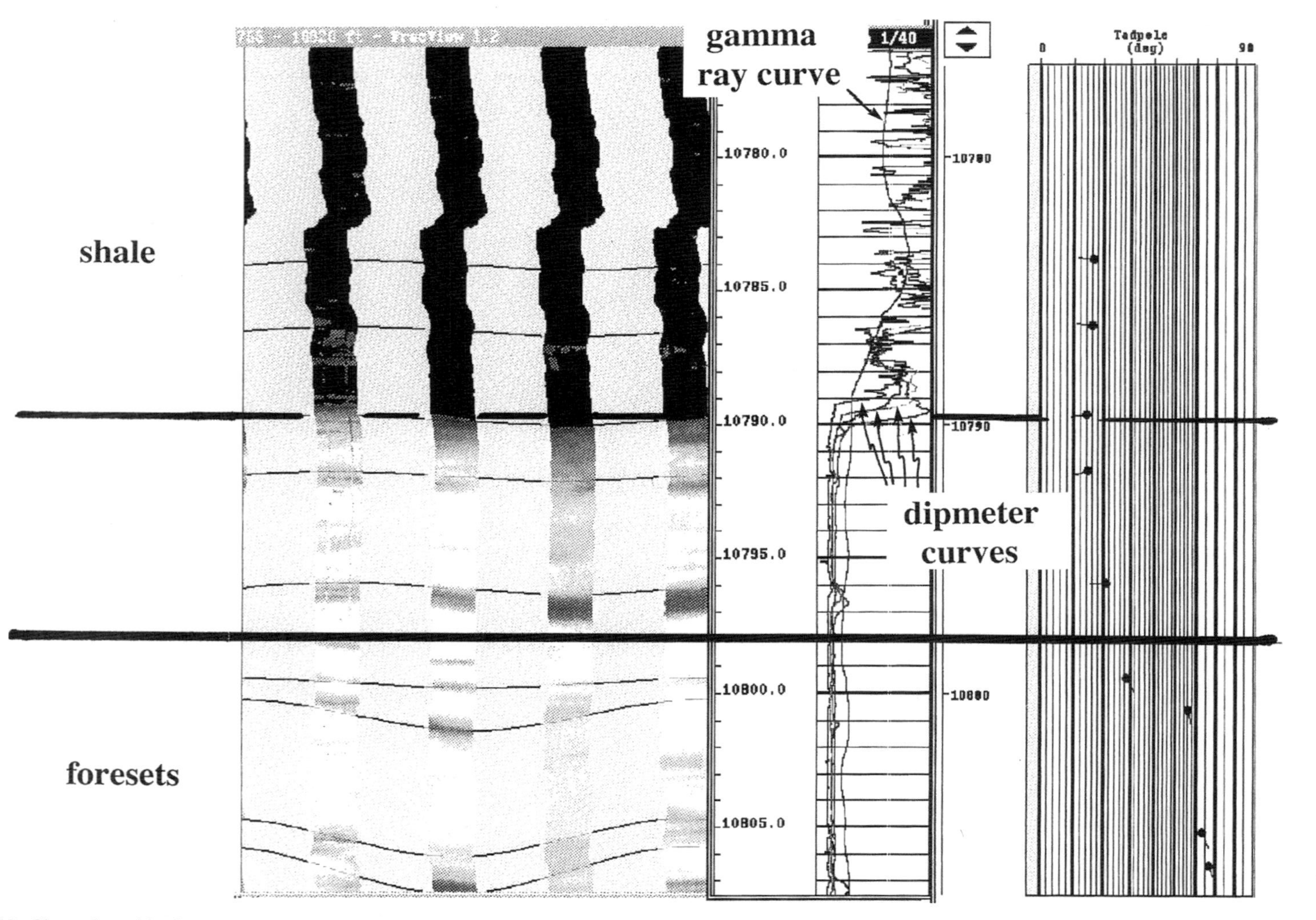

Fig. 13. Channel sand in the Oso 18 well. Image data show foreset beds (below 10 800 ft) and lateral accretion channel fill sediments (between 10 800 and 10 790 ft), but dipmeter curves (the four 'active' curves in the central column) miss the detail.

interval does consist of two amalgamated channel sands underlain by inter-distributary silts and shale. When the channels are composed of very coarse-grained sand, even the grainy texture is imaged by the FMS (Fig. 11). By comparison, small-scale tidal bundles (with bedforms as small as 2–3 cm) are clearly seen by the FMS (Fig. 12). Finally, shales may be contorted by soft sediment deformation, burrowed, or left intact as thinly-bedded hemipelagic sediments. All these details can be seen with the FMS but must be inferred at best by the OBDT.

When making depositional facies interpretations from dipmeter data, calibration to conventional core is often a necessity, since the four densely sampled microresistivity curves may not provide sufficient details of the sediments. The Oso 18 well provides an example of this limitation (Fig. 13). The image data indicate channel foreset beds (10 800 ft and below) and lateral accretion channel fill features present from 10 800–10 790 ft. The gamma ray curve (the sparsely-sampled 'lazy' curve in the central column) shows only the presence of the channel sand and overlying shale. The dipmeter data (the four 'active' curves in the middle column) show the sand and shale, but miss the subtleties necessary to distinguish the channel forest beds from

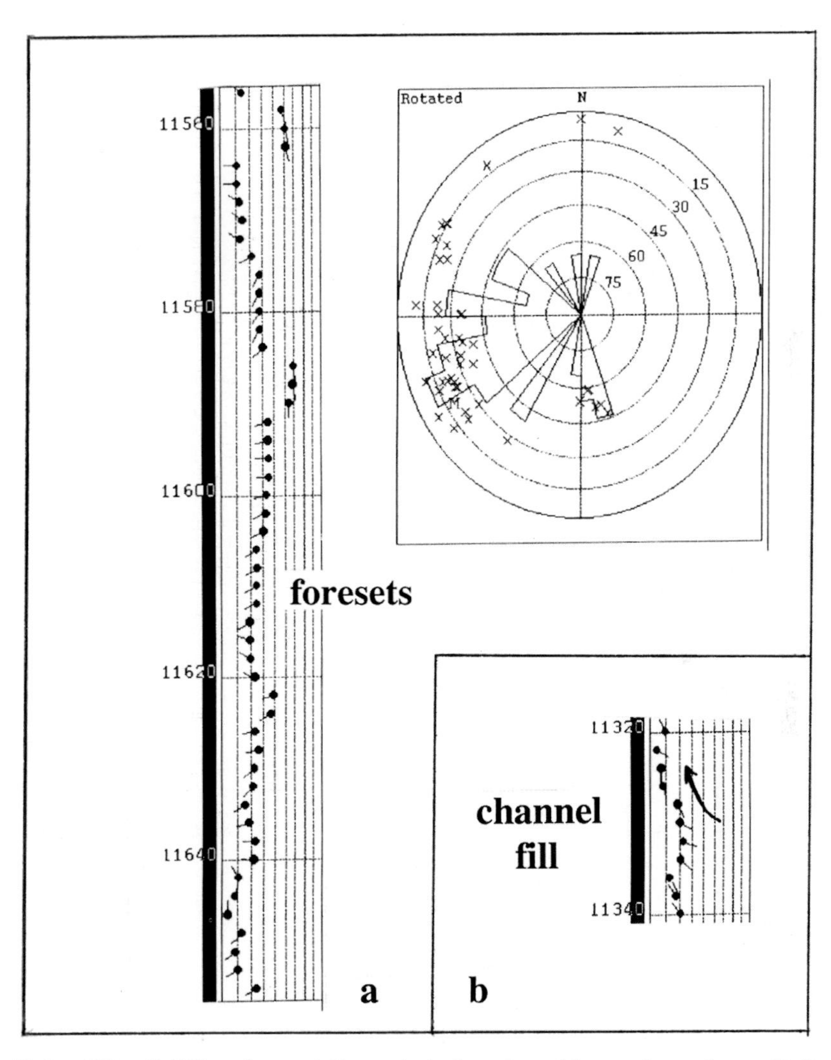

Fig. 14. (A) Oso 31D well, 1Y1 sand, zone 6. Foreset beds show channel flow to the southwest. (B) Oso 31D well, 1Y1 sand, zone 2. Channel fill pattern between 11 320 and 11 335 ft shows channel flow direction of southwest. Note that 8° of structural dip have been removed.

the channel fill features. Thus, while the dipmeter microresistivity curve data often yield a more detailed picture of the sediments than gamma ray data, this is not universally true.

Sand body geometries

Besides helping to recognize depositional facies, dipmeter and image data also allow one to make palaeocurrent observations and thus predict sand body geometry. The most commonly preserved palaeocurrent indicators are the channel trough crossbeds, which show downstream direction, and the lateral accretion bedding associated with channel abandonment, which show a cross-stream direction. FMS images show both current indicators (Figs 6 & 13). The lower portion of the channel sands contains cross-bed foresets seen as brighter (less-silty) intervals with higher angle bed dips (represented by higher amplitude sinusoids on the images and also seen in the dip tadpoles). These beds show the direction of flow in the channel, while the upper portion of each channel section shows a cross-channel pattern of fining-upward sediments (increasingly silty images with lower angle sinusoids and dip tadpoles which decrease in amplitude upward and are orthogonal to the foresets).

Dipmeter microresistivity curves under-sample both channel foreset and channel features, but may gather enough data to allow for an accurate interpretation fill (Fig. 14). There are

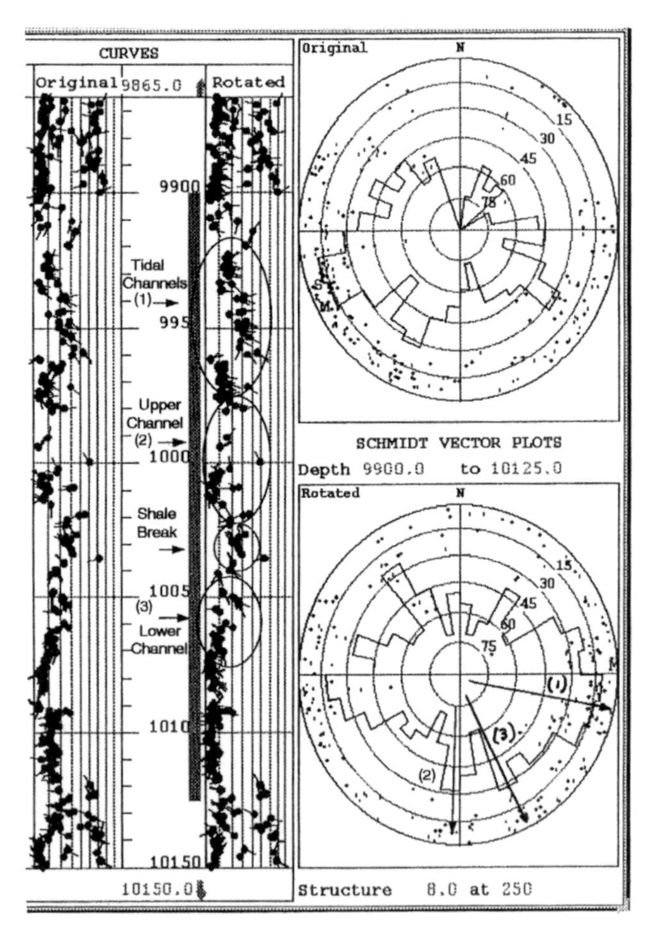

Fig. 15. 'Classic' dipmeter interpretation in the Oso 8 well. A structural dip of 8° magnitude, azimuth 250° is removed from the rotated Schmidt vector plot. Tidal creek channels (1) above the distributary channels are shown oriented in an ESE direction. The upper channel (2) flowed SSW and shows a good channel fill pattern at the top. The lower channel (3) was flowing to the SSE.

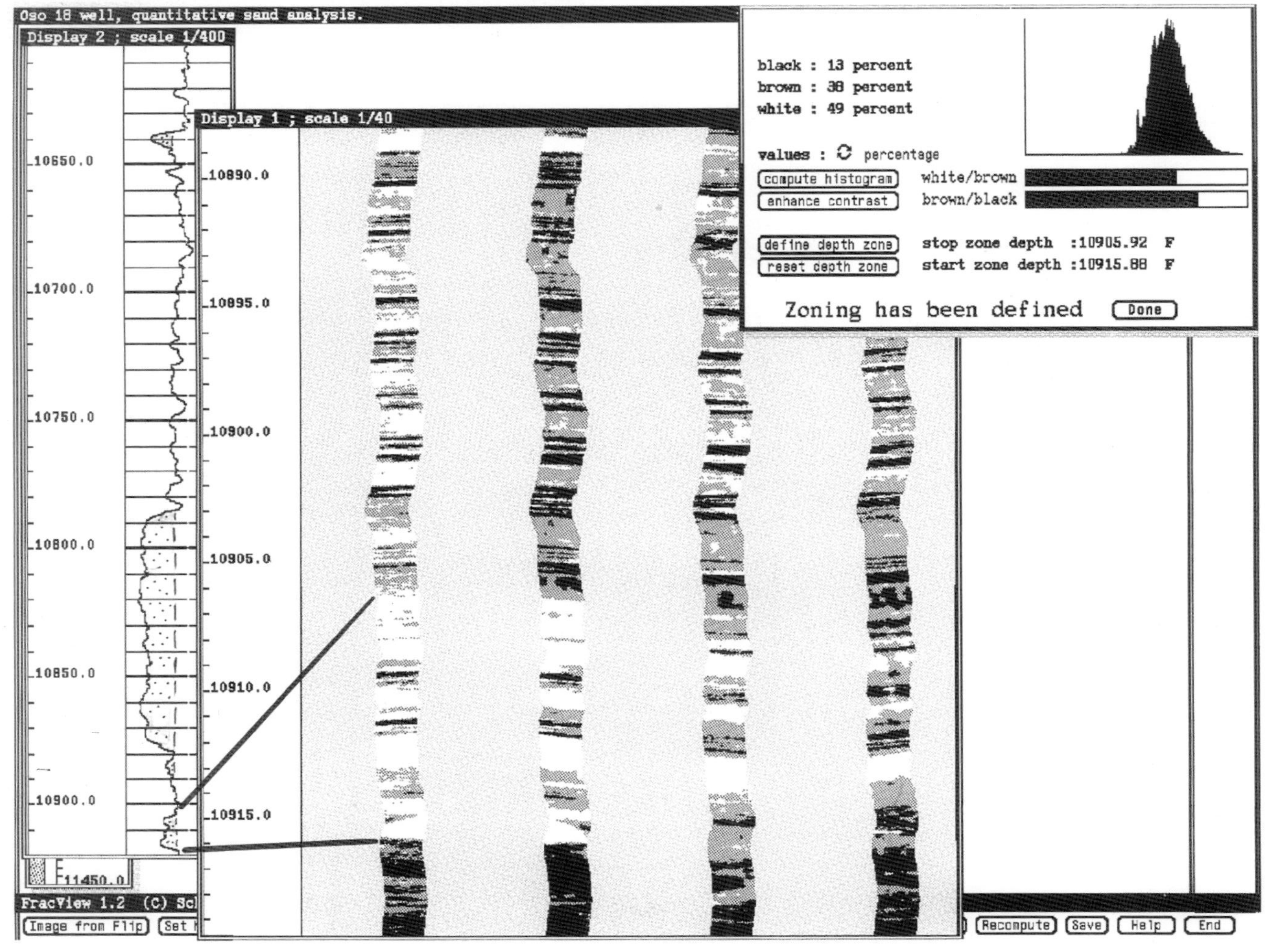

Fig. 16. Net sand calculations from FMS data, Oso 18 well. See text for details.

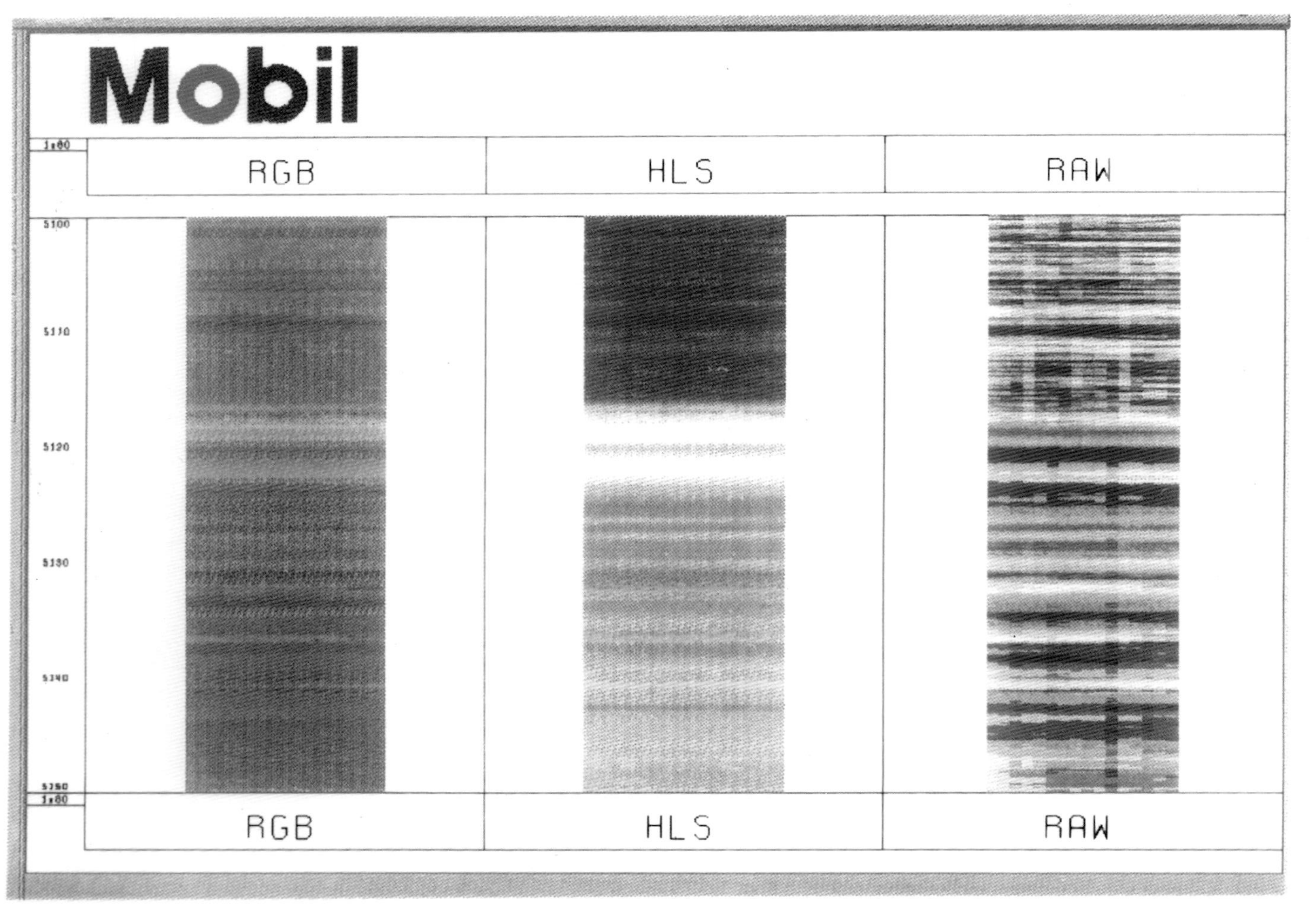

Fig. 17. Red-Green-Blue (RGB) and Hue-Lightness-Saturation (HLS) images of a single FMS pad convolved with gamma ray and density data. See text for details.

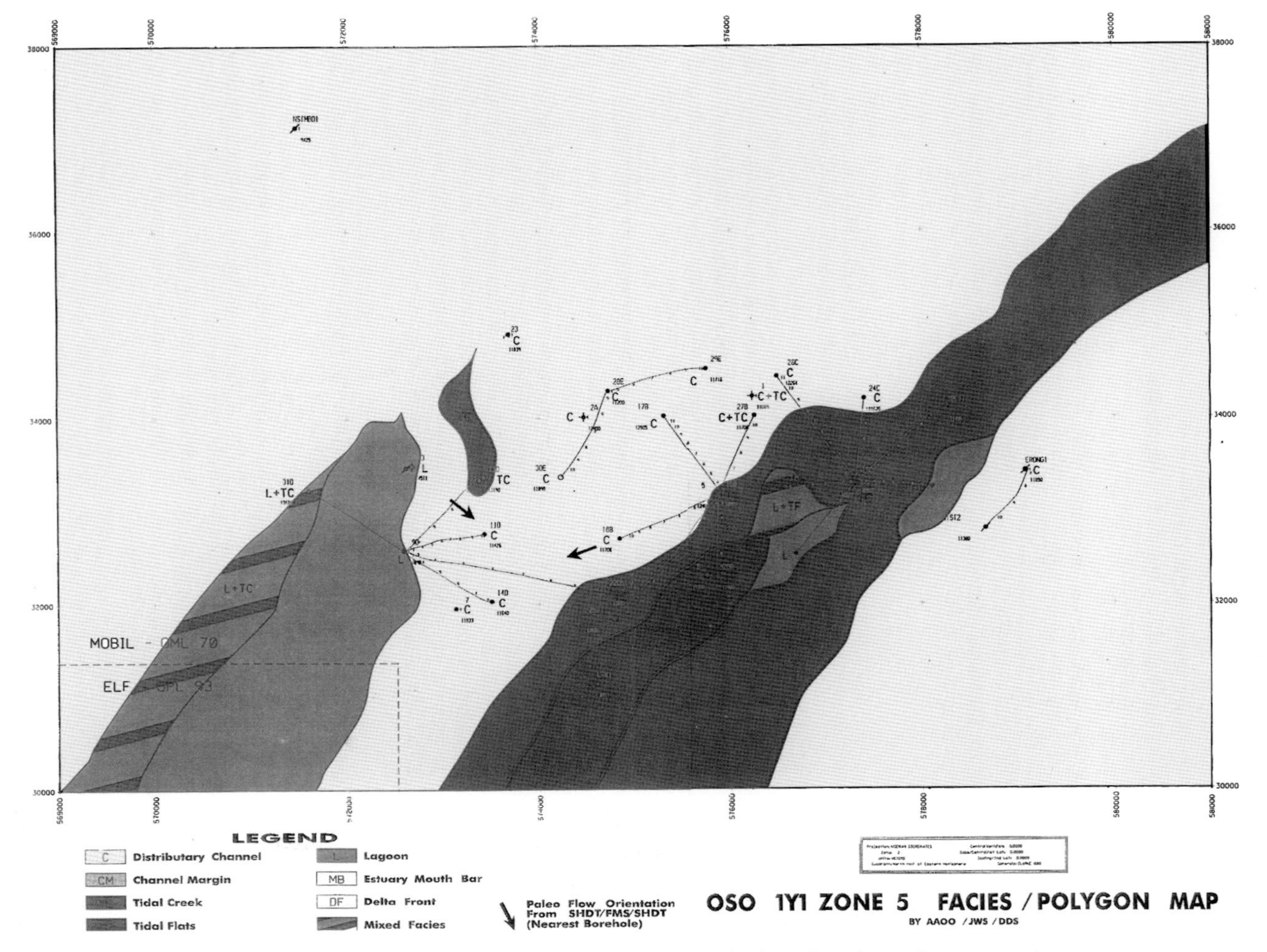

Fig. 18. Map of the Oso Field, 1Y1 reservoir, zone 5 showing facies distributions and channel orientations (arrows).

other depositional features which can produce the same dip patterns, so detailed interpretations such as these should be done with great care when only presented with dipmeter data. In the Oso Field, conventional core was available, and specific correlation of dips to core sediments in the Oso delta also allowed for orienting depositional direction – the tidal creeks with their thin-bedded tidal sediments (Fig. 12). Figure 15 shows a 'classic' dipmeter interpretation of river channel and tidal creek dip signatures. Note that the tidal creek dip signature consists of good quality dips all with the same orientation reflecting the thin-bedded, consistent nature of the sediments. The tidal channels are often roughly orthogonal to the main distributary channel sands.

Quantitative reservoir analysis

The dipmeter and FMS/FMI tools collect relative microresistivity data, not true resistivity. These data are not consistent from well to well, and are sometimes not consistent within one well or even between pads on a single tool. If manipulated carefully, however, these data can be calibrated, resampled, or both, in order to generate quantitative information for reservoir analysis.

Figure 16 shows a portion of the Oso 18 well. An apparent channel sand is shown on the gamma ray curve (on the left of the figure) from 10 785–10 880 ft. The sand present in the 10 905–10 915 interval is not as clean as the channel, but a 10 ft thickness is indicated. A more detailed image of that sand is seen on the FMS data. By setting cutoffs on the image data to model sand (white), silt (gray), and shale (black; see insert at upper right), the sandy interval is documented to contain only 49% net sand over the 10 ft. This technique is best done by calibrating the FMS image to conventional core to get the most accurate cutoffs, then extrapolating away from the cored intervals.

Another technique to quantify physical properties using image logs is to use satellite false color technology (Fig. 17). Every wireline lot curve consists of a set of values (e.g. 0–150 api units for a gamma ray log). Any set of log values can be used to create a false color band (say, of red color). By convolving three such data sets (for example, the microresistivity data of an FMS with gamma ray, density, or sonic log) into one image, one can create a false color display called an RGB (for Red, Green, and Blue) image or an HLS image (for Hue, Lightness, and Saturation). Such images can also be created from derived curves such as water saturation or porosity. By creating this presentation with the image, a more accurate picture and measure of the small-scale distribution of the attribute can be measured.

Summary

Analysis of depositional environments coupled with palaeocurrent information to constrain sand body geometry allows one to construct a detailed stratigraphic model of the Oso delta. Maps of each stratigraphic layer are constructed using the image and dipmeter data, conventional core data, and wireline data available at the wells (Fig. 18). These layer maps show a set of facies 'polygons' existing in a distribution which is similar to the modern Niger deltaic system (cf. NEDECO 1961). Channels border tidal flats and tidal creeks which split and subdivide downstream.

In addition, measurement of the channel-fill dimensions demonstrates that they have widths and thicknesses falling in the range expected for straight to moderately sinuous channel systems of the modern and Miocene Niger delta (cf. Weber 1971). Most 1Y1 channel-fills have thicknesses of 30–90 ft (TVT), which approximates the depth of modern distributary channels of the Niger delta (NEDECO 1959; 1961). Mapped (non-amalgamated) sand body widths typically range from 1000–10 000 ft.

This detailed depositional model has been coupled with petrophysical and engineering data for each facies to quantitatively map volumetrics, to predict flow connectivity within the reservoir, and to guide producibility by locating gas injectors at optimum positions (Fig. 1).

Thus, the FMS/FMI image data serve as the connecting link between core and wireline curve data to allow for analysis of disparate data sets on a common basis. The image data bridge a broad range of scales for reservoir analysis – from thin-section to basin in scope. Emerging techniques of using the images quantitatively further their utility in reservoir characterization.

The authors would like to acknowledge the aid and insights of several individuals: A. K. Bhatia, E. B. Ogiamien, and R. M. Vaught, all of MPN; T. W. Cooley and B. E. Welton (MEPTEC). The previous work of R. D. Kreisa and others of MEPTEC provided a firm basis for this reservoir characterization effort. The support of G. K. Baker and Mike Croft (MEPTEC) and V. K. Oyofo, D. O. Lambert-Aikhionbare, and J. Y. K. Blevins of MPN is also gratefully acknowledged.

References

DOUST, H. & OMATSOLA, E. 1990. Niger Delta. *In*: EDWARDS, B. D. & SANTOGROSSI, W. (eds) *American Association of Petroleum Geologists Memoir*, **48**, 201–329.

NEDECO (Netherlands Engineering Consultants). 1959. *River studies and recommendations on improvement of Nigeria and Benue*. North Holland Publishing, Amsterdam.

——1961. *The waters of the Niger Delta, Report on an Investigation*. Unpublished company report, 316.

ODIOR, G. E. 1992. Ubit Field. *Nigerian Association of Petroleum Explorationists, Transactions*.

OOMKENS, E. 1974. Lithofacies relations in the Late Quaternary Niger delta complex. *Sedimentology*, **21**, 195–222.

WEBER, K. J. 1971. Sedimentologic aspects of oilfields in the Niger Delta. *Geologie en Mijnbow*, **50**(3). 559–576.

Contributions of acoustic imaging to the development of the Bruce Field, Northern North Sea

D. N. PARKINSON,[1,4] R. J. DIXON[2] & E. J. JOLLEY[3]

[1] *Western Atlas Logging Services, 455 London Road, Isleworth, Middlesex TW7 5AB, UK*
[2] *BP Research Centre, Chertsey Road, Sunbury-on-Thames, Middlesex TW16 7LN, UK*
[3] *BP Exploration, Farburn Industrial Estate, Dyce, Aberdeen AB2 0PD, UK*
[4] *Present Address: Parkinson Geoscience Ltd, 39 Sydney Road, Richmond, Surrey TW9 1UB, UK (e-mail: neil.parkinson@dial.pipex.com)*

Abstract: An interpretation of four acoustic images from the Bruce Field is presented. It provides an example of what can be achieved using moderate-quality data from a North Sea siliciclastic reservoir. Faults have been detected and oriented using evidence from large-scale block rotations, small-scale drag, direct imaging of fault planes, identification of fault damage zones and breakout rotation. Fracture frequency is highly variable and is dominantly controlled by fault-related damage, though rock properties may also be important. Damage zones are some 25–50 m thick and appear to be concentrated on the hanging walls of their controlling faults. These damage zones are important barriers to flow. Fractures are orientated NNE–SSW. A stress-field map has been produced combining image and dip-log data. The dominant orientation of the minimum horizontal principal stress is parallel to the fractures, though it is influenced locally by structure. The combination of stress field and fracture studies predict the dominance of east-west fluid flow in the reservoir, 'baffled' by north-south fracture barriers. Palaeocurrent information has been used to help construct a suite of time-slice palaeogeographies which describe the sedimentary heterogeneity.

Good acoustic images are difficult to acquire, and are generally of inferior resolution to resistivity images of the same rock. They do have unique strengths: their 360° borehole coverage and sensitivity to borehole wall shape are particularly useful for the characterization of breakouts and near borehole-parallel fractures. They also have potential as a high-resolution porosity measurement (Hansen & Parkinson 1999). However, the principal reason for choosing acoustic rather than resistivity imaging in the North Sea, and in many other areas, is as an imaging technology for wells drilled with oil-based or other highly-resistive drilling fluids, where galvanic resistivity devices simply cannot operate.

Given the problems of acquisition, and the often indifferent character of the well-site result, purchasers of acoustic image data are entitled to ask whether the effort is worthwhile. The objective of this paper is to address this question by illustrating the range of information which has been extracted from acoustic images of the Bruce Field reservoir in the United Kingdom North Sea. As will be seen, Bruce does not present an ideal acquisition environment. Hence, the examples here can be taken as what is commonly possible with an acoustic image, rather than what can be achieved under ideal conditions.

Bruce is a giant gas-condensate field which lies on the western side of the Viking Graben. The area is structurally complex, possibly due to the transfer of extension between two major graben-bounding faults which set up a tectonic and palaeogeographic province known as the Bruce-Beryl Embayment (Fig. 1).

The principal reservoir rocks of the Bruce Field are sandstones of Aalenian to Bathonian age (Dixon *et al.* 1996). They were thus deposited contemporaneously with the Brent Group which forms many of the reservoirs of the North Sea. However, facies in the Bruce-Beryl Embayment contrast somewhat with the Brent Group, being ascribed to a range of fluvial and tidal, as well as shallow-marine, settings.

Image logs were, and are, seen as relevant to the following critical issues:

- Despite a high quality 3D seismic grid, the structural complexity of the area means that there continues to be much ambiguity in

From: LOVELL, M. A., WILLIAMSON, G. & HARVEY, P. K. (eds) 1999. Borehole Imaging: applications and case histories. Geological Society, London, Special Publications, **159**, 259–270. 1-86239-043-6/99/$15.00.

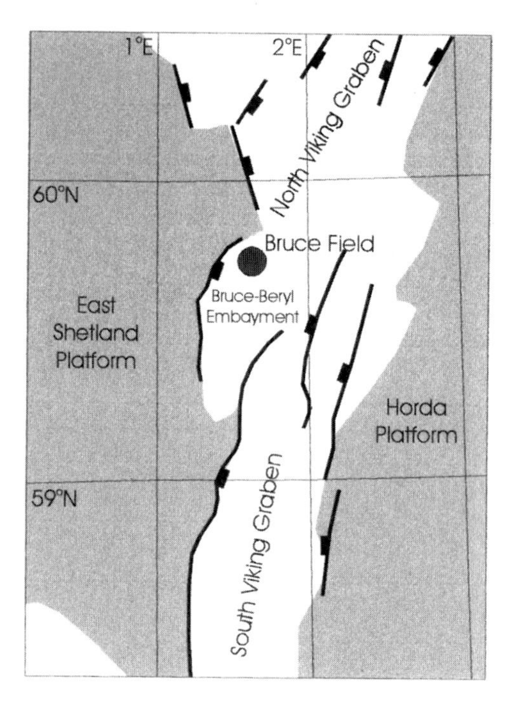

the structural interpretation, with obvious impact upon reservoir modeling and the siting of individual wells. Image logs have the potential to provide local structural dip control and information on the presence and orientation of faults.

- Fracture frequency is very variable. Fractures tend to impede fluid flow in Bruce, and at their highest frequency can eliminate wells from economic production or injection. Image logs have the potential to improve our understanding of the distribution, orientation and character of fractures.
- There is increasing awareness of the influence of the sub-surface stress field on fluid flow (e.g. Heffer & Koutsabeloulis 1995). Image logs have the potential to provide information on the orientation and magnitudes of *in situ* stresses.
- The facies and range of depositional environments in Bruce leads to considerable sedimentological heterogeneity. Image logs have the potential to provide palaeocurrent information to constrain the time-slice paleogeographic maps which are key building blocks of the reservoir model.

As we hope to demonstrate, the success of acoustic imaging logs in addressing these issues has resulted in their adoption as a key part of the Bruce logging suite. As we gain confidence, these logs are also influencing operational decisions: fracture frequency, as measured from image logs, has been used to influence completion strategy; a quick-look structural interpretation of one of the wells described here had a material impact on the planning of a side-track; and image logs have been used to locate and orientate rotary side-wall coring points.

Data acquisition

Most acoustic imaging tools (sometimes known as 'Borehole Televiewers') operate on similar principles. The data for this study were acquired using Western Atlas' Circumferential Borehole Imaging Log (CBIL^{TM}). This comprises a single acoustic transceiver rotating at six revolutions per second, acquiring 250 data samples per revolution. At normal logging speeds of 10 ft/min the result is a spiral of data with a pitch of approximately 8 mm (Fig. 2). Each 'sample' comprises two measurements: the amplitude of the returning signal, and the two-way travel-time of the acoustic pulse to the borehole wall and back to the transducer. From these two measurements, two images can be constructed, each giving an 'unwrapped' 360° picture of the inside of the borehole (Fig. 3).

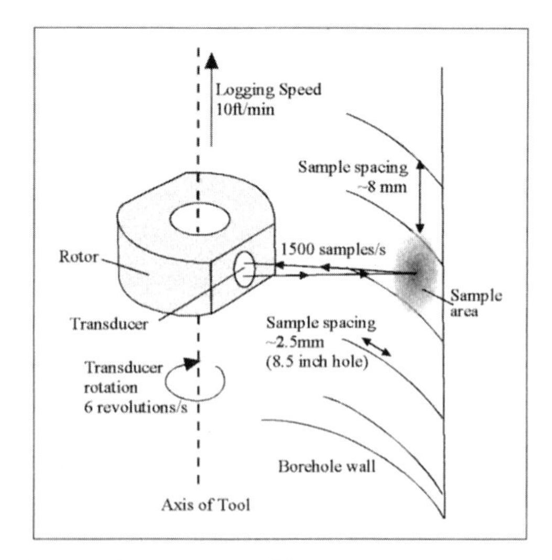

Fig. 2. Schematic of CBIL^{TM} operation (after Western Atlas, 1991). An acoustic pulse is transmitted from a rotating transducer. This is reflected from the borehole wall and the travel time and amplitude of the reflected pulse are measured. The combination of transducer rotation and translation of the logging tool creates a spiral of data from which amplitude and travel-time images can be created.

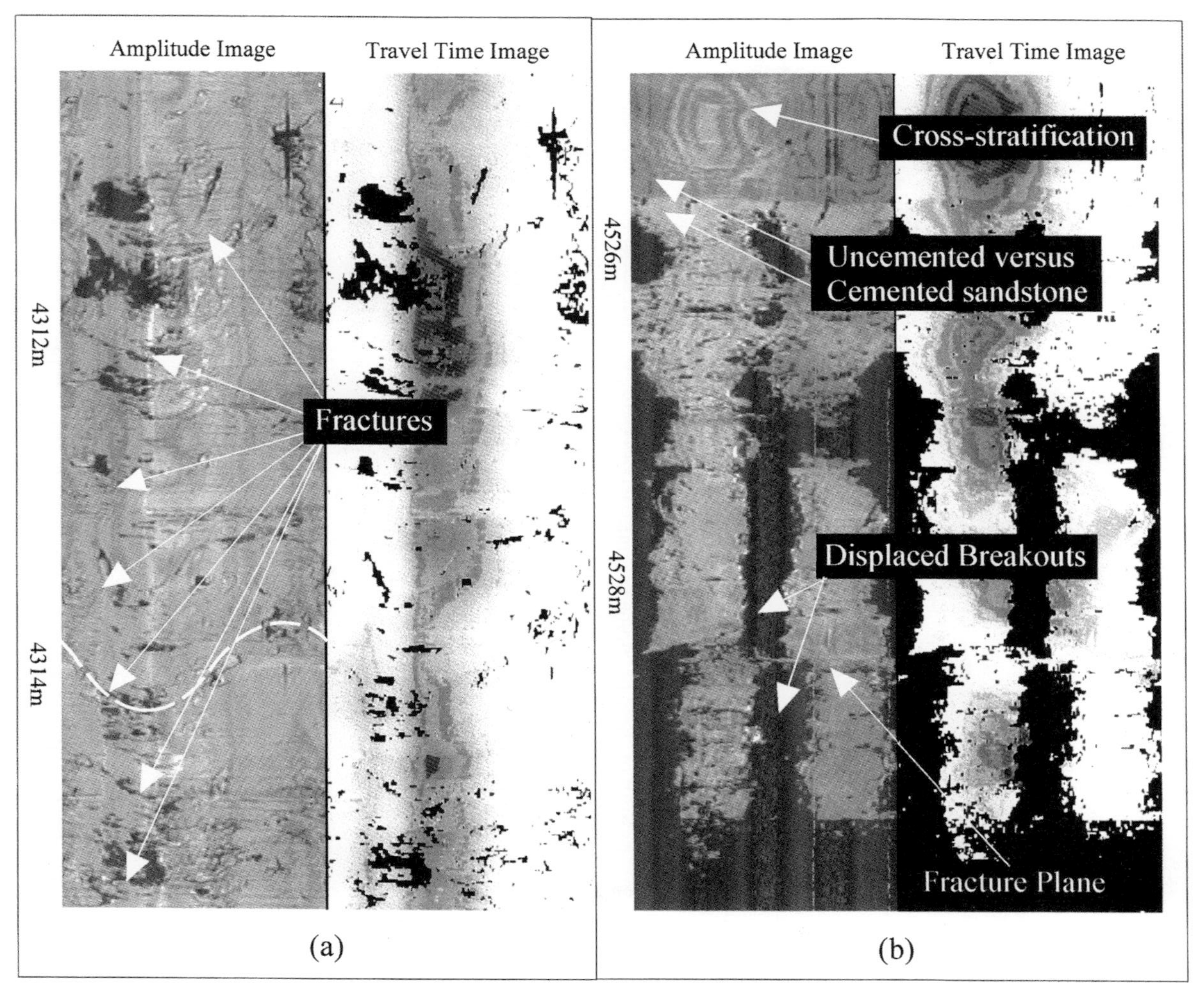

Fig. 3. Amplitude and travel-time images of inferred fault zones in well 9/9a-A17Z.

The amplitude image is of the greatest geological significance. Low amplitudes (conventionally shown dark on the image) result primarily from low acoustic impedance contrast between the drilling fluid and the borehole wall. Hence low velocity, low density, high porosity rocks tend to appear dark on the image. High velocity, high density, low porosity rocks tend to appear light. Returned acoustic amplitudes will also be low where the borehole wall is rough and acoustic energy is dispersed. This may be a result of lithology and/or of drilling damage, and may or may not be of geological significance.

The travel-time image may be thought of as a high-resolution caliper. It is used in the calculation of dip and in the measurement of 'breakout' geometry for stress-field studies. It may provide insights into the character of fractures (see below).

It is an unfortunate truism that the best image logs are obtained from poor reservoirs. Low porosity fractured limestones yield excellent images. In the North Sea, the Rotliegendes yields outstanding images showing the internal structure of aeolian bedforms in great detail. Conversely, poorly consolidated, high porosity rocks give rise to both low acoustic impedance contrasts between the borehole fluid and the rock, and rugose boreholes. Both of these phenomena tend to reduce the amplitude of the returning acoustic signal. In addition, the internal structure of high-permeability sandstones may be masked by mud-cake build-up.

Bruce tends to fall towards the 'good reservoir-poor image' end of the spectrum, with particular problems being experienced with mud cake, which impact our ability to undertake detailed sedimentology. In addition, problems have occasionally been encountered with spirally-drilled borehole, and images from the more deviated wells have been impaired by tool eccentralization.

Structure, faults and unconformities

A primary aim of image interpretation on Bruce is to contribute to the structural model by providing localized dip information, and identifying and orientating faults. Well 9/9a-A17Z provides examples of several approaches to fault characterization.

Detection of large-scale block rotation

We have found that continuous plots of dip azimuth provide a useful initial overview of the data, though these plots should be used with care as the lack of a uniform depth scale can give a misleading view of the data, and dip angle information is not represented. The plots are constructed by plotting arrows of unit length end-to-end in the direction of dip azimuth, so that a series of westerly dips would yield a line running from right to left on the page. If followed by northerly dips, the line would turn to run towards the top of the page. In the case of 9/9a-A17Z, distinct large-scale changes in dip azimuth were detectable at around 4528 m and 4313 m (Fig. 4). These changes suggest the presence of faults or unconformities and one would wish to investigate the log in more detail around these depths.

Direct imaging of the fault plane

It is rarely possible to demonstrate displacement unequivocally on images: we normally *describe*

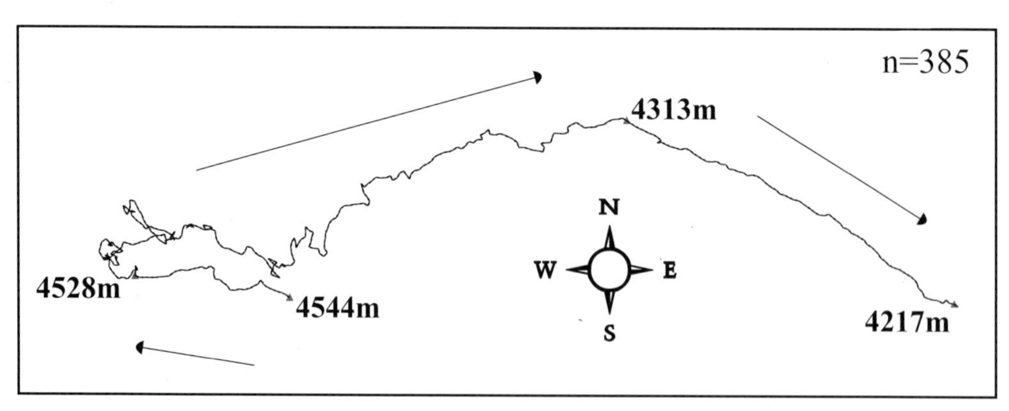

Fig. 4. Continuous plot of bedding dip azimuth data for well 9/9a-A17Z, showing change from westerly to easterly dips around 4530 m and change from north-easterly to south-easterly dips around 4313 m. Note that data are plotted from the base of the well, up-hole.

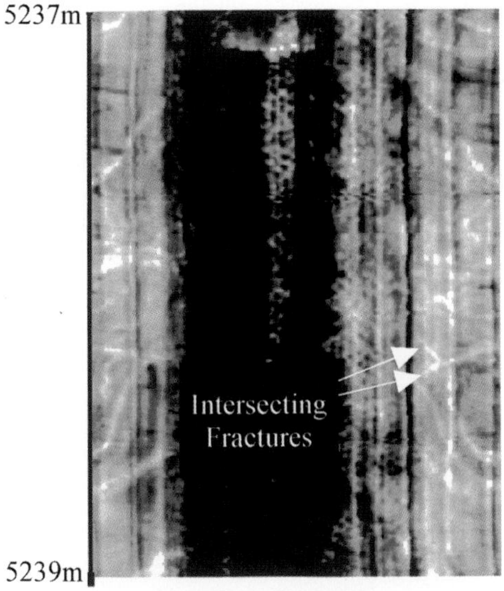

Fig. 5. Typical appearance of Bruce fractures on an amplitude image. This image (from Well 9/9a-A15) shows very high fracture densities, though image quality suffers badly from tool eccentering which causes the dark vertical area near the centre of the image.

fractures and only *infer* faults. A17Z provides examples of the type of inference which is common. Fractures occur at both of the depths highlighted on the dip azimuth plot (Fig. 3). The fractures around 4313 m are different in appearance from 'normal' Bruce fractures. 'Normal' fractures (discussed in more detail below) are 'bright' on the amplitude image and do not appear on the travel time image (Fig. 5). The fractures around 4313 m are dark on the amplitude image and also appear discontinuously on the travel time image (Fig. 3a). We may infer from this that the fractures around 4313 m are either open or have been excavated in some way by the drilling process (such as by the wash-out of fault-plane gouge). The coincidence between these distinctive fractures and large-scale changes in structural dip azimuth strongly suggests that the fractures represent a fault zone. If this is true, then we have information on the fault plane orientations directly from the image: the mean dip of the fractures in the vicinity of 4313 m is 113/71°, with a spherical dispersion of 13° (calculations according to the method of Fisher 1953).

Fault-related deformation and roll-over

The interval above 4313 m shows a systematic downwards increase in dip angle (Fig. 6a). This pattern is commonly generated by fault-plane drag or by roll-over into listric faults. In this case we believe that we have directly imaged the fault zone and that this dips ESE, in the same direction as structural dip. Geometrically, this implies a normal fault with the downwards-increase in dip associated with drag. If we assume that bedding dip angle is constant with constant distance from the fault plane, then we can use the drag pattern to place a minimum constraint on fault displacement (Ozkaya & Mattner 1996). This is in the region of 6–20 m depending upon assumptions about 'regional' bedding dip.

Rotation of breakouts

Acoustic images commonly reveal broad zones of borehole failure known as 'breakouts'. They are usually borehole-parallel, occur on opposite sides of the hole and show some relationship to lithology, i.e. they are related to rock strength. There is a growing consensus that breakouts result from compressive failure of the borehole wall, and that they form in the direction of minimum effective compressional stress on the borehole (Bell & Gough 1979; Zoback *et al.* 1985). For vertical wells where one of the principal stresses is also vertical (overburden), breakouts form in the direction of minimum horizontal principal stress.

In both A17Z and 9/8a-18 we have observed sharp, localized changes in breakout orientation across fractures (Fig. 3b). This type of observation has been attributed to stress field perturbation, either simply due to the lithological anisotropy of the fault plane (Vernik & Zoback 1992) or actual slip on faults. The latter either historically ('fossil earthquakes') or induced by the drilling operation (Barton & Zoback 1994). The fracture in A17 is located at 4528 m where the dip azimuth plot suggests a large-scale change in azimuth (Fig. 4). As will be seen, there is also an increase in fracture concentration towards this depth. The combination of breakout rotation, dip azimuth change and fracture concentration suggests the presence of a fault. If the fracture at 4528 m does represent this fault plane, then the fault plane dips westwards at a low angle (281/25°).

Identification of fault damage zones

As our work on Bruce has developed, the evidence for faulting as the dominant control on fracturing has become more robust (see below). We now feel able to use this in a predictive sense, i.e. we would suspect that observed

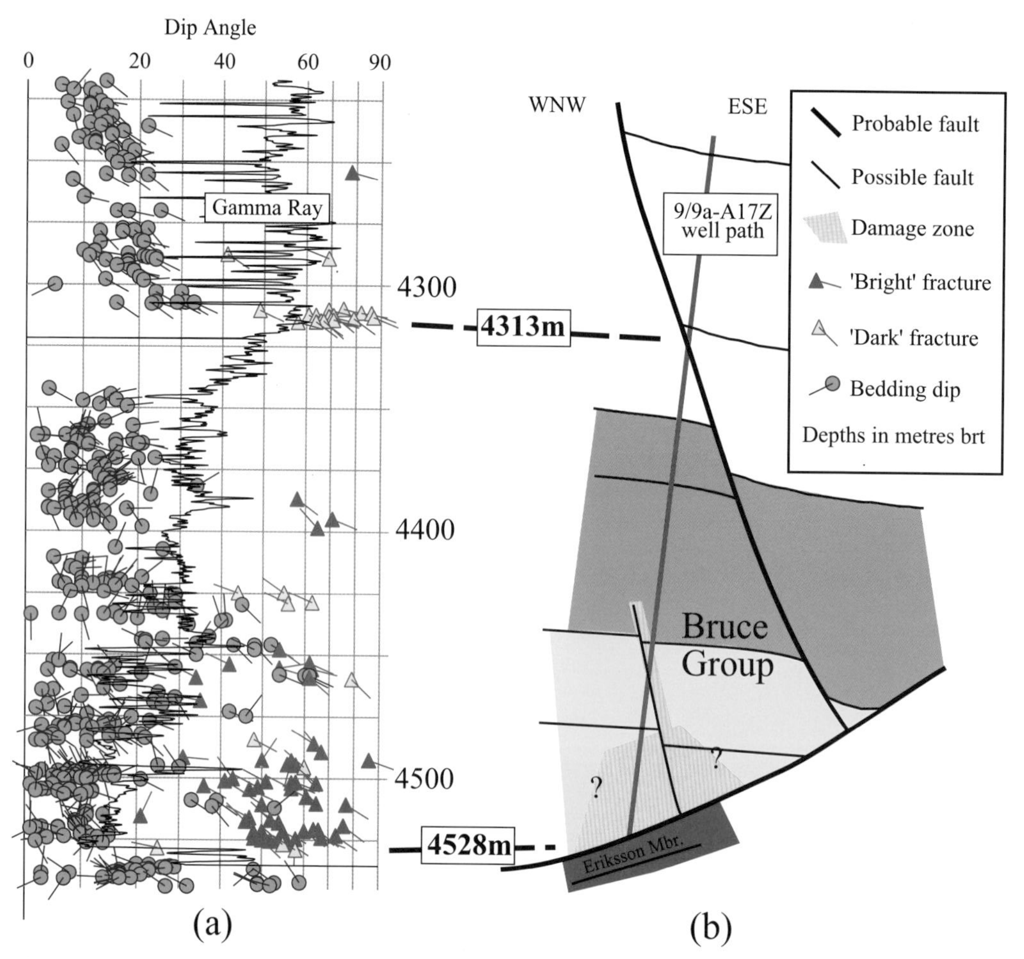

Fig. 6. (**a**) 'Tadpole Plot' of bedding and fracture dip data for well 9/9a-A17Z, and (**b**) structural interpretation of these data. Note the downwards-increase in dip towards the fault zone at 4313 m and the increase in fracture frequency towards the inferred low-angle fault at 4528 m.

concentrations of fractures are fault related. A17 is largely un-fractured, but there is a marked increase in the number of 'bright' fractures (fractures of a style common in the Bruce reservoir) towards the base, consistent with the interpretation of a fault at 4528 m (Fig. 6a). The peak in 'dark' fractures at 4313 m is interpreted as a steep eastwards-dipping fault plane. Other more minor fracture peaks may also be associated with faulting and appear on our final structural interpretation of the well (Fig. 6b).

Fractures & stress field

Image logs enable us to measure three aspects of a fracture system: fracture character, fracture orientation, and fracture frequency.

Bruce fractures commonly form fine, bright sinusoids on the amplitude image (Fig. 5) but are unseen on the travel-time image. This appearance is entirely consonant with core data, which shows the fractures to be filled with high density material: either close-packed grains, granulation seams or mineralization. The fractures generally have very low transmissivities and are believed to seriously damage reservoir performance. Occasionally, distinctive dark fractures are seen on the amplitude image. As noted above, these commonly turn out, on other evidence, to be faults.

Fracture strike is consistently NNE–SSW in all the studied wells (Fig. 7). Most fractures dip at angles greater than 60°, though the fractures near the base of A17Z, discussed above, dip at somewhat lower angles commensurate with their interpreted relationship to low-angle faulting.

Before discussing fracture frequency, a caveat is necessary. Because of the low acoustic impedance contrast presented by typical Bruce fractures, there is a marked difference between the true fracture frequency, observed in core, and the apparent fracture frequency, observed on the images. Direct comparison has only been possible in 9/9a-A15, which is a relatively poor image due to tool eccentering in a deviated well (Fig. 5). This comparison suggests a ratio of image to core fractures of the order 1:5. However, there is no doubt that high fracture frequencies on the image correlate with high fracture frequencies on the ground, for example well 9/9a-A18 (Fig. 7) was so badly damaged as to be almost unproducible.

The distribution of fractures from four wells in relation to stratigraphy and faulting is summarized in Fig. 8. In three out of four wells the fracture distribution can be related directly to known faults. No faults were detected in A15, though this may be a function of image quality. The upper fault in A17Z is not associated with a significant observed damage zone, which may relate to the relatively fine-grained lithology cut by the fault. In all cases, if our interpreted fault orientations are correct then it appears that the damage zones are concentrated in the hanging-walls of their associated faults (note that 9/9a-A18 is a highly deviated well so that the well bore passes downhole from foot-wall to hanging-wall).

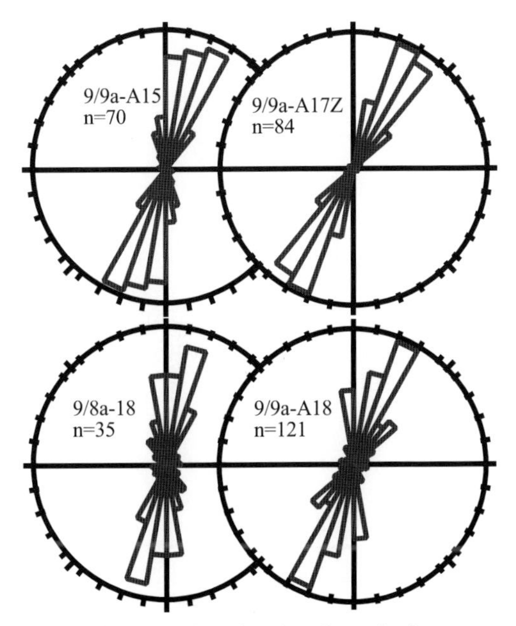

Fig. 7. Fracture orientation data from the four studied wells.

Well bore breakouts are common on the Bruce images. Assuming that one of the principal stresses is vertical then, as discussed above, breakout orientation for the near-vertical wells (9/8a-18 and 9/9a-A17Z) can be read directly as the orientation of the minimum horizontal principal stress. These observations have been combined with diplog evidence from the early field appraisal wells (using methods described by Bell & Gough 1979) to produce a stress map for the Bruce Field (Fig. 9). The interpretation of breakout orientation in deviated wells, such as 9/9a-A15 and -A18, requires further constraints (Mastin 1988) and data from these wells has not been incorporated into the map.

The map shows a dominant NNE–SSW orientation for the minimum horizontal principal stress, and a strong relationship to the mapped structural elements, which appear to perturb the regional stress field: note for example how breakout orientation follows the mapped fault shapes on the eastern flank of the field.

The observed NNE–SSW fractures would clearly have little tendency to fail in the mapped stress field and indeed would be held firmly closed. We would expect that this would emphasize their significance as flow barriers. Conversely, the body of evidence for preferential flow sub-parallel to the maximum horizontal principal stress in oilfields (e.g. Heffer & Koutsabeloulis 1995) would suggest preferential flow in an east–west direction.

Sedimentary heterogeneities must clearly be superimposed on our model, but our fracture and stress-field studies predict preferred east–west flow, 'baffled' by a variable density of low-transmissivity NNE–SSW oriented faults and fault-related fractures.

Palaeogeography

The Bruce Group represents a range of fluvial, tidal and shallow marine settings. Cross-stratification is common in many of the sand bodies but unevenly represented on the images, tending to be best-seen in the toesets (which may have high clay content and thus contrasts of acoustic impedance) and less well seen in foresets (which are cleaner with little impedance contrast).

As always with palaeocurrent studies, the key to using the image information has been close

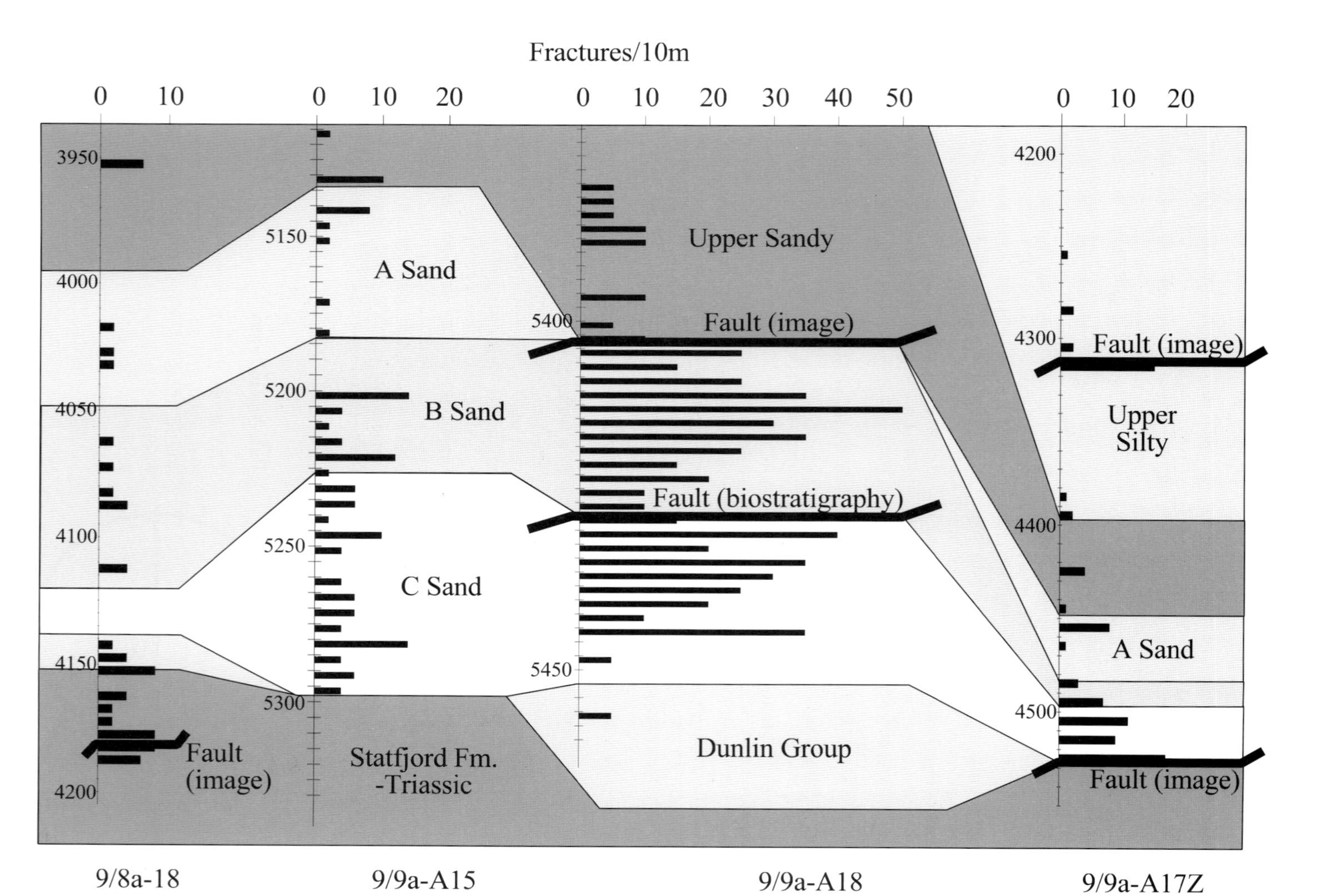

Fig. 8. Distribution of fractures in the four studied wells in relation to stratigraphy and known faulting. Fractures appear to be concentrated in damage zones on the hanging walls of faults (note that 9/9a-A18 is highly deviated).

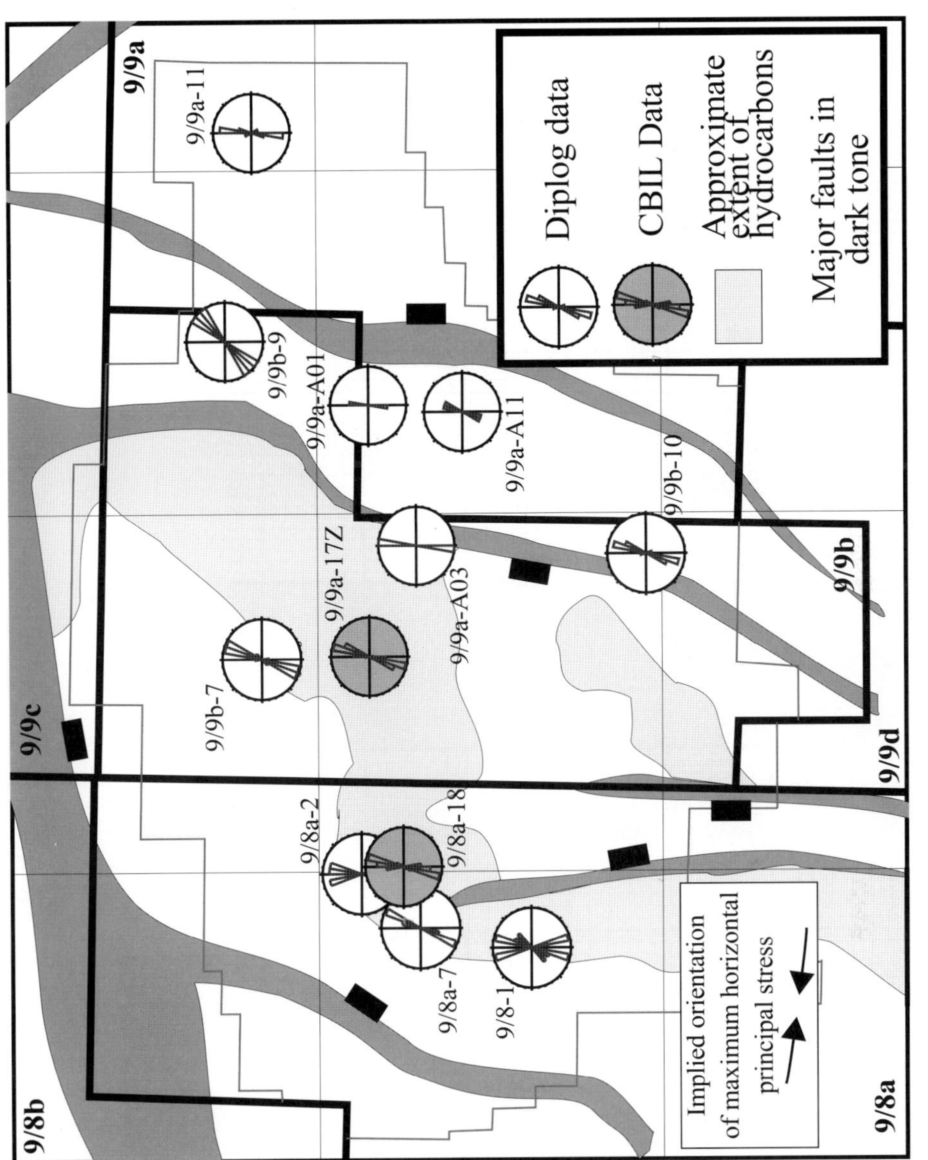

Fig. 9. Orientations of the minimum horizontal principal stress interpreted from image- and dip-log data over the Bruce Field. Note the strong relationship to major structural elements.

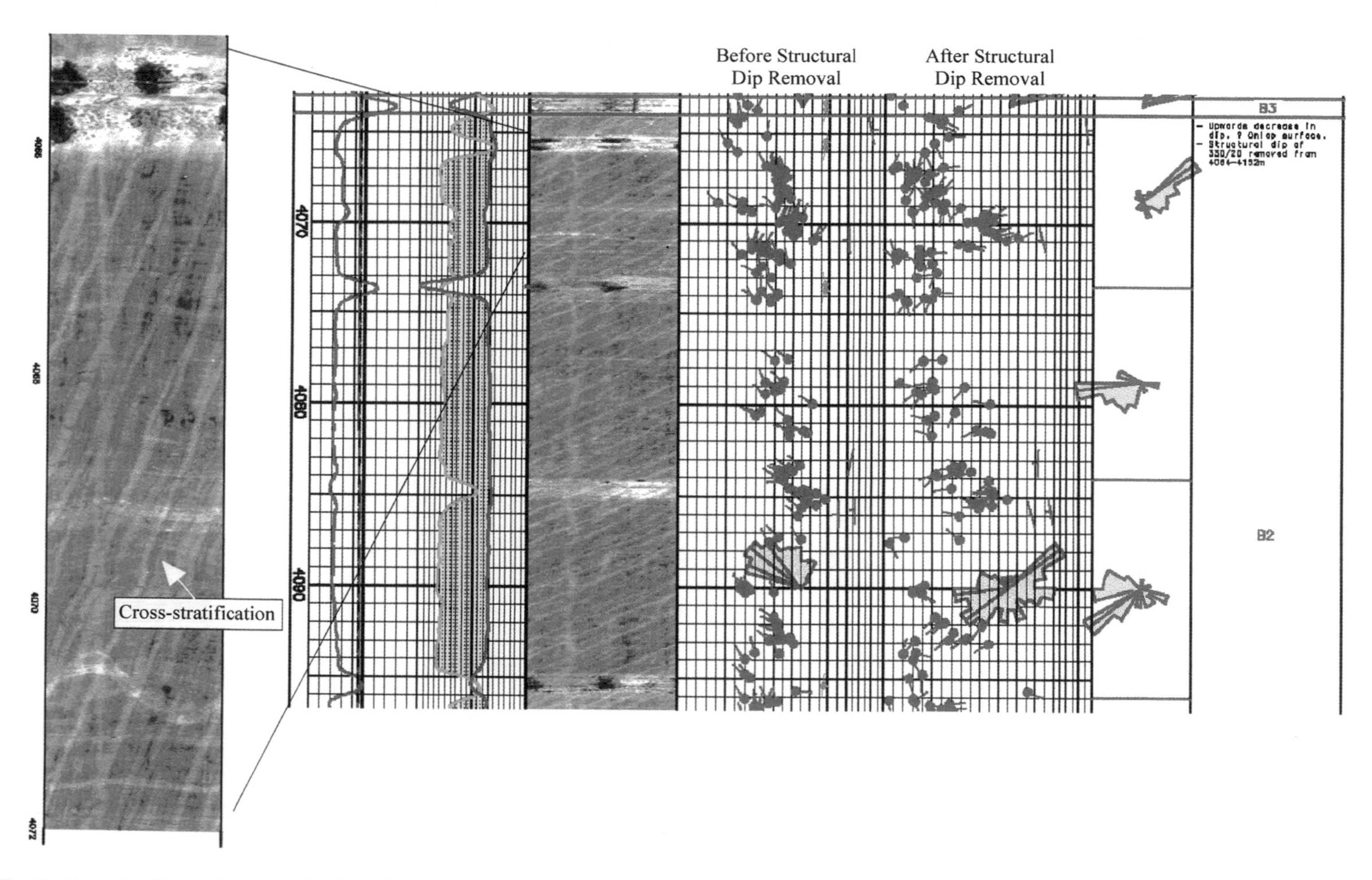

Fig. 10. Example of image interpretation for palaeocurrent analysis. Well 9/8a-18. Note also breakout above 4066 m, bright fracture at 4071 m, and diagonal striping due to scratching of the borehole wall by other logging tools and/or centralizers above the CBIL transducer.

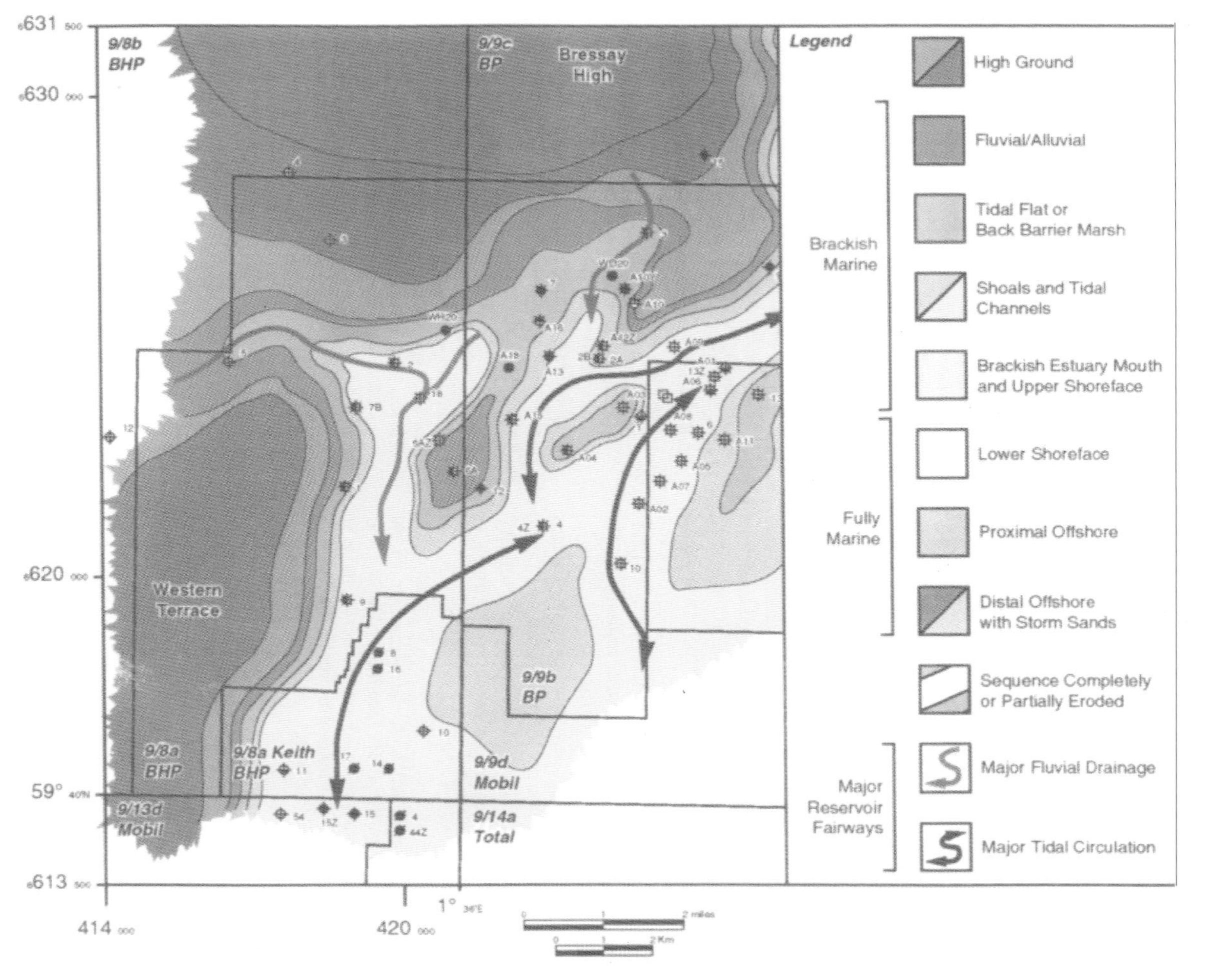

Fig. 11. Example palaeogeographic map for Bruce B2 Sand times (Bajocian), incorporating the results of palaeocurrent information and well correlation.

integration with the sedimentological description of the reservoir, from which it has been possible to identify and orientate individual tidal sandbodies (Fig. 10). Note the variability in sediment transport direction at the bed and parasequence scale, such that 'lumping' of data at coarser resolution, or using arbitrary subdivisions of the stratigraphy would be meaningless.

We have extracted palaeocurrent information from all the available images and have supplemented the data set using dip-log data from the appraisal drilling programme. All of the dip-log data, much of it from poor-quality logs acquired in oil-based mud, was re-picked by hand, rather than machine-processed, in order to extract the maximum useable information. Figure 11 is typical of the maps which have resulted from this work.

Conclusions

Despite the difficulties of the measurement, acoustic image logging can be expected to deliver important information on the fault, fracture, stress and depositional anisotropies in a reservoir. In the majority of cases, where core coverage is limited and un-oriented, and drilling is with oil-based mud, acoustic imaging is the only way to obtain this information.

As our experience develops, we are finding that quick-look image interpretation can also be an important part of operational activity, for example in planning completion strategy or side-tracks.

This paper is published with the permission of the Bruce Partnership: BP, BHP, Deminex, Elf and Total. The interpretations presented are, however, those of the authors and do not necessarily represent the views of all or any of the Partners.

References

BARTON, C. A. & ZOBACK, M. D. 1994. Stress perturbations associated with achieve faults penetrated by boreholes: Possible evidence for near-complete stress drop and a new technique for stress magnitude measurement. *Journal of Geophysical Research*, **99**, 9373–9390.

BELL, J. S. & GOUGH, D. I. 1979. Northeast–southwest compressive stress in Alberta: evidence from oil wells. *Earth and Planetary Science Letters*, **45**, 475–482

DIXON, R. J., BECKLY, A., DODD, C. & LOS, A. 1996. Reservoir Geology of the Bruce Field; Bruce–Beryl Embayment, Quadrant 9, UKCS. *In*: OAKMAN, C. D., MARTIN, J. H. & CORBETT, P. W. N. (eds) *Cores from the North West European Hydrocarbon Province: An Illustration of Geological Applications from Exploration & Production*. Memoir. Geological Society London.

FISHER, R. 1953. Dispersion on a sphere. *Proceedings of the Royal Society*, Series A, **217**, 295–306.

HANSEN, T. & PARKINSON, D. N. 1999. *This Volume*.

HEFFER, K. J. & KOUTSABELOULIS, N. C. 1995. Stress effects on reservoir flow: Numerical modeling used to reproduce field data. *In*: DE HAAN, H. J. (ed.) *New Developments in Improved Oil Recovery*, Geological Society Special Publication, **84**, 81–88.

MASTIN, L. 1988. Effect of Borehole Deviation on Breakout Orientations. *Journal of Geophysical Research*, **93**, 9187–9195.

OZKAYA, S. I. & MATTNER, J. 1996. DRAG – An EXCEL Visual basic program for modeling fault drag using cubic splines and calculation of minimum dip and strike separation. *Computers & Geosciences*, **22**(5), 512–524.

VERNICK, L. & ZOBACK, M. D. 1992. Estimation of maximum horizontal principal stress magnitude from stress-induced well bore breakouts in the Cajon Pass Scientific Research Borehole. *Journal of Geophysical Research*, **97**, 5109–5119.

WESTERN ATLAS 1991. *Circumferential Borehole Imaging Log (CBIL)*. Marketing Literature.

ZOBACK, M. D., MOOS, D., MASTIN, L. & ANDERSON, R. N. 1985. Well bore breakouts and *in situ* stress. *Journal of Geophysical Research*, **90**, 5523–5530.

Fault visualization from borehole images for sidetrack optimization

L. M. GRACE,[1] B. M. NEWBERRY[2] & J. H. HARPER[3]

[1] *Schlumberger Wireline and Testing, 4100 Spring Valley Road, Suite 600, Dallas, TX 75244 (e-mail: grace@dallas.wireline.slb.com)*

[2] *Schlumberger-GeoQuest, 5599 San Felipe, Suite 1700, Houston, TX 77056-2752 (e-mail: bnewberry@houston.geoquest.slb.com)*

[3] *Schlumberger-GeoQuest, 5444 Westheimer, Suite 800, Houston, TX 77056 (e-mail: harper@houston.geoquest.slb.com)*

Abstract: Missing pay sections in faulted areas is a common occurrence. An accurate measurement of the fault plane and surrounding bed distortion can substantially reduce the risk involved in the sidetrack decision. The measurement of the fault plane includes the dip and strike as well as the sealing nature of the fault plane. Our experience indicates mineral-filled fault planes are always sealing. This has lead to several interesting sidetracks, many of which are against conventional wisdom. Case studies included show an extensional fault sidetrack to the downthrown block; a perforating determination based on the location of the actual sealing fault; an extensional fault sidetrack based on the non-sealing properties of the fault plane. In these cases the distortion of the surrounding blocks are modeled using enhanced computer techniques (Etchecopar 1992). The input of the actual fault plane dip and strike is shown to be the key to an accurate cross-sectional projection. The visualization of the fault model is critical to the resolution of the sidetrack azimuth and distance. Whether a fault plane is sealing is also an important contribution to any sidetrack analysis. This technique has accurately predicted the sidetrack location for many successful wells.

It is not an infrequent event that a well planned borehole will not penetrate the objective zone. Although there are numerous geological reasons for a pay zone to be absent, one of the most common occurs when the trajectory of the well-bore has crossed a previously unknown fault. What then ensues is usually a tense period during which decisions have to be made as to the best course of action. Is it better to consider the well a dry hole and move on to other projects? Was the well trajectory a near miss to the objective zone? If so, then is it possible to sidetrack the well into the proper location or is another bore-hole required? The answers to such questions normally involve significant financial risk and the above considerations must be addressed in a very short time frame. Whenever a fault was suspected, from correlated missing or repeated section, the traditional approach was to use dip data obtained from a dipmeter logging device to attempt to locate and orient the fault Although frequently successful, this technique has several potential shortcomings. In order to execute a sidetrack successfully, the fault position and orientation must be accurately determined. Fault plane analysis techniques from dip data usually rely almost exclusively on examination of the deformation observed in surfaces adjacent or near to the actual fault plane. Many faults do not have any associated distortion (Billings 1972). If distortion is present then the fault model must be known and the primary fault displacement must be dip-slip. Most faults have a component of rotational or oblique slip. A far more limiting factor in using only dip data for fault plane analysis is that often there is little or no discernible deformation of surfaces near the fault plane. The authors have observed hundreds of such faults which were undetected on the dip plots. The introduction of high-resolution, borehole imaging devices, particularly the electrical logging tools, brought about a major advancement in the analysis of faults (Nurmi 1991). The borehole images provided, for the first time, the opportunity to clearly visualize the fault plane itself rather than to have to infer it from other data. The ability to observe the mineral filling of the fault plane is critical to the sealing properties of the fault. If the fault plane is conductive, it is not possible from electrical images alone to determine whether the fault is sealing or non-sealing. However all resistive, hence mineralized, fault planes have proven to be sealing

From: LOVELL, M. A., WILLIAMSON, G. & HARVEY, P. K. (eds) 1999. Borehole Imaging: applications and case histories. Geological Society, London, Special Publications, **159**, 271–281. 1-86239-043-6/99/$15.00.

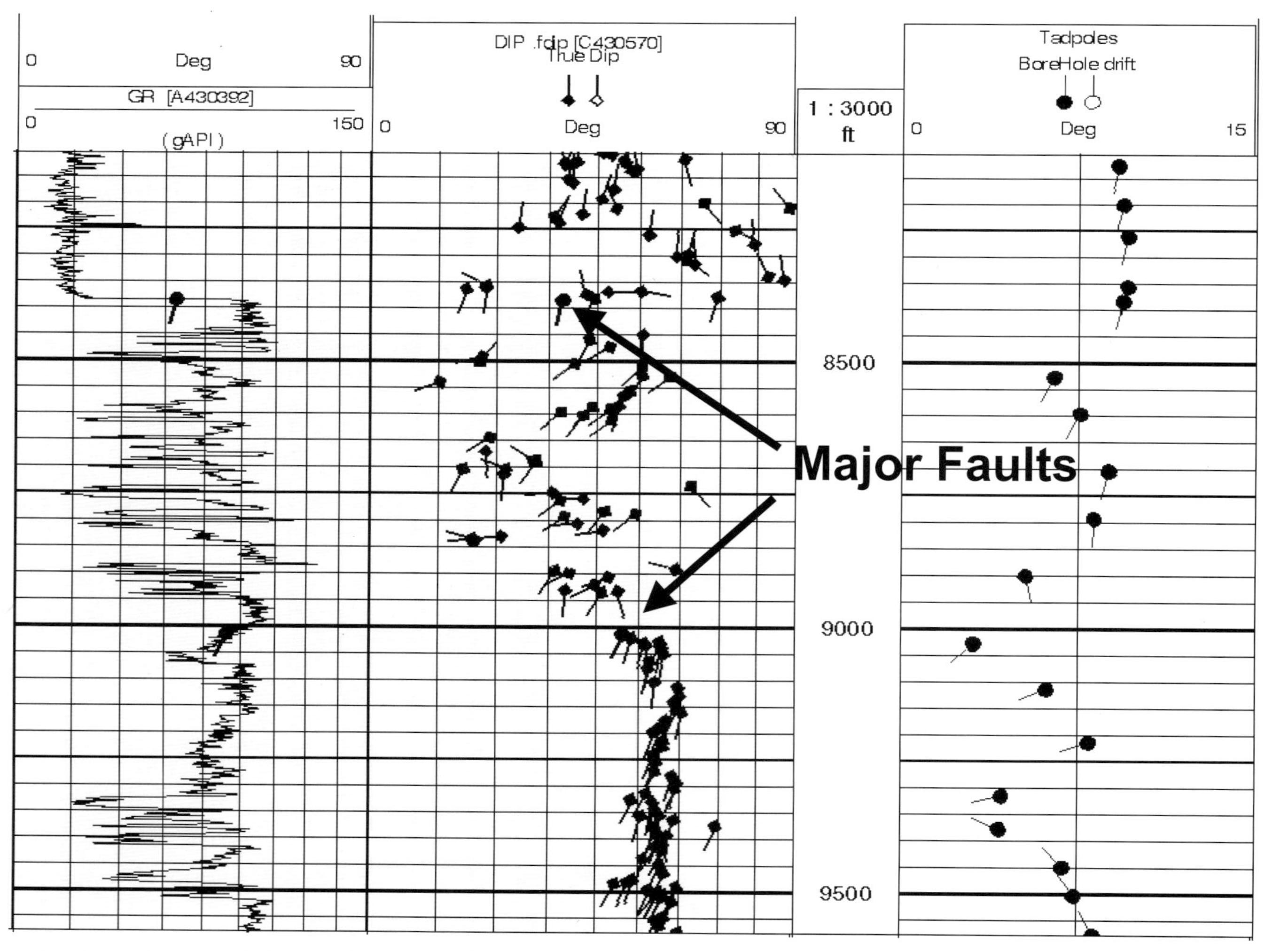

Fig. 1. Computed dip results over faulted section (reproduced with permission from Grace *et al.* 1997).

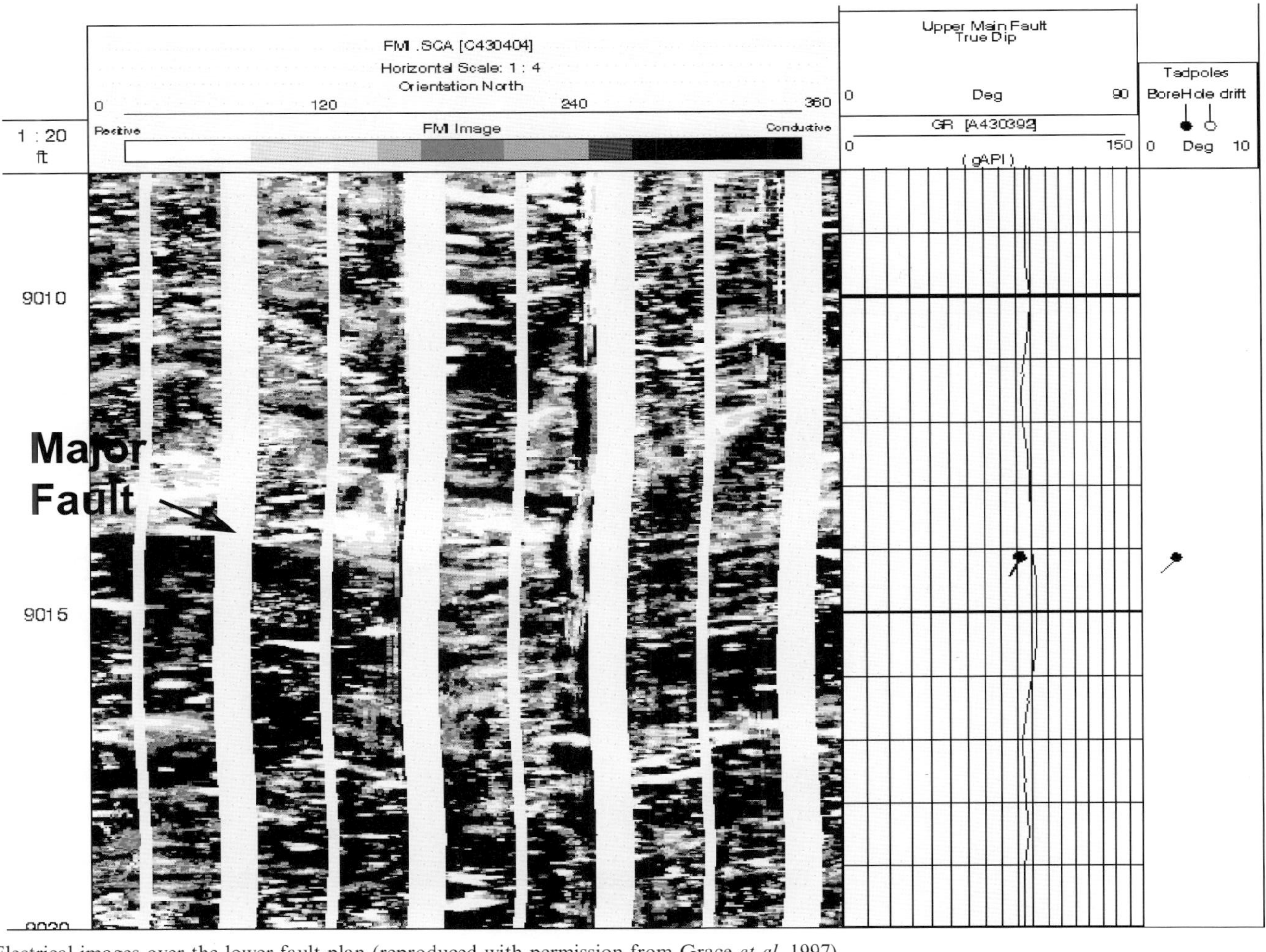

Fig. 2. Electrical images over the lower fault plan (reproduced with permission from Grace *et al.* 1997).

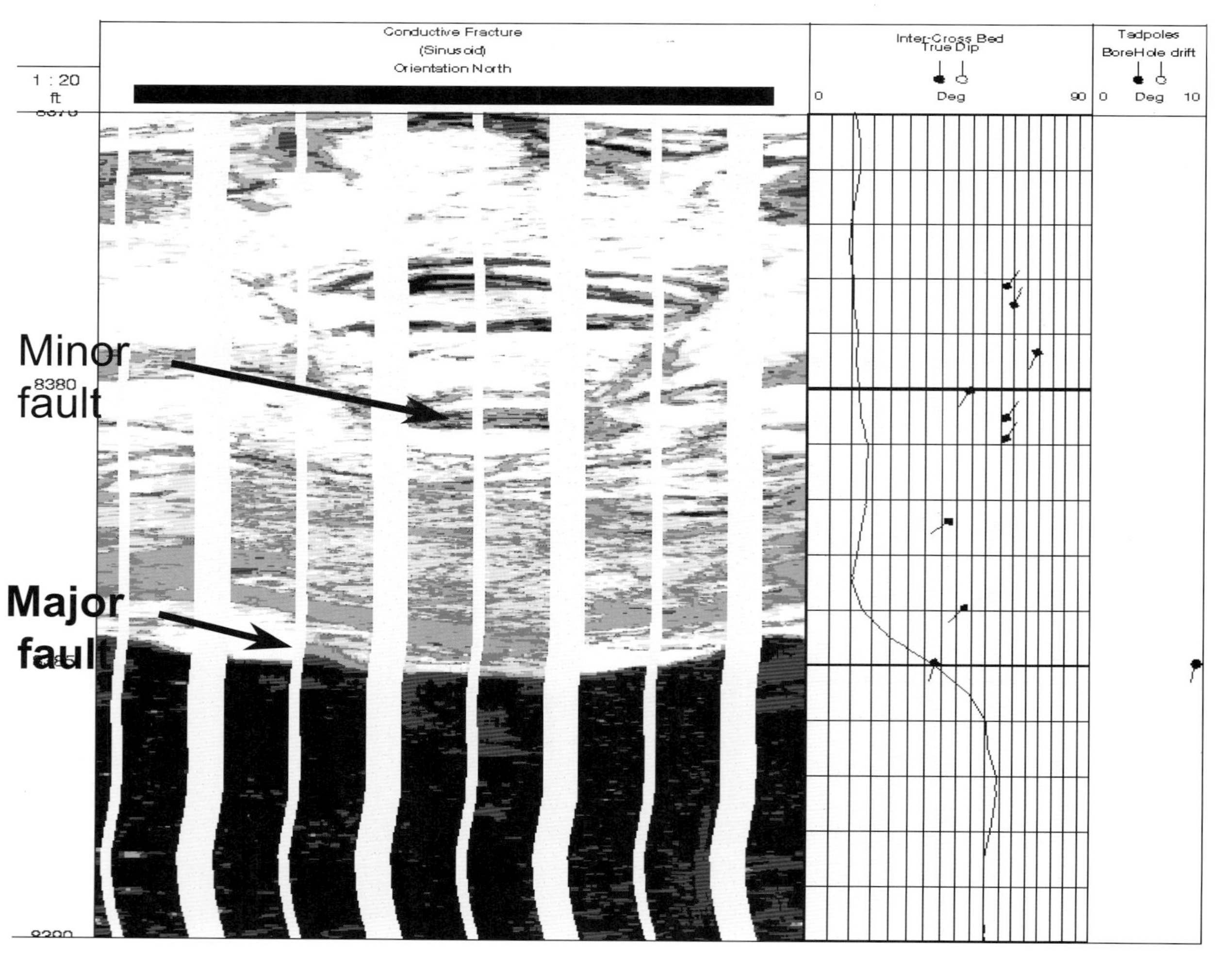

Fig. 3. Electrical images over the upper fault plane (reproduced with permission from Grace *et al.* 1997).

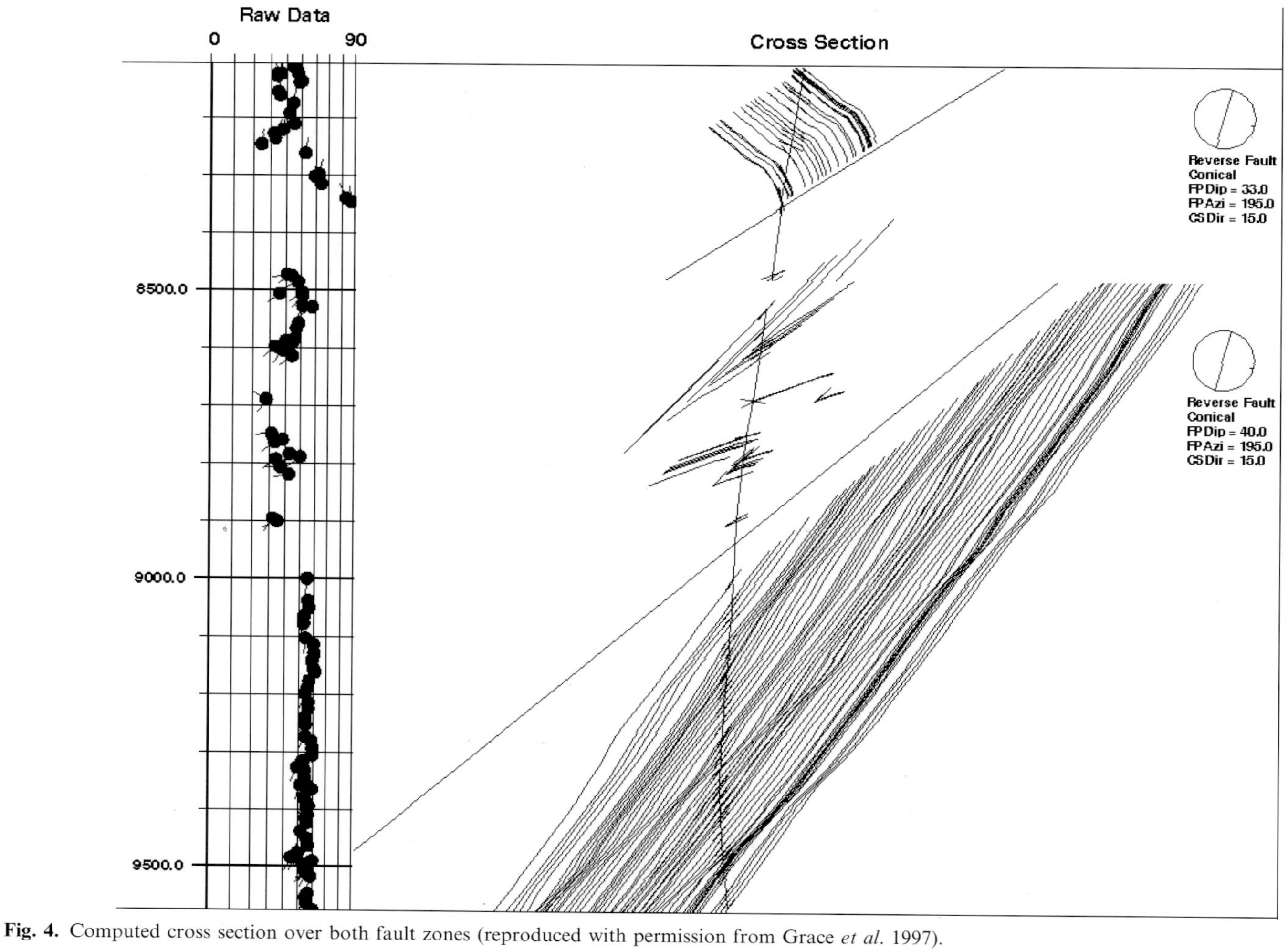

Fig. 4. Computed cross section over both fault zones (reproduced with permission from Grace *et al.* 1997).

in the author's experience. This observation has been instrumental in many successful offset recommendations.

Technique

As viewed on electrical images, fault planes will appear as more or less planar events against which features from both sides, such as bedding sets, truncate. Although the thickness of an individual fault zone can range from the indeterminately thin to many meters in thickness, the most common value is in the order of 10 cm. The most precise method of fault orientation is by interactive software on a computer workstation. This allows the interpreter to delimit very precisely the points which define the fault plane and immediately obtain the calculated azimuth and dip magnitude of the fault. Another useful feature of electrical images is that it is often possible to conclude with a high degree of certainty whether a fault is sealing. When the fault zone is seen to be resistive, it is generally safe to conclude that the fault is 'healed' or sealing. Unfortunately, the case of a conductive fault is not so straightforward and it is usually not possible to give a definitive answer. The interval could be filled with conductive minerals and be sealing or it could be filled with conductive drilling fluid and be non-sealing. Visualization of the fault plane however does not convey any reliable information regarding the relative vertical displacement of the fault blocks. This piece of the puzzle must come from other sources such as log correlation with offset wells. Once the accurate location, orientation and vertical displacement of the fault have been determined, the solution to the sidetrack problem becomes merely an exercise in geometry to determine the needed borehole trajectory to land the well at the optimum position.

Case study 1

A complex thrust-faulted province provides the first case study. The target sand penetrated by the borehole, immediately below an obvious thrust fault, exhibited unexpectedly low porosity and low permeability. Note the large gamma ray change (Fig. 1) at 8385 ft (2556.4 m) and the

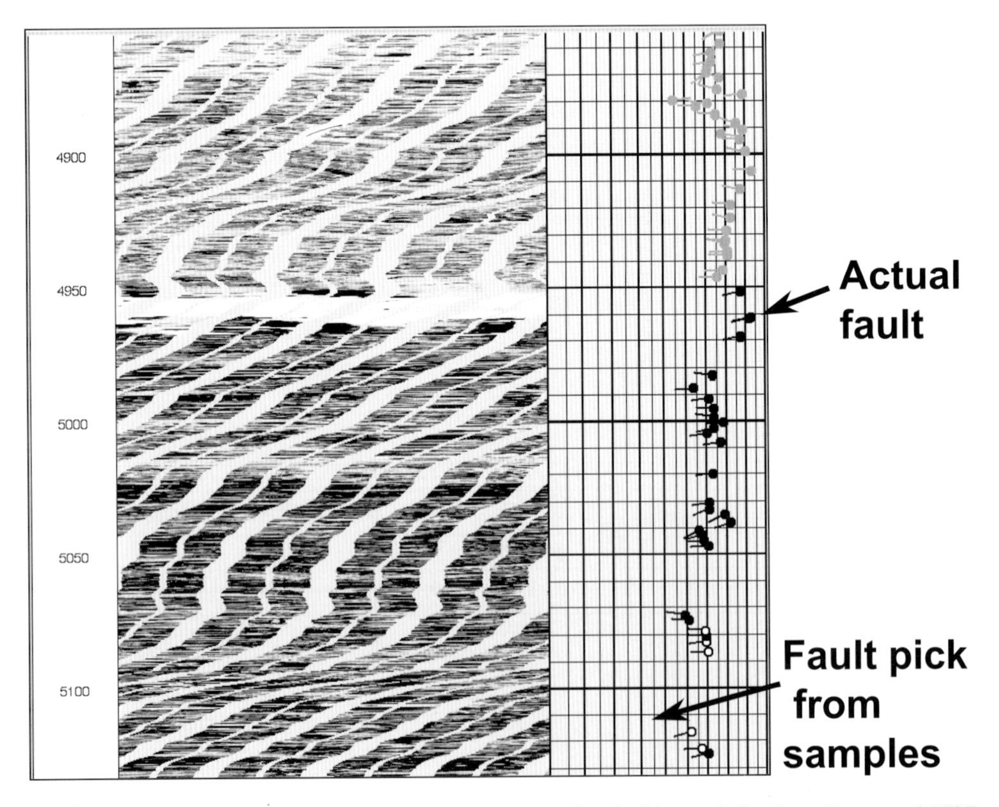

Fig. 5. Electrical images and dips over thrust fault section (reproduced with permission from Grace *et al.* 1997).

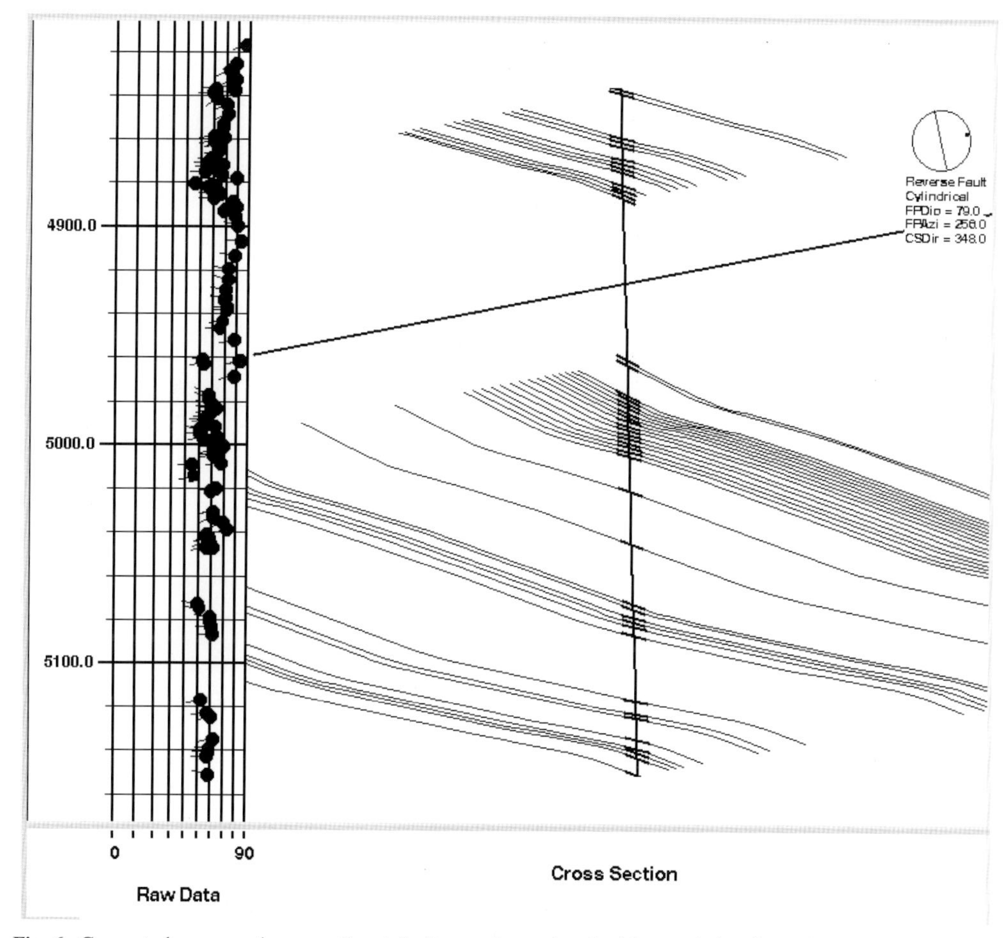

Fig. 6. Computed cross section over thrust fault zone (reproduced with permission from Grace *et al.* 1997).

corresponding change of image character shown in Fig. 3. The first inference is that the sand observed on the logs is not the target sand and a sidetrack under the upper thrust fault is in order. A close inspection of the electrical images (Fig. 2) reveals a second thrust fault at 9014 ft (2748.2 m). Fault breccia and some drill cuttings remaining on the low side of the borehole wall tend to intermittently obscure the fault plane as seen on the electrical images. Note also the abrupt changes in borehole drift at these depths. In consolidated formations, abrupt changes in borehole trajectory are often observed. Thus the pay sand section is bounded by two fault planes (Fig. 4). This suggests the pay sand may have been diagenetically altered which resulted in the lowering of the porosity and permeability. Stratigraphic analysis of the sands indicates the depositional facies of the sand to be a series of proximal delta fans. Given this set of circumstances, the final conclusion is that a sidetrack under the upper thrust fault and parallel to the fault plane would have little chance of success.

Case study 2

Another thrust fault example (Figs 5 & 6) shows how electrical images dramatically influenced the production of a well. The primary geological objective in this area is a series of sands in the sub-thrust section. Analysis of cutting samples suggested the fault was located at approximately 5110 ft (1557.9 m). The subsequent perforating decision was based upon this fault pick. An analysis of the images clearly indicated a mineralized fault at 4950 ft (1509.1 m). This is considerably higher in the section and added several sands for potential completion. Perforating this

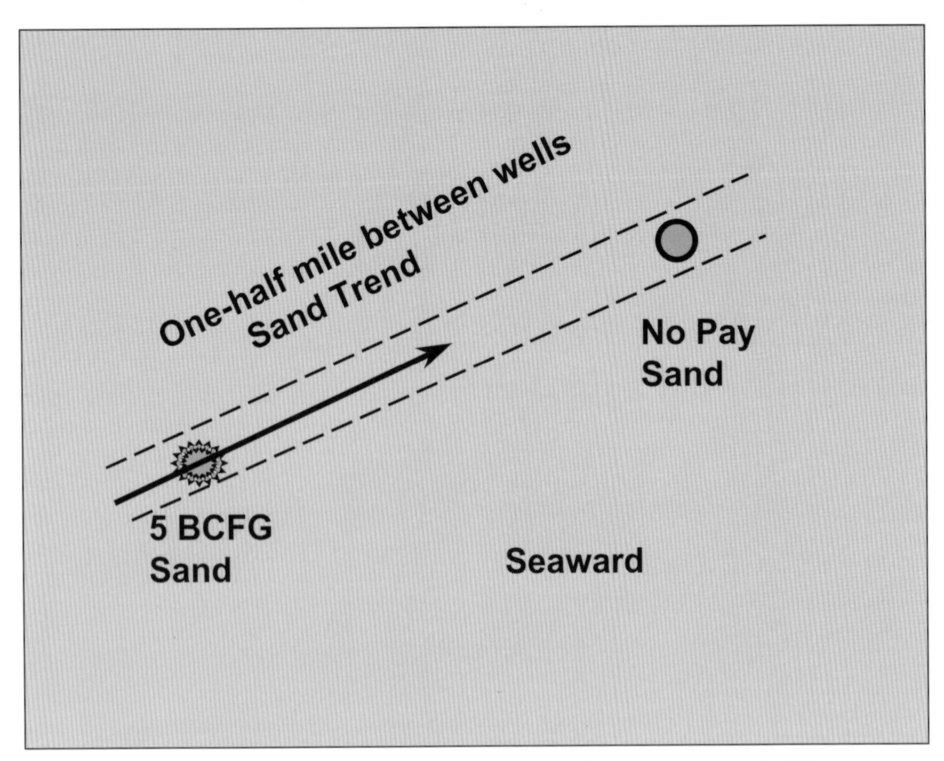

Fig. 7. Orientation of longshore sand bar (reproduced with permission from Grace *et al.* 1997).

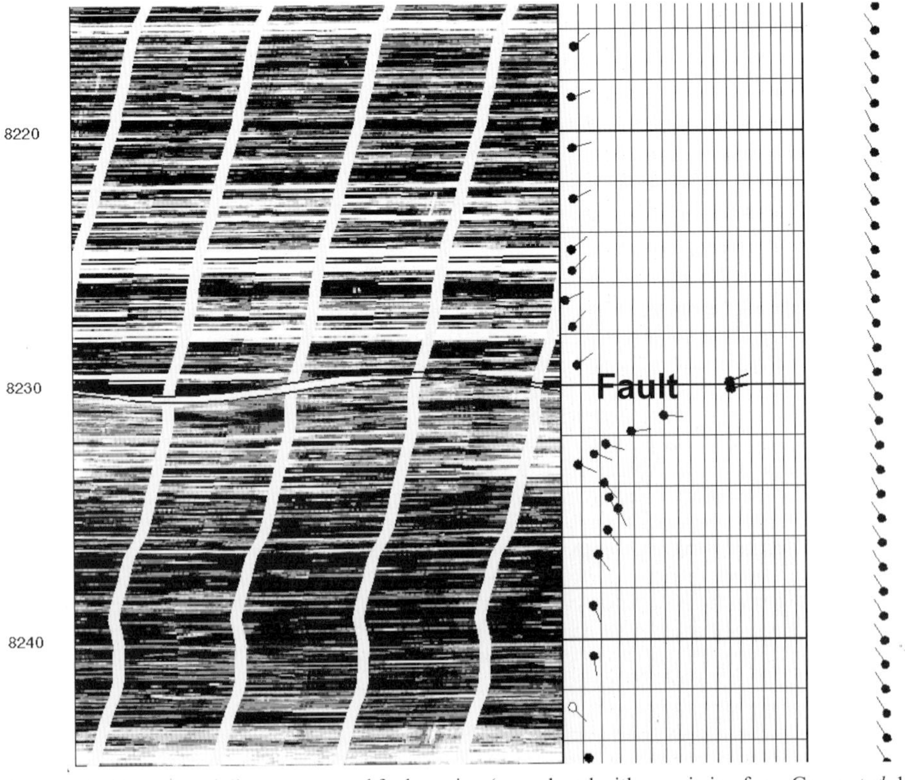

Fig. 8. Electrical images and dips over normal fault section (reproduced with permission from Grace *et al.* 1997).

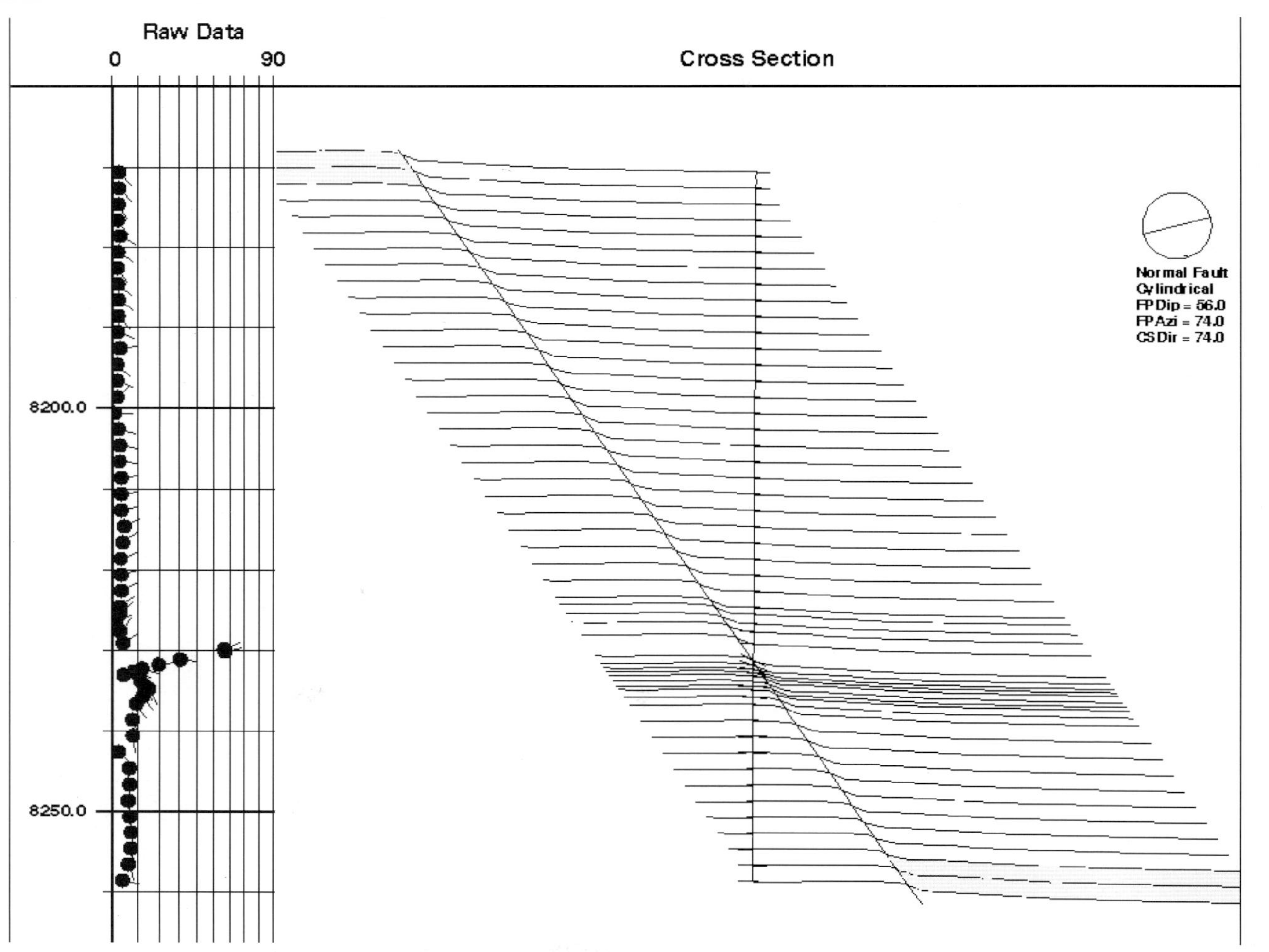

Fig. 9. Computed cross section over normal fault zone (reproduced with permission from Grace *et al.* 1997).

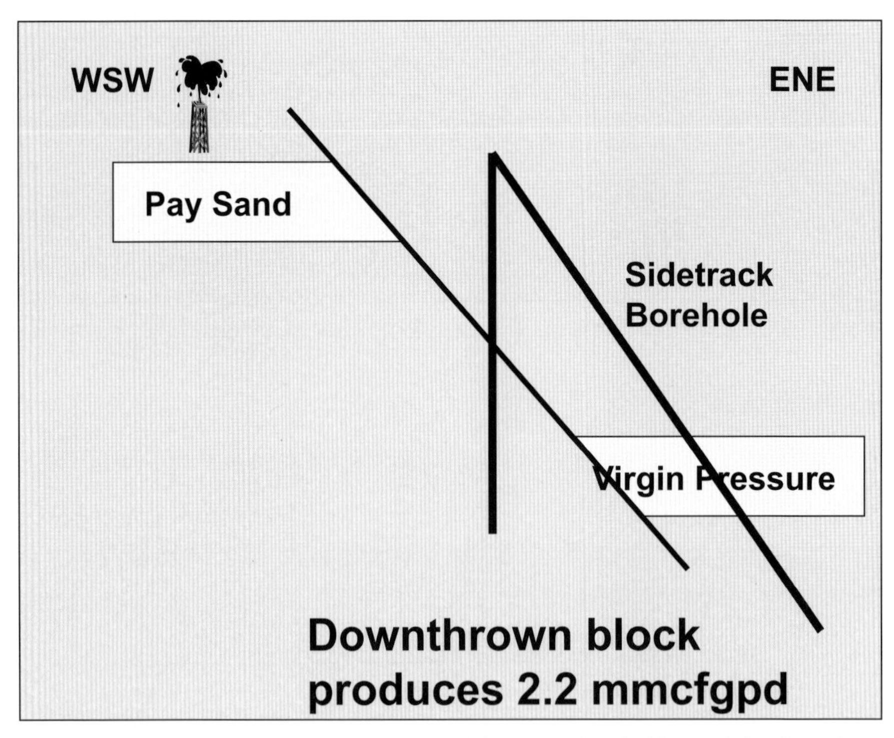

Fig. 10. Structural cross section of faulted longshore sand bar (reproduced with permission from Grace *et al.* 1997).

additional section added over 5 MMCFGPD to the total production.

Case study 3

This well is a Tertiary objective from the Texas Gulf Coast. The discovery well was drilled and produced 5 BCFG from a longshore shoaling sequence. The new well was located one-half mile ENE along the trend of the shoal (Fig. 7). No indication of the target sand was observed on the logs. Correlation of logs with the producing well showed 95 ft (29 m) of missing section even though seismic did not indicate a fault. The computed dip results strongly suggested a fault in the vicinity of 8230 ft (2509.1 m) but even with advanced dip processing mostly low angle dips with considerable azimuthal scatter were found. It would have been impossible to accurately infer a fault orientation from such data. The borehole electrical images clearly displayed a resistive fault (Fig. 8), which is a normal fault due to the missing section (Fig. 9), at 8230 ft (2509.1 m). The downdip direction of the fault was calculated to be N74°E, indicating a strike of N16°W–S16°E with a dip magnitude of 56°.

A sidetrack decision involved several factors:

- the discovery well should have partially depleted the reservoir
- the fault plane is resistive; thus a sealing fault
- sidetracks to the downthrown block are contrary to conventional wisdom because of the potential for water production
- a sidetrack to the downthrown block has virgin pressure potential

A sidetrack well was drilled into the downthrown block, encountering the pay sand as expected. This interval produced 2.2 MMCFGPD at virgin reservoir pressure (Fig. 10).

Conclusions

Any method that relies entirely on near wellbore data to determine the orientation of geological features is inherently less reliable as the features are extrapolated further from the borehole. Fault plane location and orientation are often inferred from projections of dips from bedding features that have been distorted during the faulting process. For a large number of faults, the distortion patterns observed in the adjacent beds are insufficient for this technique to provide reliable results. Fault analysis from electrical images permits an accurate determination of fault orientation at the borehole. Although all faults are certainly not truly planar features, extrapolation of the fault plane as observed at the borehole has proven to be a far more reliable method, particularly when dealing with low-distortion or no distortion faults. At worst, this technique provides a good crosscheck of the reasonability of results from other methods.

References

BILLINGS, M. P. 1972. *Structural Geology*, 3rd ed. Englewood Cliffs, Prentice-Hall, Inc.

ETCHECOPAR, A. 1992. Looking at dips from a different angle. *In*: CHARNOCK, G. (ed.) *Charismatique*. Paris, Schlumberger Data Services.

GRACE, L. M. & NEWBERRY, B. M. *et al.* 1997. *Geologic Applications of Dipmeter and Borehole Electrical Images*. Schlumberger Oilfield Services.

NURMI, R. 1991. The 3 forces of faulting. *In*: CHARNOCK, G. & GAMBLE, S. (eds) *Middle East Well Evaluation Review*. Dubai, Schlumberger Technical Services, Inc.

Borehole images of the ocean crust: case histories from the Ocean Drilling Program

T. S. BREWER,[1] P. K. HARVEY,[1] S. HAGGAS,[1]
P. A. PEZARD[2] & D. GOLDBERG[3]

[1] *Leicester University Borehole Research Group, Department of Geology, University of Leicester, University Road, Leicester, LE1 7RH, UK (e-mail: tsb5@le.ac.uk)*
[2] *CEREGE, Aix-en-Provence, France*
[3] *Lamont Doherty Earth Observatory, Columbia University, New York, USA*

Abstract: Downhole logging is an integral part of the Ocean Drilling Program (ODP), although the choice of tool deployment is primarily a function of the scientific objectives of an individual leg. Following the successful deployment of the Formation Microscanner (FMS) during Leg 126, FMS images have been used to constrain a variety of sedimentary, igneous, metamorphic and tectonic questions in the ocean. However, it is only recently, that the full potential of FMS images and other logging data has been employed in the study of volcanology, which provides important constraints on the evolution of the volcanic pile when fully integrated with core based measurements. Furthermore, many of the holes drilled into the volcanics often have incomplete core recovery and here the logging results can be used to constrain the lithological, chemical and mineralogical variations in such intervals. Using the log based stratigraphy from crust created at slow (Hole 395A) and intermediate (Holes 504B and 896A) spreading centres, it is evident that the proportions of volcanic rock types (pillows/flows/breccias) correlate with the spreading rate. This relationship is suggested from submersible observations, but the ODP data provide additional data to test such hypotheses and to also extended and test it within the third dimension.

Processes responsible for the construction of ocean lithosphere are some of the most fundamental within the earth, and in their present form they have been operating for at least 2 Ga. The rate of oceanic crust generation influences both large-scale tectonic processes, the distribution of continental crust, and major geochemical processes, such as the composition of the ocean water. Models for ocean crust generation, which have been supplemented by extensive research on ophiolites, have traditionally proposed a layered structure (Hill 1957, Raith 1963), within which Layer 2 is composed of volcanic rocks (Layer 2A) and sheeted dykes (Layer 2B). Recent work by Goldberg & Sun (1997) has focused attention on this subdivision of Layer 2 and has demonstrated that the thickness of the sub-layers (Layers 2A/2B) is probably a function of the spreading rate of the ridge system. Crustal architecture of the volcanic layer can be broadly defined in terms of the proportions, distribution and spatial relationships of the various lithologies; the main rock types being pillow lavas, flows and breccias. The first order control on crustal architecture is primarily a function of the spreading rate, whereas localized variations in any crustal segment reflect the dynamics of the magma chambers (e.g. eruption rate, timing and amount of recharge).

Throughout the 1970s and 1980s the Deep Sea Drilling Project (DSDP) and the subsequent Ocean Drilling Program drilled and cored a number of holes which penetrated layer 2A. In many of these holes core recovery was generally low (<30%) and the recovered material was frequently biased towards the rheologically more competent units (e.g. pillow cores, massive centres of flows, Brewer *et al.* 1998). Low recovery makes accurate core location difficult (or produces potentially large errors) and also the identification of the various volcanic lithologies (e.g. pillows, flows, breccias) becomes extremely difficult and often subjective, due to the lack of the characteristic features (Table 1) or the inability to recover delicate structures (e.g. pillow rims or breccias).

In this paper we review core and logging data from three boreholes which penetrate significant sections of ocean floor volcanics (Layer 2A)

From: Lovell, M. A., Williamson, G. & Harvey, P. K. (eds) 1999. Borehole Imaging: applications and case histories. Geological Society, London, Special Publications, **159**, 283–294. 1-86239-043-6/99/$15.00.

Table 1. *Criteria used for the visual core during the drilling of Holes 504B, 896A and 395A. In Fig. 8 massive units, thin flows and sheet flows are all grouped together as flows*

Hole 504B (after Adamson 1985)

- Massive unit: two types (a) medium to coarse grained, little affected by drilling; core in long intact pieces. (b) medium to coarse grained basalt; cores generally break into short pieces.
- Dykes: a unit with one or two chilled intrusive margins.
- Sheet Flow: characterized by parallel closely spaced (<0.5 m), mainly horizontal flat glassy selvages.
- Thin Flow: homogeneous areas of core >1 m thick, composed of fine to medium grained basalt.
- Pillows: all material remaining after classification into above types. Characterized by presence of chilled margins (curved or inclined), hyaloclastic breccias, fine grained and often highly fractured rock.

Hole 896A (after Alt *et al.* 1993)

- Massive Units: lack of curved glassy or chilled margins, the presence of well-developed brown oxidative alteration, generally higher recovery and longer sections of core than in pillows or breccias. Grain size varies from microcrystalline to fine grained.
- Dykes: lack of glass and variolitic textures in hand specimen. Presence of veins or breccia zones separating the chilled margin from the host rock.
- Pillow lavas: characterized by curved to planar or irregular chilled and/or glassy margins, an interior variolitic zone, poorly developed oxidative alteration. Abundant fracturing and veining (veins generally thick, >1 mm) and are relatively fine grained.
- Breccias: only recorded as a unit where two or three pieces occurred together. Two types of breccia, hyaloclastic and matrix supported were identified by not differentiated in the visual core descriptions.

Hole 395A

No attempt was made during Leg 45 (Melson *et al.* 1979) to discriminate the various rock types and in the basement stratigraphy only pillow lavas and breccias are shown (Melson *et al.* 1979).

formed at intermediate (Holes 896A, 504B) and slow (Hole 396A) spreading centres in order to establish the lithostratigraphy at each drill site. From this approach it is possible to more accurately constrain the proportions of the various lithologies in Layer 2A, to test if spreading rate is a first order control on crustal architecture, and to establish some of the second order controls on crustal architecture in an individual borehole.

Geological setting

ODP Holes 504B and 896A are situated in the equatorial East Pacific, approximately 200 km to the south of the Costa Rica spreading centre (Fig. 1). Hole 504B is the deepest hole so far drilled into oceanic basement, penetrating 2.111 km of *c.* 5.9 Ma crust (Alt *et al.* 1993). In this hole, Layer 2A is 570 m thick and is composed predominantly of pillow lavas and flows, which then pass through an 310 m thick transition zone into the sheeted dykes of Layer 2B. Within Layer 2A, previous studies have identified a number of flows and pillow lavas (Adamson 1985, Pezard 1990). In the drilling of the volcanics, a number of different types of breccia were recovered from Hole 504B descriptions (Cann *et al.* 1983), but these pieces were excluded from the core lithostratigraphy.

Hole 896A is located approximately 1 km south–east of Hole 504B (Fig. 1), in *c.* 6.1 Ma crust (Alt *et al.* 1993). This hole penetrates *c.* 290 m of basement, with core recovery averaging 26.9%, and the main lithologies being flows, pillow lavas and minor breccias (Alt *et al.* 1993). Recently this core-based lithostratigraphy has been revised by integrating core and logging data to give a more representative record (Brewer *et al.* 1998).

Hole 395A was drilled at 22°N on the western edge of the mid-Atlantic Ridge (Fig. 2), in 7 Ma old crust, during DSDP Leg 45 (Melson *et al.* 1979). The drill site is located in a isolated sediment pond in a water depth of 4482 m, which has low heat flow (Hussong *et al.* 1979, Langseth *et al.* 1992); this hole penetrates to a depth of 609 mbsf, through a sediment cover of 93 m (Melson *et al.* 1979). Following the drilling of this site, the hole has been re-entered four times to conduct downhole experiments and logging (Hyndman *et al.* 1984, Bryan *et al.* 1988, Gable *et al.* 1992, Becker *et al.* in press). Logging results used in this study are those collected during DSDP Leg 78B (Hyndman & Salisbury 1984), which have been validated by the recent logging during ODP Leg 174B (Becker *et al.* in press).

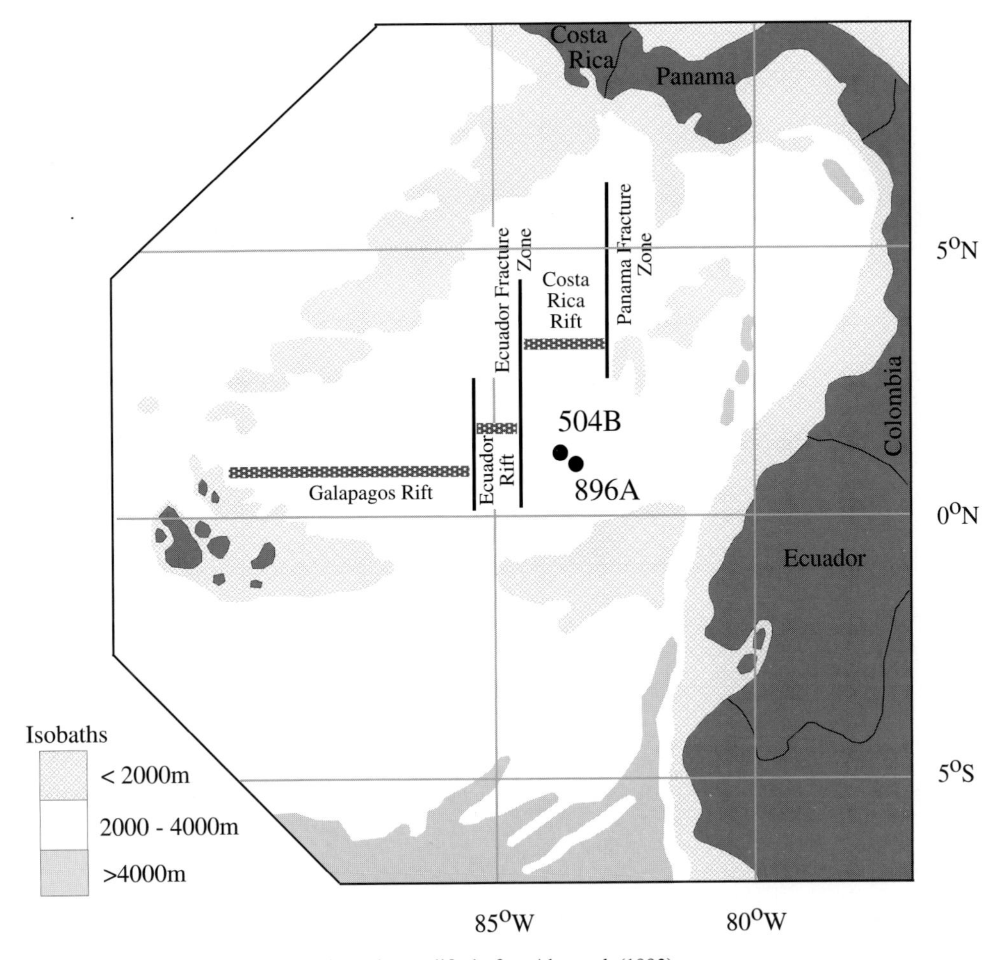

Fig 1. Location of Holes 504B and 896A, modified after Alt *et al.* (1993).

Previously, by use of the resistivity logs, Hyndman & Salisbury (1984) were able to identify a series of massive units, while the remainder of the lavas were classified as pillow lavas.

Basement lithologies

Criteria for identification of pillow lavas, flows and breccias in each hole is summarized in Table 1. These criteria are often ambiguous and represent a non-robust classification scheme. In each of the studied holes, much of the core consists of small pieces (<5 cm), often lacking structures (destroyed during drilling), leading to potentially large errors in the estimates of the lithological types and correspounding depths. It is therefore, advantageous to develop a classification scheme that uses a wider set of data, in particular wireline logging results.

Downhole logging

Downhole logging results often provide near continuous records of the chemical and/or physical properties of lithologies in the borehole wall (Goldberg 1997). The types and coverage of the different logs available from each hole is summarized in Fig. 3 details of the data processing and quality of the individual logs is described in Goldberg (1997) and Brewer *et al.* (1998).

Caliper log

The caliper log is a record of hole size and highlights the zones of oversizing, this is particularly important since many of the logging tools are affected to varying degrees by tool stand-off and borehole size (Hertzog *et al.* 1987, Bristow & deMenocal 1992, Kerr *et al.* 1992, Brewer *et al.*

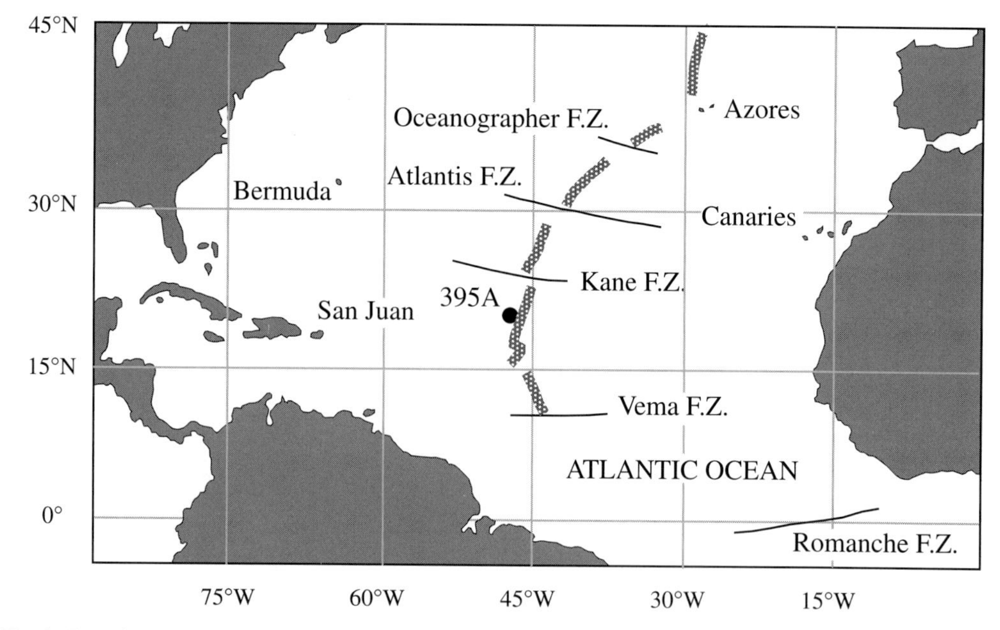

Fig. 2. Location of Hole 395A, modified from Hyndaman & Salisbury (1983).

1997), which may lead to erroneous results due to an attenuated signal from the formation (Fig. 4). The caliper log can also be used to aid lithology identification in layer 2A (Brewer *et al.* 1998). Using borehole geometry from the caliper log, the following rock types can be identified:

- flows: generally in-gauge hole size, smooth caliper log
- pillow lavas: variable hole size, characterized by micro-relief with occasional relatively large breakouts
- breccia: general uniform caliper log with localized micro-relief

Geochemical Logging Tool (GLT^{TM})

The GLT provides in-situ measurements of selected oxides (Si, Al, Ca, Fe, Ti, K) and trace elements (U, Th, Gd, S), where the data quality is strongly dependent upon the borehole size (Hertzog *et al.* 1987, Bristow & deMenocal 1992, Kerr *et al.* 1992, Brewer *et al.* 1997, Harvey *et al.* 1997). The reduction in data quality due to stand-off is primarily a function of:

(a) the sensitivity of the Aluminium Activation Tool (Al measurements) to borehole enlargements and

(b) subsequent data processing that applies a closure model, assuming a constant sum for the major element oxides. The applicability of such a closure model to ocean floor volcanics is particularly problematic because two major elements (Mg and Na) are not determined, but in the processing their concentrations are usually set to 10% (the value in fresh basalt, Brewer *et al.* 1997).

Clearly, such an assumption is not valid throughout Layer 2A, where ($MgO + Na_2O$) can reach up to 20% in some alteration minerals (Laverne *et al.* 1996, Teagle *et al.* 1996). In zones of intense alteration or enlargement (very low Al values) anomalous GLT derived compositions can be produced due to this artefact of processing (Fig. 4).

However, by careful screening of the data (removal of data from break-outs) both sampling and processing artefacts can be removed, producing usable data sets. It must, however, be stressed that even in the best hole conditions, the quality of the log-derived geochemistry is still lower than that of geochemical data measured from core samples by traditional methods (e.g. XRF (X-ray flourescence spectrometry), Brewer *et al.* 1992, Harvey & Lovell 1992, Kerr *et al.* 1992). Therefore, a simple comparison of GLT data with core-derived chemistry from the same

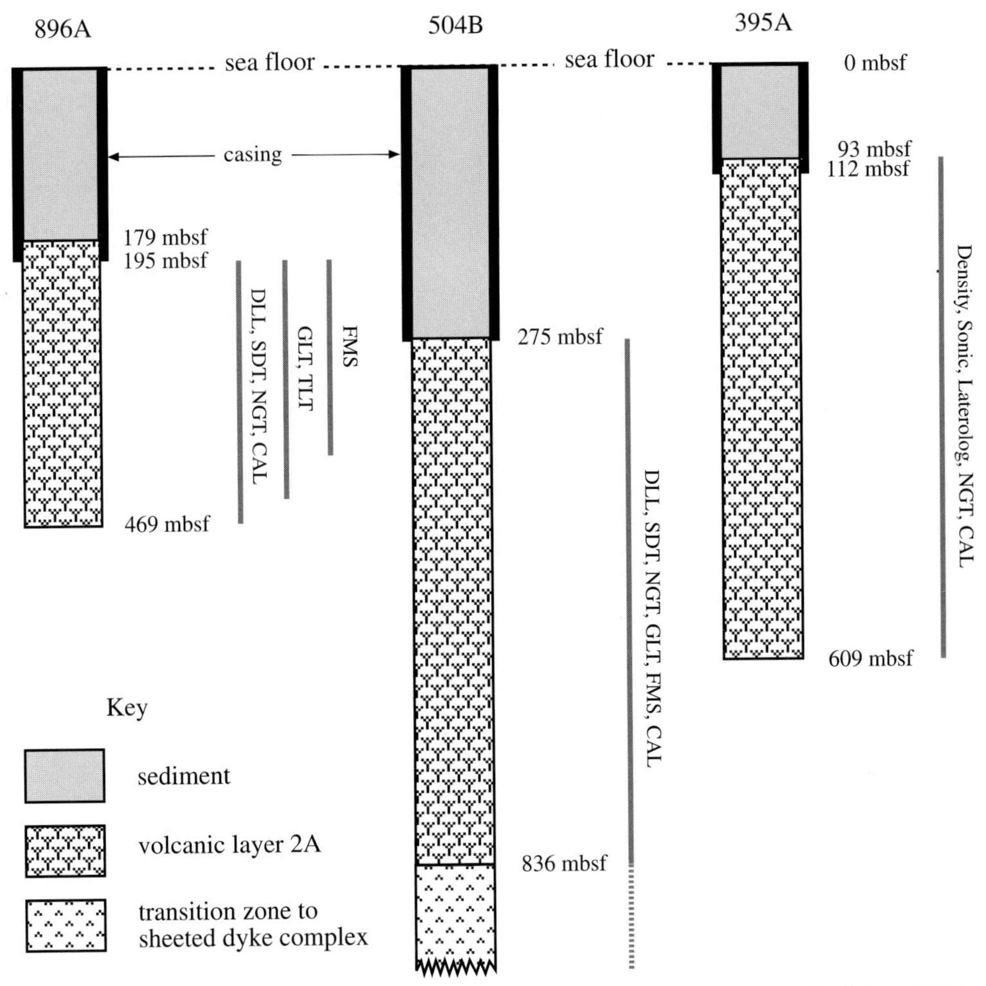

Fig. 3. Location of various logging tools in Holes 896A (Alt *et al.* 1993), 504B (Dick *et al.* 1992) and 395A (Hyndaman & Salisbury 1983). FMS: Formation Microscanner, GLT: Geochemical Logging Tool, DLL: Dual Laterolog, SDT: Sonic Tool, NGT: Natural Gamma Tool, CAL: Caliper.

interval does not provide a means of calibration, and this in part reflects the different volumes sampled by the two data sets (Brewer *et al.* 1992, Brewer *et al.* 1997). In Fig. 5 two sections from Hole 896A are shown; these have similar hole size (note the similar caliper logs), but there is a marked difference between the GLT curves. Figure 5a is a section through a sheet flow and is characterized by relatively smooth Al, Fe and Ca concentrations, which reflect the generally uniform structure of the flow. In contrast, in Fig. 5b the caliper log has microrelief and spiky Ca, Al and Fe curves. This section covers a zone of carbonate cemented breccia (basalt clasts), and therefore the spiking of GLT curves reflects the heterogeneous nature of this rock type.

Formation Microscanner (FMS^{TM})

Several studies have now demonstrated the potential for using FMS images to discriminate different volcanic lithologies in ocean basement holes (Davis *et al.* 1992, Langseth & Becker 1994, Brewer *et al.* 1995, Brewer *et al.* 1998). An advantage of the FMS images is that they may be viewed as resistivity maps of partial sections of the borehole wall (in the ODP *c.* 20% coverage), which allows for the mapping of lithological boundaries, the determination of the internal structure of breccias and pillow lavas and the mapping of the distribution of secondary low temperature vein arrays. This mapping reflects the conductivity contrast between the

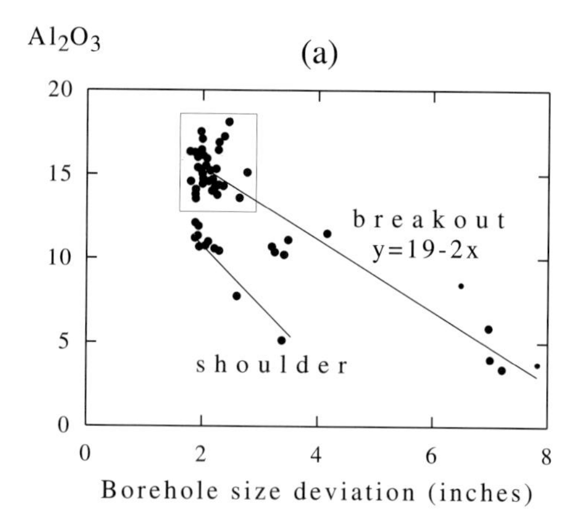

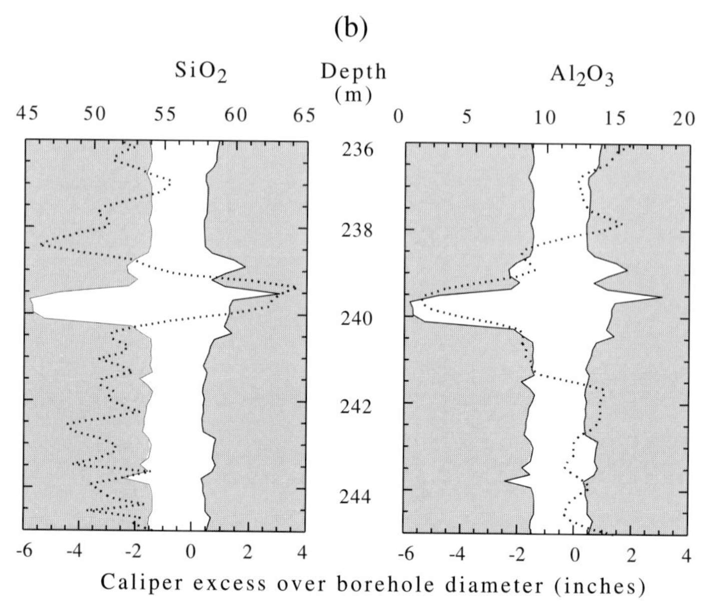

Fig. 4. (**a**) GLT derived Al_2O_3 concentration versus hole size, showing the variation induced by the breakout and the shoulder affect from a breakout in Hole 896A. (**b**) Variation of GLT derived SiO_2 and Al_2O_3 versus depth in the vicinity of a major breakout in Hole 896A.

basaltic clasts or pillows and the breccia matrix or the interpillow material respectively (Fig. 6). Examples of FMS images from submarine breccias, pillows and flows are shown in Fig. 6. The critical features are as follows:

- Flows: images dominated by largely uniform resistivity. Individual flows truncated by discrete fractures (conductive infills, e.g. clays) of variable orientations (Fig. 6a).
- Pillow lavas: elliptical shaped sacks of resistive basalt, with variously sized conductive intercalations of interpillow material (Fig. 6b).
- Breccias: generally mottled images within which both veined and unveined clasts can be identified (Fig. 6c). Rapid semi-quantitative variations in resistivity reflect size, shape and distribution of clasts and the composition of the matrix material.

Frequently in basement holes the distinction between different lithologies is acheived by a combination of different logging results (Brewer

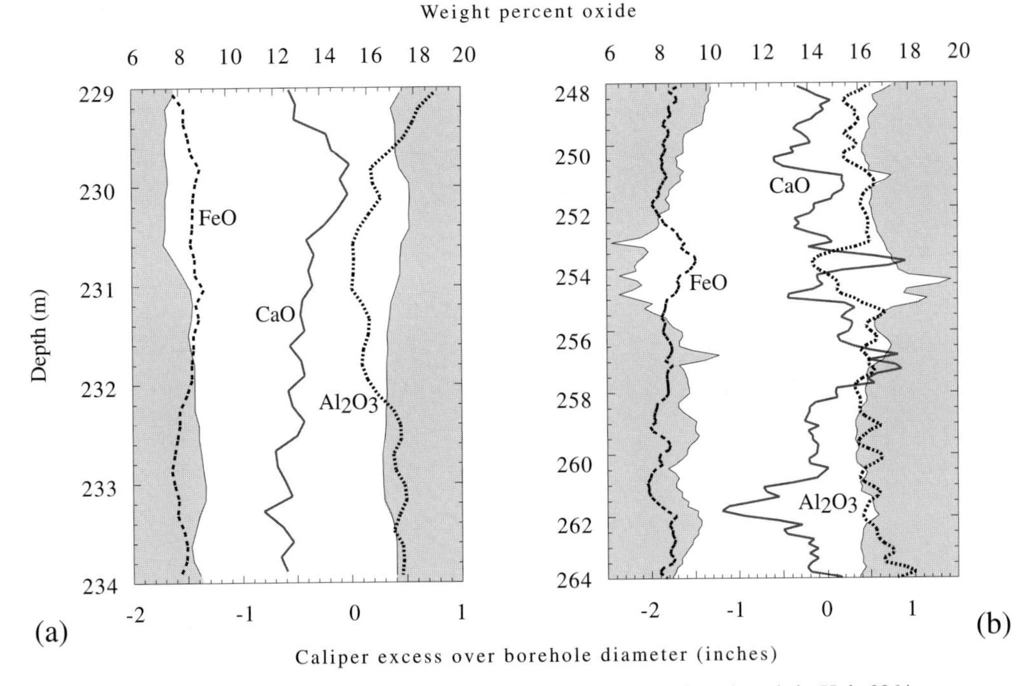

Fig. 5. Variations of GLT derived Ca, Fe and Al in (**a**) a flow, and (**b**) a breccia unit in Hole 896A.

et al. 1998), although the FMS provides the 'classic visual image' which can be considered as a snap-shot of the outcrop.

Integration of logging results

Brewer *et al.* (1998) provide a detailed account of core-log integration as applied to ODP Hole 896A, and the criteria used to discriminate the different lithologies are summarized in Fig. 7. Using these criteria the lithostratigraphy in ODP Holes 504B and 396A has been constructed from petrophysical data. A limitation of this approach concerns the availability and quality of the logging data in both holes. Logging data for Hole 395A comes from ODP Leg 78A, when both the FMS and GLT tools were not available (Hyndman & Salisbury 1984). In Hole 504, the quality of the FMS is moderate to poor, due to the failure of some of the resistivity pads during the logging run and also to oversizing in the upper part of the hole (Dick *et al.* 1992, Alt *et al.* 1993). However, if these limitations are accepted it is possible to reconstruct the lithostratigraphy from the downhole measurements and to estimate the different proportions of the lithologies present (Fig. 7). For comparison, in Fig. 7 the core-based lithological proportions are also presented and the following points are evident:

- core-based estimations of breccias in the three holes are always low
- core-based estimations are always dominated by pillow lavas.

Part of the variation between the two data sets probably reflects:

(a) inability to recover breccia, and
(b) the core classification schemes used to identify the various lithologies.

The robustness of the classification scheme is improved by integration of additional logging data and core measurements, so extending the range of parameters used to discriminate lithologies. Further substantiation of the log-based approach is that different operators working in isolation have also been able to discriminated the same lithologies from identical data sets (Langseth & Becker 1994, Brewer *et al.* 1995).

An application of this work relates to the correlation between volcanic architecture and the morphology and structure of individual ridges. Topographic mapping of ridges has established the following bathymetric profiles:

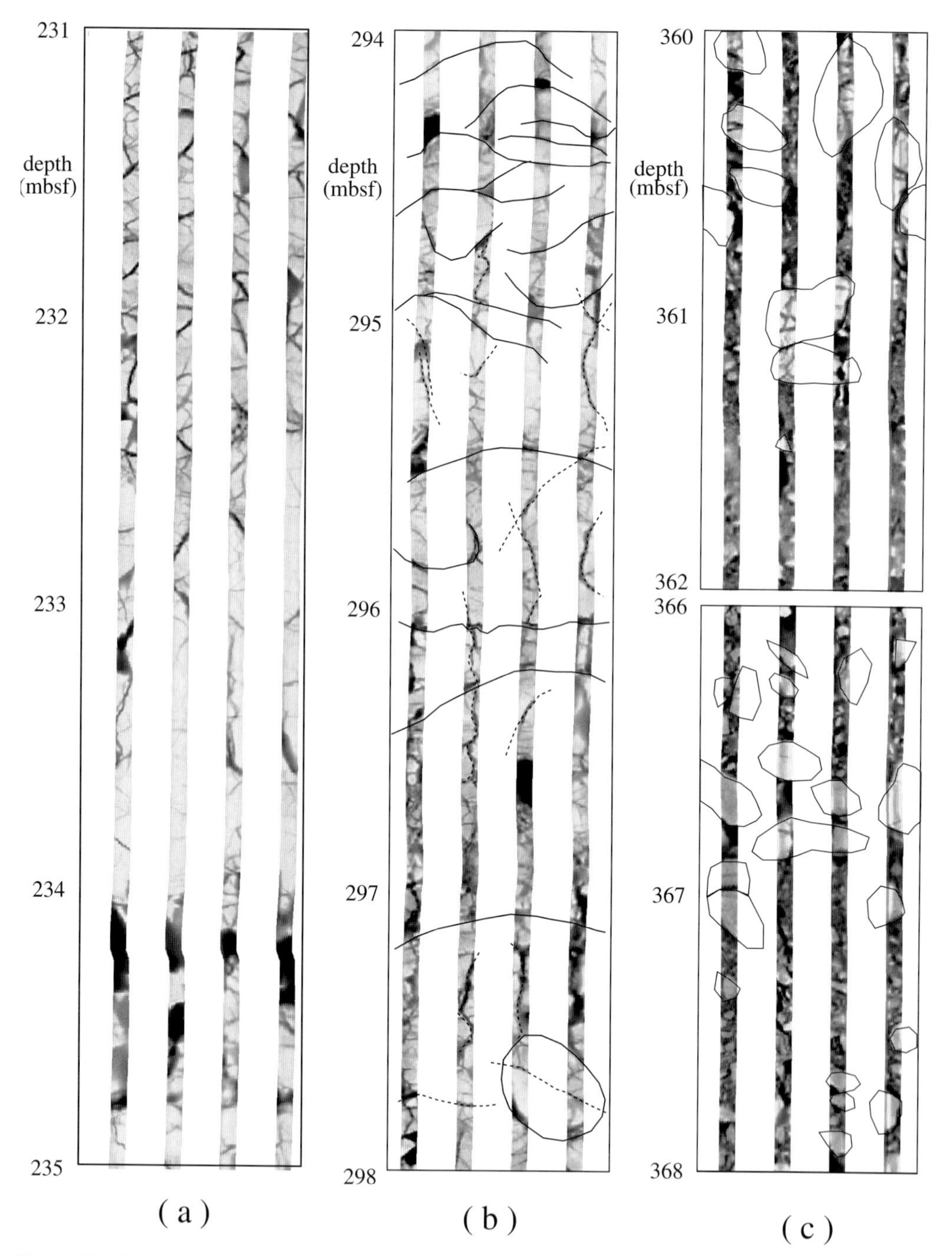

Fig. 6. FMS images from (**a**) a flow, (**b**) a pillow lava unit, and (**c**) a breccia unit in Hole 896A.

- Slow spreading (0.7–1.4 cm/a, e.g. Mid Atlantic Ridge, Heirtzler & van Andel 1977, Ballard & van Andel 1977): rift mountains on each side of the rift. The rift valley floor slopes gently towards the axis.
- Intermediate spreading (6 cm/a, 21°N East Pacific, Francheteau & Ballard 1983): A shallow axial depression (comparable width to Mid-Atlantic Ridge) situated on central swell of approximately 7 km width.

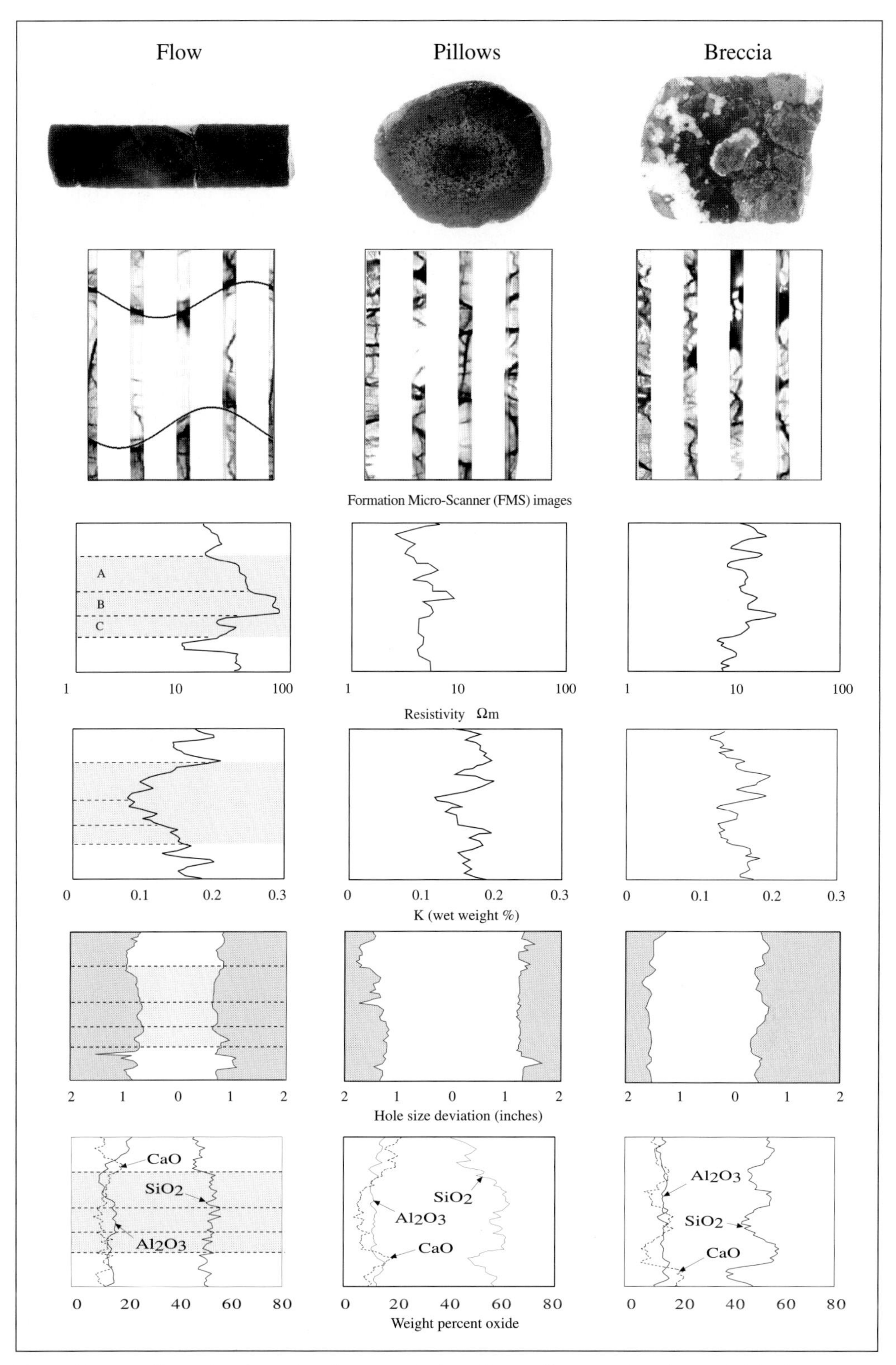

Fig 7. Schematic illustrating the various log response from flows, pillows and breccias. Resistivity from LLD, K from NGT, hole size from caliper and oxides from GLT tool.

- Intermediate to fast spreading (10.2 cm/a, e.g. 13°N East Pacific ridge Francheteau & Ballard 1983): a very small axial graben (200 m wide, 50 m deep) situated on a central swell.
- Fast Spreading (16 cm/a, e.g. East Pacific Ridge 20°S Francheteau & Ballard 1983): no graben present, a prominent axial ridge situated on central swell (*c.* 7 km wide).

A key question is does the style of volcanism change with different spreading rates. This can be answered from submersible and geophysical data (Bonatti & Harrison 1988, Smith & Cann 1992), but drilling also provides another dimension (i.e. time). Unfortunately, ultra-fast spreading ridges have yet to be drilled and so we are limited to an initial comparison between slow (e.g. Hole 395A) and intermediate (Holes 896A and 504B) ridges. However, even at this preliminary stage, testable hypotheses can be formulated. Using the log-core derived lithostratigraphies for each of the three holes, it is evident that the proportion of flows increases and that of pillow lavas decreases with increased spreading rate (Fig. 8a). This variation probably reflects enhanced effusion rates (Bonatti & Harrison 1988, Wilson & Head 1997), allowing more axially focused sheet flows to develop. It is also evident from Hole 896A that pillow lavas are more abundant in the upper parts of the stratigraphy (i.e. youngest volcanism, Fig. 8b), suggesting a change in volcanic style within the neovolcanic zone (i.e. axial rift). This change in the style of volcanism probably reflects small volumes of magma (i.e. small discrete magma chambers) and effusion rates in an off-axis setting which would favour pillow lava eruptions.

Conclusions

Imaging and traditional downhole logging data provide a novel data set with which to study ocean floor volcanism. The application of these different techniques is presently at a juvenile stage due to the limited number of data sets. However, the potential advantage of the log based classification is that in each hole the quality of the logging data is both very good (with the exception of some FMS in Hole 504B) and that it provides a near continuous record within the borehole. However, a cautionary note is that these data sets should not be treated independently from the core-derived data since accurate and precise core-log integration is critical to any successful study.

Further upscaling of the results from such studies can only then be attempted with data derived from submersible and marine geophysics (e.g. seismic surveys and deep towed devices). From such a integrated approach it will then be possible to address the fundamental questions of process and dynamics of ridge systems.

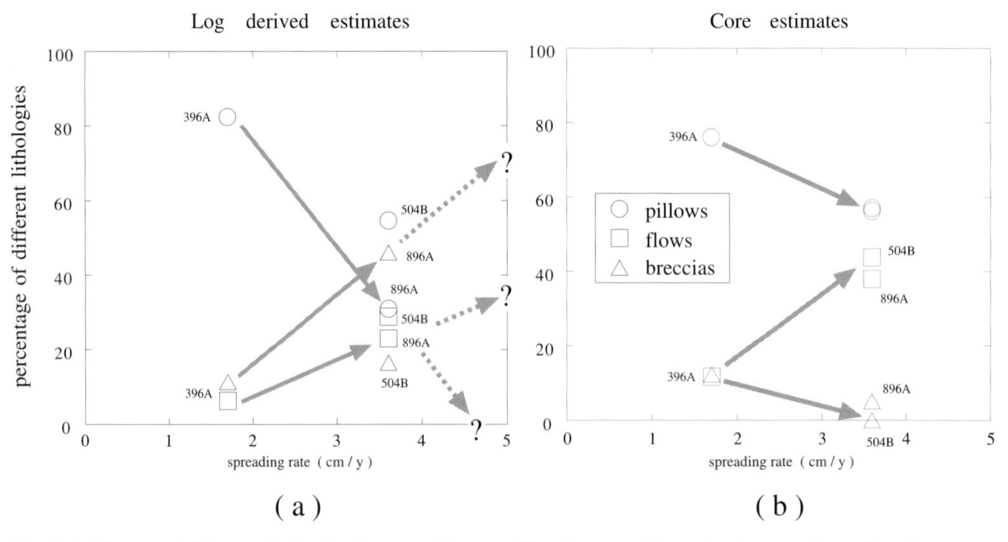

Fig. 8. (**a**) Log, and (**b**) core derived estimates of flows, pillow lavas and breccias in crust formed at slow spreading centre (Hole 395A), an intermediate spreading (Holes 504B and 896A) and a linear extrapolation for fast spreading.

ODP log interpretation facilities at Leicester are supported by a grant from the Natural Environmental Research Council (GST/02/684).

References

ADAMSON, A. C. 1985. *Basement lithostratigraphy, Deep Sea Drilling Project Hole 504B.* Initial Reports of the Deep Sea Drilling Project, **83**, 121–127

ALT, J. C., KINOSHITA, H. STOKKING, L. B. *et al.* 1993. *Proceedings of the Ocean Drilling Program, Initial Reports*, **148**, College Station, TX (Ocean Drilling Program).

BALLARD, R. D. & VAN ANDEL, T. H. 1977. Morphology and tectonics of the inner rift valley at latitude 36°56′N on the mid-Atlantic Ridge. *Bulletin of the Geological Society of America*, **88**, 507–530.

BECKER, K., MALONE, M. J. & SHIPBOARD PARTY. *Initial Reports of the Ocean Drilling Program*, 174B, in press.

BONATTI, E. & HARRISON, C. G. A. 1988. Eruption style of basalt in oceanic spreading ridges and seamounts: effect of magma temperature and viscosity. *Journal of Geophysical Research*, **93**(B4), 2967–2980.

BREWER, T. S., PELLING, R., LOVELL, M. A. & HARVEY, P. K. 1992. The validity of whole-rock geochemistry in the study of oceanic crust: a case study from ODP Hole 504B. *In*: PARSON, L. M., MURTON, B. J. & BROWNING, P. (eds) *Ophilolites and their Modern Oceanic Analogues.* Geological Society of London Special Publication, **60**, 263–278.

——, LOVELL, M. A., HARVEY, P. K. & WILLIAMSON, G. 1995. *Stratigraphy of the ocean crust in Hole 896A from FMS images.* Scientific Drilling, **5**, 87–92.

——, HARVEY, P. K. & LOVELL, M. A. 1997. Ocean crust evolution: constraints arising from the detailed analysis of downhole logs. *In*: PAWLOWSKY-GLAHN, V. (ed.) *Proceedings of IMAG'97, The Third Annual Conference of the International Association of Mathematical Geology*, **1**, 221–226.

——, ——, ——, HAGGAS, S., WILLIAMSON, G. & PEZARD, P. 1998. *Ocean floor volcanism: constraints from core-log data.* Special Publication of the Geological Society of London, **136**, 341–362.

BRISTOW, J. F. & DEMENOCAL, P. B. 1992. *Evaluation of the quality of geochemical logging data in hole 798B.* Proceedings of the Ocean Drilling Program, Scientific Results, **127/128**, 1021–1035.

BRYAN, W. B., JUTEAU, T. & SHIPBOARD SCIENTIFIC PARTY 1988. *Proceedings of the Ocean Drilling Program, Initial Reports*, **109**, College Station, TX (Ocean Drilling Program).

CANN, J. R., LANGSETH, M. G., HONOREZ, J., VON HERZEN, R. P., WHITE, S. M. & SHIPBOARD SCIENTIFIC PARTY 1983. *Initial Reports of the Deep Sea Drilling Project*, **69**, Washington (US Government Printing Office).

DAVIS, E. E., MOTTL, M. J. & SHIPBOARD SCIENTIFIC PARTY 1992. *Proceedings of the Ocean Drilling Program, Initial Reports*, **139**, College Station, TX (Ocean Drilling Program).

DICK, H. J. B., ERZINGER, J. & STOKKING, L. B. 1992. *Proceedings of the Ocean Drilling Program, Initial Reports*, **140**, College Station, TX (Ocean Drilling Program).

FRANCHETEAU, J. & BALLARD, R. D. 1983. The East Pacific Rise near 21°N, 13°N and 20°S: inferences for along strike variability of axial processes of the mid-ocean ridge. *Earth and Planetary Science Letters*, **64**, 93–116.

GABLE, R., MORIN, R. H. & BECKER, K. 1992. Geothermal state of DSDP Holes 333A, 395A and 534A: results of the DIANAUT. *Geophysical Research Letters*, **19**, 513–516.

GOLDBERG, D. 1997. The role of downhole logging measurements in marine geology and geophysics. *Reviews of Geophysics*, **35**, 315–342.

—— & SUN, Y. F. 1997. Attenuation differences in layer 2A in intermediate- and slow-spreading oceanic crust. *Earth and Planetary Science Letters*, **150**, 221–231.

HARVEY, P. K. & LOVELL, M. A. 1992. Downhole mineralogy logs: mineral inversion and the problem of compositional colinearity. *In*: HURST, A. C., GRIFFITHS, C. M. & WORTHINGTON, P. F. (eds) *Geological Applications of Wireline Logs II.* Geological Society of London Special Publication, **66**, 361–368.

——, ——, BREWER, T. S. & LOFTS, J. C. 1997. Modelling strategy in the generation of mineral logs. *In*: PAWLOWSKY-GLAHN, V. (ed.) *Proceedings of IMAGU97, The 3rd Annual Conference of the International Association of Mathematical Geology, part* **1**, 201–206.

HEIRTZLER & VAN ANDEL, T. J. 1977. Project Famous: its origin programs and settings, *Bulletin Geological Society of America*, **88**, 481–487.

HERTZOG, R., COLSON, L., SEEMAN, B. *et al.* 1987. *Geochemical Logging with Spectrometry Tools.* 62nd Annual Technical Conference and Exhibition of the Society of Petroleum Engineers, Dallas, September 27–30, 1–13.

HILL, M. N. 1957. Recent geophysical exploration of the Ocean floor. *Physics and Chemistry of the Earth*, **2**, 129–163.

HYNDMAN, R. D. & SALISBURY, M. H. 1984. *The physical nature of the oceanic crust on the Mid-Atlantic Ridge, DSDP hole 395A.* Initial Reports of the Deep Sea Drilling Project, **78B**, 839–848.

——, —— & SHIPBOARD PARTY 1984. *Initial Reports of the Deep Sea Drilling Project*, **78B**, Washington (US Government Printing Office).

HUSSONG, D. M., FRYER, P. B., TUTHILL, J. D. & WIPPERMAN, L. K. 1979. The geological and geophysical setting near DSDP site 395, north Atlantic Ocean. *In*: MELSON, W. G., RABINOWITZ, P. D. *et al.*, *Initial Reports of the DSDP*, **45**, 23–27 (US Government Printing Office).

KERR, S. A., GRAU, J. A. & SCHWEITZER, J. S. 1992. A comparison between elemental logs and core data. *Nuclear Geophysics*, **3**, 303–323.

LANGSETH, M. G. & BECKER, K. 1994. Structure of igneous basement at Sites 857 and 858 based on Leg 139 downhole logging. *In*: MOTTL, M. J., DAVIS, E. E., FISHER, A. T. & SLACK, J. F. (eds) *Proceedings of the Ocean Drilling Program, Scientific Results*, **139**, 573–584. College Station, TX (Ocean Drilling Program).

——, ——, VON HERZEN, R. P. & SCHULTHEISS, P. 1992. Heat and fluid flux through sediment on the western flank of the mid-Atlantic Ridge: a hydrogeological study of North Pond. *Geophysical Reserach Letters*, **19**, 517–520.

LAVERNE, C., BELAROCHI, A. & HONNOREZ, J. 1966. Alteration mineralogy and chemistry of the upper oceanic crust from Hole 896A, Costa Rica Rift. *In*: ALT, J. C., KINOSHITA, H., STOKKING, L. B. & MICHAEL, P. J. (eds) *Proceedings of the Ocean Drilling Program*, **148**, 151–170.

MELSON, W. G., RABINOWITZ, P. D. & SHIPBOARD PARTY 1979. *Initial Reports of the Deep Sea Drilling Project*, **45**, Washington (Government Printing Office).

PEZARD, P. A. 1990. Electrical properties of mid-ocean ridge basalts and implications for the structure of the upper ocean crust in Hole 504B. *Journal of Geophysical Research*, **95**(B6), 9237–9264.

RAITH, R. W. 1963. The crustal rocks. *In*: HILL, M. N. (ed.) *The Sea*, **3**, 85–102, Interscience, New York.

SMITH, D. K. & CANN, J. R. 1992. The role of seamount volcanism in crustal construction at the Mid-Atlantic Ridge (24°–30°N). *Journal of Geophysical Research*, **97**(B2), 1645–1658.

TEAGLE, D. A. H., ALT, J. C., BACH, W., HALLIADY, A. N. & ERZINGER, J. 1996. Alteration of upper ocean crust in a ridge-flank hydrothermal upflow zone: mineral, chemistry and isotopic constraints from Hole 896A. *In*: ALT, J. C., KINOSHITA, H., STOKKING, L. B. & MICHAEL, P. J. (eds) *Proceedings of the Ocean Drilling Program*, **148**, 119–150.

Index

Page numbers in *italic* refer to Tables

Index of Artefacts